new	media
old	media

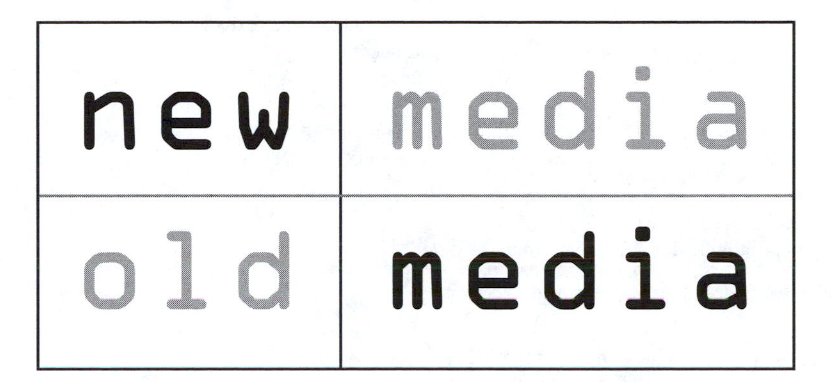

A History and Theory Reader

Edited by

Wendy Hui Kyong Chun &
Thomas Keenan

Published in 2006 by
Routledge
Taylor & Francis Group
270 Madison Avenue
New York, NY 10016

Published in Great Britain by
Routledge
Taylor & Francis Group
2 Park Square
Milton Park, Abingdon
Oxon OX14 4RN

Printed in the United States of America on acid-free paper
10 9 8 7 6 5 4 3

International Standard Book Number-10: 0-415-94223-3 (Hardcover) 0-415-94224-1 (Softcover)
International Standard Book Number-13: 978-0-415-94223-2 (Hardcover) 978-0-415-94224-9 (Softcover)
Library of Congress Card Number 2005014486

Library of Congress Cataloging-in-Publication Data

New media, old media : a history and theory reader / edited by Wendy Hui Kyong Chun, Thomas W. Keenan.
p. cm.
Includes bibliographical references and index.
ISBN 0-415-94223-3 (hardback : alk. paper) -- ISBN 0-415-94224-1 (pbk. : alk. paper)
1. Mass media--History. I. Chun, Wendy Hui Kyong, 1969- II. Keenan, Thomas, 1959-

P90.N52 2005
302.23'09--dc22 2005014486

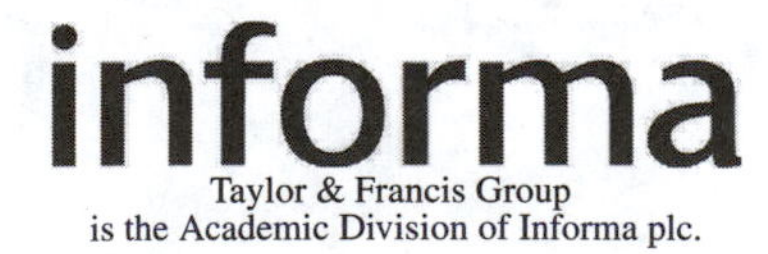

Visit the Taylor & Francis Web site at
http://www.taylorandfrancis.com

and the Routledge Web site at
http://www.routledge-ny.com

Contents

Acknowledgments

The idea for this volume stemmed from an international conference entitled *The Archaeology of Multi-Media* at Brown University in November 2000, which was generously supported by the Malcolm S. Forbes Center for Research in Culture and Media Studies, the Department of Modern Culture and Media, the Office of the Provost, the Pembroke Center for Teaching and Research on Women, the Scholarly Technology Group, the Multimedia Lab, the Watson Institute for International Studies, and the Dean of the College. We thank all of the sponsors, as well as Alexander Russo, Susan McNeil, and Liza Hebert, without whose hard work the conference would never have happened. We are also grateful to James Der Derian, Julia Flanders, and the entire faculty of the Department of Modern Culture and Media for their invaluable contributions to the discussion.

Thanks to them and to Melanie Kohnen for her excellent work in indexing and formatting the text. The path from the conference to this collection has been facilitated by grants by the Malcolm S. Forbes Center and Brown University's Office of the Vice President for Research, as well as a fellowship from the Radcliffe Institute for Advanced Study at Harvard University and a Henry Merritt Wriston Fellowship from Brown. Thanks to Peter Krapp for his translations. We also thank our editor at Routledge, Matthew Byrnie, for his interest in this project and for his supreme patience and support.

Some of the chapters in this volume have been previously published, and we are grateful to the original publishers and copyright holders for their permission to reprint them.

Thomas Y. Levin, "'Tones from out of Nowhere': Rudolph Pfenniger and the Archeology of Synthetic Sound," *Grey Room* 12 (Summer 2003). ©2003 Thomas Y. Levin. Used by permission.

Vannevar Bush, "Memex Revisited," *Science Is Not Enough* (New York: William Morrow & Co., 1967). ©1967, William Morrow & Co. Used by permission.

Cornelia Vismann, "Out of File, Out of Mind," *Interarchive: Archivarische Praktiken und Handlungsräume im zeitgenössischen Umfeld/ Archival Practices and Sites in the Contemporary Art Field,* Ed. Beatrice von Bismarck et al. (Köln: Verlag der Buchhandlung Walther König, 2002). ©2002 Verlag der Buchhandlung Walther König. Used by permission.

Wolfgang Hagen, "The Style of Sources: Remarks on the Theory and History of Programming Languages," first published in German as "Der Stil der Sourcen: Anmerkungen zur Theorie und Geschichte der Programmiersprachen." In *Hyperkult,* Ed. W. Coy, G. C. Tholen, and R. Warnke (Basel u.a.: Stroemfeld 1997). ©Wolfgang Hagen. Used by permission.

Alexander R. Galloway, "Protocol vs. Institutionalization," *Protocol* (Cambridge: MIT, 2004). ©2004, MIT. Used by permission.

Tara McPherson, "Reload: Liveness, Mobility, and the Web," *Visual Culture Reader 2.0*, Ed. Nicholas Mirzoeff (New York: Routledge, 2003). ©2003, Routledge. Used by permission.

Lev Manovich, "Generation Flash," "Generation Flash," *rhizome* (www.rhizome.org), 2002. ©2002 Lev Manovich. Used by permission.

Julian Dibbell, "Viruses Are Good For You," *Wired* 3.02 (February 1995). ©1995, Julian Dibbell. Used by permission.

Anders Michelson, "The Imaginary of the Artificial: Automata, Models, Machinics — On promiscuous modeling as precondition for poststructuralist ontology," in *Reality/Simulacra/Artificial: Ontologies of Postmodernity*. Ed. Enrique Larreta (Rio de Janeiro: Universida de Candido Mendes, 2003). ©2003, Universida de Candido Mendes. Used by permission.

An earlier version of Mary Ann Doane, "Information, Crisis, and Catastrophe" appeared in *Logics of Television: Essays in Cultural Criticism*. Ed. Patricia Mellencamp. Bloomington: Indiana UP, 1990. ©1990, Indiana UP. Used by permission.

A longer version of Geert Lovink, "Deep Europe: A History of the Syndicate Network" appeared in *My First Recession* (Rotterdam: V2_Publishing/NAi Publishers, 2003). ©V_2Publishing/NAi Publishers. Used by permission.

Vicente Rafael, "The Cell Phone and the Crowd: Messianic Politics in the Contemporary Philippines," *Public Culture* 15:3 (2003). ©2003, Duke UP. Used by permission.

Lisa Nakamura, "Re-mastering the Internet: the Work of Race in the Age of Digital Reproduction," *Cybertypes* (NY: Routledge, 2003). ©2003, Routledge. Used by permission.

Mark Wigley, "Network Fever," *Grey Room* 4 (Summer 2001). ©2001, Mark Wigley. Used by permission.

Introduction
Did Somebody Say New Media?

Wendy Hui Kyong Chun

When the first encounter with some object surprises us, and we judge it to be new, or very different from what we formerly knew, or from what we supposed that it ought to be, that causes us to wonder and be surprised; and because that may happen before we in any way know whether this object is agreeable to us or is not so, it appears to me that wonder is the first of all the passions; and it has no opposites, because if the object which presents itself has nothing in it that surprises us, we are in nowise moved regarding it, and we consider it without passion.

—Rene Descartes, *The Passions of the Soul,* article 53

To be new is peculiar to the world that has become picture.

—Martin Heidegger[1]

Emergence is always produced through a particular stage of forces.

—Michel Foucault[2]

The term "new media" came into prominence in the mid-1990s, usurping the place of "multimedia" in the fields of business and art. Unlike its predecessor, the term "new media" was not accommodating: it portrayed other media as old or dead; it converged rather than multiplied; it did not efface itself in favor of a happy if redundant plurality.[3] The singular plurality of the phrase ("new media" is a plural noun treated as a singular subject) stemmed from its negative definition: it was not mass media, specifically television. It was fluid, individualized connectivity, a medium to distribute control and freedom. Although new media depended heavily on computerization, new media was not simply "digital media": that is, it was not digitized forms of other media (photography, video, text), but rather an interactive medium or form of distribution as independent as the information it relayed.

Although the term "new media" has been used since the 1960s, it rose (and arguably fell) with dotcom mania, cyberspace, and interactive television. The signs of new media's difficult times: the

New York New Media Association folded in 2003, its assets purchased by the Software & Information Industry Association and the address newmedia@aol.com given back to Mark Stahlman (who claims to have coined "new media"—in 2004 he was pushing the phrase "3-space"); clickz.com bought newmedia.com; many new media groups within corporate structures (Apple, Gannett, etc.) and many new media companies disappeared. Importantly, this demise does not coincide with the demise of media once called new, but rather with industry's quest to survive and thrive after the "new economy" bubble and after new media's wide acceptance. Does it, after all, make sense to have a New Media Group within Apple after 2003? New media's decline in academia has been less precipitous, although the slippery term "emerging media" has gained momentum. From the start, new media studies sought a critical middle ground between commercial propaganda and intellectual conservatism. Film and television scholars, artists and humanities scholars eager to explore the potential of networked computation without necessarily engaging prior traditions of hypertext or humanities computing supported the term "new media."[4] Also, as the utopianism or dystopianism of early net studies became painfully clear, some scholars further distanced themselves by separating new media studies from "cyberstudies" (thus the rapid disappearance of William Gibson's fiction from new media courses and readers). Cyberspace, not new media, was the mistaken term. Most importantly, new media has traction because of programs and jobs perpetuated in its name—it is a field with its own emerging canon and institutional space.

Much critical debate within new media studies has centered on: What is/are new media? Is new media new? What is new about new media?—questions arguably precipitated by the widespread acceptance of the term itself.[5] Regardless, these debates produced many insightful histories for and theories of new media, which redrew disciplinary borders. For instance, Jay Bolter and Richard Grusin in *Remediation: Understanding New Media* linked all media from the Renaissance to Virtual Reality through "remediation," "immediacy," and "hypermediacy."[6] Others focused more closely on the "new" to establish historical continuity. Lisa Gitelman and Geoffrey B. Pingree in *New Media 1740–1915* (part of David Thorburn et al.'s *Media in Transition* series, which seeks to understand the aesthetics of cross-historical media transition) argued, "all media were once 'new media'" and "emergent media may be seen as instances of both risk and potential."[7] Still others, such as Lev Manovich in *The Language of New Media*, expanded the definition of new media through formalist principles indebted to historical analysis. *The Language of New Media* emphasized the importance of programmability rather than computer display and distribution, while at the same time viewing new media as the product of the merging of computation with media storage (most importantly film). Following Manovich, Noah Wardrip-Fruin and Nick Montfort have compiled the comprehensive and definitively titled *The New Media Reader*, documenting and indeed creating new media history as the progressive marriage of computation and art, a marriage that produced the computer as an expressive medium.

All these texts are important and have influenced many of this collection's chapters, but they all—inadvertently, purposefully, or ironically—grant computation, new or media a strange stability and obscure new media's commercial history. Computation may be key to new media, but computation does not automatically lead to new media or to software. No one, as Wolfgang Hagen argues in "The Style of Sources: Remarks on the Theory and History of Programming Languages," meant to create the computer as we know it, and the computer emerged as a media machine because of language-based software. This "communicative demand," he argues, came from all sides: economic, organization of labor, symbolic manipulation. In terms of media, histories that reach from the Renaissance to the present day elide the fact that: one, although the word medium does stretch across this time period, its meaning differs significantly throughout; two, the plural-singular term "media" marks a significant discontinuity. According to the *Oxford English Dictionary* (OED), media stems from the Latin *medium* meaning middle, center, midst, intermediate course, intermediary (hence medium/average height and spiritual medium). In the fifteenth century, medium emerged as an intervening substance in English, stemming from the post-classical Latin phrase *per medium* (through the medium of) in use in British sources since the thirteenth century. The term "media" (as opposed to mediums or medium) is linked to mass media: in the eighteenth century, paper was a medium of circulation, as was money; in the nineteenth century, electricity was a medium; in the late nineteenth and twentieth centuries, media emerged as the term to describe inexpensive newspapers and magazines and, in an affront to English and Latin, became a singular noun. The rise of media coincided with its portrayal as transparent rather than intervening, and although Friedrich Kittler himself does not engage the etymology of media, his argument in *Gramophone, Film, Typewriter* that these media displaced writing as the universal storage medium maps nicely onto the emergence of the term "media."[8] To be clear, to claim that media is an important discontinuity that calls into question fluid histories from the Renaissance printing press or perspectival painting to the present is not to claim that no overarching argument can ever be made about mediums or media. It is to say that any such argument must grapple with the ways that mediums have changed, rather than concentrating on the remarkable yet overdetermined similarities between entities now considered media.

The term "new" is also surprisingly uninterrogated. Those debunking the newness of new media often write as if we could all agree on or know the new, as if the new were not itself a historical category linked to the rise of modernity. The new should have no precedent, should break with the everyday, and thus should be difficult, if not impossible, to describe. If something is new—that is known or made for the first time—then we should, according to Descartes in his influential definition of the new, fall into a passionate state of wonder or surprise. The "new," however, is described and explained all the time and describing something as "new" seems a way to dispel surprise or to create it *before* an actual encounter (actually using the Internet, for instance, is banal in comparison to its pre-mass usage filmic, televisual, and print representations). To call X "new" is to categorize it, to describe and prescribe it, while at the same time to insist that X is wonderful, singular, without opposite or precedent. This insistence more often than not erases X's previous existence (case in point, the "discovery" of the "new world"). The Internet was not new in 1995, the year it arguably became new. Its moment of "newness" coincided less with its "invention" or its mass usage (in 1995 significantly more Americans had heard about the Internet than actually been on it), but rather with a political move to deregulate it and with increased coverage of it in other mass media. We accepted the Internet or new media as new because of a concerted effort to make it new, because of novels, films, television news programs, advertisements, and political debates that portrayed it as new, wondrous, strange.[9]

To be new, however, is not simply to be singular. The new contains within itself repetition: one of the OED definitions of it is "coming as a resumption or repetition of some previous act or thing; starting afresh" (this notion of repetition is also contained in the word revolution). "Make it new" is a stock modernist phrase and it exemplifies the type of repetition enabled by the new—the transformation of something already known and familiar into something wonderful. The new is "fresh, further, additional," "restored after demolition, decay, disappearance, etc." (OED). Along these lines, the Internet seemed to make old theories, dreams, and structures new again, revitalizing Athenian democracy, the bourgeois public sphere, deconstruction and capitalism. The Internet seemed to renew the new, and technology, with its endless upgrades, is relentlessly new. This "making new" reveals the importance of interrogating the forces behind any emergence, the importance of shifting from "what is new" to analyzing what work the new does. What enables anything to be called new and How does the new affect other fields, which it simultaneously draws from and repudiates?

To answer these questions, this collection brings together scholars working in new media, media archaeology, film, television, cultural and literary studies to investigate new media and the political, cultural, economic, and epistemological forces necessary to its emergence. Divided into five sections—Archaeology of Multi-Media, Archives, Power-Code, Network Events and Theorizing "New" Media—it argues that these forces cut across fields of race and sexuality, create new global political events, and impact, rather than solve, political problems. The texts in the "Archaeology of Multi-Media" section re-think histories of "older" media, such as film, photography, sound and physical space in light of the computer screen, while at the same time analyzing the importance of these media to the emergence "new media." The texts in the second section, "Archives," examine the continuing relevance (or not) of archives to digitized media. The chapters in "Power-Code" analyze code and its relationship to the circulation knowledge and "empowerment," for new media depends on the computer's transformation from a calculator into a programmable communications medium. "Network Events" further questions knowledge and power, but rather than focusing on code or the computer, looks more broadly at the uses of networked media and transformations in media events. The texts in the final section, "Theorizing 'New' Media," address the theoretical challenges posed by new media.

Rather than present a unified theoretical front or create an inevitable historical trajectory, this book connects forms of media analysis that have usually been separated. It does so not for the sake of diversity, but rather to map the field of new media studies, for this mapping necessitates bringing together continental European media archaeologists, who have tended to concentrate on the logics and physics of hardware and software, and Anglo-speaking critics, who have focused on the subjective and cultural effects of media, or on the transformative possibilities of interfaces. Media Archaeology, indebted to the German scholar Friedrich Kittler, as well as the French Michel Foucault and the Canadian Marshall McLuhan, excavates the technological conditions of the sayable and thinkable and strongly critiques narrative media history. As Wolfgang Ernst explains, "media archaeology describes the non-discursive practices specified in the elements of the techno-cultural archive. Media archaeology is confronted with Cartesian objects, which are mathematisable things..."[10] However, if cultural studies has been criticized for not engaging technology rigorously, media archaeologists often appear as "hardware-maniac, assembler-devoted and anti-interface ascetics, fixed to a (military) history of media without regard to the present media culture."[11] They often seem blind to content and user practices. British, U.S., and Australian cultural/media studies' insistence on technology as experienced by users highlights the importance of economics, politics, and culture and relentlessly critiques technological determinism. Refusing to adjudicate this debate, this book brings together the significant texts of both approaches to chart their surprising agreements and disagreements, common assumptions and uncommon insights, and through these map the field's possibilities and blindnesses.

Approaches to the Multi-Media Archive

> *The archive is the first law of what can be said, the system that governs the appearance of statements as unique events. But the archive is also that which determines that all these things said do not accumulate endlessly in an amorphous mass, nor are they inscribed in an unbroken linearity, nor do they disappear at the mercy of chance external accidents; but they are grouped together in distinct figures, composed together in accordance with multiple relations, maintained or blurred in accordance with specific regularities... it is that which differentiates discourses in their multiple existence and specifies them in their own duration....*
>
> *This term [archaeology] does not imply the search for a beginning; it does not relate analysis to a geological excavation. It designates the general theme of a description that questions the already-said at the level of its existence: of the enunciative function that operates within it, of the discursive formation, and the general archive system to which it belongs. Archaeology describes discourses as practices specified in the element of the archive.*
>
> —Michel Foucault[12]

The "Archaeology of Multi-Media" and "Archives" sections take on Michel Foucault's influential archaeology of knowledge. Treating knowledge-power as a grid, Foucault's archaeology explores the ties between elements of knowledge and power.[13] It seeks to defuse the effects of legitimacy by revealing what makes something legitimate and what allows for its acceptance. Archaeology examines the enunciative functions of the "already-said" and its relationship to the general archive, where the archive is "the system that governs the appearance of statements as unique events" that "differentiates discourses in their multiple existence and specifies them in their own duration." Discourses are thus objects and practices that obey particular rules. Nothing, Foucault argues, can appear as knowledge if it does not conform to the rules and the constraints of a given discourse in a given epoch; and nothing functions as power unless its exertion complies with the procedures, instruments, means or objectives valid in more or less coherent systems of knowledge.[14] Archaeology, as a systematic description of the discourse-object, focuses on regularities rather than moments of "originality."[15] It does not wholly ignore the unique, the original or the moment of "discovery," but rather, even within these statements, it reveals the regularity that enables them and their differentiation. Archaeology is also fundamentally anti-humanist: it decenters consciousness by refusing a history of continuity, by refusing anthropology.[16]

Following Foucault, to pose the question of the archaeology of multi-media or multi-media as archive is to question the relationship between multi-media and knowledge, multi-media and power. However, it is also to question Foucault's privileging of documents and discourse (Foucault argues that the emergence of this new history coincides with a crisis of the "document." Instead of treating documents as mute but decipherable traces of consciousness, history now treats documents as *monuments*), for media, as Kittler has argued, limit Foucault's project: "all of his [Foucault's] analyses end immediately before that point in time at which other media penetrated the library's stacks [because] Discourse analyses cannot be applied to sound archives or towers of film rolls."[17] Multi-media, through its simulacral multiplicity, arguably dis- or re-places documents (treated as monuments or otherwise); yet documents (as non-digitally manufactured texts) both disappear *and* proliferate (as heuristic devices). These simulacral differences also displace archival distinctions and perhaps archive the term "archive." Thus, to put these sections under the rubric of "media archaeology" and to address this in writing is perhaps already too limited. However, rather than simply extending Foucault or Kittler (even though extension nicely implies distortion and disfiguration), these chapters use scholarly, popular, and technical notions of archaeology and archives as a point of departure in order to examine the relationship between memory and media, storage and mass dissemination, past and present. As well, these chapters register the signs and clues of our media and critical situation, as computers seem to be emerging as a new universal medium, changing power-knowledge within universities and beyond.[18] So, even given Kittler's critique, the "return" to archaeology seems itself overdetermined: archaeology's privileging of rules and statements dovetails nicely with the operation of higher-level software languages—computers and archaeology reinforce each other's truths.

The articles in the first section, "Archaeology of Multi-Media," rethink the archaeology of "older media," such as film, photography and sound, while also investigating the importance of these media to the emergence of the digital as multiple. In "Early Film History and Multi-Media: An Archaeology of Possible Futures?" Thomas Elsaesser uses digitization as an impossible zero degree from which to displace himself from habitual ways of thinking and interrogate the ways in which early cinema challenges film history's "from . . . to" narratives. With multi-media, he argues, the history of the cinema looks more like the archaeology of the Panopticon. Geoffrey Batchen in "Electricity Made Visible" argues that new media has a history as old as modernity itself. Computation and media storage met in the nineteenth century through the intersection of photography, Babbage's difference engine, and telegraphy. Thomas Levin in "'Tones From Out of Nowhere': Rudolph Pfenninger and the Archaeology of Synthetic Sound" argues, through a reading of the early twentieth century synthetic sound projects of Rudolf Pfenninger, that the loss of indexicality, which many associate with digitization, has a longer, analog history.

The "Archives" section explores more closely the possibilities and limitations of a multi-media archive, focusing on the relationship between archives, power and narratives of progress. It moves from Vannevar Bush's optimistic post-World War II view in "Memex Revisited" of future information processing technology as saving us from our ever-expanding archives (its unconsulted records threaten to bury us and our "civilization") to Cornelia Vismann's critical post-reunification assessment of such emancipatory dreams and of the physics and the symbolics of bureaucratic files in "Out of File, Out of Mind." The next chapter, "Dis/continuities: Does The Archive Become Metaphorical in Multi-Media Space?", contains Wolfgang Ernst's plea to archive the term "archive." According to Ernst, the computer has "an *arché*, a (archeo-)logics of its own" and does not order itself according to human perception: the term "multi-media" is a conceit produced for humans. In contrast, Richard Dienst in "Breaking Down: Godard's Histories" offers a materialist analysis of the human perception of images, digital or otherwise, through a reading of Jean-Luc Godard's *Histoire(s) du Cinéma*. To see an image as an image, Dienst argues, requires an enormous collective and cumulative effort over many millennia: images remain to be seen and it is our task to use images in the work of remembrance, critique and imagination in order to change the scope of life. Lastly, Lynne Joyrich in "Ordering Law, Judging History: Deliberations on Court TV" examines the way in which television can serve as a "mass" archive that scandalously spreads scandalous knowledge. Concentrating on Court TV (its time and its myriad parallels to law and soap operas), she argues that it can help us understand how "through various cultural and media forums . . . processes of knowing are offered and refused."

Power-Code-Network

Rather than focusing on the term "archaeology," the next three sections of the collection, "Power-Code," "Network Events," and "Theorizing 'New' Media" further examine the term "knowledge," for the rise of new media is intimately linked to the conflation of information with knowledge. Although the term "information revolution" preceded the Internet, information as revolutionizing capitalist society was not entirely regularized—popularized and accepted as true—until the Internet emerged as the mass medium to end mass media.[19] This regularization made banal and perverted Foucault's own insights. If once the coupling of knowledge with power seemed critical or insightful, "knowledge *is* power" (different, as Thomas Keenan has argued, from knowledge-power) became the motto for Etrade.com and for the "knowledge economy" more generally.[20] "Knowledge *is* power" posits information as a commodity, but what is information and how did it gain such significance?

What is information? The only quantifiable definition of information stems from telecommunications engineering and seemingly has no relation to meaning and knowledge. Claude Shannon defines information as the entropy of a discrete set of probabilities; Warren Weaver, interpreting Shannon's work for a lay public, defines information as a measure of "freedom of choice," for information is the degree of choice (possible number of messages) within a system. As such, information is essential to determining the wire capacity necessary for relatively error-free transmission. As N. Katherine Hayles has argued in *How We Became Posthuman*, through this engineering definition, "information lost its body"—it became "extractable" from actual things.[21] Of course, defining information in this manner also *created* information, transforming its meaning from the process of forming a person or a thing to something that can be transferred and processed (hence, although information lost a body, it was/is never entirely disembodied, since it always exists in a material form). But, we are still some ways from information as a *meaningful* non-exclusive commodity that defies laws of exchange and retroactively defines all storable knowledge as commodities.

Information transmission does, however, get us to modern, stored-program computers and thus to the rise of software, algorithm-based data-analysis, and information as potentially meaningful stored-data rather than entropy. The mostly unquestioned relationship between computers and information stems from the necessity to transport data from one location to another within a computational device. The coming together, Wolfgang Hagen argues, of von Neumann and Shannon. From this transmission or metaphorization (a metaphor is literally a transfer), software and information have become portable entities and computers (human or otherwise) information processors. Software—this thing extracted from hardware that Kittler has argued does not exist (everything comes down in the end to voltage differences)—has been crucial to the creation of the information society, to the new economy, to workers as knowledge workers or symbolic analysts who manipulate information. Software as commodity is key to knowledge as power: as the power to earn a good wage in emerging markets, if no longer necessarily in developed ones. Moreover, without computers understood as software-hardware hybrids, information would not be valuable: without the ability to process "information" efficiently, information would languish as so many factors to confuse human analysis (hence the promise and limitations of Bush's analog memex). Software/information as a commodity has depended on expanded intellectual property rights and encryption. If information's rampant reproducibility (a computer reads by writing elsewhere) once seemed to render intellectual property obsolete, new laws and technology make "fair use" almost impossible. Against these phenomena, free and open source software movements have emerged, movements that Kittler, in "Science as Open Source Process," sees as key to the ongoing survival of the university.[22] By emphasizing the free circulation of information, the Free Software Movement moves knowledge towards what Jean François Lyotard predicted it would be in a society of freely accessible information: the creative use of information. Information itself, Lyotard argued, is only valuable in a zero-sum game.[23]

The chapters in "Power-Code" take on "knowledge-power," offering parts of its grid, analyzing the rise of code and its relationship to the circulation knowledge and "empowerment"—issues posed in the previous section. Wolfgang Hagen, in "The Style of Sources: Remarks on the Theory and History of Programming Languages," stresses the importance of unarchivable and unforeseen programming languages to the transformation of the computer into a media machine. Friedrich Kittler, in "Science as Open Source Process" and "Cold War Networks or Kaiserstr. 2, Neubabelsberg," examines the institutional structures necessary for the emergence of software and cold war information networks. Tracing the relationship between power and code, Kittler provocatively argues that academic freedom will fall or stand with open source, for the free circulation of knowledge—without patents and copyrights—has always been crucial to universities. Hardware, on the other hand, is allied with secrecy, the military, and control.

The next five chapters debate the question of control, specifically the relationship between programming and agency, surfing and using. Lev Manovich in "Generation Flash" argues that programming in the early 2000s moved a new generation of artists away from the old and tired act of postmodern citation towards a new romanticism and a new modernist aesthetic of clean lines and transparent causality. In contrast to this vision of romantic creation, Alexander Galloway in "Protocol vs. Institutionalization" examines the control structures necessary for the so-called open circulation of knowledge, from theoretically open organizations comprising members of a relatively homogenous social class of techno-elites to TCP/IP, the protocol driving the Internet. The net, he argues, is founded on control, not freedom. Tara McPherson in "Reload: Liveness, Mobility, and the Web" weighs in on this debate by emphasizing the web as a technology of experience, rather than simply an effect of software. While critiquing the overblown promises made by commercial prophets of "convergence" during dotcom mania, McPherson argues that "choice," "presence," "movement," and "possibility" are all terms that could describe the experience of web surfing. Julian Dibbell, writing during the heyday of artificial life, returns us to the question of code, but through alien code: viruses whose assertive presence drives fear in the heart of users who believe they control their machines. Viruses, he argues, operate both as a virus-maker's signature and as a self-replicating program that denies authorship. Lastly, Anders Michelson in "The Imaginary of the Artificial: Automata, Models, Machinics—On Promiscuous Modeling as Precondition for Post-structuralist Ontology" argues that although the computer is based on "the image of man," it leads elsewhere. The "machinic" is now creative. It consitutes what he calls *the imaginary of the artificial*, "an inexplicit and poorly understood impetus for the creative articulation of the artificial."

The next section, "Network Events" further pursues knowledge-power, but rather than focusing on code or the computer, looks more broadly at global information flows. Transmission and "knowledge is power," it stresses, are not limited to computer buses or high-speed data networks. Concentrating on catastrophic media events and on the ways in which the media create a "we" and a "they," this section examines the possibilities and limitations of global mass media. It also delves into the various temporalities of media and mediated life, from Mary Ann Doane's analysis of television's reliance on the catastrophe (catastrophe allows U.S. television to mimic the experience of colliding with the real and to deny its reliance on capitalist economics) to McKenzie Wark's analysis of the limits of time-consuming traditional scholarship in "The Weird Global Media Event and the Tactical Intellectual [version 3.0]." According to Wark, catastrophic images are weird global media events: sudden irruptions of raw facticity that can redraw boundaries and reveal the time and power of the uneven media space in which they take place.

The next three chapters focus on the "communities" or audiences created by global media, as well as on popular and critical assumptions about the nature of technology and technological power. "We" may be unable to recognize the power of technology precisely because "we" want to see it as a direct cause and because "we" are formed in response to technology: "we" essentialize and fetishize technology, rather than examine the ways it amplifies forms of power with which "we" are already familiar. Arvind Rajagopal makes this point in "Imperceptible Perceptions in Our Technological Modernity," arguing that technology has become fetishized as the cause of racial and cultural difference in popular rhetoric and critical theory; but, as the 9/11 airplane flying terrorists and more positively activism on the part of "untouchables" in India reveals, global technology leaves no outside, leaves no one untouchable. Geert Lovink in "Deep Europe: A History of the Syndicate Network" exposes the fallacies of global communications as naturally solving the problems of history through a reading of Syndicate, an email list that sought to bridge East-West (Europe) through the notion of a "Deep Europe." Vicente Rafael in "The Cell Phone and the Crowd: Messianic Politics in the Contemporary Philippines" also interrogates media essentialism, power and dreams of contact, but through a reading of People Power II. Contemporary Filipino middle-class fantasies of the cell phone and the crowd, he argues, render the masses voiceless by viewing the cell phone and the crowd as simple transmitters of bourgeois justice.

The last section "Theorizing 'New' Media" pursues knowledge-power by investigating new media's impact on scholarly knowledge. Each author in this section either offers new theories or terms in light of "new media," or argues against their necessity. Together, these chapters map out the disciplinary challenges posed by "new media" to disciplines from Asian American Studies to literary studies; from queer to architectural theory. Lisa Nakamura begins this section with "Cybertyping and the Work of Race in the Age of Digital Reproduction," which introduces the term "cybertypes" to describe the ways in which race and ethnicity proliferated in mainstream new media during the late 1990s. Cybertypes, she argues, alleviate white anxiety in the face of fluid and uncertain identity by concealing the West's colonization of global media and its domestic racist practices; cybertypes, however, are also after/images—a mind's eye projection of the real—and thus open the possibility of seeing differently. Nicholas Mirzoeff in "Network Subjects: or, The Ghost is the Message" similarly contends that new media changes visual subjects' relationship to their media. In an analysis that moves from the Enlightenment to the present day, Mirzoeff argues that the medium itself has become the object and subject of desire, and that the endless repetition of visual selves leads to indifferent surveillance and indifference to surveillance. Ken Hillis in "Modes of Digital Identification: Virtual Technologies and Webcam Cultures" also addresses identity and desire from the Enlightenment to the present, but through the rubric of virtual reality and queer webcams. VR, he argues, blurs the boundary between the virtual and the real, leading us to reside not in the desert of the real, but rather in a magical world designed by humans for humans.

The next two chapters offer historical analyses that question the newness of new media, as well as various intellectual histories of it. According to Peter Krapp, many theories of new media portray it both as a radical departure and as a long awaited development, turning much of what new media has supposedly superseded into new media *avant la lettre*. This hindsight, Krapp argues in "Hypertext avant la lettre," is *the* symptom of new media. Mark Wigley in "Network Fever" similarly interrogates the newness of network analysis, arguing that we are at the end, rather than the beginning, of network logic. Tracing the complex web of interrelations between architecture and information theory, Wigley argues that contemporary discourse about the net realizes nineteenth-century fantasies that were acted out throughout the twentieth century.

Did Somebody Say New Media?

Slavoj Žižek in his introduction to *Did Somebody Say Totalitarianism* argues that totalitarianism serves as an ideological antioxidant, taming free radicals in order to help the social body maintain its politico-ideological good health. Totalitarianism has been used to dismiss Leftist critique of liberal democracy as the "twin" of Rightist fascist dictatorship: "the moment one accepts 'totalitarianism,' one is firmly located within the liberal-democratic horizon."[24] Thus, Žižek argues, totalitarianism "is a kind of *stopgap*: instead of enabling us to think, forcing us to acquire a new insight into the historical reality it describes, it relieves us of the duty to think, or even actively *prevents* us from thinking."[25] Although new media is clearly different from totalitarianism, it too can function as a stopgap. The moment one accepts new media, one is firmly located within a technological progressivism that thrives on obsolescence and that prevents active thinking about technology-knowledge-power. The term itself has circumscribed debate to Is new media new, or What makes it new? As a whole, this collection refuses new media as a stopgap, probing into the historical reality it describes. These essays, with considerable cohesion and integration across a disparate set of fields, provide new points of reference for evaluating all those claims—political, social, ethical—made about the digital age. They share a prejudice against representations of digital media as rendering obsolete or converging all other forms of media; as solving or perpetuating various sorts of social and political discriminations and oppressions; as economic miracle, nightmare, or fraud. They also share a common prejudice against simply dismissing those utopian promises made on behalf of new media, choosing instead to analyze the import and effect of those promises. Committed to historical research and to theoretical innovation and themselves historically located, they suggest that in the light of digital programmability, seemingly forgotten moments in the history of the media we glibly call "old" can be rediscovered and transformed.

This collection thus seeks to shake loose current intellectual trajectories and common sense understandings of new media—what it was, what caused it to be, what it will be. It challenges its status as new *or* old, as converging *or* diverging, as revolutionary *or* reactionary, concentrating instead on what—culturally, technologically, ideologically—enabled such adjectives to be applied to the Internet and other media classed as new. It also concentrates on the actualities of the media itself—its hardware, its software, its user interface—and on the experience of using it, of being entangled within it. Most importantly, it refuses to see new media as a simple cause and its effects as limited to those who use it on a daily basis. We thus offer this collection of theoretical and historical texts not to settle, but to unsettle, the question of the relationship between knowledge, information, code and power.

Notes

1. "Age of the World Picture." *The Question Concerning Technology and Other Essays.* Trans. William Lovitt. (New York: Harper and Row, 1977), 132.
2. "Nietzsche, Genealogy, History," *language, counter-memory, practice: selected essays and interviews by Michel Foucault.* Ed. Donald F. Bouchard. (Ithaca, NY: Cornell UP, 1977), 148–9.
3. Of course the term "multi-media" itself erased the multiplicity inherent to film, television, etc.
4. Anna Everett and John T. Caldwell, for instance, in their introduction to *New Media: Theories and Practices of Digitextuality* (New York: Routledge, 2003, xi–xxx) write, "When we consider the far-reaching impact of ascendant digital media systems and what their increasing corporatization augurs for individuals' technology access and technologized social processes alike, then the essential role of media theorists, scholars, and practitioners in helping to ensure the humanistic values prevail in the new digital order is clear" (xi).
5. See *Nettime-l* October 2003 "What *ARE* New Media?" http://www.nettime.org, the first section of Lev Manovich's *The Language of New Media* (Cambridge: MIT, 2001) and Lev Manovich's and Janet H. Murray's introduction to *The New Media Reader* (Cambridge: MIT 2003).
6. See Jay Bolter and Richard Grusin, *Remediation: Understanding New Media* (Cambridge: MIT, 1999).
7. Lisa Gitelman and Geoffrey B. Pingree, "Introduction" *New Media 1740–1915* (Cambridge: MIT, 2003, xi–xxii), xi, xv.
8. See Friedrich Kittler, *Gramophone, Film, Typewriter* (Palo Alto: Stanford University Press, 1999).
9. As Tom Gunning argues in "Re-Newing Old Technologies: Astonishment, Second Nature, and the Uncanny in Technology from the Previous Turn-of-the-Century" (in *Rethinking Media Change: An Aesthetics of Transition* edited by David Thorburn and Henry Jenkins [Cambridge: MIT, 2003, 39–60] the wonder associated with the "new" is renewed whenever technology fails.
10. Geert Lovink, "Archive Rumblings: Interview with German media archeologist Wolfgang Ernst," *Nettime-l*, 26 February 2003, http://amsterdam.nettime.org/Lists-Archives/nettime-l-0302/msg00132.html.
11. Ibid.
12. Foucault, *The Archaeology of Knowledge* (New York: Tavistock, 1972), 129, 131.
13. Knowledge [*savoir*] here refers "to all the procedures and all the effects of knowledge [*connaissance*] that are acceptable at a given moment and in a defined domain ... the term 'power' ... [covers] a whole series of particular mechanisms, definable and defined, that seem capable of inducing behaviors or discourses" (*Archaeology* 394).
14. Foucault, "What is Critique" in *What is Enlightenment?: Eighteenth-Century Answers and Twentieth-Century Questions.* Ed. James Schmidt (Berkeley: University of California Press, 1996: 382–98), 94.
15. Foucault, *The Archaeology of Knowledge*, 138, 144–5.
16. Ibid., 16.
17. Friedrich Kittler, *Gramophone, Film, Typewriter*, 5.
18. Friedrich Kittler in *Gramophone, Film, Typewriter* most forcefully claims: "Increasingly, data flows once confined to books and later to records and films are disappearing into black holes and boxes that, as artificial intelligences, are bidding us farewell on their way to nameless high commands. In this situation we are left only with reminiscences, that is to say, with stories. How that which is written in no book came to pass may still be for books to record. Pushed to their margins even obsolete media may become sensitive enough to register the signs and clues of a situation." (xl)
19. Manuel Castells, in *The Rise of the Network Society*, revised edition (Oxford: Blackwell Publishers, 2000), argues that information technologies (and thus the information technology revolution) first diffused widely in the 1970s. Although the 1970s are certainly important, it is not until the Internet emerged as a mass medium that the information revolution became part of everyday language. For more on the emergence of the Internet as a mass medium, see Wendy Hui Kyong Chun, *Control and Freedom: Power and Paranoia in the Age of Fiber Optics* (Cambridge, MA: MIT Press, 2005).
20. Thomas Keenan, *Fables of Responsibility: Aberrations and Predicaments in Ethics and Politics* (Stanford: Stanford University Press, 1997), 146–7.
21. N. Katherine Hayles, *How We Became Posthuman: Virtual Bodies in Cybernetics, Literature, and Informatics* (Chicago: University of Chicago, 1999).
22. The Open Source Movement insists that source code should be available to everyone, which does not mean that software should have no price or that open source software cannot be bundled with proprietary "add-ons." The Free Software Movement (which uses the GNU software license) believes that source code should be always be freely distributed—that everyone should be able to improve on it and these improvements should be free in turn (although again, this does not mean that software should have no price). Free software plays on a recognized hacker ethic: information should be free. Richard Stallman, pioneer of the Free Software movement, is unflinching in his belief that this is an ethical stance that has little to do with benefits of open source production.
23. Jean François Lyotard, *The Postmodern Condition: A Report on Learning*, trans. by Geoff Bennington and Brian Massumi (Minneapolis: University of Minnesota, 1984).
24. Slavoj Žižek, *Did Somebody Say Totalitarianism?* (London: Verso, 2001), 3.
25. Ibid.

Part I

Archaeology of Multi-Media

1

Early Film History and Multi-Media
An Archaeology of Possible Futures?

Thomas Elsaesser

Can Film History Go Digital?

The spectre stalking film history is that of its own obsolescence. It is widely assumed that the digital convergence between image-, audio- and print media—and thus the practice of multi-media—must inevitably modify and eventually overturn our traditional notions of film history. But this assumption rests on several unstated premises both about this convergence and about film history. What is evident is that the electronic media do not fit neatly into a linear or chronologically conceived film history, focussed on film as text or artifact. However, it is not at all obvious that digitization is the reason why the new media present such a challenge, historically as well as theoretically, to cinema studies.[1] Perhaps it merely forces into the open inherent flaws and contradictions, shortcomings and misconceptions in our current picture? Does the digital image constitute a radical break in the practice of imaging, or is it just the logical-technological continuation of a long and complex history of mechanical vision, which traditional film theory has never fully tried to encompass? Is film history vulnerable, because it has operated with notions of origins and teleology that even on their own terms are untenable in the light of what we know, for instance, about early cinema? This paper wants to put the latter question as its working hypothesis, and in order to do so, I want to start with identifying a number of what I take to be typical attitudes among film scholars when it comes to responding to the (digital) multi-media.

We Have to Draw a Line in the Silicone Sand

To some, the electronic media do not belong to the history of cinema at all. On this side of the divide are above all those for whom the photographic image is sacred, and for whom celluloid is the baseline of a 150-year visual heritage that must not be plundered, devalued, faked or forged. Jean Douchet, a respected critic in the tradition of André Bazin, thinks the loss of the indexical link with the real in the digital image presents a major threat to mankind's pictorial patrimony, as well as to a cinephile universe, of which he feels himself to be guardian:

the business of testing "content" for its acceptability across the different platforms, and their promotion of synergy is old Hollywood's vertical integration by another name: the takeovers and mergers hardly disguise the move towards monopolies, and an anti-trust case like the one brought against Microsoft indicates just how difficult, but also how necessary it is to monitor such cartels, if there is to be diversity and (some would say) innovation, which could counter the current convergence of multi-media towards mono-content.

Not so long ago, but before the high-tech bubble burst, London's *The Economist* ran a sobering survey about the IT revolution.[12] While it was true that the computer and modern telephony had brought a massive fall in the cost of communication and thus had increased the flow of information through the economy, it was not yet proven whether the "new economy" would be remembered as a revolution, in the same way as the invention of the steam engine had been a revolution, which—via the railways—created the mass-market. Or that of electricity, which—via the assembly-line, the extension of the working day, the invention of leisure and entertainment—brought about not only new and more efficient ways of making things, but led to the creation of new things altogether. The cinema, as we know, is very much a consequence of both these revolutions, of urbanization and electrification. According to *The Economist*, besides the cost of information, it is the cost of energy that is the real variable in a major, epochal social transformation, which is why it suggests that the development of new fuel cells may well be a bigger breakthrough on a global scale (when we consider also the political priority of "developed nations" to shed their dependence on oil) than either the computer or the mobile phone: a prediction that seems hard to believe from our present vantage point, not to mention from within our own discipline of film and media studies, except that the push towards miniaturization and mobility of our information and entertainment devices (e.g., laptop, mobile phone) also implies new and more efficient sources of energy.

Beyond the Post: Archaeology of a Media Revolution?

Where, in these different stances towards the digital does one locate oneself as a film historian? To be susceptible to the argument that only the silver-based photographic image counts, is to recognize the optico-chemical image's special historical value as a record with evidentiary as well as archival status. Film archivists, for instance, are convinced that celluloid is still a more durable and reliable material support of audiovisual data than digital storage media. On the other hand, to hold to the position that the photographic mode, from the vantage point of the post-photographic era, is merely a historically special instance of the graphic mode, is to acknowledge that photography, cinema, and the digital media merely reflect the respectively current technological state of this graphic mode. In such a perspective, the photographic mode (heavily fetishized in our culture because of its "realism," i.e., the seemingly unique combination of iconic and indexical reference) is merely one possible articulation, whose truth-claims are spurious and whose special evidentiary status much exaggerated. This is an argument which, at the height of the semiological turn and thus within a different vocabulary, was forcefully put by Umberto Eco when he deconstructed the indexical level of the photographic image into a dozen or so iconic and symbolic codes.[13] The Czech media historian Vilem Flusser also pointed out, some thirty years ago, that in any photograph, the distribution of the grain already prefigures both the dots of the video-image and the numerical grid of the digital image.[14] Other scholars and filmmakers have likewise drawn analogies between the mechanized loom of Jacquard in the eighteenth century, the Hollerith cards that made the fortune of IBM in the late nineteenth century, and the television image of the de Forester cathode ray tube in the twentieth century.[15]

If one therefore positions oneself, regarding the indexical nature of the photographic image, not in the past, but in the post, one tends to regard digitization less as a technical standard (important though this is, of course), but more like a zero-degree that allows one to reflect upon one's understanding of both film history and cinema theory. As a zero degree, it is, necessarily, an imaginary or impossible place from which one speaks when examining either "the new"or "the now." Digitization, at this early point in time, may for historians of the cinema be no more than the name of this impossible place, serving as a heuristic device, which helps them displace themselves in relation to a number of habitual ways of thinking. They need not decide whether digitization is, technically speaking, a moment of progress, but aesthetically speaking a step backward; whether it is, economically speaking, a risk, and politically speaking the tool of a new totalitarianism.

Instead, it permits a look at multi-media across a number of other, more abstract or general parameters, such as: fixed and/or mobile perceiver; image and/or text; distance and/or proximity; passive reception and/or interactive participation; two-dimensional "flat" image and/or three-dimensional virtual environment; looking through a "window on the world" and/or "immersed in a horizonless space." If these are some of the characteristics of the debate around multi-media, film scholars can once more find their bearings, since they are also the parameters familiar to any student of early cinema and of modern art.

Rather than pursue these aesthetic parameters, I want to sketch instead an archaeological agenda, taken from Michel Foucault's *Archaeology of Knowledge*, which, for instance, states: "archaeology does not imply the search for a beginning, [. . . it] questions the already-said at the level of existence [. . .] and it describes discourses as practices." It is easy to translate these three propositions into terms that echo the preoccupations of scholarship in early cinema: No search for beginnings: what early cinema has taught us is that the cinema has several origins, and therefore also no specific essence: in fact, at the limit, it has yet to be "invented."

Questioning the already-said at the level of existence: film history is best described as a series of discontinuous snapshots that illuminate a whole topography: the task is to map this field as a network, rather than as discrete units. More specifically, I am struck by the existence of what could be called (but finally are not) the S/M "perversions" of the cinematic apparatus. Among these normally-abnormal *dispositifs* one could name: **s**cience and **m**edicine, **s**urveillance and the **m**ilitary, **s**ensory-**m**otor coordination in the "movement image," and maybe I should add "GMS" and "MMS," to include the mobile phone alluded to above.

Discourse as practice: what does an archaeology of the discourses that constitute "cinema" tell us about it as a medium, and its relation to other media practices? Several scholars, notably Laurent Mannoni and Deac Rossell, have shown that the ideas and experiments of the so-called "losers" or "also-rans" in the race for being "first" in making moving images a viable reality have much to tell us about our present state of multi-media.[16]

I am only too aware of Friedrich Kittler's critique of Foucault: Kittler argues that Foucault's archive is the "entropy of the post-office," and that Foucault (along with Derrida) still sets writing and script as the default value of all communication and storage.[17] Foucault's mistake, according to Kittler, is that he does not see writing, too, as a technical medium, which means his notion of archaeology stops short prior to the modern recording media of gramophone, film, typewriter.[18] Kittler preferred to go to Lacan, but a Lacan read across Alan Turing, John von Neumann and Shannon-Weaver's information theory, in order to arrive at the appropriate theory of the "materialities of communication."[19]

I have elsewhere tried to look at what such a critique means for understanding the relation between distinct (multi-) media in their chronological succession, that is, the question of convergence, divergence, deferral, and difference.[20] It complicates the somewhat tongue-in-cheek position of George Lucas, quoted above, when he suggests that using digital equipment makes no difference to his *métier* as a director. For even when executing the same tasks, the change of medium alters forever the status of these tasks. In the case of the new digital media, we are as much subject to Marshall McLuhan's notion that the content of a medium is the form of the previous medium, as to Walter Benjamin's remark that art-forms often aspire to effects that can only be realized with the introduction of a "changed technical standard."[21] This is especially intriguing, seeing that the computer (as currently deployed in the generation of visuals) is not (yet) a technology of inscription and simulation, as much as it is one of transcription and emulation (of the effects of previous media-practices, from typewriter to camera, from newspaper to television, from radio to tape-recorder). Bolter and Grusin's notion of "Remediation" tries to address this issue, and Lev Manovich, too, has argued that the technically more advanced and historically more recent modes of media-practice do not oppose the previous ones, but in their organization subsume them, making their content and properties into mere "effects" that can be reproduced, usually faster, cheaper and in automated fashion. What has hitherto been thought of as the dominant mode or the default value of the cinematic system, namely live-action photography, now becomes a mere local instance of a practice or performance which the new medium organizes at a higher plane of generality. Thus the digital image, understood as a graphic mode, includes the photographic mode as only one among a range of modes or effects it is capable of. Rick Altman makes a similar point to that of McLuhan and Manovich when he argues that each successive technology is charged not to represent "reality," but the version of reality established and naturalized by a previously dominant technology.[22]

Archaeology I: The Cinema Has No Origins

How might this help us answer the point I began with, namely that digital media do not fit into tradition concepts of film history? A first step might be to deconstruct not only chronological uni-linear accounts, but also to put a question mark behind the "genealogical" approach to the cinema. Among film historians it is now generally accepted that the cinema has too many origins, none of which adds up to a history. For instance, if one goes back to the genealogies of the cinema reprinted in the textbooks of only twenty years ago, one can observe the kind of self-evidence that today seems startling for its blind spots. There, the history of photography, the history of projection, and the "discovery" of persistence of vision are listed as the triple pillars that sustain the temple of the Seventh Art. Or, to change the metaphor: they appear as the three major tributaries that finally—miraculously but also inevitably—join up around 1895 to become the mighty river we know as the cinema. But as we also know, an archaeology is the opposite of genealogy: the latter tries to trace back a continuous line of descent from the present to the past, the former knows that only the presumption of discontinuity and the synecdoche of the fragment can hope to give a present access to its pasts.

A media archaeologist would therefore notice above all what is missing or has been suppressed and left out in our genealogical chart. Sound, for instance, since we now know the silent cinema was rarely if ever silent, in which case: why is the history of the phonograph not listed as another tributary? Or what about the telephone as an indispensable element of what we would now understand by the cinema in the multi-media environment? Radio-waves? The wave and particle-theories of light? Electro-magnetic fields? The history of aviation? Do we not need Babbage's difference engine ranged parallel to his friend's William Henry Fox Talbott's Calotypes or Louis Daguerre's sensitized copper plates? Here, our media-archaeologist might begin to protest, arguing that we are simply being additive, factoring in the "missing links," while still operating within basically mono-medial teleologies, except that we have inverted them, since we are now guilty of a kind of hind-sight history, unrolling the whole story backwards from our own—no doubt equally limited and partial—contemporary perspective of the computer-phone-Internet-satellite configuration.

If we were to time-travel, and place ourselves at the end of the nineteenth century, we could see the cinematograph in 1895, depending on the vantage point, both as a Johnny-come-lately and a perilously premature birth. A latecomer, in that the Lumières' invention was no more than a mechanized slide-show, whose special effects for a long time were inferior to any twin or triple-turret magic lantern, worked by a singer-lecturer assisting the skilled lanternist-operator, which could supply sound and image, verbal commentary and color, abstractly moving designs and representations from life. Premature, as we shall see, because the late nineteenth century might have been poised on the brink of a quite different imaging technology, which the popularity of the cinema in some ways "delayed."

Few now recall that many of the so-called pioneers—among them Pierre Jules César Janssen, Ottomar Anschütz, Etienne-Jules Marey, Edweard Muybridge and even the Lumière Brothers—were either not at all, or not primarily interested in the entertainment uses and storytelling possibilities of the cinematograph, thinking of it in the first instance as a scientific instrument or toy. Were they blind to the economic potential of entertainment and its social role in the late nineteenth century, or did they have something in mind that only the emergence of an entirely different technology nearly a hundred years later could bring to light? A media archaeologist faces any number of such questions that need to be put to film history. The answers are likely to lead to even more revisions in our conception not only of early cinema, but of the cinema in general.[23] So much so, that today, near-forgotten figures such as Marey or his assistant Georges Demenÿ look as interesting as the Lumière Brothers,[24] and Oskar Messter seems as emblematic for an archaeology of multi-media as Thomas Alva Edison used to be for the history of the cinema and the film industry.[25] Never very well-known outside Germany, Messter and his Alabastra 3-D projections of 1900, his synchronized sound pictures from 1902, his medical films from 1904, or his airborne surveillance cameras from 1914 nonetheless strike one as sometimes more fantastic than Jules Verne's novels, and much more prescient, because nearly all his ideas were implemented. Messter's indefatigable search for applications of the moving image parallel to its entertainment uses testify to such a pragmatic understanding of the different potentials of the cinematic apparatus that he stands at the intersection of several histories, many of which we are only now recognizing as having *histories*: those configurations and applications of the basic apparatus I earlier listed as its S/M practices.

Thanks to Paul Virilio and Friedrich Kittler, (but also thanks to CNN, Iraq, Serbia, Kosovo, Afghanistan . . .), we know a good deal more about the complex War and Cinema—or "surveillance and the military"—than even two decades ago.[26] In other words, it is the very practical and urgent impact of satellite technology, space exploration, and airborne or terrestrial surveillance that has sensitized us to a continuous, if submerged alternative history of cinema, which is now being recovered in the form of an "archaeology" of the present.[27]

Yet it is worth recalling also the opposite: that much of what we now consider as belonging to early film and thus to the history of cinema was not initially intended or indeed suited to performance in a movie-theater: scientific films, medical films, or training films, for instance. At the same time, such staples of early cinema programming as the view, the actualities, and many other types of films or genres, did initially rely on techniques of vision and on a habitus of observation that had to be "disciplined," in order to fit into the movie theater and become suitable for collective, public reception. Think of the landscape view, or the painted panorama: prior to the cinema, they relied on the mobile observer, optimizing his varying point of view; think of the stereoscope, or the so-called "Claude glass" and a multitude of other devices: they were in everyday use, but usually in the privacy of the home, in the artist's studio, or handled by a solitary spectator.[28] Yet the cinema borrowed from all these genres and practices, adapting them and significantly transforming their cultural meaning. In the process, both the mode of presentation and the audiences had to be "adjusted"—to fit into the movie-theater and its program format.

What this suggests is that the different ways in which the moving image in its multi-medial electronic form is today "breaking the frame" and exceeding, if not altogether exiting the movie theater (giant display screens in airport lounges or railway stations, monitors in all walks of life, from gallery spaces to museum video art, from installation pieces to football stadiums, from London's Hyde Park during Lady Diana's funeral service in Westminster Abbey to DVD-movies on laptop computer screens) indicate that we may be "returning" to early cinema practice,[29] or we may be on the threshold of another powerful surge of "disciplining" and normatively prioritizing one particular standard of the multi-media image over others. However, the instability of the current configuration is by no means novel. For instance, audiences seem to have been there before, if less dramatically, when the drive-in cinema was competing with the television screen, converting the automobile into a living room, combining the erotic intimacy of "staying home" with the giant outdoor screen of "an evening at the movies." More generally, and going back to the "origins" of the cinema, it will be remembered how unstable, around 1895, were the definitions and minimal conditions that eventually led to exactly dating the cinema's invention. Some of the questions were: does chronophotography qualify as cinema, or do we require the Maltese cross to give the illusion of continuous motion? Why was Emile Raynaud's continuously moving strip of paper, with painted images projected on to a screen not good enough as the birth of cinema? Why should only images taken with a camera and fixed on celluloid qualify? If photographic images, why not Edison's peephole device instead of the Lumières (later and derivative) device for projecting images on a screen? Did it make a difference if these moving images were first shown to a scientific community or before a paying public? As we know, it was decided that only the latter audience "really" counted, with the result that in the end it took four or five different (some would say, arbitrarily selected) qualifiers or limiting conditions, in order to make December 28th, 1895 the date, and the Lumière Brothers the authors of the "invention" of the cinema.[30] In this sense, the history of the cinema responds not so much to the Bazinian inquiry "what is cinema," but has to start from the question: "when is cinema"?

Archaeology II: Film in the Expanded Field, or "When is Cinema?"

In other words, were one to construct the "origins" of (digital) multi-media backwards in the manner of the new film historians, trying to date the "birth" of the cinema, one would face some hard choices. I mentioned factoring in Babbage's difference engine and Bell's telephone. But nearer home, i.e., today's digital world, necessary additions and adjustments might include the Morse code or the radar screen. For an archeological approach, on the other hand, it may be a matter not only of broadening the range of questions considered pertinent, but once more to shift the angle of inquiry and revise one's historiographic premises, by taking in the discontinuities, the so-called dead-ends, and by taking seriously the possibility of the astonishing otherness of the past. That the case for a wider agenda in film history, as well as for a different focus, is a compelling one, has not been an insight exclusively owed to the new media. Even before the advent of digitization, it was obvious that the cinema had always also existed in what one might call an expanded field.[31] "Expanded field" in the sense indicated above, namely that there have been very distinct uses of the cinematograph and the moving image, as well as of the recording and reproducing technologies associated with them, other than in the entertainment industries. What is new—and perhaps a consequence of the new digital media—is that we are now willing to grant these uses the status of parallel or parallax cinema histories.[32]

For a sense of this expanded field in the context of alternative histories, an anecdote once told to me by Vivian Sobchack might illustrate the point. One day, she was driving on a San Francisco freeway behind a van with the words "Pullman's Underground Film" written on the back. Being a film scholar with catholic interests, she became curious, since in all her years of teaching the American avant-garde, she had never come across a filmmaker or a collective by that name. As she accelerated and leveled with the van, in order to see whether she recognized anyone inside, she read, neatly stenciled across the driver's door: "Pullman's Underground Film: The Bay Area's Specialists in Electronic Sewer Inspection."

Perhaps only in the state and the region that is home to the Pacific Film Archive and to Silicon Valley could the industrial users of cameras salute the artistic film community with such a handsome tribute. But as the case of the so-called pioneers shows: the non-entertainment and nonart uses of the cinematic apparatus at the turn of the nineteenth to the twentieth century did not disappear with the institution of narrative cinema as the norm, or the emergence of the full-length feature film around 1907, they merely went underground. But this underground was in many instances contiguous to the above ground, and in several cases the very condition of possibility for the developments in the cinema's entertainment uses, certainly when we recall once more how many of the technical innovations in the fields of photography, the cinema, and the new media were financed and first tested for warfare and military objectives (to name just a few of the best-known: the powerful searchlights of WWI, the 16mm portable camera, the Ampex (audio- and video-) recording tape, the television camera, the computer, the Internet). Hence my suggestion of the different S/M registers of the cinematic apparatus: surveillance and the military, science and medicine, sensoring and monitoring—to which, in a Deleuzian spirit, I added a fourth: the sensory-motor coordination of the human body in classical cinema.

It would take me too far to pursue these practices and their *dispositifs* in detail here, or to construct around them the kind of film history of image-interference that would open up to surprising connections even the cinema-history we think we know so well. Jean-Luc Godard, in his *Histoire(s) du cinéma* draws strong conclusions of complicity and disavowal from similar historical montage-effects, when on footage taken from George Stevens' 16mm color film of the U.S. Army's liberation of Nazi camps he superimposes a scene featuring Elizabeth Taylor and Montgomery Clift from the same director's (black-and-white) studio-production *A Place in the Sun*.[33]

Archaeology III: Discourses in Default: The Dog That Did Not Bark

I suggested earlier that the cinema was not only a late-comer, if we consider that most of the technologies necessary for its implementation had been known for some fifty years previously. Judged by its effects, it was also a bit of a changeling, having had to compete with much grander spectacles like panoramas, phantasmagorias, and the skilful suggestion of motion, of dissolves and superimpositions done with magic lanterns. Yet there is even a sense in which the cinema was not only a bastard, but an unwanted child altogether. According to some scholars, neither Edison's peep show nor Lumière's public projection was what the nineteenth century had been waiting for. What it was imagining for its technotopic future was domestic television, and preferably two-way television.[34] And the Victorians not only dreamt of television. They were as hungry for instantaneity, for simultaneity, and interactivity as we are today, and they also had a good idea of what it would mean to be connected to an internet: after all, they had developed the telegraph-system![35]

This puts me in mind of the well-known Sherlock Holmes story of "the dog that did not bark," which turns on Holmes' ingenious deduction that the burglar could not have been a stranger, since the house was guarded by a dog—that did not bark. The story makes a point, useful for historians and heartening to the media archaeologist, namely that the vital determinant might be the one you have overlooked, because its significance lies in its absence. For instance, years ago, I finally grasped the editing principle of Edwin S. Porter's *Life of an American Fireman*, when it was pointed out to me that in order to explain the overlap of the rescue scene (which is shown successively, from outside the house and then again from the inside), one only had to think of it as early cinema's version of television's "action replay" mode. After all, when a goal is being scored during a televised soccer game, it is shown repeatedly from different angles, and at different speeds. Likewise, a dramatic rescue of a woman and her child from the raging flames deserves an action replay, too. The dog that did not bark in *Life of an American Fireman*, in other words, was the lecturer, the *bonimenteur*, whom Edwin Porter could assume to have commented the action when his film was being shown.[36]

More generally, the dog that did not bark for generations of early cinema scholars, was, of course, sound. Only recently have we begun to realize not only the importance of sound-effects, but also the huge variety of musical accompaniments, the different kinds of off-screen sound, in-house commentary, and even "the silences" of early cinema.[37] Thus, some of the most interesting work on the multi-media aspects of early cinema in a historical perspective that illuminates our present situation comes from scholars who, for the last decade or two, have radically revised our notion of sound and cinema. We can now inform ourselves about the Gaumont sound systems, the Messter sound system, the Lloyd Lachmann system, the Beck system, the Noto-system and countless others, most of them very ingenious (and some of them even successful) in providing constant if not permanent synchronisation well before 1927.[38] Equally intriguing is the fact that systems were developed, where synchronization was not the only aim of marrying or combining sound, music, text, and image. The exhibition context, the contact space of live audiences and what could be called the "performative imperative" also played an important role. The history of sound prior to 1927 is also the story of the auditorium space as a multi-medial space, just as the history of early sound film up to the mid-1930s, at least in Europe, is incomprehensible if one does not factor in radio as an institution, and the gramophone as the key home entertainment gadget, with hit songs and theme tunes—then, as now—a major selling point for the products of the film industry.[39]

Yet why, until two decades ago, was this knowledge deemed irrelevant? Perhaps in order to obtain the neatly linear film history we have been accustomed to, instead of having to trace the crooked dog-leg logic that the cinema did in fact follow (and which we still only partly understand)? It follows from this that the cinema's traditional telos of greater and greater realism, or the classic evolutionary scheme from silent to sound, from black and white to color, from the flat, two-dimensional screen surface to 3-D, from peephole to IMAX-screen just does not hold up: all the "from... to" histories have for too long been, as we now realize, deeply flawed. They seem factually so inaccurate as to make one wonder what kind of intellectual sleight of hand, or acts of censorship must have taken place for so much knowledge about early cinema and so many discourses about color, sound, and the many experiments with giant screens or 3-D glasses to have been "forgotten." What secret wish, what mixture of belief and disavowal has been attached to the dominant teleological narrative to make it gain such wide circulation, to give it the credibility of a doxa and the unquestioned certainty of the commonplace?

From "the dog that did not bark" in cinema history, to the "dog-leg logic" of its actual development (to which we might add the "wagging the dog" logic of its inverted cause-and-effect relationships): such might be an alternative agenda for "revisionist" film historiography in order to integrate, rather than merely accommodate, the cinema's relation to digital multi-media. Their reliance on what I have called the parallel histories or S/M practices of the cinematic apparatus are so much more evident that we can now see these histories as discourses and these discourses as practices; it would even be inaccurate to say that they went underground.

Perhaps it was us, the film historians who have been underground. For the history of early cinema in the expanded field can, as indicated, provide many names of inventors, showmen, and *bricoleurs* whose ways of thinking about moving images, about sound-and-image combinations, about simultaneity and interactivity landed them in dead-ends, at least from the retrospective teleology of the traditional "birth" of cinema. An archaeology of multi-media, by contrast, gives a glimpse of the different balance sheet of winners and losers, losers as winners. It puts one in mind of another of Walter Benjamin's sayings—that history is usually written by the winners: in the new film history, the losers can once more have a place. For what an archaeological practice very quickly teaches one is not only that it is hard to tell winners from losers at this stage in the game, but that we are constantly rediscovering losers in the past who turn out to have become if not winners, then the great-grandfathers of winners.

As so often in the history of inventions, some of the most influential or momentous ones were the by-products of quite other discoveries, or turned out quite differently from what their makers had intended: technical "progress" has rarely the eureka-experience and more often a knight's move logic as its basis. If the history of the cinematic apparatus is a good example of this, the film projector to this day is its perfect image: apart from being a mechanized magic lantern, it still shows quite clearly that what allowed this magic lantern to be mechanized were the treadle sewing machine, the perforated Morse telegraph tape and the Gatling machine gun. All three have disappeared in their respective areas of applications, but they are miraculously preserved in the retrofitted adaptation still to be found in every projection room (though probably not for much longer). A media-archaeologist of "virtual reality" might well be prepared to trade the history of the camera obscura and the stereoscope (so crucial to the historian of cinema), for learning more about Messter's Alabastra projections, Mesdag's panorama in Scheveningen, or Robertson's phantasmagorias. To which an archaeologically minded art-historian might add: why go to Eadweard Muybridge, if you can learn all you need to know about the late nineteenth century's obsession with fixing and recording the fleeting moment not from chronophotography, but from studying Manet's brushstrokes and the folds in his female figures' dresses?[40]

It was indeed the film historians, who have perhaps been in the dark too long: we had not noticed—maybe because we did not want to notice—how, for instance, the military tail had been wagging the entertainment dog all along, or how the Orwellian nightmare of surveillance had probably also all along been the mask and mimicry of the performative pleasure of being seen, of being looked at, and of being looked after. We may have to welcome the multi-media as not so much the emulation of cinema, or as the "content" of its form. Rather, while the industry is waiting for a "killer application," historians might consider the multi-media in Benjamin's sense, as the realization of those effects that the cinema could not itself deliver, however much the Lacanian "stade du mirrior" paradigm and its subsequent look/gaze theoretical elaborations in film studies had tried to extend it in this a direction. With the multi-media, another age-old dream seems to be coming true: *esse est percipi*—to be is to be perceived. That, too, is of course a thought in the spirit of Foucault. It would make the history of the cinema more like the archaeology of the panopticon, and in the Nietzschean absence of God, the dream would no longer be for humankind's immortal double, but for someone to—once again—watch over you: a specter is, after all, stalking film history—the absence of "God" as the loss of faith in perception.[41]

Either way, one conclusion might be that the new digital media's relation to cinema is neither a matter of opposition to classical cinema (in the form of a "return" of a cinema of attraction), nor as its McLuhanite subsumption or emulation. Early cinema, classical cinema and contemporary post-cinema can also be seen on another, if even more complex line of development, where each marks a step in the severance of images from their material referents—a story that could take us at least as far back as the Renaissance. If in the transition from early to classical cinema, it was narrative as the logic of implication and inference that both "translated" and "preserved" the image's "here" and "now," the switch from the photographic to the post-photographic or digital mode allows moving images to "represent" time in ways not encompassed by narrative, hitherto the cinema's most familiar spatio-temporal support and indexical register. In which case, the moving image will have lent itself to the culture of telling stories only for a short while, a mere hundred years or so, before it began to move on. No doubt, once we know where it is heading, a new "archaeology" will also have to be at hand.

Notes

1. This paper extends some of the ideas first put forward by me in "Convergence, Divergence, Difference" in Thomas Elsaesser, Kay Hoffmann (eds.), *Cinema Futures – Cain, Abel or Cable? The Screen Arts in the Digital Age* (Amsterdam: Amsterdam University Press, 1998), 9–23.
2. Le Cinéma: Vers son deuxième siècle, conference held at the Odéon, Paris, 20 March 1995. Press handout of Jean Douchet's lecture, in English, 1.
3. "In their imagination the [pioneers] saw the cinema as a total and complete representation of reality; they saw in a trice the reconstruction of a perfect illusion of the outside world in sound, colour and relief." André Bazin, "The Myth of Total Cinema," in *What is Cinema: Vol. I* (Berkeley: University of California Press, 1967), 20. See also Warren Buckland, "Between Science Fact and Science Fiction: Spielberg's Digital Dinosaurs, Possible Worlds, and the New Aesthetic Realism," *Screen* vol. 40, no. 2 (Summer 1999,), 177–192. The link between the aesthetics of neorealism and immersive virtual reality is made implicitly in the opening section ("The Logic of Transparent Immediacy") of Jay David Bolter and Richard Grusin, *Remediation* (Cambridge: MIT Press, 1999), 21–31.
4. Kevin Kelly and Paula Parisi, "Beyond Star Wars What's Next for George Lucas," in *Wired* 5.02 (February 1997): 164.

5. See the most recent edition of *Film Art*, sporting a still from *The Matrix* on the cover, and Kristin Thompson, *Hollywood Storytelling* (Cambridge: Harvard University Press, 1998).
6. Tom Gunning, "An Aesthetic of Astonishment. Early Film and the (In)credulous Spectator," *Art & Text* 34 (1989): 31–45.
7. Ann Friedberg, "The End of Cinema: Multi-media and Technological Change," in C. Gledhill and L. Williams (eds.), *Re-inventing Film Studies* (London: Arnold, 2000): 438–452.
8. Lev Manovich, "Prologue: Vertov's Dataset." In *The Language of New Media* (Cambridge: MIT Press, 2001), XIV–XXIV.
9. When it first appeared in the summer of 2001, the Microsoft gamebox was specifically designed and marketed as a "convergence" device, in order to bring together the computer, the television set and internet access around a video-game console. *The Economist* called it a "Trojan horse" in the home (*The Economist*, October 20, 2001).
10. Justine Cassell and Henry Jenkins (eds.), *From Barbie to Mortal Kombat. Gender and Compter Games* (Cambridge: MIT Press, 2000).
11. Earlier examples of an "inferior" technology winning because of a better distribution infrastructure are the Morse telegraph which won against superior European *apparati* in imposing its standard, and the Bell telephone, widely regarded as initially inferior to Elisha Gray's machine.
12. "A Survey of Innovation in Industry," Special Supplement, *The Economist*, February 20, 1999, 5–8.
13. Umberto Eco, "Zu einer Semiotik der visuellen Codes," in *Einführung in die Semiotik* (Munich: Wilhelm Fink, 1972), 195–292, esp. 214–230.
14. Some of Flusser's essays have now been collected and posthumously edited as *Villem Flusser, Ins Universum der technischen Bilder* (Munich: European Photography, 2000).
15. Among the filmmakers, one could name Harun Farocki, whose *Wie Man sieht* (Germany, 1986, 16 mm, color, 72 min.) provides such an "archaeology" of the links between the television image and the computer.
16. See also Geoffrey Batchen's "Electricity Made Visible," where he makes a strong case for Morse, Lenoir and Bidwell to be reinscribed into the prehistory of audiovisual digital media, even if they do not belong to the prehistory of cinema.
17. "Foucault's idea of the archive—in his practice as a scholar if not in his theory—is identical with the library—always proceeded from the historical a-priori of writing. Which is why discourse analysis has had trouble only with periods whose modes of data-processing exceeded the alphabetic monopoly of storage and transcription. [No wonder that] Foucault's historical work stops around 1850." Friedrich Kittler, "Nachwort," in *Aufschreibsysteme 1800/1900* (2nd ed.), (Munich: Wilhelm Fink, 1987), 429.
18. Friedrich Kittler, *Discourse Networks* (Stanford: University of Stanford Press, 1996).
19. Friedrich Kittler, "Die Welt des Symbolischen, eine Welt der Maschine," in *Draculas Vermächtnis* (Leipzig: Reclam, 1993), 58–80. See also Hans Ulrich Gumbrecht and K. Ludwig Pfeiffer (eds.), *Materialities of Communication* (Stanford: Stanford University Press, 1994).
20. "Digital Cinema: Delivery, Event, Time" in T. Elsaesser, K. Hofmann (eds.), *Cinema Futures. Cain Abel or Cable* (Amsterdam: Amsterdam University Press, 1996), 201–222.
21. "One of the foremost tasks of art has always been the creation of a demand which could be fully satisfied only later. The history of every art form shows critical epochs in which a certain art form aspires to effects which could be fully obtained only with a changed technical standard." Walter Benjamin, *The Work of Art in the Age of Mechanical Reproduction, Illuminations* (New York: Schocken, 1969), 237.
22. Altman calls this the "representation of representation" and he draws the following conclusion: "This new approach considers that every ideological force must by necessity grapple with the residue of another ideological impetus embodied in competing representational modes." Rick Altman, "Representational Technologies," *Iris*, vol.2, no.2, (1984), 16.
23. Tom Gunning, *D. W. Griffith and the Origins of American Narrative Film* (Urbana: University of Illinois Press, 1991); Noel Burch, *Life to Those Shadows* (Berkeley: University of California Press, 1990); Charles Musser, *Before the Nickelodeon: Edwin S. Porter and the Edison Manufaturing Company* (Berkeley: University of California Press, 1991) and William Uricchio, "Cinema as Detour?" in, K. Hicketier, E. Müller, & R. Rother (eds.), *Der Film in der Geschichte* (Berlin: Sigma, 1997), 19–25.
24. On Marey, see Marta Braun, *Picturing Time: The Work of Etienne-Jules Marey* (Chicago: University of Chicago Press, 1992).
25. For a shift in the evaluation not only of the French pioneers, see Laurent Mannoni, *Le grand art de la lumière et de l'ombre* (Paris: Nathan, 1994). The reassessment of Oskar Messter is due largely to Martin Loiperdinger's archival work. See M. Loiperdinger (ed.), *Oskar Messter: ein Filmpionier der Kaiserzeit* (Frankfurt: Strömfeld/Roter Stern, 1994).
26. Paul Virilio, *War and Cinema: Logistics of Perception* (London: Verso, 1997). Friedrich Kittler, "Gramophone, Film, Typewriter," October 41 (Summer 1987): 101–118.
27. See Thomas Y. Levin, Ursula Frohne, and Peter Weibel (eds.), *CTRL [SPACE]: Rhetorics of Surveillance from Bentham to Big Brother* (Cambridge: MIT Press 2002).
28. For the presence of optical toys and precision instruments in the artist's studio, see David Hockney, *Secret Knowledge* (London: Phaidon, 2001).
29. Extrapolating from Tom Gunning's concept of the "cinema of attraction," several scholars have argued for such a "return." See Vivian Sobchack's *Screening Space. The American Science Fiction Film* (2nd ed.) (New York: 1987); Scott Bukatman, *Terminal Identity. The Virtual Subject in Postmodern Science Fiction* (Durham, NC and London: Duke University Press, 1993); Miriam Hansen, "Early cinema, late cinema: permutations of the public sphere" *Screen*, 34/3 (Fall 1993): 197–210.
30. Deac Rossell, *Living Pictures. The Origins of the Movies* (Albany: University of New York Press, 1998).
31. The idea of "expanded cinema" originated in the avant-garde of the 1960, which successfully reconstructed for itself a pedigree and a tradition. See Gene Youngblood, *Expanded Cinema* (New York: E.P. Dutton, 1970).

32. For the concept of parallax histories, see Catherine Russell, *Narrative Mortality. Death Closure and New Waves Cinemas* (Minneapolis: Minnesota University Press, 1995), 186–187.
33. The sequence is discussed by Alan Wright "Elizabeth Taylor at Auschwitz: JLG and the real object of montage," in Michael Temple and James Williams (eds.), *The Cinema Alone* (Amsterdam: University of Amsterdam Press, 2000), 51–60.
34. Siegfried Zielinski, in *Audiovisions* (Amsterdam: Amsterdam University Press, 1999), argues that the cinema has been no more than an "intermezzo" in the history of audio-visions. See also the widely reproduced cartoon from Punch, 1879, which shows grandparents sitting in front of a fireplace, and above, instead of a mirror, a two-way screen with attached telephone that allows them to see and speak to their daughter and grand-children in the colonies.
35. Tom Standage, *The Victorian Internet* (New York: Walker & Co, 1998).
36. André Gaudreault, "Le retour du bonimenteur refoulé…(ou serait-ce le bonisseur-conférencier, le commentateur, le conférencier, le présentateur ou le 'speacher')," *Iris* 22 (Fall 1996): 17–32.
37. Rick Altman, "The Silence of the Silents," *The Musical Quarterly* 80/4 (Winter 1996): 648–718.
38. Michael Wedel, "Messter's 'silent' Heirs: Synch Systems of the German Music Film 1914–1929." *Film History* 11/4 (1999): 464–476.
39. Thomas Elsaesser, "Going Live: Body and Voice in Early Sound Film," in G. Krenn and A. Loacker (eds.), *Zauber der Boheme* (Vienna: Film Archiv Austria, 2002), 271–298.
40. On the instant in Manet and its relation to cinema, see Jacques Aumont, *The Image* (London: BFI, 1998) and Jonathan Crary, *Suspensions of Perception* (Cambridge: MIT Press, 1998).
41. "We have all had enough, hearing about the death of God […]. What has happened was simply the progressive disintegration of a faith in perception […]; the zero degree of representation merely fulfilled the prophecy voiced a thousand years earlier by Nicephorus, Patriarche of Constantinople during the quarrel with the iconoclasts: 'If we remove the image not only Christ, but the whole universe disappears.'" Paul Virilio, *The Vision Machine* (London: BFI Publishing, 1994), 16–17.

2

Electricity Made Visible

Geoffrey Batchen

> *"if... electricity can be made visible... I see no reason why intelligence might not be instantaneously transmitted by electricity to any distance."*
>
> —Samuel Morse, 1837

In his recent book, *The Language of New Media*, perhaps the most intelligent yet written on the subject, Lev Manovich attempts to provide a genealogy for the language of the computer and therefore of new media in general. Manovich defines "language" in somewhat formal terms—"the emergent conventions, recurrent design patterns, and key forms of new media"—even while he is concerned to locate these conventions, patterns and forms within a relevant cultural and conceptual history. And as with all histories, this concern periodically touches on questions of origin and essence. As he puts it, "if we construct an archaeology connecting new computer-based techniques of media creation with previous techniques of representation and simulation, where should we locate the essential historical breaks?"

Where indeed? Manovich himself decides to use a theory and history of cinema as the "key conceptual lens" through which he will look at this question. This is despite his concession that two important moments in his genealogy—the concurrent inventions of photography and computing—precede the emergence of cinema by seventy years or so. He explains this temporal gap by arguing that "the two trajectories [photo-media and computing] ran in parallel without ever crossing paths."[1] Until, apparently, the "key year" of 1936, when a German engineer named Konrad Zuse began building a digital computer (the Z1) in his parents' living room that used punched tape made from discarded 35mm movie film.[2] "Zuse's film, with its strange superimposition of binary over iconic code, anticipates the convergence that will follow half a century later. The two separate historical trajectories finally meet. Media and computer—Daguerre's daguerreotype and Babbage's Analytical Engine, the Lumiére Cinématographie and Hollerith's tabulator—merge into one. All existing media are translated into numerical data accessible for the computer."[3]

Zuse's machine is a wonderfully concrete metaphor for Manovich's origin story, and he quite appropriately repeats its conceptual architecture as the cover design for his book. But the plausibility of this particular historical metaphor depends on two provocative claims: that computing and photo-media have no interaction until the 1930s and that cinema is the key to any understanding of the forms and development of new media. Such claims represent a challenge to all historians of visual culture, asking us to address in more detail the genealogy of new media and to articulate the nuanced history that it deserves. This essay aims to be one more, necessarily small, contribution to this task. In the process it will extend Manovich's narrative back about one hundred years in order to look at two further artifacts of metaphoric import for new media: a photogenic drawing of a piece of lace sent by Henry Talbot to Charles Babbage in 1839, and Samuel Morse's first electric telegraph instrument, made in 1837.

Not that either of these rather modest-looking objects tells us very much on its own (each represents, in fact, the intersection of a number of other communication systems and technologies). In any case, as Michel Foucault has insisted, "archaeology is not in search of inventions... What it seeks... is to uncover the regularity of a discursive practice."[4] So my examination of these two artifacts will seek to place them within a broader set of discursive practices that I will argue provide the foundations for another reading of the history of both "new media" and its logics.

It's strange that Manovich identifies the beginnings of photography with the work of Frenchman Louis Daguerre and his metallic daguerreotype process rather than with the paper-based experiments of Englishman William Henry Fox Talbot. Strange, because Talbot was a close friend of Charles Babbage, the inventor of the computer. Both being expert mathematicians, there was considerable exchange between the two men about their respective experiments. I have written about the extent of their interactions elsewhere, but it seems worth repeating some of that here.[5] Prompted by the announcement in France on January 7, 1839 of the invention of Daguerre's photographic system, Talbot hurriedly presented a selection of his own prints to the Royal Institution in London on January 25. The title of an essay by Talbot released a week later begins by posing the problem of photography's identity. Photography is, he tells us, "the art of photogenic drawing," but then he goes on to insist that, through this same process, "natural objects may be able to delineate themselves without the aid of the artist's pencil."[6]

So, for Talbot, photography apparently both is and is not a mode of drawing; it combines a faithful reflection of nature with nature's production of itself as a picture, somehow incorporating the actions of both the artist *and* that artist's object of study. With this conundrum in place, he goes on in his text to posit yet another. Never quite able to decide whether the origins of photography are to be found in nature or in culture, Talbot comes up with a descriptive phrase that contains elements of each: "the art of fixing a shadow." In adopting such a phrase he recognises that photography is actually about recording the absence of light, or at least the differential effects of its absence or presence. To put it in more contemporary terms, photography is a binary (and therefore numerical) system of representation involving the transmutation of luminous information into on/off tonal patterns made visible by light-sensitive chemistry. As Roland Barthes has argued, then, the emergence of photography represents, among other things, a "decisive mutation of informational economies."[7]

This is never so clearly expressed as in Talbot's many contact prints of pieces of lace. To make such a contact print or photogram, the lace first had to be placed directly on photographic paper, paper designed to register this differential play of light. Here object and image, reality and representation, come face to face, literally touching each other. Only when the lace has been removed can its photographic trace be seen, a trace composed of just dark spaces and white lines (no shading or tonal range here). By this means, photography allows Talbot's lace samples to be present as image even when they are absent as objects. In other words, a piece of lace is transformed by photography into a *sign* of lace, into a ghostly doubling of the lace's identity. This doubling is doubled again when, as in the vast majority of cases, Talbot presents this sign to us in its negative state (so that what was black in reality is white in the image, and so on). As an overt simulation, then, the photogram's persuasive power depends on a lingering spectre of the total entity, a continual re-presentation of the initial coming together of image and lace on the photographic paper. Accordingly there is always this prior moment, this something other than itself, to which the photogram (and photography in general) must continually defer in order to be itself.

Figure 2.1 William Henry Fox Talbot, *Lace*, December 1845 (Plate XX from *The Pencil of Nature*), photogenic drawing contact print negative, collection of the J. Paul Getty Museum, Los Angeles (84.XM.478.14).

Featured amongst the earliest of his photographs, lace was a very common subject for Talbot's contact prints, allowing him to demonstrate the exact, indexical copying of "small delicate threads" that his photography could provide.[8] And using a starkly-patterned piece of lace as a matrix was a convenient way to produce high-contrast images with his still-primitive chemistry. But it also allowed him to demonstrate the strange implosion of representation and reality (again, culture and nature) that made photography of any kind possible. In his first paper on photography, dated January 31, 1839, Talbot tells the story of showing a photograph of lace to a group of friends and asking them whether it was a "good representation." They replied that they were not so easily fooled, for it "was evidently no picture, but the piece of lace itself."[9] This gratifying story demonstrated that contact printing was able to present the lace as a kind of "true illusion" of itself.

When Talbot included one of these lace negatives in *The Pencil of Nature* in December 1845, his accompanying text carefully explained the difference between a contact print ("directly taken from the lace itself") and the positive copies that could be taken from this first print (in which case "the lace would be represented *black* upon a *white* ground"). However, as he suggests, a negative image of lace is perfectly acceptable, "black lace being as familiar to the eye as white lace, and the object being only to exhibit the pattern with accuracy."[10] So this is a photograph not so much of lace as of its *patterning*, of its numerical, regular repetitions of smaller geometric units in order to make up a whole.[11] It's as if Talbot wants to show us that the photograph too is made up of a series of smaller units (in his magnified examples we see nothing but these geometric pixels). In these pictures, the units that make up the meaning ("lace") also make up the medium ("photography").[12] Moreover Talbot recognizes from the outset that while photography always provides an indexical truth-to-presence, it doesn't necessarily offer a truth-to-appearance. Photography involves, in other words, an abstraction of visual data; it's a fledgling form of information culture.

In February and May of 1839, shortly after his announcement of photography, Talbot sent Babbage first a copy of his privately-printed *Some Account of the Art of Photogenic Drawing*, and then, as if to illustrate its arguments, eight examples of his prints. One of these prints was a contact print of two pieces of lace, now titled *Samples of Lace* (c. 1839).[13] Like all contact prints, this image is a one-to-one copy of its referents, an exact visual replica of their original lace patterns. Seemingly unmediated by the human hand, this replication is here rendered taxonomic according to the dispassionate methods of modern science. The lace samples appear to float in the fathomless depth of a flattened pictorial space, or on an otherwise blank tabula rasa. This sense of flatness is increased by the fact that, as photogenic drawings, the lace images are right *in*, rather than merely on, the paper which holds them. Figure and ground, image and support, fibres and tone, touchable reality and optical simulation, are here all collapsed into the same visual experience.

This particular contact print involves what is for Talbot an unusually complex composition. It comprises the imprints of two pieces of lace, the first of them elaborated along one edge with a floral design (very similar to the piece later reproduced in *The Pencil of Nature*), and the other featuring a more simple pattern repeated along both edges. This second piece is allowed to extend right across the picture plane, cut off at each end by the edges of Talbot's paper in a way that leaves no visible defect in the inexorable flow of its patterning. The other piece has been placed on the

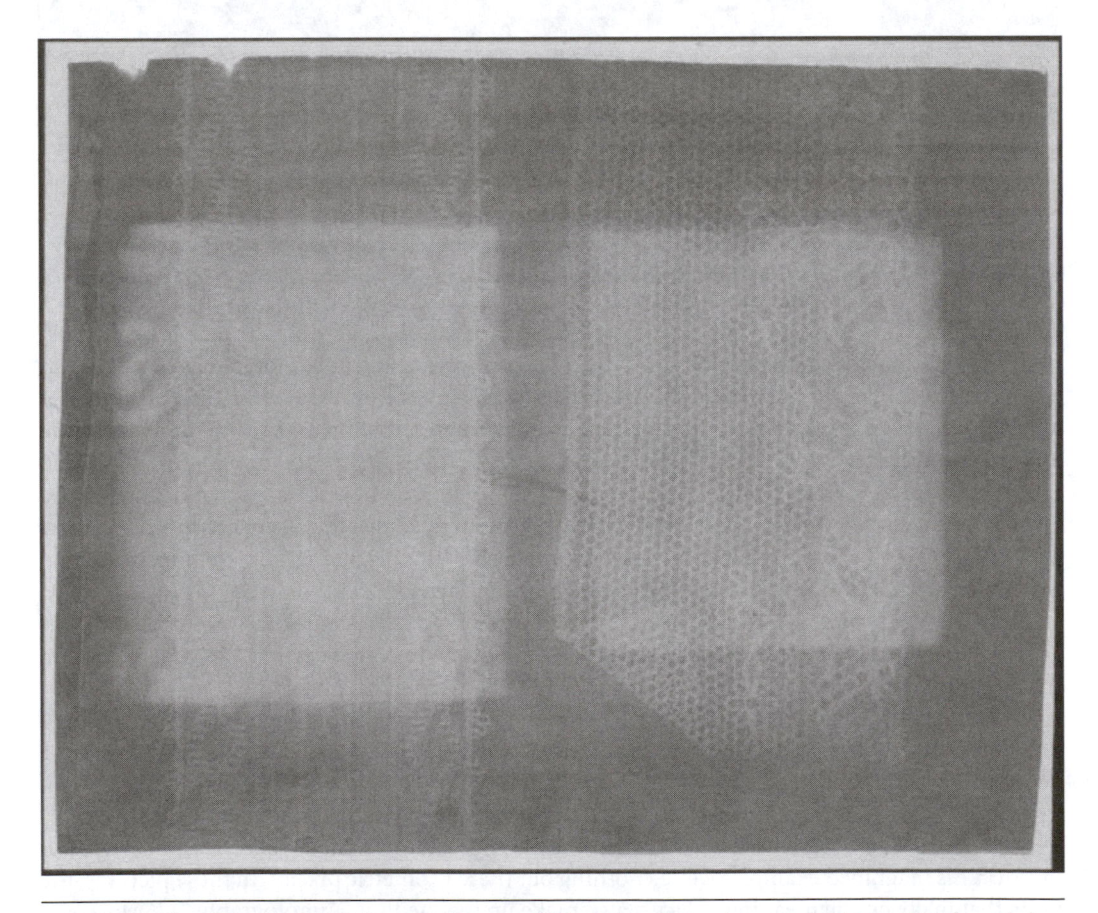

Figure 2.2 William Henry Fox Talbot, *Samples of Lace*, c. 1839, photogenic drawing contact print negative, formerly from the collection of Charles Babbage; now in private collection of Dr. Walter Knysz, Jr.

photographic paper so that it comes in from one side, but stops short of the other. So this second piece of lace is presented as an independent object, an object that extends outside the picture plane as well as into it. In the first case, the lace pattern *is* the picture while in the other it is simply *in* the picture. It's a disconcerting use of the available picture plane, as if acknowledging once again this medium's schizophrenic implosion of nature and culture. This is of course a demonstration picture and it seems that aesthetic concerns like symmetry are not as important as depictive ones. Again, what matters is the evidence this example provides of photography's ability to exactly reproduce patterns. This photogram is about making mathematics visible.[14] Perhaps that is why there is so little embellishment on the part of the maker, except for one corner that has been allowed to turn back and fold over itself (also a feature of the example chosen for *The Pencil of Nature*). This intimation of depth in an otherwise flat pictorial scene works to remind us of the lace's physicality, of the fact that lace does take up space in the real world, even if not in the photogram.

Although the two pieces of lace do not actually overlap on this sheet, there is also a suggestion here of the possibility of montage, of the juxtaposition or even superimposition of two unlike images within a single photographic surface. Talbot was in fact already familiar with this kind of practice. In 1839 a German experimenter named Johann Carl Enslen sent Talbot a photomontage of a drawing of the head of Christ transposed onto a contact print of a leaf. On February 26, 1839, Talbot's friend John Herschel produced a similar type of photograph showing another leaf with a calligraphic character superimposed over it. Need I point out that both these montages feature an other-worldly juxtaposition of elements from both nature and culture, thus reinacting the same implosion that makes photography of any sort possible? All the disruptive/productive techniques of photomontage, so familiar to us now from both the later history of photography and the ubiquitous products of Adobe Photoshop, are right there from photo-media's beginnings.

Babbage might also have seen another significance in Talbot's choice of lace as his subject matter. As Douglas Nickel has suggested, "behind Talbot's presentation of lace images lay the development of the machine-made lace industry in England."[15] In 1837, so-called "Jacquard cards" had been introduced into English lace-making machines for the first time, signalling the relegation of hand-made lace to the luxury market. And Mark Haworth-Booth has recently reported that the lace Talbot used for his picture in *The Pencil of Nature* was indeed machine-made.[16] Apparently it was manufactured in Nottingham by a Pusher machine, which produced the two kinds of mesh ground onto which was sewn machine-made Picot edging. The embroidery was hand-done by women or girls.[17] Talbot's lace matrix was therefore a proudly English artefact, as was its photographic replica. But it was also a demonstration of the further expansion of industrialisation into everyday life, and with it a significant change in labor practices (female labor in this case), changes to which photography of course contributed. It certainly didn't take long for Talbot to target lace manufacturers as potential customers for his new process. On January 23, 1839, he sent a photogenic drawing of lace to Sir William Jackson Hooker to show to manufacturers in Glasgow. Hooker wrote back on March 20, 1839 to report that "your specimen of Photogenic drawing... has interested the Glasgow people very much, especially the *Muslin* Manufacturers—& also excited great attention at a Scientific Meeting."[18]

Babbage, inventor of several automatic computing devices, himself owned a mechanically-woven silk portrait of Joseph Marie Jacquard, the Frenchman who in 1804 had completed the building of a loom directed by a train of punched cards. The portrait shows Jacquard holding a compass, sign of mathematical calculation, sitting in front of a small model of a Jacquard loom. When Babbage writes the history of his own thinking about computing, he specifically refers us to the development of this loom.[19] For by early 1836 Babbage had adopted Jacquard's system of cards into his plans for a computing Analytical Engine. A picture of a piece of lace must therefore have had particular meaning for him in 1839. It's ironic then that, thanks in part to Babbage's own pioneering work, we now look back at Talbot's lace pictures with eyes accustomed to seeing the world through the equally pixellated screen of a computer.

Figure 2.3 Dider Petit et Cie, *Portrait of J. M. Jacquard*, c. 1839, machine woven silk 51 × 36 cm, collection of Science Museum, London.

When Ada Lovelace comes to write about the Analytical Engine in 1843 she conjures its effects (never otherwise made visible, for the machine remained unfinished) in terms of an image that closely resembles one of Talbot's flowery lace contact prints. As she says, the Analytical Engine "*weaves algebraic patterns* just as the Jacquard-loom weaves flowers and leaves."[20] Babbage called Lovelace, the daughter of the poet Lord Byron, his "Enchantress of Numbers." More recent commentators have been eager to point out the relatively rudimentary nature of her grasp of mathematics.[21] But the presence of Lovelace is important in this story because she points to both the poetic and the metaphysical implications of working in this field. Indeed, in keeping with her Romantic heritage, she saw them all (mathematics, invention, poetry, theology) as part of the same grand endeavour. "The effects of the study [of mathematics include an]...immense development of *imagination*: so much so, that I feel no doubt if I continue my studies I shall in due time be a *Poet*. This effect may *seem* strange but it is not strange, to *me*. I believe I see its causes & connection clearly." Her ambition, she goes on to say, is "to add my mite to the accumulated & accumulating knowledge of the world especially in some way more particularly tending to illustrate the wisdom & ways of God!"[22] She repeats the idea in a later letter to Babbage; "I do *not* believe that my father was (or ever could have been) such a *Poet* as I *shall* be an *Analyst*, (& Metaphysician); for with me the two go together indissolubly."[23]

Before pursuing this question of metaphysics a little further, it is well to remember that Talbot's contact prints of lace have at least one more significant aspect. For they also conjure the imminent transference of the photograph from one medium to another via photo-mechanical printing, and, following that, the electronic flow of data that the photographic image has become today. Indeed the first images photographically impressed on a woodblock to allow the printing of exact facsimiles were reproduced in *The Magazine of Science, And School of Arts* on April 27, 1839 and included a contact print of a piece of lace very similar to Talbot's. As early as 1847 Talbot was writing notes to himself about the theoretical possibilities of "transferring photography to steel engraving" by means of electro-chemistry. He went on to patent a photo-engraving process in October 1852 that used a piece of lace, actually some "black crape or gauze," to decompose a given image. This image was formed photographically on a plate of metal by contact printing "an opaque leaf of a plant," so that it, having been pixellated, could be turned into an etched plate and printed in ink on paper.[24] Using what Talbot called his "photographic veils," everything that was so reproduced came striated with a pattern of threads; turned back, in a sense, into a piece of shaped lace.[25]

Not much has been made of the surreal quality of some of Talbot's early photomechanical images. Take *View of Edinburgh and fern* (c.1853), for example. This photoglyphic engraving on paper presents a camera-view of an Edinburgh street and, above it, almost overlapping the street scene, is a reproduction of a contact print of a twig of fern. Both are visibly fixed in place with five pieces of tape. The picture's means of production are laid bare, and all attempts to create a visual illusion, a window-onto-the-world, are abandoned in favor of the sheer wonder of mechanical reproduction. Flatness and depth, looking down and looking in, touch and sight, the natural and the cultural, here and there, domesticity and travel, the unique and the multiple, collage and montage, photography and mechanical printing: all are merged into a single image screen. With this technology, truly multiple reproduction of all sorts of photographic images would soon be possible, as would the transfer of these images from world to photographic paper to metal plate to inked paper.[26] Photographs could now travel far and wide and so could those who looked at them (the placeless quality of the digital image is here prefigured). No wonder that one acquaintance commented in 1867, upon examining one of Talbot's photo-engravings, that now "he should not despair of being able to *fly*."[27]

Others had actually already equated photography with flying. Talbot's friend David Brewster had come up with a practical form of stereoscopy in the1830s, before the announcement of photography. When you look through an instrument at one of these doubled images, a scene appears to be three-dimensional, receding back into virtual space as a series of overlapping planes. As early as December 1840, Talbot made some pairs of photogenic drawings of statuettes, "at a somewhat wide angle," for use in Charles Wheatstone's competing reflecting stereoscope; these are the earliest known stereo photographs.[28] Wheatstone also organised for Henry Collen to take the first stereo photo-portrait on August 17, 1841, using Talbot's calotype process. Its subject was none other than Charles Babbage, who thus became the first cybernaut, the first subject to be transformed into a photo-induced virtual reality.[29]

In June 1859 the American cultural commentator Oliver Wendell Holmes wrote in wonder that, when looking through his own stereo viewer, "I pass, in a moment, from the banks of the Charles to the ford of the Jordan, and leave my outward frame in the arm-chair at my table, while in spirit I am looking down upon Jerusalem from the Mount of Olives." He then goes even further: "we will venture on a few glimpses at a conceivable, if not a possible future," a future in which Holmes envisioned no less than "the divorce of form and substance."

"*Form is henceforth divorced from matter*. In fact, matter as a visible object is of no great use any longer, except as the mould on which form is shaped. Give us a few negatives of a thing worth seeing, taken from different points of view, and that is all we want of it. Pull it down or burn it up, if you please.... Matter in large masses must always be fixed and dear; form is cheap and transportable.... Every conceivable object of Nature and Art will soon scale off its surface for us. . . . The consequence of this will soon be such an enormous collection of forms that they will have to be classified and arranged in vast libraries, as books are now."[30]

Speaking (as we are once again) of the conjunction of Nature and Art in photography, two other "objects" that Talbot often used in order to make contact prints were botanical specimens and samples of handwriting. It's no surprise then to find him sometimes combining all three elements in the same print. In at least one undated example, *Lace and Grasses, with an Alphabet*, he included a scrap of lace pattern, some tiny plant forms, and a complete alphabet in his own hand, all on the one piece of paper. In a 1985 exhibition catalogue, Judith Petite offers the following commentary on the lace imprints made in the 1850s by Victor Hugo. "Musing on these impressions as their author urges us to do, we may . . . recall that text and textile have the same common origin, and that ever since antiquity—see Plato's *Politicus*—the interweaving of threads has been compared to that of words."[31] Talbot, a noted scholar of both Greek and English etymology, must surely have reflected on this same association, especially given the eventual adoption of the Greek-derived word "photography" (light-writing) for his process. The photograph of lace he sent to Babbage therefore also imbricates a vast range of other representational systems, including weaving, mechanical reproduction, and linguistics.

This reminds us in turn of one of Talbot's other great passions: translation, especially of hieroglyphics and cuneiform. He published a photographically-illustrated booklet on a hieroglyphic translation in 1846 and one of his last photoglyphic engravings in 1874 featured a transliteration and translation of Assyrian cuneiform. This interest in the problem of translation, in inventing and cracking codes and designing solutions to coded problems, was shared by two of Talbot's friends; you guessed it, Charles Wheatstone and Charles Babbage! In 1854, for example, Babbage used his vast mathematical knowledge to decipher a coded message previously thought to be unbreakable, and he and Wheatstone not only devised their own cipher system but also spent their Sunday mornings deciphering secret messages sent by lovers in code through personal ads in British newspapers.[32] Such an interest was obviously also relevant to Babbage's ongoing work on a coded system for his computing machines.

So far my shorthand history of this moment of emergence has touched on four inter-related technologies and their conceptual apparatuses—photography, mechanical weaving, computing, and photo-mechanical printing. Conceived around 1800, each of these multi-media developments is therefore synonymous with modernity itself, and thus with capitalism, industrialization, colonialism, patriarchy, and all of modernity's other attributes. Devised more or less simultaneously, each also shares a desire to automate the act of representation and to thereby displace the human body from an active to a relatively passive role. And each recognizes representation itself as involving the transmission of visual information from one place to another, or from one form into another, information that has first been turned into an abstract mode of data. Already then, we seem to have identified the emergence of all of the attributes Manovich argues are specific to 'new media': "numerical representation, modularity, automation, variability, and cultural transcoding."[33]

What relationship, though, did photography have to the actual development of the computer? Some contemporary commentators not only recognized their conjunction but also saw them as being of the same order, representing together the incursion of a new kind of voracious and all-inclusive cyberculture. American writer Nathaniel Willis, for example, referred his readers to the work of Babbage when announcing the discovery of photography in an essay published in *The Corsair* on April 13, 1839. Willis is anxious to make the point that existing art forms are now under threat, given that "all nature shall paint herself—fields, rivers, trees, houses, plains, mountains, cities, shall all paint themselves at a bidding, and at a few moments notice . . . Talk no more of 'holding the mirror up to nature'—she will hold it up to herself." Nature, it seems, has acquired the means to make her own pictographic notations. And Willis sees such an achievement as synonymous with the thinking of Babbage from two years before. "Mr Babbage in his (miscalled ninth Bridgwater) Treatise announces the astounding fact, as a very sublime truth, that every word uttered from the creation of the world has registered itself, and is still speaking, and will speak for ever in vibration. In fact, there is a great album of Babel. But what too, if the great business of the sun be to act register likewise, and to give impressions of our looks, and pictures of our actions...the whole universal nature being nothing more than phonetic and photogenic structures."[34] The conception of Babbage's calculating engines, a key element of his *Treatise*, thus becomes not only a part of the history of computing but also of the then-disintegrating field of natural philosophy—and is therefore closely related not only to photography but also to the Romantic poetry and painting produced in this same period.[35]

As it happened, Babbage displayed a number of examples of Talbot's photogenic drawings and calotypes at his famous London soirées ("for the decoration of my drawing room and the delight of my friends"), intellectual gatherings that Talbot and his family occasionally attended in person.[36] Between 1833 and 1842, among the other entertainments at such gatherings was a working model of a portion of Babbage's first computing machine, the Difference Engine he had built in 1832. It seems likely then that visitors to Babbage's drawing room between 1839 and 1842 encountered photography and computing together, for the first time at the same time.[37] Lady Annabella Byron and her daughter Ada were among those who visited Babbage's drawing room (this visit was what inspired Ada to go on to study mathematics and eventually become Babbage's assistant and interpreter). Lady Byron described her first viewing of the Difference Engine in a letter dated June 21, 1833, exclaiming that "there was a sublimity in the views thus opened of the ultimate results of intellectual power."[38] On November 28, 1834 Lady Byron further records in her diary that Babbage explicitly "alleged that the engine could show that miracles were not only possible but probable."[39]

We're back, it seems, to the question of metaphysics. In September 1839, the same year in which he announced his photographic experiments, Talbot published a tract titled *The Antiquity of the Book of Genesis*, pursuing a theme (the origins of the world, the origins of our account of this origin) already canvassed by Babbage. For Babbage too had been exploring the relationship of culture and nature, in that same *Ninth Bridgewater Treatise* of May 1837 already mentioned by Willis. In this particular tract, Babbage attempted to reconcile biblical belief and evolutionary evidence, and he did so by pointing to the creative, even miraculous, possibilities of God's "natural laws," i.e., mathematics. And he explicitly based this argument on the algorithmic feedback functions calculated by his Difference Engine. In other words, Babbage conceived of his computer as a cultural artefact that enabled nature (and therefore God) to represent itself in the form of mathematical equations (just as Talbot saw photography as enabling nature to represent itself according to the natural laws of physics and chemistry). Thus, each of Babbage's calculating machines was perceived as proof incarnate of the possibility of "natural" miracles and therefore a confirmation of the existence of a still-active and present God; this was the sublimity, the "ultimate results of intellectual power," to which Lady Byron refers above. Might they both have thought similarly about the photographs that Babbage exhibited beside his calculating machine?

Others certainly did. For example, the concurrent discoveries of photography and another important mechanical invention, telegraphy, were often compared during this period as confirmations of natural theology. Drawing in part on the arguments in Babbage's *Treatise*, Edward Hitchcock, Professor of Geology and Natural Theology at Amherst College, saw them both as evidence of what in the 1840s he called the "Telegraphic System of the Universe." "The discoveries of modern science...show us that there is a literal sense in which the material creation receives an impression from all our words and actions that can never be effaced; and that nature, through all time, is ever ready to bear testimony of what we have said and done." He goes on to suggest that, "thrown into a poetic form, this principal converts creation:

Into a vast sounding gallery;
Into a vast picture gallery;
And into a universal telegraph."[40]

Strange again that Manovich makes no mention of the electric telegraph in his genealogy for new media. For "universal telegraphy" was something imagined as early as the mid-eighteenth century and made manifest in the 1820s and 1830s—at the same time, then, as photo-media and computing also emerged. The aim was to harness the properties of electricity to send images of every kind—sounds, letters, words, and even pictures—through wires and from place to place. A number of people worked on this grand idea, the idea of "the world itself rolling through the air" as Walt Whitman put it in 1850.[41] A key breakthrough came in July 1838, when the Englishman Edward Davy was granted a patent for an electric telegraph system in which a current being received is passed through a moving paper tape soaked in potassium iodide, thus leaving a colored mark with each flow. Electricity was thereby turned into a legible image, moreover a kind of image produced very much like a photograph (automatically, as a chemical reaction to received energy). With this example in mind, in 1842 Alexander Bain, a Scotsman, devised a telegraphy system that could transmit simple line drawings as well as text, "an arrangement for taking copies of surfaces at distant places by means of electricity." This primitive facsimile machine included an "endless silk ribbon," which he saturated in printers' ink and against which a metal rod would press to leave a mark on the paper beneath, apparently producing an image "in a series of small dots."[42]

However, perhaps the most intriguing experimenter with electric telegraphy was the American painter Samuel Morse. After attending Yale between 1805 and 1810, Morse had gone on to a career as a prominent painter and occasional inventor. In 1821, for example, he had attempted to invent a photographic process, but finding "that light produced dark, and dark light, I presumed the production of a true image to be impracticable, and gave up the attempt."[43] This experience made him immediately responsive to Daguerre's announcement of his photographic process in January 1839; he met with the Frenchman in Paris on March 7, and Daguerre returned the compliment on March 8 in order to examine Morse's telegraphic invention in the American's apartment. In May 1839, back in the United States, Morse had Daguerre elected an Honorary Member of the National Academy of Design. By September, having acquired and translated a copy of Daguerre's *Manual*, Morse had made his first daguerreotype (a view of a Unitarian church opposite New York University) and in the following month attempted to take portraits. Shortly thereafter he opened a commercial studio with John Draper and began taking in pupils.[44]

Morse and Draper produced at least one remarkable daguerreotype, a still life very reminiscent in its composition, backdrop, and cornucopia of constituent elements of Daguerre's own early still life images. Morse and Draper's "photographic painting" (as Morse called it) shows four overlapping figurative images (some of them copies of other people's work, in a kind of mini version of his 1832 painting *The Gallery of the Louvre*) drawn by Morse (one of them bears his reversed signature) and haphazardly pinned against a textured piece of cloth. The composition also includes a shelf bearing some glass and ceramic vessels, scientific instruments, a chemistry book (with a label reading "Hare's Chem," Hare being Draper's chemistry teacher), and a statuette. The shallowness of the depicted space and the uncentered, seemingly arbitrary array of images, both two and three-dimensional, encourages the viewer's eye to scroll back and forth across the whole picture plane without resting on any one spot. Symbolizing the collaboration of art and science (and of Morse and Draper), this photograph also speaks to a new kind of visual culture in which everything is soon going to be transformed into a seamless, multi-directional flow of reproductions.[45]

Yet another representational system was to occupy Morse between his ventures into the world of painting and photography. During 1832, Morse conceived of a telegraphic system that would harness electricity to transmit messages along wires between any two points. He later remembered remarking to friends, "if...the presence of electricity can be made visible...I see no reason why

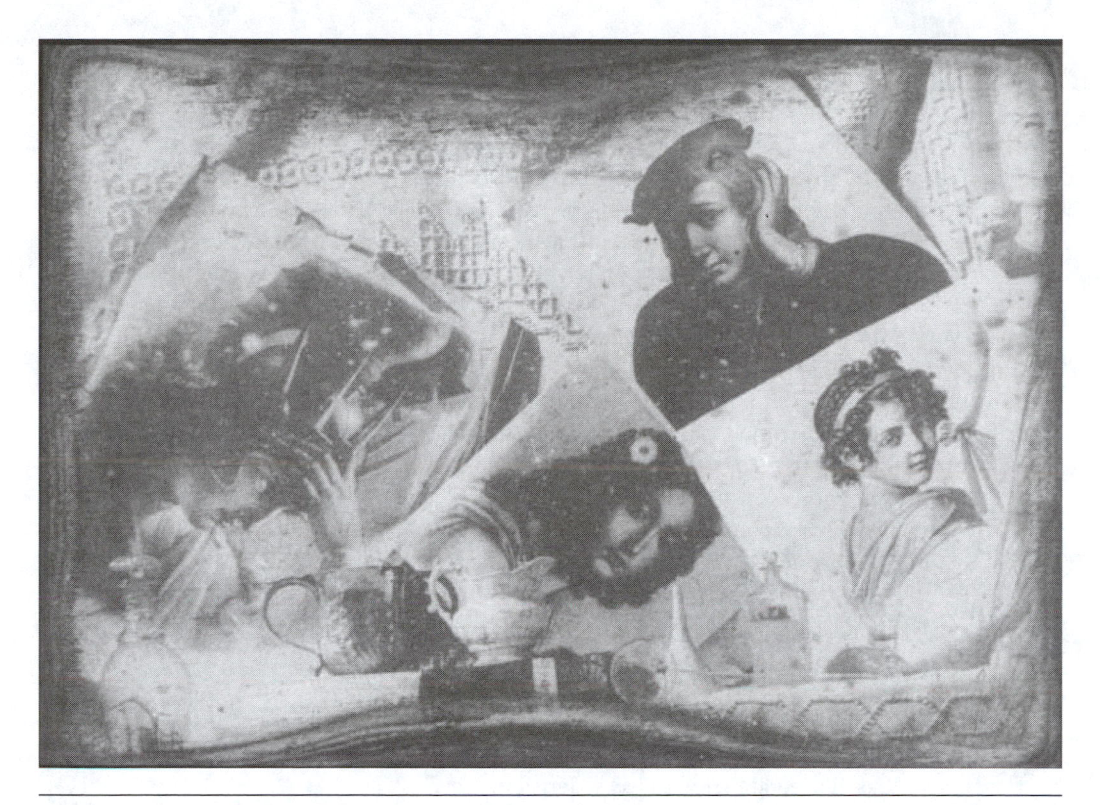

Figure 2.4 *Samuel Morse & John Draper,* Still life, 1839–40, daguerreotype. Collection Photographic History National Museum of American History, Smithsonian Institution.

intelligence might not be instantaneously transmitted by electricity to any distance."[46] He imagined fulfilling this bold prophecy of a new media by translating the alphabet into a numerical code and then transmitting these numbers as breaks in the flow of electricity, as dots, spaces, and dashes. At various moments he experimented with a system like Davy's in which the electricity would automatically leave a mark on some chemically prepared paper, but eventually decided instead on an apparatus in which two electromagnets would work in concert to mechanically mark the paper with a pencil.

Poverty and other discouragements delayed the building of this apparatus until 1837, when he was able to make a crude prototype in his studio in New York. As Morse recalled, this first instrument (which still exists) was comprised of, among other components, "an old picture or canvas frame fastened to a table" and "the wheels of an old wooden clock moved by a weight to carry the paper forward."[47] Time, painting, drawing, mathematics, and electricity are combined to transmit and reconstitute images (but also sounds and textures) in coded numerical form as a series of binary electrical pulses, and all this a hundred years before Zuse built his digital computer. Manovich reads Zuse's machine as a dramatic discarding of cinema and its conventions by new media: "a son murders his father," he declares.[48] Morse's apparatus also incorporates the death of an earlier form of representation. For it was in this same year, 1837, that Morse completed one of his finest and final paintings, a full-length portrait of his daughter titled *The Muse: Susan Walker Morse* (1836–37). A young woman sits with pencil in hand and sketch paper in lap, ready to make her first mark, her face turned up as if searching for divine inspiration.[49] The picture is, says Paul Staiti, "unique in the stress placed on depicting the anxious threshold of representation.... poised in an expanded moment of epistemological crisis."[50] Morse himself was living out that crisis. For in 1837 he also learnt that he had not been among those chosen to paint pictures for the interior of the Capitol building, and this, he later said, "killed" him as a painter.[51] "I did not abandon her, she abandoned me," he recalled, although in fact the demands of his telegraphic apparatus made further concentration on his first love, painting, impossible.[52] Once again we are witness to a deadly Oedipal moment in apparatus form, with Morse's discarded canvas frame being stripped of its painted picture to make way for the abstract pencil markings of an electrical pulse. The birth of telegraphy in 1837 is at the cost of the death of painting (or at least of its iconic pretenses).[53]

At one point Morse imagined that telegraphy might also overcome the advent of photography. While in Paris to promote his invention, he wrote back to his business partner Francis Smith. "I am told every hour that the two great wonders of Paris just now, about which everybody is conversing, are Daguerre's wonderful results in fixing permanently the image of the *camera obscura*, and Morse's Electro-Magnetic Telegraph, and they do not hesitate to add that, beautiful as are the results of Daguerre's experiments, the invention of the Electro-Magnetic Telegraph is that which will surpass, in the greatness of the revolution to be effected, all other inventions."[54] However, as we've heard, Morse soon took up photography himself, apparently seeing this most modern of representational systems as compatible with his thinking about telegraphy. Others also recognized this compatibility, as evidenced in this anonymous American poem of 1852:

> FRANKLIN brought down the lightning from the clouds,
> MORSE bade it act along the trembling wire;
> The trump of Fame their praises gave aloud,
> And others with the same high thoughts inspire.
> DAGUERRE arose—his visionary scheme
> Was viewed at first with jeers, derision, scorn,
> Conquered at last by the grand power supreme
> Of god-like mind—another art was born.[55]

Indeed, it wasn't long before Morse's telegraphic data network was being used as a vehicle for the transfer of photographic images. In 1867 the Frenchman Jean Lenoir proposed the telegraphic transmission of photographic images by reducing them to stark contrasts of black and white, to a matter of presence and absence; that is, to a kind of digital image. In fact the discovery that would make possible the transmission of continuous-tone images had already occurred back in 1839, the same year photography was announced, when the French physicist Alexandre Edmond Becquerel noticed that the voltage output of a metal-acid battery changes with exposure to light. The direct relationship between voltage output and exposure to light that Becquerel had noticed was ascribed to bars of crystalline selenium in 1873 by an Englishman, Willoughby Smith. He was conducting tests for the first transatlantic cable and discovered that the electrical resistance of selenium depends on the amount of light that falls on it. With this discovery in place, images could potentially be sent from one place to another using electricity in concert with two matching selenium converters.

By 1878 Alexander Graham Bell was suggesting in a lecture that it was possible to "hear a shadow" fall on a piece of selenium connected in circuit with his telephone; in a clear reference to Talbot's "art of fixing a shadow," he called his new invention the photophone. In the following year, Britain's *Punch* magazine published a cartoon about Thomas Edison's imagined Telephonoscope, picturing it as an "electric camera-obscura" which can apparently transmit light and sound in real time from Ceylon to England. The cartoon shows a tennis match in progress between some young English colonists, while one of their number speaks to her father back home in Wilton Place. In the left foreground a dark Ceylonese woman, seen sitting next to the family dog, nurses a white child. Much like photography, the Telephonoscope indiscriminately transmits whatever data comes within its scope, including the signs of class difference and racial hierarchies.[56]

Meanwhile, work was still being done on methods of transmitting photographic images via a telegraphic wire. In 1880 a Portuguese professor named Adriano de Paiva suggested, in a treatise titled *La télescopie électrique*, coating the receptive surface of his camera obscura with selenium to allow the images formed to be transmitted using telegraph lines. In March 1881 an Englishman named Shelford Bidwell demonstrated a new apparatus, called a Telephotograph Device, capable of transmitting any kind of picture, including photographs. Basically, he had come up with a method of scanning an image, breaking it up into smaller elements that could be transmitted as a linear stream of electrical impulses and then reassembling them, using the differential response of selenium to these impulses, as a two-dimensional image. As he modestly speculated in a paper delivered to the Royal Society in London in 1881, "I cannot but think that it is capable of indefinite development, and should there ever be a demand for telephotography, it may turn out to be a useful member of society."[57] Scientists in Germany and France improved on the idea to the point that by 1908, phototelegraphy was being used to send images of all kinds over telephone lines. In 1907, for example, *Scientific American* published a photographic reproduction of Germany's Crown Prince sent by Arthur Korn from Berlin over a telegraph wire. This kind of picture took about twelve minutes to transmit, although by having a coarser scan and lines wider apart the transmission of a full-sized picture could take place in six minutes.[58]

So how might this unexpected interaction of photography, telegraphy and computing inform our understanding of the history of new media? Well, first and foremost it demonstrates that these three representational systems were never separate or opposed to each other but in fact had a common chronological, philosophical and representational trajectory (and, of course, a common social, political and economic context). It also shows that there was, at the least, a conceptual convergence of photo-media and computing in the 1830s, a full century before Zuse cluttered up his parents' living room with his film-directed calculating device. By the 1880s, photographic images were being converted into numerical data, transmitted by binary electrical impulses to another place, and reconstituted as images. This would seem to fulfill most of the conditions for new media, except for the actual, physical involvement of a computer (although one could argue, based on the history just given, that the logics of computing are already inscribed in the practice of phototelegraphy). And all this is taking place amidst a "regular discourse" in which many of the practices, themes and concepts of the digital age are already being widely canvassed.

What this suggests is that new media has a surprisingly long history, a history as old as modernity itself. The "new" in new media might therefore best be sought, not in the formal qualities of its "language," but in that language's contemporary reception and meanings. This would shift our history from a concern with how images are technically made and transmitted, to political and social questions about their past and current contexts of production, dissemination and interpretation.[59] What world view, what assumptions about the way life ought to be lived now, are embodied and reproduced in the visual culture of today's electronic media? How can we engage, and, if necessary, contest these assumptions? These sorts of questions bring us back to the "archaeology" that Manovich seeks to construct for new media, for history is, as always, a good place from which to begin any answer. But now the word "archaeology" must conjure, not so much a vertical excavation of developments in imaging technologies, but rather Michel Foucault's more troublesome effort to relate particular apparatuses to "the body of rules that enable them to form as objects of a discourse and thus constitute the conditions of their historical appearance."[60] The identification of these "rules," of what Foucault calls "a *positive unconscious* of knowledge," turns such a history into a necessarily political enterprise.[61] For in identifying new media's various rules of formation, our history must also identify its (its subject's, but also its own) imbrication within broader social issues, and thus its relationship to particular deployments of power.

What my own brief history has argued is that photography is present in new media, even when it's not, just as new media has always been imbricated in the genealogical fabric of what is supposed to be its predecessor.[62] Of course, my discussion has concentrated only on the fate of the photograph in this story, and has thus ignored the amazing breadth of other image-types and means of image-formation and dissemination that Manovich identifies with the world of new media. But this breadth is precisely why any single "conceptual lens," whether derived from photography or cinema, is going to be inadequate to an analysis of new media as a total phenomenon. What a singular focus on the photograph can do, in the face of this difficulty, is identify new media with a certain type of historical economy that does seem true to its multifarious character.[63] Belying linear chronology in favor of a three-dimensional network of connections and nodules, Foucault's version of archaeology is the historical equivalent of a hypertext document (the history it produces is *thick* with unpredictable connections). With it comes a more complex rendition of the relations of past and present, and of the "new" and the "old." It also comes with a difficult set of political challenges for the writing of history itself, for the *way* one writes that history. Indeed, as my own text has demonstrated, this kind of history "produces what it forbids, making possible the very thing it makes impossible."[64] But what better description could there be for the 'language' of this strange and convoluted entity called new media?

Notes

1. Lev Manovich, *The Language of New Media* (Cambridge, MA: The MIT Press, 2001), 12, 8, 9, 23. See also our exchange in Geoffrey Batchen, "Voiceover" (a conversation with Lev Manovich and Rachel Greene), *Afterimage* 29: 4 (January/February 2002): 11.
2. Relying on an illustration from an introductory book by Charles and Ray Eames, titled *A Computer Perspective* (Cambridge, MA: Harvard University Press, 1973), Manovich remarks that the film stock used by Zuse shows "a typical movie scene...two people involved in a room in some action." (25) He neglects to mention that this film stock is actually a negative; punching holes through it therefore represents this film's complete obliteration. In any case, this particular footage is probably a quite arbitrary choice on the Eames' part, for the original Z1 was totally destroyed by Allied bombing during the war (Zuse rebuilt it only in 1986). The most detailed account of the development of the Z1 in English is to be found in Paul E. Ceruzzi, "The Early Computers of Konrad Zuse, 1935 to 1945," *Annals of the History of Computing* 3: 3 (July 1981): 241–262. Ceruzzi reports that it was Zuse's friend and co-worker Helmut Schreyer who suggested they use discarded movie stock, based on his experience as a movie projectionist (248). But Zuse already had an interest in cinema. Another historian, Friedrich L. Bauer, tells us that Zuse attempted to build a computer only after "juvenile inventor dreams" of city planning, photography and moon rockets (the last inspired by Fritz Lang's 1929 film *Frau im Mond*). See Friedrich L. Bauer, "Between Zuse and Rutishauser: The Early Development of Digital Computing in Central Europe." In *A History of Computing in the Twentieth Century*, edited by N. Metropolis, J. Howlett and Gian-Carlo Rota (New York: Academic Press, 1980), 507.
3. Manovich, *The Language of New Media*, 25. Given his theme, an historical coupling of computing and cinema from which sprang new media, it is odd that Manovich never mentions a famous fictionalized version of this same copulation. In *The Difference Engine*, a 1991 sci-fi novel by William Gibson and Bruce Sterling, the authors imagine the nineteenth century as if Babbage's steam-driven computing engine had in fact been finished and distributed in the 1830s. One result is the production of cartes-de-visite with an "Engine-stipled portrait" (7); another is the invention of "kinotropes," an engine (or computer) driven animation projector. See William Gibson and Bruce Sterling, *The Difference Engine* (New York: Bantam Books, 1991). For a commentary on this book, and in particular on its representation of Charles Babbage and Ada Lovelace, see Jay Clayton, "Hacking the Nineteenth Century." In *Victorian Afterlife: Postmodern Culture Rewrites the Nineteenth Century*, edited by John Kucich and Dianne F. Sadoff (Minneapolis: University of Minnesota Press, 2000), 186–210.
4. Michel Foucault, *The Archaeology of Knowledge*, trans. A.M. Sheridan Smith (New York: Pantheon Books, 1972), 144.
5. Geoffrey Batchen, "Obediant Numbers, Soft Delight" (1998), *Each Wild Idea: Writing, Photography, History* (Cambridge, MA: The MIT Press, 2001), 164–174, 223–226. Talbot appears to have first met Babbage at a breakfast held on June 26, 1831. Also present was Talbot's friend John Herschel, one of Babbage's closest friends, and the man who went on to popularise the word "photography." See Larry Schaaf, *Out of the Shadows: Herschel, Talbot & the Invention of Photography* (New Haven & London: Yale University Press, 1992), 33. Another who attended that breakfast was David Brewster, a close friend of both Talbot and Babbage. In 1834, for example, Brewster published his *Letters on Natural Magic* and devoted several pages to "Babbage's Calculating Machinery," which he described, after having witnessed its operation, as a "stupendous undertaking" (265–266).
6. William Henry Fox Talbot, "Some Account of the Art of Photogenic Drawing, or The Process by Which Natural Objects May Be Made to Delineate Themselves without the Aid of the Artist's Pencil" (January 31, 1839), as reproduced in Beaumont Newhall ed., *Photography: Essays and Images* (London: Secker & Warburg, 1980), 23–30.
7. Roland Barthes, "Rhetoric of the Image" (1964), *Image-Music-Text* (Hill and Wang, 1977), 45. Speaking of informational economies, Australian Aboriginal culture has been organized around information exchange for thousands of years; we might do well to look to that culture for ways to manage this exchange in a more sophisticated and flexible manner

than we do now. Indeed, we need to write a history of information that takes into account a full range of cultural attitudes and practices. See Eric Michaels, "For a Cultural Future: Francis Jupurrurla Makes TV at Yuendumu" (1987), *Bad Aboriginal Art: Tradition, Media, and Technological Horizons* (Minneapolis: University of Minnesota Press, 1994), 98–125, 189.

8. William Henry Fox Talbot, *The Pencil of Nature*, Plate XX. In *Henry Fox Talbot: Selected texts and bibliography*, edited by Mike Weaver (Oxford: Clio Press, 1992), 101.
9. Talbot, "Some Account of the Art of Photogenic Drawing," *Photography: Essays and Images*, 24.
10. Talbot, "Plate XX," *The Pencil of Nature*, in Weaver, *Henry Fox Talbot*, 101.
11. This distinction between lace and its patterning is repeated frequently in the contemporary literature about Talbot's contact prints. Talbot himself referred to them as "a pattern of lace" in a letter to the editor of *The Literary Gazette* dated January 30, 1839. He also exhibited "copies of Lace, of various patterns" at the British Association Meeting in August 1839. When reviewing *The Pencil of Nature* on January 10, 1846, *The Literary Gazette* praises his publication of "an absolutely perfect pattern of lace." See Larry Schaaf, *H. Fox Talbot's The Pencil of Nature Anniversary Facsimile: Introductory Volume* (New York: Hans P. Kraus, 1989), 61. For more on Talbot's lace pictures, see my "Dibujos de encaje" (Patterns of Lace). In *Huellas de Luz: El Arte y los Experimentos de William Henry Fox Talbot*, edited by Catherine Coleman (Traces of Light: The Art and Experiments of William Henry Fox Talbot) (exhibition catalogue, Madrid: Museo Nacional Centro de Arte Reina Sofia/Aldeasa, 2001), 53–59, 354–357.
12. Talbot thereby underlines one of the photographic image's most distinctive properties, its apparent transparency to its referent. This is a property also commented on by, among others, Siegfried Kracauer. "If one could look through a magnifying glass one could make out the grain, the millions of little dots that constitute the diva, the waves and the hotel. The picture, however, does not refer to the dot matrix but to the living diva on the Lido." Siegfried Kracauer, "Photography" (1927). In *The Mass Ornament: Weimar Essays*, edited by Thomas Y. Levin (Cambridge, MA: Harvard University Press, 1995), 47.
13. This print has recently been sold at a Sotheby's auction in London. See Lot 39 in the Sotheby's catalogue *Fine Photographs from the Collection of Paul E. Walter* (May 10, 2001), 34–35.
14. Like many scholars of his generation, Talbot was trained to see the world in mathematical terms, and this at a time when mathematics itself was undergoing fundamental changes, especially at Cambridge, where Talbot studied. In Britain these changes were initiated by a group led by two men who were to become among Talbot's closest friends, Charles Babbage and John Herschel. Talbot published a number of papers on mathematical problems in the 1830s and in 1838 received the Queen's Royal medal, after being recommended by the Royal Society for his work on integral calculus. In the later 1850s he again began publishing in mathematics, looking specifically at number theory. He published his final mathematical paper, on integer roots, in 1875, two years before his death. These interests could not help but inform his work with photography. For more on Talbot's mathematical interests, see H.J.P. Arnold, *William Henry Fox Talbot: Pioneer of photography and man of science* (London: Hutchinson Benham, 1977). I have written elsewhere about how Talbot asked his contemporaries to examine one of his earliest pictures, *Latticed Window (with the Camera Obscura)* of August 1835, through a magnifying lens so that the panes of glass in the window can be counted. Looking here is regarded as the equivalent of counting, and photography is presented as just one more way to translate the world into numbers. See Geoffrey Batchen, "A Philosophical Window," *History of Photography* 26: 2 (Summer 2002): 100–112. Indeed, camera-generated photographs in general were assumed to be instances of mathematics at work, translating the world into a picture via those ideal geometries that are embodied in the camera apparatus itself. Whatever the kind of photography, the camera's geometrical workings guarantee the veracity of the image produced. This is the assumption behind the claim of Edmond de Valicourt de Séranvillers in his *New Manual of Photography* in 1853: "In the daguerreotype, when we have acquired a certain method; success is, so to speak, mathematical, and the uniformity of the means employed yields an identity of result with the precision and regularity of a machine." He thereby repeats an idea first presented by François Arago in his announcement to the French Chamber of Deputies on June 15, 1839, arguing that photography produced "drawings, in which the objects preserve their mathematical delineation in its most minute details." See *An Historical and Descriptive Account of the various Processes of the Daguerréotype and the Diorama, by Daguerre* (London: McLean, 1839), 1.
15. Douglas R. Nickel, "Nature's Supernaturalism: William Henry Fox Talbot and Botanical Illustration." In *Intersections: Lithography, Photography and the Traditions of Printmaking*, edited by Kathleen Howe (Albuquerque: University of New Mexico Press, 1998), 19.
16. Mark Haworth-Booth, speaking through the Electronic Information Unit, Canon Photography Gallery at the Victoria & Albert Museum, London.
17. This information was generously provided by Clare Browne, Assistant Curator in the Department of Textiles and Dress at the Victoria & Albert Museum, London.
18. See Schaaf, *Out of the Shadows*, 47.
19. Charles Babbage, "Of the Analytical Engine," *Passages from the Life of a Philosopher* (1864), edited by Martin Campbell-Kelly (New Brunswick, NJ: Rutgers University Press, 1994), 88–89. Another inspiration was Babbage's early encounter with the automatons made by John Merlin, one of which he later purchased, repaired and exhibited in his drawing room. See Anne French, *John Joseph Merlin: The Ingenious Mechanick* (exhibition catalogue, London: The Greater London Council, 1985), and Babbage, *Passages*, 273–274.
20. Ada Augusta [sic], Countess of Lovelace, "Sketch of the Analytical Engine: Notes by the Translator." In *Charles Babbage and his Calculating Engines*, edited by Philip Morrison and Emily Morrison (New York: Dover, 1961), 252. In this same essay, Lovelace declares that mathematics alone can "adequately express the great facts of the natural world" (272).
21. See, for example, Benjamin Woolley, *The Bride of Science: Romance, Reason, and Byron's Daughter* (New York: McGraw-Hill, 2001), and Jim Holt, "The Ada Perplex," *The New Yorker* (March 5, 2001), 88–93.

22. Ada Lovelace, in a letter to her mother Lady Byron, dated January 11, 1841; as quoted in Betty Alexander Toole, *Ada, The Enchantress of Numbers: Prophet of the Computer Age* (Mill Valley, CA: Strawberry Press, 1998), 96.
23. As quoted in Holt, "The Ada Perplex," 92. Lovelace's heady description of the pleasures of mathematical analysis is matched in more recent times by Vladimir Nabokov, when he speaks of composing chess problems. "Inspiration of a quasi-musical type, quasi-poetical, or to be quite exact, poetico-mathematical type, attends the process of thinking up a chess composition of that sort....it belonged to an especially exhilarating order of sensation." Vladimir Nabokov, *Speak, Memory: An Autobiography Revisited* (New York: Vintage Books, 1967), 288-289. Talbot was also interested in chess, making at least ten pictures of people playing the game between 1840 and 1842. One of these shows Nicolaas Henneman contemplating his next move, a subject that is in effect an attempt to make visible the act of mathematical thinking. Perhaps it was posing for these pictures that inspired Henneman to make his own versions, this time featuring Antoine Claudet as the chess player caught in mid-thought. See Larry Schaaf, *The Photographic Art of William Henry Fox Talbot* (Princeton and Oxford: Princeton University Press, 2000), 124–125.
24. See Arnold, *William Henry Fox Talbot*, 272. For Talbot's own description of this process, see his "Photographic Engraving," *Journal of the Photographic Society of London* 1 (1854): 42–44. "The objects most easily and successfully engraved are those which can be placed in contact with the metallic plate,—such as the leaf of fern, the light feathery flowers of a grass, a piece of lace, etc. In such cases the engraving is precisely like the object; so that it would almost seem to any one, before the process was explained to him, as if the shadow of the object had itself corroded the metal,—so true is the engraving to the object" (43).
25. See the detail of such a specimen reproduced in both Eugene Ostroff, "Etching, Engraving and Photography: History of Photomechanical Reproduction," *The Photographic Journal* 109: 10 (October 1969): 569, and Larry Schaaf, "The Talbot Collection: National Museum of American History," *History of Photography* 24: 1 (Spring 2000): 12.
26. In 1854, for example, Viennese photographer Paul Pretsch patented a printing system he called "Photogalvanography, Engraving by Light and Electricity"; in 1856 he founded the Photogalvanographic Company in London, employing Roger Fenton as his chief photographer. By the 1880s the half-tone process, which involves using a simple dot or crossline screen to break up the continuous tone of the photograph, enabled photographic images to be mechanically printed with text. The first half-tone reproduction of a photograph, labelled as a "Reproduction Direct from Nature," appeared in the *New York Daily Graphic* of March 4, 1880. Its photographer, Stephen Horgan, wrote that the plate was produced "in such a manner that an artist could add figures or other changes as he wished, the photographic effect being retained in horizontal lines through the picture." See Roy Flukinger, Larry Schaaf, and Standish Meacham, *Paul Martin: Victorian Photographer* (Austin, University of Texas Press, 1977), 13.
27. Harriet Mundy, writing to Talbot on February 25, 1867, as quoted in Arnold, *William Henry Fox Talbot*, 293.
28. For the story of the making of these images, see Stephen F. Joseph, "Wheatstone's Double Vision," *History of Photography* 8: 4 (October-December 1984): 329–332, and Stephen F. Joseph, "Wheatstone and Fenton: A Vision Shared," *History of Photography* 9: 4 (October–December 1985): 305–309.
29. Babbage apparently kept these stereo portraits in his own collection. See Henry Collen's 1854 recollection in Joseph, "Wheatstone's Double Vision," ibid, 330. In 1841 Wheatstone also commissioned daguerreotype stereo-portraits from the London studios of rivals Richard Beard and Antoine Claudet. Interestingly, in 1853 Claudet patented a sliding stereoscope apparatus that could show cinematic-type movement. See Arthur T. Gill, "The First Movie?," *The Photographic Journal*, 109: 1 (January 1969): 26–29, and Arthur T. Gill, "The Rise and Fall of the Daguerreotype," *The Photographic Journal*, 114: 3 (March 1974): 131. See also my "Spectres of Cyberspace" (1996). In *The Visual Culture Reader*, edited by Nicholas Mirzoeff (London & New York: Routledge, 1998), 273–278.
30. Oliver Wendell Holmes, "The Stereoscope and the Stereograph" (1859), in Newhall, *Photography: Essays and Images*, 59, 60.
31. Petite is quoted in Ann Philbin and Florian Rodari eds., *Shadows of a Hand: The Drawings of Victor Hugo* (exhibition catalogue, New York: The Drawing Center, 1998), 32.
32. See the discussion of Babbage's code-breaking in Simon Singh, *The Code Book: The Science of Secrecy from Ancient Egypt to Quantum Cryptography* (Anchor Books, 2000).
33. Manovich, *The Language of New Media*, 20.
34. Nathaniel Willis, "The Pencil of Nature: A New Discovery," *The Corsair: A Gazette of Literature, Art, Dramatic Criticism, Fashion and Novelty*, 1: 5 (New York, April 13, 1839), 70–72.
35. The links between the invention of photography and contemporary developments in Romantic poetry and painting are discussed in my *Burning with Desire: The Conception of Photography* (Cambridge, MA: The MIT Press, 1997). The same connections can be traced with regards to the conception of the computer. As already mentioned, Babbage was greatly assisted in the articulation of his ideas about the Analytical Engine by Lord Byron's daughter, Ada Lovelace. John Murray, the publisher of Babbage's *Ninth Bridgewater Treatise*, was also Byron's publisher and Babbage opens his autobiographical *Passages* with a quotation from Byron's *Don Juan*. Talbot was himself a poet, as well as a noted scholar of the English language. See, for example, his Romantic ballad *The Magic Mirror* (1830), reproduced in Weaver, *Henry Fox Talbot*, 37–39. Talbot's neighbor was Byron's friend and biographer, the Irish poet Thomas Moore. In late February 1840, possibly thinking of a planned memorial publication on Byron ("The Tribute of Science to Poetry" Talbot called it in his research *Notebook P*), Talbot made three photogenic drawing negative contact prints of a handwritten manuscript page from Byron's *Ode to Napoleon*. See Larry Schaaf, *Sun Pictures: Photogenic Drawings by William Henry Fox Talbot* (Catalogue Seven, New York: Hans Krauss Jr., 1995), 32–39. He also made a photograph of some lines of one of Moore's own poems. See Schaaf, *Facsimile: Introductory Volume*, 35–36. In a letter to Talbot's mother Lady Elisabeth Feilding dated May 15, 1839, Thomas Moore had this to say after a visit to Lacock Abbey: "Both Talbot and his collaborateur, the Sun, were in high force & splendour, and I promised to write something about their joint doings, if I could but get paper sensitive enough for the purpose." See Arnold, *Talbot*, 120, 334.
36. Babbage to Lady Elisabeth Fielding, March 18, 1841 (Lacock Abbey 40–35).

37. For example, in February 1840 Babbage hosted a soirée for Talbot, in which were presented a selection of framed and glazed photographs. See Schaaf, *The Photographic Art of William Henry Fox Talbot*, 84. We also know that Babbage gave Lady Byron a "treat," as he put it, by lending her five of Talbot's calotypes in February 1844. "Many thanks for the loan of those beautiful photographs. They were much admired last Saturday Evg.... In the meantime I gave Lady Byron a treat to whom I lent them for a few hours." Charles Babbage, in a letter to Talbot dated February 26, 1844, held by the Fox Talbot Museum, Lacock Abbey.
38. Lady Byron, in a letter to Dr William King, dated June 21, 1833, as quoted in Toole, *Ada, The Enchantress of Numbers*, 38–39.
39. Ibid, 50.
40. Edward Hitchcock, *The Religion of Geology and Its Connected Sciences* (1851), as quoted in Richard Rudisill, *Mirror Image: The Influence of the Daguerreotype on American Society* (Albuquerque: University of New Mexico Press, 1971), 89–92.
41. Walt Whitman, *Poem of Pictures* (c.1850), as quoted in ibid, 92.
42. Alexander Bain, *Electric Time Pieces and Telegraphs* (London: George E. Eyre and William Spottiswoode, 1856). This publication describes and illustrates Bain's 1843 specifications for what he claimed were "improvements in producing and regulating electric currents and improvements in timepieces and in electric printing and signal telegraphs." By 1848 an Englishman named Frederick Bakewell had designed another facsimile system capable of transmitting handwriting and simple line drawings along telegraph wires, and this was followed in 1856 by Giovanni Caselli's Pantelegraph system, which was able to transmit all kinds of information, drawings as well as text, across telegraph lines; in 1865, Paris was connected to other cities by this system.
43. For more on Morse's photographic experiments and their timing, see my "'Some Experiments of Mine': The early photographic experiments of Samuel Morse," *History of Photography* 15: 1 (Spring, 1991): 37–42. It should be noted that, contrary to most published accounts of the history of photography, there is no evidence that Morse undertook these experiments in 1812 while a student at Yale.
44. See William Welling, *Photography in America: The Formative Years 1839–1900* (Albuquerque: University of New Mexico Press, 1978), 7–11.
45. See Bates Lowry and Isobel Barrett Lowry, *The Silver Canvas: Daguerreotype Masterpieces from the J. Paul Getty Museum* (Los Angeles: The J. Paul Getty Museum, 1998), 140–142. The drawing or lithograph on the left of the composition might well represent the war-like Minerva, goddess of the useful and ornamental arts.
46. Morse, writing to C.T. Jackson, September 18, 1837, as quoted in Carleton Mabee, *The American Leonardo: A Life of Samuel F.B. Morse* (New York: Alfred A. Knoff, 1944), 149. Morse later remembered the date of his first conception of an electric telegraph system as October 19, 1832. For a history of telegraphy before Morse, see the aptly titled book by Gerald J. Holzmann and Björn Pehrson, *The Early History of Data Networks* (Los Alamitos: IEEE Computer Society Press, 1995).
47. As quoted in Edward Lind Morse ed., *Samuel F.B. Morse: His Letters and Journal*, Vol. 2 (Boston: Houghton Mifflin Co., 1914), 38–39.
48. Manovich, *The Language of New Media*, 25.
49. Morse, like Talbot and Babbage, saw his creative efforts, both his painting and his invention of the electric telegraph, as God's work. See the discussion in Paul J. Staiti, *Samuel F.B. Morse* (Cambridge: Cambridge University Press, 1989), 221–224. It was fitting therefore that the first message sent by Morse from one city to another, transmitted from Washington to Baltimore on May 24, 1844 as a demonstration for members of Congress, was the phrase "What hath God wrought." Equally fitting, given the argument of this essay, was that the Biblical phrase was taken from Numbers 23: 23.
50. Staiti, *Samuel F.B. Morse*, 219–220.
51. Morse, writing in 1848, as quoted in Mabee, *The American Leonardo*, 185.
52. Morse, 1849, ibid 188.
53. Nor was this the only death associated with telegraphy's introduction. As one observer exclaimed of Morse in 1868, "our guest has annihilated both space and time in the transmission of intelligence." William Cullen Bryant, as quoted in Staiti, *Samuel F.B. Morse*, 230. In a coincidence reminiscent of Talbot's pre-photographic poem about fading images, Morse wrote an illustrated poem in March 1815 titled *Knight in Armor* that he projected through space and time by explicitly addressing it to "the possessor of this Book in the year 1960." Staiti reproduces this manuscript on page 28 of his book.
54. Morse to Smith, March 2, 1839, as quoted in Edward Lind Morse ed.,*Samuel F.B. Morse*, Vol. 2, 124–125.
55. Author unknown, "Untitled Poem" (1852). In *Light Verse on Victorian Photography*, edited by Bill Jay (Tempe: Limner Press, 1982).
56. See Steven Lubar, *Infoculture: The Smithsonian Book of Information Age Inventions* (Boston: Houghton Mifflin Co., 1993), 244. In May, 1924 the English inventor John Logie Baird was granted a patent for his "System of Transmitting Views, Portraits, and Scenes by Telegraphy or Wireless Telegraphy," later known as television. By 1928 he was able to transmit the photographic cover of a magazine being held up to his camera.
57. Shelford Bidwell, "Tele-Photography," *Nature* (February 10, 1881): 344–346.
58. T. Thorne Baker, "Photo-Telegraphy," *The Photographic Journal* Vol. XLVIII, No. 4 (April 1908): 179–186. "Photo-Telegraphy is the term which has been accepted to signify the transmission of a photograph from one place to another by electrical means. Although the quality of the pictures telegraphed is by no means perfect at present, it is at the same time sufficiently good to admit of the use of the process for commercial journalistic purposes; the present time, moreover, will be a memorable one in the future history of photo-telegraphy, for the science is now on the verge of considerable development, and many improvements of a substantial kind may be looked for in the near future." Baker reports that, depending on the system adopted, the transmission of photographs, including landscapes and portraits,

takes only about six minutes, with the resulting scanning lines reproducing themselves at 4mm apart. He goes on to speculate that the results will soon be sufficiently good to be publishable. He also reports on the possible use of photo-telegraphy in criminal investigation: "I read only a week ago that the German police were proposing to have Korn instruments in various parts of the country as a means of assisting their detective force. I should imagine, however, that in the present condition of the system the adoption of such a plan would be likely to increase the possibilities of wrongful arrest." By 1925 AT&T had opened a Telephoto service, the first commercial system of transmitting photographs over telephone lines. Even the fashion industry found a use for this novel technology. In September 1934, for example, *Harper's Bazaar* published five pages of fashion photographs by Man Ray, "radioed from Paris." All this occurred before Zuse's computer (1936), or Alan Turing's abstract for a universal computing machine (1937), or the conception of binary switching circuits (1940), or the building of the first ENIAC computer (1943), each of which were important, but later, contributions to the computer culture we know today. Speaking of that culture, it was as recently as 1957 that Russell Kirsch and colleagues at the National Bureau of Standards in Washington first constructed a drum scanner that could translate a photographic image into digital code and thereby store the information on a computer. It is worth noting that Kirsch chose as his first digital subject a family portrait of his baby son, thus allying one recent act of reproduction with his other, machinic one. See Russell A. Kirsch, "SEAC and the Start of Image Processing at the National Bureau of Standards," *IEEE Annals of the History of Computing* 20: 2 (1998): 9–13. In making such a choice, Kirsch also invests his new technology with a familiar view of the photograph as a vehicle for parental memory and family values, a view long promoted by Eastman Kodak and its many sub-contractors (such as Edward Steichen, who designed and photographed a series of advertisements for Kodak on this theme in 1934). It would be interesting to know what percentage of digital photographs are now devoted to the distribution of similar family snapshots and the ideology they represent. Images such as these can of course now be sent almost anywhere on earth, so that no grandparent can any longer escape their family obligation to admire new progeny. In 1991, for example, the Associated Press introduced its PhotoStream picture transmission system, using satellites and digital storage and manipulation capabilities (and reducing the transmission time of a black and white photograph to one minute, and a color photograph to three). By 1993 the transmission time of a color photograph sent via PhotoStream had been further reduced to less than ten seconds. Now it's virtually instantaneous and the image produced at the other end is to all intents and purposes indistinguishable from a traditional photograph. Indeed, in 2002 a California company called Foveon announced a new type of digital image sensor; with approximately 3.53 million pixels, it is the first to match or surpass the photographic capabilities of 35-millimeter film.

59. See, for example, Rachel Greene's comments in her "Voiceover" (a conversation with Lev Manovich and Geoffrey Batchen), *Afterimage* 29: 4 (January/February 2002): 11.
60. Michel Foucault, *The Archaeology of Knowledge*, 47–48. Manovich makes only a passing reference to the work of Foucault in his book (and no reference to any of Foucault's "archaeological" texts). His use of the word "archaeology" is perhaps meant more colloquially, similar to the use of the word (as in, to "make order out of a vast amount of material") adopted by C.W. Ceram in his *Archaeology of the Cinema* (London: Thames and Hudson, 1965), 9. Interestingly, Ceram dates the "pre-history" of cinema to 1832 (the same year that saw the conception of computing and electric telegraphy)—and there, of course, begins another line of response to Manovich's second major claim about the history of new media. Although again only mentioned in passing in Manovich's book, Erkki Huhtamo is another historian of new media who has taken to using the word "archaeology" in this context. However he uses it in two different ways. His 1994 essay "From Kaleidoscopomaniac to Cybernerd: Notes Toward an Archeology of Media" (also the title of his 1994 book), opens with a discussion of Ceram's *Archaeology of the Cinema* but goes on to engage the work of Foucault, Tom Gunning and Friedrich Kittler as part of an effort to articulate a suitable historical method for "a wider and more multifaceted social and cultural frame of reference" for a "media archeological approach." See Erkki Huhtamo, "From Kaleidoscopomaniac to Cybernerd: Notes Toward an Archeology of Media" (1994). In *Electronic Culture: Technology and Visual Representation*, edited by Timothy Druckrey (New York: Aperture, 1996), 296–303, 425–427. However, in his 1996 essay, "Time Traveling in the Gallery: An Archeological Approach in Media Art," Huhtamo uses the word in its more vernacular sense, refering to the work of artists who borrow technologies and ideas from the past. See Erkki Huhtamo, "Time Traveling in the Gallery: An Archeological Approach in Media Art." In *Immersed in Technology: Art and Virtual Environments*, edited by Mary Anne Moser, with Douglas MacLeod (Cambridge, MA: MIT Press, 1996), 233–268.
61. Michel Foucault, *The Order of Things: An Archaeology of the Human Sciences* (New York: Vintage, 1966, 1970), xi, xxii. For a suggestive commentary on Foucault's elliptical use of rhetoric, see Hayden White, "Michel Foucault." In *Structuralism and Since*, edited by John Sturrock (Oxford: Oxford University Press, 1979), 81–115.
62. I have elsewhere argued that photography, having been turned into an apparition of itself by the advent of digital imaging, has now attained the condition of its own subjects; it haunts the digital image with its noisy absence, making this image possible, but only at the cost of photography's own continual conjuring and erasure. See my "Carnal Knowledge," *Art Journal* 60: 1 (Spring 2001): 21–23.
63. As I have written in an earlier account: "A Foucauldian history of photography does not so much replace the idea of continuity with that of discontinuity as problematize the assumed distinction between the two. At the heart of both Foucault's method and photography's historical identity is once again this tantalizing undecidability, this play of a difference that is always differing from itself." See *Burning with Desire*, 186.
64. Jacques Derrida, *Of Grammatology*, trans. Gayatri Spivak (Baltimore: The Johns Hopkins University Press, 1976), 143.

3

"Tones from out of Nowhere"
Rudolf Pfenninger and the Archaeology of Synthetic Sound

Thomas Y. Levin

> *4.014 The gramophone record, the musical idea, the written notes, the sound waves, all stand in the same internal representational relationship to one another that obtains between language and the world.*
>
> —Ludwig Wittgenstein, *Tractatus logico-philosophicus* (1921)

"All-of-a-tremble": The Birth of Robotic Speech

On February 16, 1931, the *New York Times* ran a story on a curious development that had just taken place in England: "Synthetic Speech Demonstrated in London: Engineer Creates Voice which Never Existed" read the headline.[1] The day before, so the article began, "a robot voice spoke for the first time in a darkened room in London . . . uttering words which had never passed human lips." According to the accounts of this event in numerous European papers, a young British physicist named E.A. Humphries was working as a sound engineer for the British International Film Co. when the studio ran into a serious problem. A synchronized sound film (then still quite a novelty) starring Constance Bennett had just been completed in which the name of a rather unsavory criminal character happened to be the same as that of a certain aristocratic British family. This noble clan was either unable or unwilling to countenance the irreducible—even if seemingly paradoxical—polysemy of the proper name (so powerful, perhaps, was the new experience of hearing it actually uttered in the cinema) and threatened a libel suit if "their" name was not excised. As the film had already been shot, however, eliminating it would have involved huge reshooting costs and equally expensive production delays. Consequently, the producers supposedly decided to explore an innovative alternative: unable to get their star back into the studio to simply rerecord and postsynchronize an alternative moniker—the journalistic accounts are uniformly vague as to why—a print of the film was given instead to Humphries, who used his extensive experience as an

Figure 3.1 Rudolf Pfenninger in his laboratory with hand-drawn sound strips, 1932. Reproduced by permission of the Pfenninger Archive, Munich.

acoustic engineer to make the necessary changes to the soundtrack *by hand,* substituting in each case an alternative name *in Bennett's "own" voice.*

This curious artisanal intervention had become possible because the first widely adopted synchronized sound-on-film system—developed and marketed by the Tri-Ergon and the Tobis-Klangfilm concerns—was an *optical* recording process. Unlike the earlier Vitaphone system that employed a separate, synchronized soundtrack on phonograph discs, the new optical recording technology translated sound waves via the microphone and a photosensitive selenium cell into patterns of light that were captured photochemically as tiny graphic traces on a small strip that ran parallel to the celluloid film images.[2] "In order to create a synthetic voice," so Humphries explains, "I had to analyze the sounds I was required to reproduce one by one from the sound tracks of real voices"; having established which wave patterns belonged to which sounds—that is, the graphic sound signatures of all the required phonetic components—Humphries proceeded to combine them into the desired new sequence and then, using a magnifying glass, painstakingly draw them onto a long cardboard strip. After one hundred hours of work this sequence of graphic sound curves was photographed such that it could function as part of the optical film soundtrack and indeed, when played back on a "talkie" projector, according to the journalist who witnessed the demonstration, "slowly and distinctly, with an impeccable English accent, it spoke: 'All-of-a-tremble,' it said. That

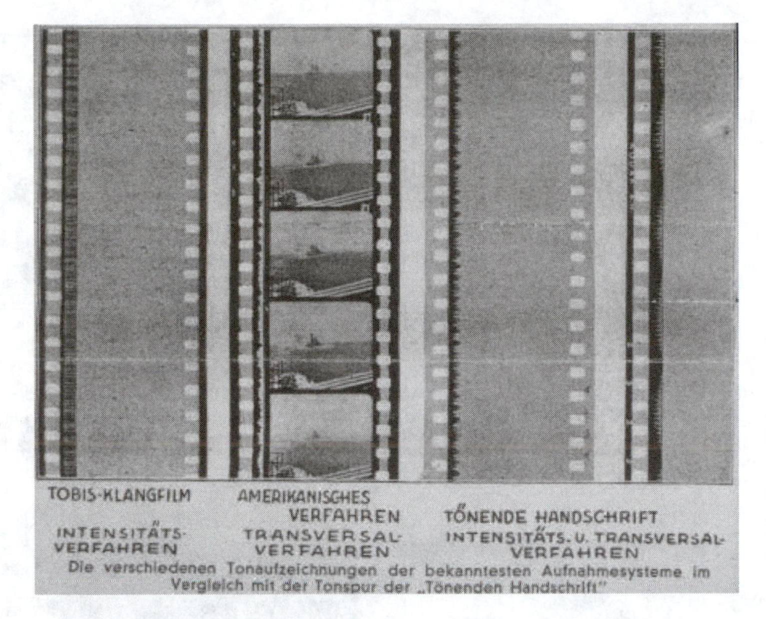

Figure 3.2 Juxtaposition of various competing optical sound systems. From left to right; the Tobis System, the American system, and two versions of the Tri-Ergon system used by Pfenninger.

was all." But these words—wonderful in their overdetermined thematization of the shiver that their status as *unheimlich* synthetic speech would provoke—were in a sense more than enough: the idea of a *synthetic sound,* of a sonic event whose origin was no longer a sounding instrument or human voice, but a graphic trace, had been conclusively transformed from an elusive theoretical fantasy dating back at least as far as Wolfgang von Kempelen's *Sprachmaschine* of 1791,[3] into what was now a technical reality.

News of the robotic utterance, of the unhuman voice, was reported widely and excitedly in the international press, betraying a nervous fascination whose theoretical stakes would only become intelligible decades later in the poststructuralist discussion of phonocentrism, of the long-standing opposition of the supposed "presence" of the voice as a guarantor of a speaker's meaning with the "fallible" and problematically "absent" status of the subject (and the resulting semantic instability) in writing. Indeed, much like the Derridian recasting of that seeming opposition that reveals writing as the very condition of possibility of speech (and, in turn, of the fullness, stability, and "presence" of the meaning subject), so too does the specter of a synthetic voice, of the technogrammatologics of Humphries's demonstration of a speaking produced not by a human agent but by a process of analysis and synthesis of acoustic data—literally by an act of inscription—profoundly change the very status of voice as such. This proleptic technological articulation of the "linguistic turn," this production of a voice by graphic means, was itself, however, the product of a long-standing project whose most recent chapter had been the invention of the phonograph and gramophone. This writing (*grame*) of sound (*phone*) had already effected a crucial dissociation, effectively making possible, through the recording and subsequent playback of the voice, the separation of speech from the seeming presence of utterance. Once, thanks to the phonograph, one's voice can resound even when one is absent—indeed even after one is dead—then voice is, as Friedrich Kittler put it so aptly, "posthum schon zu Lebzeiten" (posthumous already during [its] lifetime),[4] which is to say already of the order of writing, because to write, as Derrida once put it, is to invoke a techne that will continue to operate even during one's radical absence (i.e., one's death).

Yet while the condition of possibility of the phonographic capturing and rephenomenalization of the acoustic was indeed a kind of acoustic writing, the inscription produced by the gramophonic "pencil of nature" was barely visible, hardly readable as such. In the end, the "invention" of synthetic sound—that is, the ability to actually "write" sound as such—effectively depended on four distinct developments:

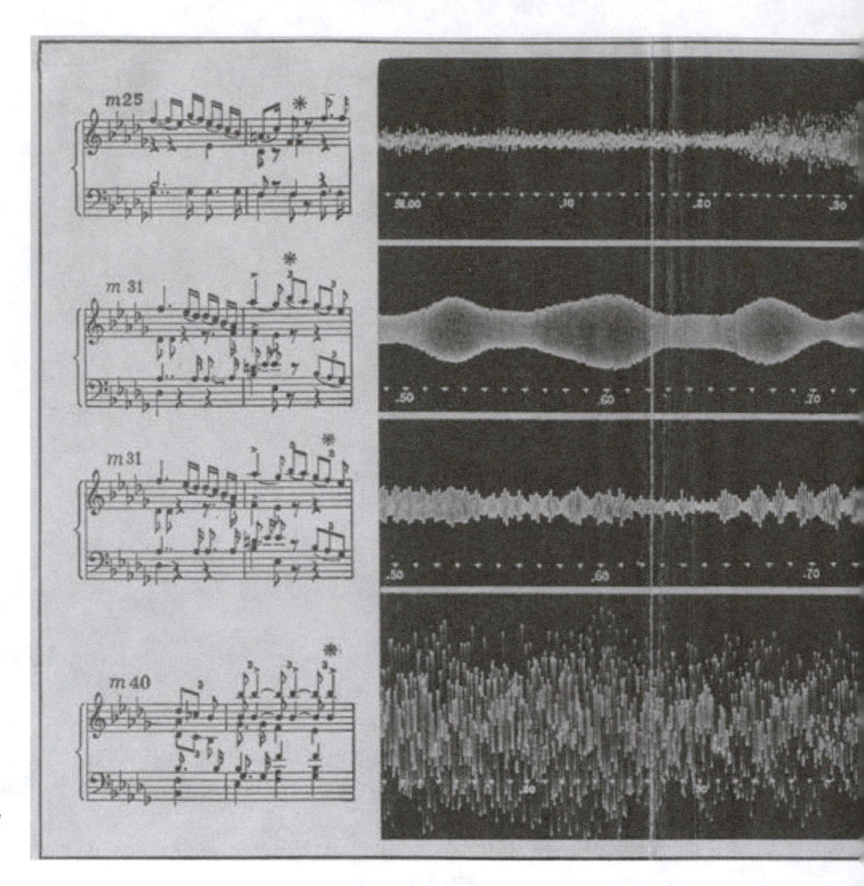

Figure 3.3 "Photographs of sound waves"—phonograph recording of the vocal sextette from "Lucia di Lammermoor" with orchestral accompaniment. Published in Dayton Clarence Miller, *The Science of Musical Sounds* (1916).

1. the initial experiments that correlated sound with graphic traces, making it possible to "see" the acoustic;
2. the invention of an acoustic writing that was not merely a graphic translation of sound but one that could also serve to reproduce it (this was the crucial contribution of the phonograph);
3. the accessibility of such acoustic inscription in a form that could be studied and manipulated *as such*; and finally
4. the systematic analysis of these now manipulatable traces such that they could be used to produce any sound at will.

The archaeology of the above-mentioned robotic speech, in turn, also involves four distinct stages:

1. the coming-into-writing (*mise-en-écriture*) of sound as mere graphic translation or transcription;
2. the functional development of that inscription as means to both trace and then rephenomenalize the inscribed sound;
3. the optical materialization of such sounding graphic traces that would render them available to artisanal interventions; and finally
4. the analytic method that would make possible a functional systematic vocabulary for generating actual sounds from simple graphematic marks (of the sort made famous by Humphries).

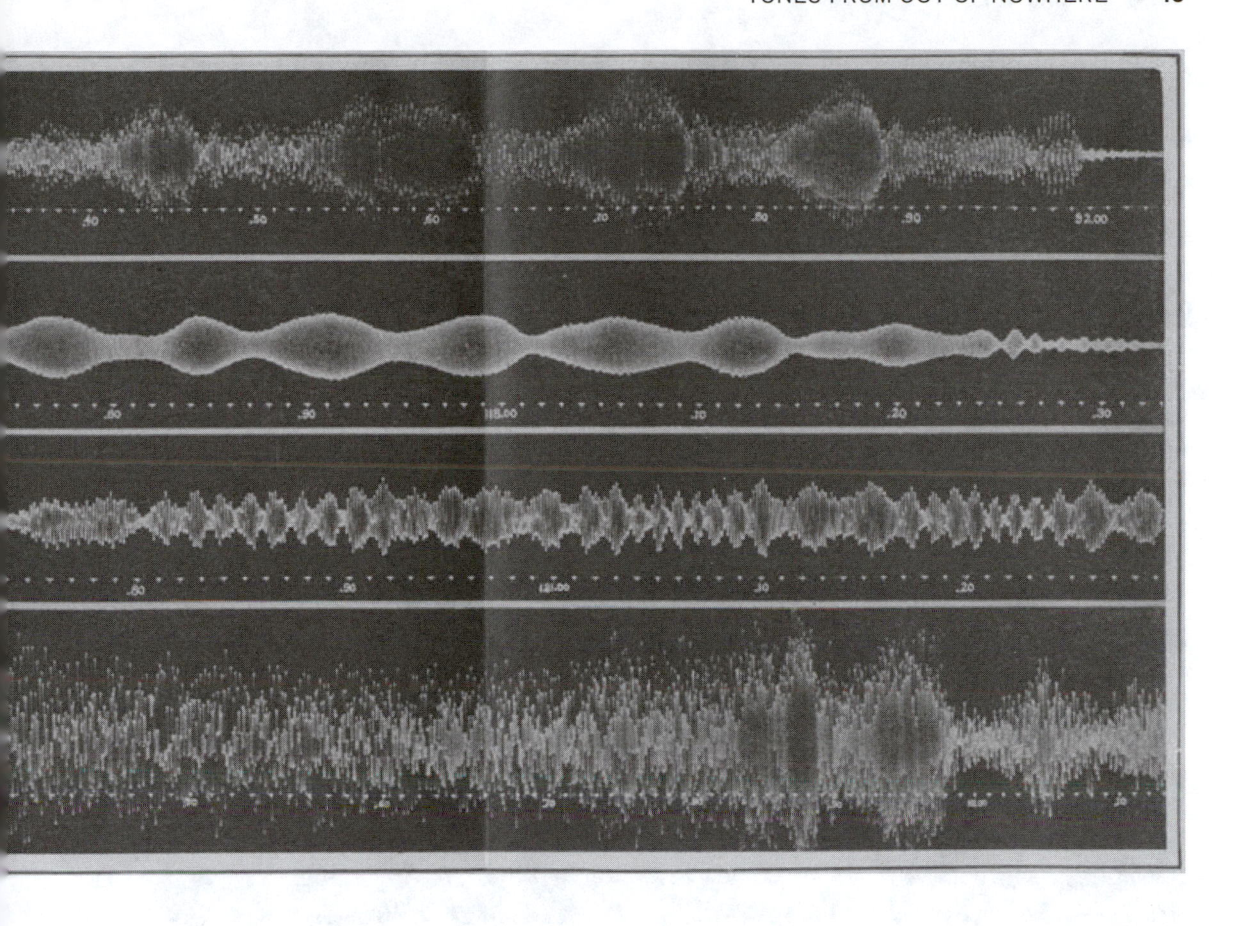

Following a brief overview of these first two, generally more well-known moments, this essay will focus on the latter, largely ignored, chapters of the fascinating story of the "discovery" of synthetic sound.

Genealogics of Acoustic Inscription

Already in the 1787 text *Entdeckungen über die Theorie des Klanges* (*Discoveries about the Theory of Sound*) by the so-called father of acoustics, Ernst Florens Friedrich Chladni, one can read about a graphic transcription of sound that, unlike all previous notational practices, was not strictly arbitrary. Chladni's discovery that a layer of quartz dust upon a sheet of glass would, when vibrated by a violin bow, form distinct and regular patterns or *Klangfiguren* (tone figures), as he called them, that correspond to specific tones, effectively demonstrated the existence of visual traces of pitches whose iconico-indexical character differentiated them in a semiotically crucial fashion from all other conventional means of notating sound. What was so exciting about these acoustic "ur-images" (as a contemporary of Chladni called them) was that they seemed to arise from the sounds themselves, requiring for their intelligibility not the hermeneutics appropriate to all other forms of musical notation but instead something more akin to an acoustic *physics*. The subsequent prehistory of the phonograph—and Chladni's practical insight into the relationship of sound, vibration, and its graphic transcriptionality points to nothing less than the inscriptional condition of possibility of the phonograph as such—is concerned initially with the rendition of sound as (visible) trace. Indeed, this task was of great interest to the nascent field of early linguistics known since the 1830s alternately as *Tonschreibekunst,* phonography, or vibrography, which both supported and profited from various protophonographic inventions.[5] Central among these were Edouard Léon Scott's wonderfully named "phon-autograph" of 1857, often described as the first oscillograph employed for the study of the human voice; the Scott-Koenig Phonautograph" of 1859, which (like its predecessor) transcribed sound waves in real time as linear squiggles; and Edward L. Nichols and Ernst George Merritt's photographic records of the flickering of Rudolph Koenig's 1862 manometric capsule, in which changes in pressure produced by sound waves are captured by the vibrations of a burning gas flame. In various ways, all these technologies were exploring the relationship of speech and inscription, as evidenced, for example, in the experiments undertaken

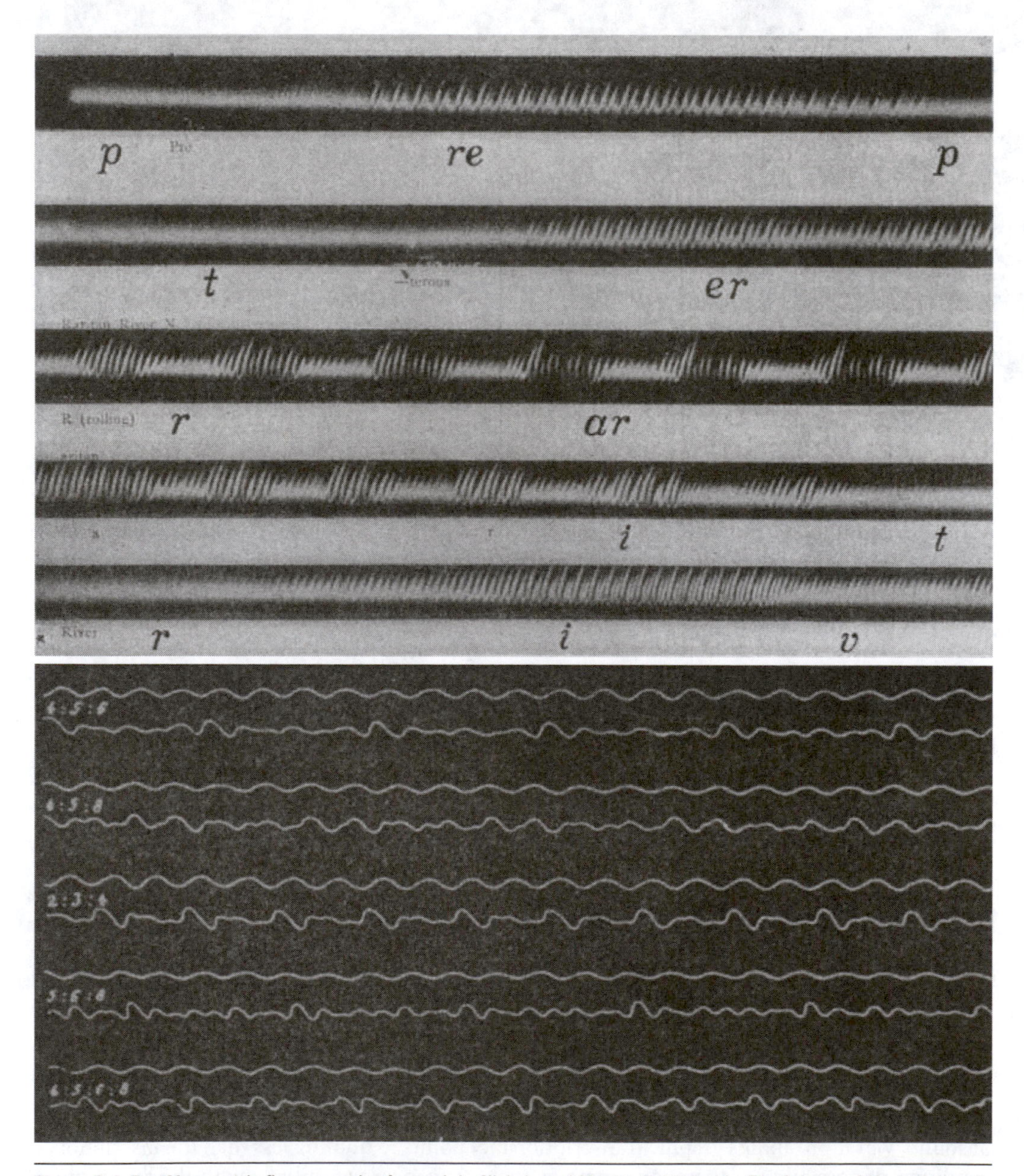

Figure 3.4 Top: Manometric flame records of speech by Nichols and Merritt. Published in *The Science of Musical Sounds.* Bottom: Phonautograph records. Published in *The Science of Musical Sounds.*

in 1874 by the Utrecht physiologist and ophthalmologist Franciscus Cornelius Donders, who is described as having used Scott's phonautograph to record the voice of the British phonetician Henry Sweet, noting next to the acoustic traces the exact letters being spoken, while a tuning fork was used to calibrate the curves.[6]

But if sound in general—and speech in particular—is here rendered visible by various means as graphic traces, this particular sort of readability (with its undeniable analytic value) is bought at the price of a certain sort of functionality: sound is literally made graphic, but in the process becomes *mute*. This changes dramatically in the next stage of this techno-historical narrative. Thomas Alva Edison's invention in 1877 of the first fully functional acoustic read/write apparatus successfully pioneered a new mode of inscription that both recorded *and* re-produced sound, albeit now at the price of the virtual invisibility of the traces involved. What had previously been a visually accessible but nonsounding graphematics of the acoustic was now capable of both tracing and rephenomenalizing sound, but by means of an inscription that—in a gesture of Media-historical coquetry—hid the secrets of its semiotic specificity in the recesses of the phonographic grooves. This invisibility not only served to foster the magical aura that surrounded the new "talking machines"—leading some early witnesses of the first demonstration of Edison's new machine at the Paris Academy of Sciences on March 11, 1878, to accuse the inventor's representative du Moncel of ventriloquistic charlatanry[7]—but also raised the question as to the status of the cylindrical traces. It was generally acknowledged that the tiny variations in the spiral groove were a writing of some sort—indeed, as Friedrich Kittler has noted, the reason why it is Edison's cylinder phonograph and not Emil Berliner's flat gramophone record that has been the repeated object of literary fascination is due to no small degree to the fact that the cylinder's "read/write" inscriptional capacity (it is both a playback *and* recording device) enables it to do what was previously impossible on paper.[8] Nevertheless, contemporaries of Edison's invention were divided as to whether one ought ever "to hope to be able to read the impressions and traces of phonographs, for these traces will vary, not alone with the quality of the voices, but also with the differently related times of starting of the harmonics of these voices, and with the different relative intensities of these harmonics."[9] Others, however, were convinced that, as a later enthusiast put it, "by studying the inscriptions

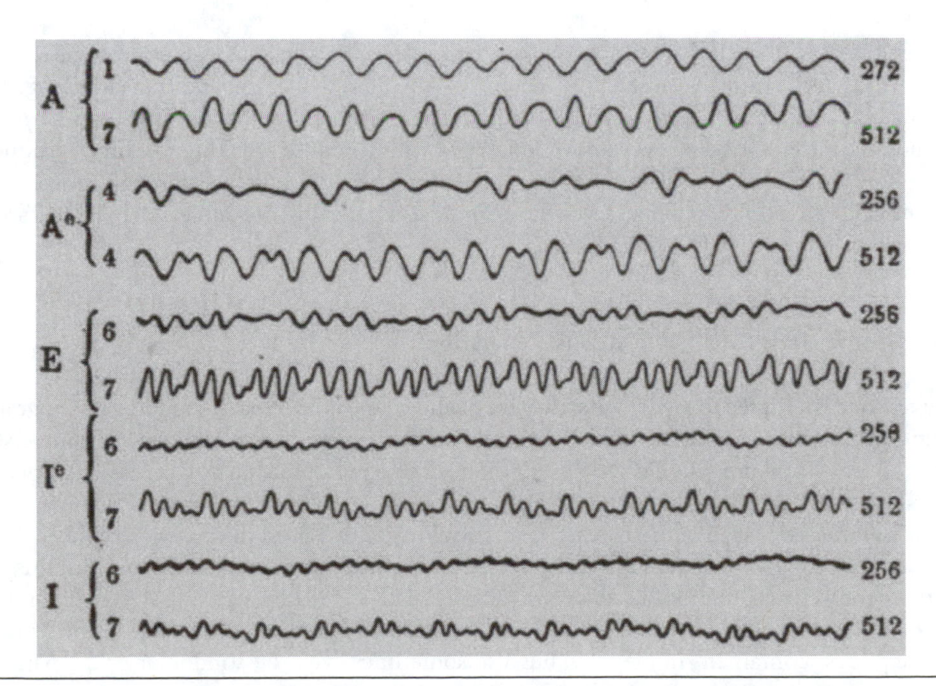

Figure 3.5 "Vowel curves enlarged from a phonographic record." Published in *The Science of Musical Sounds.*

closely one may come to an exact knowledge of these inscriptions and read them as easily as one reads musical notes for sound."[10]

For reasons whose motivations might well have been less than entirely "scientific," Edison's own position was that the gramophonic traces ought *not* be understood as writing. In the context of congressional hearings in 1906 and 1908 on the question of whether recorded sound was copyrightable, Frank L. Dyer, Edison's patent attorney, CEO, and sometime biographer, testified that recordings were not copies of "writings" because they were not legible. To support this claim he recounted how Edison had attempted in vain to make the phonograph records readable through the following laboratory strategy: having made a recording of the letter *a*, "he examined with a microscope each particular indentation and made a drawing of it, so that at the end of two or three days he had what he thought was a picture of the letter 'a.'" But when he compared different recordings of the same letter it became clear that the "two pictures were absolutely dissimilar."[11] This spurious confusion of the status of alphabetical and phonological signifiers (the two recordings of the letter *a* are different because they record both the letter and its *pronunciation*)—which seems suspiciously convenient in this economico-juridical context—does not arise in a similar debate that took place in the German court system the same year, concerning the status of recordings of Polish songs that glorified the independence struggles of the previous century. After a series of earlier decisions pro and contra, the high court decided unambiguously that these gramophonic inscriptions were indeed writing and could thus be prosecuted under paragraph 41 of the criminal code that governs illegal "writings, depictions or representations":

> The question as to whether the impressions on the records and cylinders are to be considered as written signs according to paragraph 41 of the State Legal Code must be answered in the affirmative. The sounds of the human voice are captured by the phonograph in the same fashion as they are by alphabetic writing. Both are an incorporation of the content of thought and it makes no difference that the alphabetic writing conveys this content by means of the eye while the phonograph conveys it by means of the ear since the system of writing for the blind, which conveys the content by means of touch, is a form of writing in the sense of paragraph 41.[12]

Given that the definition of writing invoked in this decision is strictly a functional one (phonographic traces are writing because they function as a medium that stores and transmits language), what remains unexamined here is the specificity of these almost invisible scribbles *as inscriptions.* Like most end users, the court was more concerned with what the speaking machines produced, but not how they did so. This latter question did however become an issue, although in an entirely different field of research—phonetics—whose foundational text is Alexander Melville Bell's 1867 opus entitled, appropriately, *Visible Speech.*[13]

From "Groove-Script" to "Opto-Acoustic Notation"

Provoked, one is tempted to say, by the script-like quality of the now actually sounding phonographic inscriptions and their migration into the invisibility of the groove, phonologists and phoneticists of various stripes—pursuing the elusive Rosetta Stone of phonographic hieroglyphics—attempted in various ways to make these functional acoustic traces visible.[14] Above and beyond their particular scientific motivation, each of these experiments also implicitly raised the question of the legibility of the semiotic logic of the gramophonic traces. Indeed, the continuing fascination with this possibility might well account for the sensation caused as late as 1981 by a certain Arthur B. Lintgen, who was able—repeatedly and reliably—to "read" unlabeled gramophone records, identifying not only the pieces "contained" in the vinyl but also sometimes even the conductor or the national-

Figure 3.6 "Writing Angel" Grammophon Logo.

ity of the orchestra of that particular recording, merely by looking at the patterns of the grooves. It matters little whether "man who sees what others hear" (as he is called in the headline of the lengthy *New York Times* account of his unusual ability[15]) was *actually* doing what he claimed: in either case his performance and its widespread reception (as evidenced, for example, by his subsequent appearance on the ABC television program *That's Incredible*) are both significant as cultural allegory, as a *mise-en-scène* of the at least *potential* readability of the still indexical gramophonic trace at the very moment that the material inscription of sound—with the advent of the compact disc and its hallmark digital encoding in the early 1980s—was becoming phenomenally even more elusive. Lintgen's *Trauerspiel* of acoustic indexicality, quite possibly the last manifestation of the long and anecdotally rich history of the readability of acoustic inscription, also confirms that not only the prehistory but also the posthistory of the phonograph can reveal what remains hidden in the depths of gramophonic grooves.[16]

Implicit in the drive to read the gramophonic traces is the notion that, once decipherable, this code could also be employed for *writing*. While the impulse to both *read and write* sound was, according to Douglas Kahn, "a desire, already quite common among technologists in the 1880s,"[17] the fascination exerted by the sheer phenomenal wonder of recorded sound (and all its equally astonishing technical consequences, such as acoustic reversibility and pitch manipulation) was—understandably—so great that for the first fifty years following the invention of the phonograph it effectively distracted attention from the various practical and theoretical questions raised by the gramophonic traces themselves, even when these were acknowledged as such. Typical in this regard is the simultaneous blindness and insight regarding gramophonic inscription in the following highly suggestive passage from Ludwig Wittgenstein's *Tractatus logico-philosophicus* of 1921:

> 4.0141 There is a general rule according to which the musician can extrapolate the symphony from the score, and according to which one can derive the symphony from the groove on the gramophone record and then, using the first rule, in turn derive the score once again. That is what constitutes the inner similarity between these seemingly so completely different constructs. And this rule is the law of projection, which projects the symphony into the language of musical notation. It is the rule for the translation of the language of musical notation into the language of the gramophone record.[18]

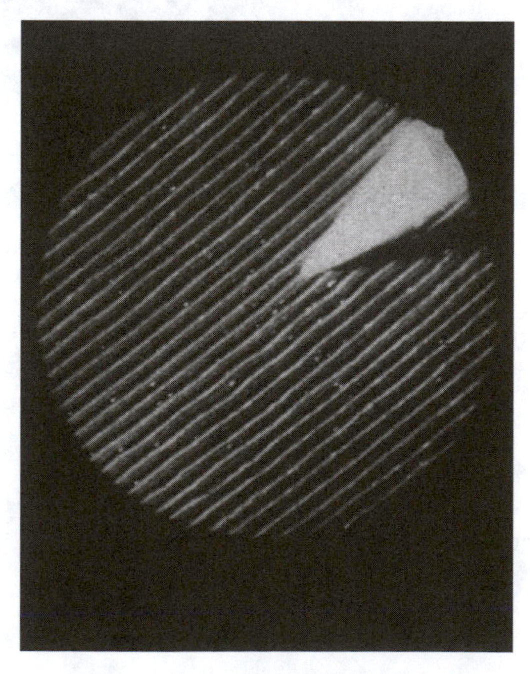

Figure 3.7 Close-up photograph of a phonograph record showing the point of the needle and the "wavy" grooves. Area shown is 1/3" in diameter.

While Wittgenstein invokes both the gramophonic trace and "the language of the gramophone record," and in the final line even effectively juxtaposes gramophonic "language" with another form of musical notation, a careful reading of the passage reveals that Wittgenstein's concern is not the character of the gramophone record's inscriptions as such but rather the technical capacity of that "language" to store and re-produce sound. Dramatically different, by comparison, and an index of an important shift in the sensibility toward the *semiotic specificity* of the gramophonic grooves as such, is the intriguing remark in Rainer Maria Rilke's famous prose piece "Ur-Geräusch" (Primal Sound), written only two years earlier, in which the young poet describes his early fascination with the new acoustic technology: "As time would tell it was not the sound from the horn that dominated my recollection, but instead it was those curious signs etched into the cylinder that remained much more significant to me."[19] Unlike Wittgenstein, for whom the gramophone is significant thanks to its capacity to re-produce a given piece of music, Rilke's concern is with the "ur-sound" that might arise from a gramophonic tracing of the cranial groove in a skull sitting on his table. This thought experiment raises the question of the gramophone's capacity to render audible sounds that were never previously recorded, or, in Kittler's apt terminological recasting, to decode an inscription that had never been previously encoded.[20] While the appeal of this seminal techno-semiotic allegory lies precisely in the nonetheless still referential fascination that informs Rilke's musings on the skull's groove as the locus of some sort of a *signal* (i.e., an inscription that, while not produced by a subject, might nevertheless be a trace of some other signifying agency), the sound that this hypothetical phonography of the cephalic suture would in fact produce would most probably resemble what we tend to call noise and as such would "refer" acoustically more to the materiality of technical mediation as such—that is, to the literal topography of the sonic groove.[21]

The stakes involved in the difference between Wittgenstein's focus on the *result* of gramophonic inscription and Rilke's insistence on the epistemological questions raised by the physical mediation *as such*, are given what is probably their most programmatic articulation in the famous essay by the pioneering avant-garde polymath László Moholy-Nagy entitled "Production-Reproduction," which appeared in 1922 in the journal *De Stijl*.[22] In this classic text of Weimar-era gramophonic modernism, Moholy-Nagy argues that because art serves to train man's sensory and other apparatuses for the reception of the new, then creative activities that hope to do justice to the imperatives of their time must explore the unknown rather than simply re-produce the familiar. Applied to the acoustic domain, this means that the gramophone must be transformed from a mere means of *re*-production (i.e., a medium that simply records, stores, and then rephenomenalizes sounds created elsewhere) into a tool of *pro*duction, an instrument in its own right; that is, a technology that will produce new, previously unheard sounds specific to its capacities. In doing so, it would realize a potential also promised (but also not always realized) by other new mechanical musical devices—such as the Trautonium, Sphaerophon, and the Atherophon or Theremin—which were all the rage in the Western musical world of the 1920s. Manifesting a focus more reminiscent of Rilke than Wittgenstein, Moholy-Nagy proposes that one undertake a scientific examination of the tiny inscriptions in the grooves of the phonograph in order to learn exactly what graphic forms corresponded to which acoustic phenomena. Through magnification, he suggests, one could discover the general formal logic that governed the relation of the acoustic to the graphematic, master it, and then be able to produce marks that, once reduced to the appropriate size and inscribed onto the record surface, would literally be acoustic writing:

> the grooves are incised by human agency into the wax plate, without any external mechanical means, which then produce sound effects that would signify—without new instruments and without an orchestra—a fundamental innovation in sound production (of new, hitherto unknown sounds and tonal relations) both in composition and in musical performance.
>
> The primary condition for such work is laboratory experiments: precise examination of the kind of grooves (as regards length, width, depth, etc.) brought about by the different

Figure 3.8 Lásló Moholy-Nagy: *Gramofonplatte* (gramophone record). Published in *Malerei Fotografie Film* (1925).

> sounds; examination of the man-made grooves; and finally mechanical-technical experiments for perfecting the groove-manuscript score. (Or perhaps the mechanical reduction of large groove-script records.)

Liberating the gramophone from the mere "photographic" re-production of prior sounds, this "groove-script alphabet"—as Moholy-Nagy called it a year later in an essay entitled "New Form in Music: Potentialities of the Phonograph"— would make the gramophone into "an overall instrument . . . which supersedes all instruments used so far," allowing one to employ the technology as a means to write sound directly, enabling composers to eliminate the intermediary of the performance by "writing" their compositions as sounding scripts, and making it possible for sound artists to express and transmit any language or sound, including previously unheard acoustic forms and works.[23]

In the mid-1920s Moholy-Nagy's challenge was taken up and further articulated by the music critic Hans Heinz Stuckenschmidt in a series of polemical interventions in numerous journals ranging from *Der Auftakt* to *Modern Music*. Enlisting the gramophone in the project of a musical *Neue Sachlichkeit* (New Objectivity), Stuckenschmidt mobilized Moholy-Nagy's arguments (both implicitly and explicitly) for debates in musical composition, interpretation, and performance, including the highly provocative claim that by means of works written specifically for the new technologies, the composer could eliminate the subjective dimensions invariably introduced both through the irreducibly ambiguous character of musical notation and the vicissitudes of "live" performance. Insisting that, thanks to machines such as the gramophone, "the role of the interpretor is a thing of the past,"[24] Stuckenschmidt's philo-gramophonic articles elicited vicious and often Luddite responses. Happily, however, there was also another dimension to the reception of his polemics—one that responded to his important claim that "the essential significance of these machines [phonographs and gramophones] lies in the possibility of writing for them in an authentic fashion."[25] Continuing what was by then almost a tradition of pieces composed expressly for new acoustic technologies—such as Ferruccio Busoni's 1908 sketch "Für die [sic] Pianola" or Igor Stravinsky's "Etude pour Pianola" of 1917 (whose 1921 premiere in London took place in the player piano company's own "Aeolian Hall")—the 1920s had witnessed a proliferation of works written for "musical machines" (as they were called at the time). These experiments were most often premiered at new music festivals such as the Donaueschingen Musiktage whose 1926 program featured works for Welte-Mignon pianola rolls composed by Paul Hindemith, Ernst Toch, and Gerhart Münch. Although Stuckenschmidt claimed as early as 1925 that "I myself carried out fundamental experiments with the gramophone at the same time that George Antheil was doing so in Paris,"[26] the earliest documented public performance of gramophone-specific music was not until 1930 at the Musikfest Neue Musik held at the Staatliche Hochschule für Musik in Berlin, where Ernst Toch presented a gramophonic montage of his four-part "Fuge aus der Geographie" and Paul Hindemith premiered his oft-invoked but only recently rediscovered experiments in "grammophonplatten-eigene Stücke" (pieces specifically for gramophone records).[27]

While one cannot ignore the very real possibility that various gramophone- specific sound experiments, of which there are few or no remaining traces, might have been undertaken in marginal venues, laboratories, and nonperformance contexts in the later 1920s, the extended interval between Stuckenschmidt's 1925 rearticulation of Moholy's 1922 proposal and the known instances of its subsequent realization might nevertheless be quite telling. In fact, it matters little whether Hindemith and Toch's 1930 gramophonic compositions were, as a contemporary critic called them, the very first of their kind.[28] What is significant is that while both explored the new sonic possibilities offered by the overlapping of multiple recordings and "live" music, as well as the variations in speed, pitch and timbre that could be achieved only by the creative "misuse" of the gramophone, neither of their compositions nor any of the other "gramophonic" works of that period, to my knowledge, actually intervened at the level of the "groove-script alphabet." Despite published journalistic accounts describing early groove-script experiments by Moholy-Nagy and Antheil,[29] Moholy-Nagy himself confirms that although he had been able to get both Stuckenschmidt and Antheil interested in exploring this possibility in the mid-1920s and although the director of the Vox Corporation, a certain Jatho, had agreed to allow them to use their laboratories, "in the end my suggestions were never fully worked out in detail."[30] According to Moholy-Nagy, this was due to various institutional circumstances: Antheil, he explains, moved to Paris where he worked on player pianos for Pleyel, and Moholy himself had to devote his attentions to his new job at the Weimar Bauhaus. The reasons might also have been more technical in nature, as suggested by Hindemith's own rather skeptical remarks on the pragmatics of groove-script composing published only a few years prior to his proto-turntablist appearance in Berlin:

> The attempts to manually etch musical events onto gramophone or phonograph records have so far remained unsuccessful. At present we have come so far as to be able to depict very simple relations such as specific vowels in conjunction with specific pitches. But it is a very long way from here to the generation of even plain musical works. I don't think that it will ever be possible to make this mode of inscription useful for musical practice.[31]

As it turns out, Hindemith was both right and wrong: as he predicted, the gramophone would *never* prove amenable to the realization of a proper groove-script alphabet; yet, contrary to his prognosis, something very akin to the possibility envisioned by Moholy-Nagy was in fact being worked out at almost exactly the same time as the Hindemith-Toch experiments, albeit in a somewhat different medium—the synchronized sound film.

Always the pragmatist, Moholy-Nagy immediately recognized in the new *optical* film sound processes being adopted in the late 1920s a means to effectively realize his long-standing groove-script vision. Here the technical difficulties posed by the miniature scale of the groove-script inscriptions were eliminated by a graphic transcription of sound that was visible to the human eye. In an essay entitled "Problems of the Modern Film" published in various versions and languages between 1928 and 1932, Moholy-Nagy laid down his gauntlet in typically polemical fashion, challenging film-makers to take up the task that had so far generally eluded (or been ignored by) composers:

> Contemporary "musicians" have so far not even attempted to develop the potential resources of the gramophone record, not to mention the wireless or ether-waves.... The sound film ought to enrich the sphere of our aural experience by giving us entirely unknown sound values, just as the silent film has already begun to enrich our vision.[32]

Calling for a "a true opto-acoustic synthesis in the sound film" Moholy-Nagy predicted the emergence of the "abstract sound film" (which would be complemented by the parallel genres of the "documentary" and the "montage" sound film) and suggested that experimentation be undertaken with the soundtrack *in isolation from the image track.* That is, Moholy-Nagy recognized optical film-sound technology as an important innovation *in sound recording as such,* not least because this new form of acoustic inscription seemed to make possible what had always been so frustratingly elusive in the gramophonic realm: access to sound as trace. Besides investigations of "acoustic realism" (i.e., recorded extant sounds), he insisted on the importance of:

> experiments in the use of sound units which are not produced by any extraneous agency, but are traced directly on to the sound track and then translated into actual sound in the process of projection. (E.g., the tri-ergon system uses parallel lines of a varying brightness, the alphabet of which must be previously mastered.)... It will not be possible to develop the creative possibilities of the talking film to the full until the acoustic alphabet of sound writing will have been mastered. Or, in other words, until we can write acoustic sequences on the

Figure 3.9 Examples of different optical sound systems; film still from *"Tönende Handschrift"- das Wunder des gezeichneten Tones* (1931). Reproduced by permission of the Filmmuseum Muenchen.

> sound track without having to record any real sound. Once this is achieved the sound-film composer will be able to create music from a counterpoint of unheard or even nonexistent sound values, merely by means of opto-acoustic notation.[33]

Moholy-Nagy's unambiguous recognition that the new optical sound techniques presented an alternative means to achieve in practice what he had initially conceived in terms of the groove script alphabet also might explain why, by the later 1920s, he was no longer pursuing his original gramophonic approach: film simply seemed to offer a better way to explore more or less the same issues.

As it turns out, Moholy-Nagy did not have to wait long for this challenge to be taken up and met successfully. Indeed, in an illustrated lecture "on the invention which signifies the revolutionizing of the sound film in its entirety" that he presented in various schools and lecture halls in Germany in 1932, Moholy-Nagy announced, with unambiguous excitement, that his earlier notion of the groove-script—now called "sound-script"—had already become a reality. Revisiting the history of his own writings on the possibilities of synthetic sound from the happy perspective of the visionary whose long-doubted speculations had at long last been proven right, Moholy-Nagy writes (in the published version of that lecture):

> Sound-script makes possible acoustic phenomena which conjure up out of nothing audible music without the previous play of any musical instrument. We are in a position today to be able to play written sounds, music written by hand, without involving an orchestra, by the use of the apparatus of the sound film. It is a great pleasure for me to be able to report on this acoustical phenomenon; inasmuch as I had already explained it in articles and lectures ten years ago, although I was not fortunate enough to be able to experiment with it then, I am very happy today to witness the successful realization of those of my suggestions previously labeled absurd. At the time, my starting point was that phonograph recordings could be made on the basis of an "etched alphabet." These recordings, without any sound having previously been played and captured by them, are inscribed exclusively on the basis of the imaginative world of the composer and would have been played only subsequently. A few years later I extended my phonograph experiments to include radio, sound film and television [*sic*]. *And today, thanks to the excellent work of Rudolf Pfenninger, these ideas have been successfully applied to the medium of sound film. In Pfenninger's sound-script, the theoretical prerequisites and the practical processes achieved perfection.*[34]

According to a contemporary review of the version of this lecture presented to a gathering of the *Bund das neue Frankfurt* in the Frankfurt Gloria-Palast on December 4, 1932,[35] Moholy-Nagy showed two films in conjunction with his talk: *Tönende Ornamente* by the German pioneer of abstract animation Oskar Fischinger, and *Tönende Handschrift* (Sounding Handwriting) by

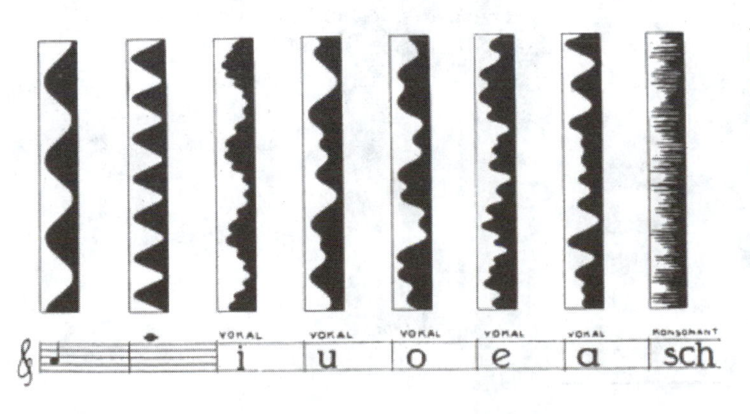

Figure 3.10 1932 Emelka publicity photograph with the caption, "This is what Rudolf Pfenninger's 'Sounding Handwriting' looks like."

a comparatively unknown Swiss-born engineer working in Munich named Rudolf Pfenninger. Given the inclusion of Fischinger in this program, and in light of the fact that his much publicized work on what he called "sounding ornaments" has led more than one film historian to credit him (implicitly or explicitly) with the invention of animated sound, why is it that Moholy-Nagy seems to insist—in an assessment later confirmed by nearly all of the historical literature—that the sole credit for the development of a functional sound script—which is to say, the invention of synthetic sound as such—belongs not to Fischinger, but to Pfenninger?[36]

The Race That Wasn't One: Fischinger, Pfenninger, and the "Discovery" of Synthetic Sound

In a classic instance of the curious simultaneity that is the repeated hallmark of the overdetermination governing the history of invention, during the early 1930s a number of people in various parts of the world were working furiously but independently on experiments in what they referred to variously as "handdrawn," "animated," "ornamental," and/or "synthetic" sound. Besides the aforementioned Humphries in England, in the Soviet Union there were, according to some accounts, no less than three separate groups of researchers working on hand-drawn sound in Leningrad and Moscow: their ranks included figures such as the composer, music theorist, and performance instigator Arsenii Avraamov; the painter, book illustrator, and animator Mikhail Tsekhanovskii; the engineer Evgenii Sholpo; the animators Nikolai Voinov and Nikolai Zhilinski; and the inventor Boris Yankovskii. While space considerations preclude anything more than a cursory treatment of these crucial Soviet contributions here, it should be noted that these groups produced some extremely important theoretical and practical results, not least being the development of a protosynthesizer called the "Variofon" and another known as the "Vibro-Eksponator."[37] At exactly the same time, and as far as I can tell without any knowledge of what was being done in the Soviet Union, similar

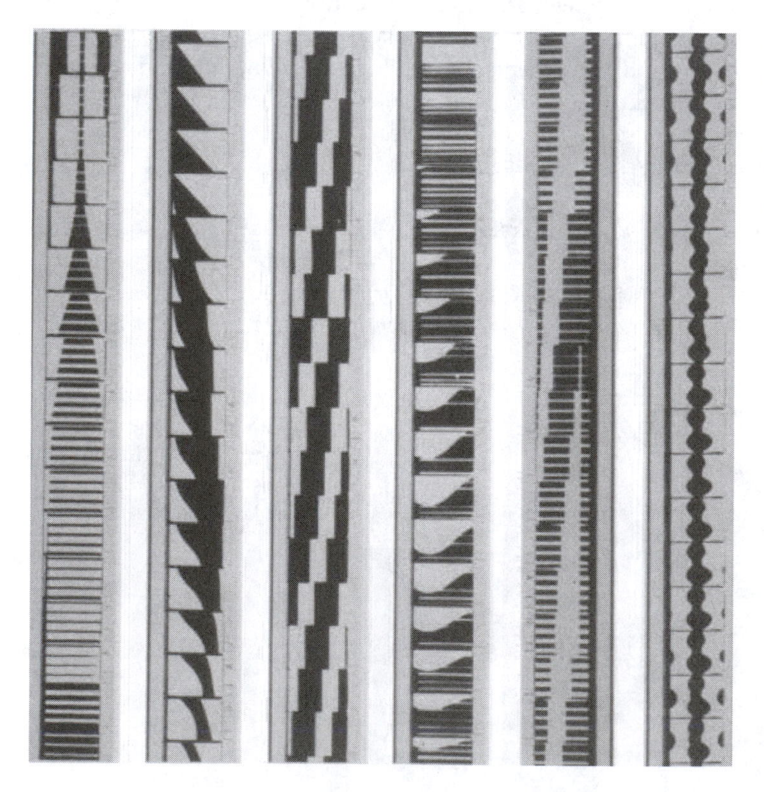

Figure 3.11 Walter Ruttmann: Strips of "musical" graphics from his early abstract "Opus" animations (1921–1925) (Centre Pompidou, Musée national d'art moderne, Paris, France).

efforts were also being undertaken in Germany by Pfenninger in Munich and, somewhat later, by Fischinger in Berlin.

Fischinger's widely discussed experiments and lectures during the years 1932–1933 grew out of his extensive earlier work in nonobjective, abstract, or, as he preferred to call it, "absolute" film, which explored the musicality of moving graphic form in the tradition of animated cinematic synesthesia established by the filmmakers Viking Eggling, Hans Richter, and Walter Ruttmann.[38] The first concrete result of these explorations in the relations between musical and graphic elements in time (which the contemporary critic Bernhard Diebold referred to with the charming neologism "Muso-Graphik"[39]) was Fischinger's compilation *Experimente mit synthetischem Ton* (Synthetic sound experiments), which was composed of "patterns, drawn on paper with pen and ink and photographed directly onto the margin of the film reserved for the sound track."[40] Fischinger's practice of making drawings on paper that would then be photographed onto the optical film sound track supposedly was inspired by his experience of hearing a key drop; struck by the fact that he recognized what he heard *as the sound of a key*, Fischinger wondered whether every shape had a corresponding sound, a sort of iconic acoustic signature.

According to William Moritz, this led Fischinger to undertake not only a series of experiments that examined the relationship between visual forms and their corresponding sonic manifestations, but also various attempts at

> drawing designs and ornaments which produced "a-musical" sounds; he found, for example, that the pattern of concentric wave-circles which was often used in cartoon and silent film iconography to represent the ringing of a door or alarm bell actually produced a buzzing clang sound when drawn in long rows and photographed onto the soundtrack area.[41]

Figure 3.12 Oskar Fischinger with rolls representing synthetic sound, circa 1932. Elfriede Fischinger Trust, courtesy Center for Visual Music.

Intrigued by the potentially far-ranging ramifications of such acoustico-visual isomorphism, Fischinger often speculated as to whether there was more than an accidental relationship between the physical shape of an object and its auditory manifestation. Might there exist some deep and previously inaccessible common structural logic that governs both the most prevalent ornamental practices of a given society and its dominant auditory patterns? Posing the question in rather explicitly nationalistic terms in a widely published 1932 essay, Fischinger states:

> Personal, national and characteristic traits naturally will also be expressed in the ornament. In terms of their vocal intonation Germans tend to make a strong attack which corresponds to a specifically jagged curve whereas the soft vocal attack of the French also manifests itself in a correspondingly different fashion in the ornament. There is thus an equally clear "mouth-writing" as there is "hand-writing."[42]

These and other related questions were the focus of investigations that Fischinger presented to great public acclaim in a lecture on synthetic sound at the Haus der Ingenieure in Berlin in the first week of August 1932.[43]

Long before the appearance of Fischinger's well-publicized explorations into the aesthetics of "tönende Ornamente," a little-known animation filmmaker and engineer named Rudolf Emil Pfenninger (1899–1976) had been busily at work in the Geiselgasteig studios of the Münchener Lichtspielkunst AG (EMELKA) perfecting what would turn out to be the first fully functioning and fully documented (i.e., not apocryphal) *systematic* technique for the entirely synthetic generation of sounds. Born in Munich as the son of the Swiss artist Emil (Rudolf) Pfenninger (1869–1936), Rudolf began studying drawing with his father, and then, after initial experiments with a self-made camera

Figure 3.13 Oskar Fischinger: *Ton Ornamente* [Sound Ornaments]. Elfriede Fischinger Trust, courtesy Center for Visual Music.

Figure 3.14 Rudolf Pfenninger holding a sound strip; film still from *"Tönende Handschrift"- das Wunder des gezeichneten Tones* (1931) (Filmmuseum Muenchen).

and an apprenticeship as set painter in the Munich Werkstätten für Bühnenkunst Hummelsheim und Romeo in 1914, worked together with Emil Pfenninger as illustrator for Gustav Hegi's multi-volume reference work on the flora of central Europe.[44] It was during this period that Pfenninger had his first contact with the movies as a projectionist at various Munich cinemas, an experience that required him to become thoroughly familiar with a wide range of film technologies (optics, mechanics, electronics). In 1921 he was discovered in Munich by the U.S. animator Louis Seel, who hired Pfenninger to draw, paint, and make animated films and text frames for silent films for the Münchener Bilderbogen. This was followed in 1925 by a new job in the Kulturfilmabteilung of the EMELKA (after UFA the second-largest film production company of the Weimar era), where he worked on films such as *Zwischen Mars und Erde* (Dir. F. Möhl, 1925). Pfenninger simultaneously pursued intensive engineering research on new radio technologies, in the course of which he developed and patented a number of improvements for loudspeakers, microphones, and so on. It was in the context of this laboratory work that he began his experiments in synthetic sound.

As with Fischinger, there is also an ur-legend surrounding the origin of what Pfenninger called his *tönende Handschrift* (Sounding Handwriting). Unlike Fischinger, however, Pfenninger seems to have been motivated less by synesthesial speculations than by economic necessity. According to the story, the poorly paid inventor Pfenninger was eager to provide a sound track for the experimental animations he was making on the side, but he could afford neither the musicians nor the studio to record them.[45] Instead, he sat down with an oscilloscope and studied the visual patterns produced by specific sounds until he was able—sometime in late 1929 or early 1930[46]—to isolate a unique graphic signature for each tone. Using the newly available optical film soundtrack to test his experimental results, he would painstakingly draw the desired curve onto a strip of paper which he then photographed in order to integrate it into the optical sound track. The resulting sound, phenomenalized by the selenium cell, was one that had never been previously recorded but was, in effect, written by hand: "hand-drawn sound," as Pfenninger called it. And indeed, the first films that Pfenninger made for EMELKA in late 1930 with an entirely synthetic sound track—an extremely labor-intensive task that involved choosing and then photographing the right paper strip of sound curves for each note—were his own undersea animation, *Pitsch und Patsch,* and a "groteskes Ballett" film directed by Heinrich Köhler and entitled *Kleine Rebellion.*

When the discovery of the "Tönende Handschrift" was first presented to journalists in a special demonstration at the Kulturlfilmabteilung of the EMELKA studios in the late spring of 1931, the numerous published accounts compared Pfenninger's breakthrough not with work by Fischinger but, instead, with the recent news of the comparable technical achievement in England by the engineer Humphries. Odo S. Matz, for example, who claims to have been one of the first to hear the results of Pfenninger's new technique, once again opens up the question of historical priority (here laced with an added dimension of national chauvinism) when he points out in his report that Pfenninger was working on his project *before* the news of Humphries's work splashed across newspapers around the world. As if this was not enough, however, Matz goes on to dismiss the achievement of the British competitor as facile techno-mimesis (why bother synthesizing the human voice when any microphone could do it better?), while the "true" pioneer Pfenninger was exploring the much more uncharted aesthetic territory of previously unheard *new* sounds: "Pfenninger, by contrast, uses similar means in order to create new sonic effects which are unknown to our ears because they cannot be generated by any instrument. Herein lies the magical quality of this invention."[47] Indeed, it may well have been the news of Pfenninger's discovery that led Fischinger to suddenly begin to explore a generative rather than simply analogic logic between graphic form and musical sounds: how else to account for the fact that, as Moritz reports, "he interrupted his work on his other projects including *Studie Nr.11* in order to produce hundreds of test images which he then recorded as images for the soundtrack."[48]

Having tantalized the public through the press accounts in 1931, very possibly so as not to be eclipsed by the stories about Humphries, EMELKA then waited over a year before announcing the first full-scale public demonstrations of Pfenninger's pioneering achievement in a multicity gala launching of a series of films with entirely synthetic sound tracks. *Die tönende Handschrift: Eine Serie gezeichneter Tonfilme eingeleitet durch ein Film-Interview* (Sounding Handwriting: A Series of Hand-Drawn Sound Films introduced by a Filmed Interview) premiered at the Munich Kammerlichtspiele on October 19, 1932, and the following day at an invitation-only matinee in the grand Marmorhaus cinema-palace in Berlin, an event also attended by Pfenninger, who personally thanked the audience for, as the *Film-Kurier* described it, "its justifiably amazed and enthusiastic response to the screening."[49] The program—which EMELKA circulated to cinemas throughout Europe in late 1932 under the title *Die tönende Handschrift*[50]—consisted of *Kleine Rebellion* and *Pitsch und Patsch,* two "groteske Puppenfilme" by the brothers Diehl entitled *Barcarole* and *Serenade,* and a "Naturfilm" entitled *Largo.* These were preceded by a fascinating pedagogical documentary entitled *Das Wunder des gezeichneten Tones* (The Wonder of Hand-Drawn Sound) (which was also released as a newsreel announcing the new discovery) and consisting of an illustrated history of sound recording followed by an on-camera interview of Pfenninger by the charismatic film personality Helmuth Renar. The journalistic response was, as one might expect, both extensive and largely enthusiastic.[51] Although generally fascinated by the technical achievement and its promise, most critics were perplexed and even annoyed by the new sounds: while some were entranced by what they felt was "very beautiful 'mechanical' music, a sort of carousel music," others wrote of its "primitive and somewhat nasal timbre," how it gave an "impression of being mechanical, almost soul-less," and that it had "a snore-like quality and (since the tones belong primarily to the realm

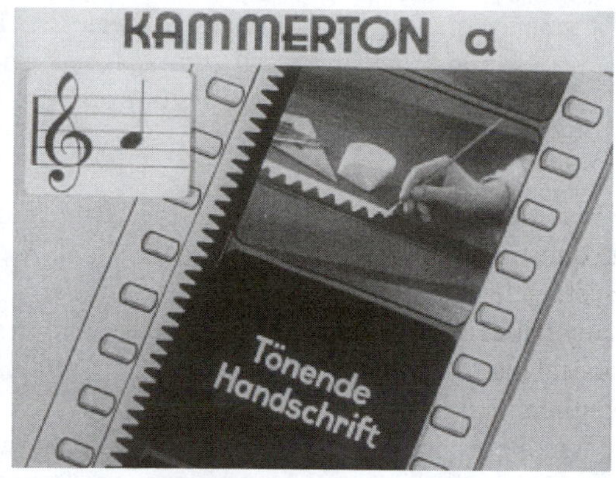

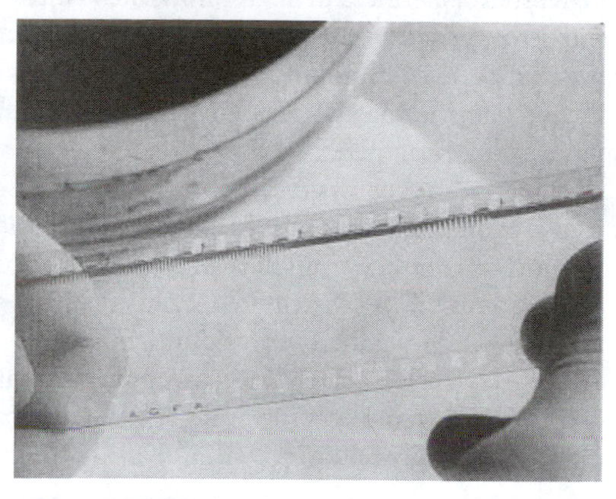

Figure 3.15 A group of four stills from *"Tönende Handschrift"- das Wunder des gezeichneten Tones* (1931) (Filmmuseum Muenchen).

of the flutes and plucking instruments), a monotone quality as well."[52] As one reviewer put it, "the sound reminds one of stopped organ pipes, muted horns, harps, xylophones. It sounds strangely unreal."[53]

In lieu of a more detailed account of the fascinating reception history, which will be undertaken elsewhere, consider the following representative account by R. Prévot published the day after the premiere in the *Münchener Neueste Nachrichten*:

> What we saw yesterday morning was more than simply initial experiments. Our technological sense was fascinated, our imagination of the future provoked!...At the same time, I must admit that our music-loving ear did go on strike, and our lively artistic consciousness was troubled.
>
> Was this still music? . . . rarely have we felt so clearly the inner difference between live art and technological construct. One heard piano and xylophone-like sounds, others which seemed to come out of a steam whistle—all of them crafted together with great precision, much as if someone were to build a tree out of a thousand pieces of wood, which can look deceptively real and yet will never bloom!...Without a doubt, this abstract, this skeletal music fit best with the animated images–here there was a sort of technical unison. But the attempt to "give life" by such musical means to the dance and mimicry of live people seemed utterly impossible. The effect was like that of a dance of the dead! Here we must give voice to a decided "halt!"
>
> ...Film has finally succeeded in creating a new "technological art" which has its own

> essence distinct from that of live theater. Perhaps the Pfenninger method will also succeed in finding tones and tonal complexes which are new and cannot be produced by natural means; i.e., a music which does not yet exist—a real music of the future? Let us hope that it turns out to be beautiful![54]

Prévot's response is typical in its combination of techno-fetishistic fascination, its concern with the question of aesthetically "appropriate" sound-image combinations, and above all in the way it registers the instinctive threat to a longstanding, supposedly a-technological conception of music. Many critics insisted that Pfenninger's invention ought to be measured against other new electronic musical instruments or technologies of the time, such as the Theremin or the Trautonium, in that, like them, its future lay in the exploitation of its capacity to make "new" sounds, not in imitating extant ones, the latter being both redundant and economically ill-advised. But this seemingly progressive openness to an unknown acoustic futurity was of course itself also a way of displacing the threat posed to the organic notion of the acoustic by synthetic sound—a tree made of wood but that can never bloom!: "*Unheimlich*," writes the critic of the *Frankfurter Zeitung*, "the degree to which technology unceasingly renders superfluous in all domains both organic creation and the natural labors of man!"[55] Nowhere is this clearer than in the simultaneous amazement and horror in response to the prospect—possibly envisioned by Pfenninger but (as far as I know) never realized—of outdoing Humphries by making a full-length "talkie" with entirely synthetic voices, a film in which, as one critic put it, "words will be spoken which belong to no person!"[56] Even critics willing to admit that all instrumental music was, as such, necessarily mechanical, had always insisted that the voice remained the residuum of the extra-technological: "Actually all music is mechanical, with the sole exception of human singing. For all music is made with machines—only the larynx is organic."[57] Pfenninger's technique effectively meant that—at least in theory—this long-standing claim was simply no longer valid.

Figure 3.16 Handwritten sound; film still from *"Tönende Handschrift" - das Wunder des gezeichneten Tones* (1931) (Filmmuseum Muenchen).

Following the Pfenninger premieres in late 1932, comparisons with Fischinger's work first begin to appear in print. While a few journalistic accounts are content merely to note the seeming similarity of the two projects, most cast the Fischinger-Pfenninger juxtaposition in terms—*basic impulse* versus *logical conclusion, decorative* versus *analytic*—that imply that it was a question not of who was the first to "discover" synthetic sound but rather of two related but, in the last analysis, very different projects.[58] This is the sense conveyed in the roundup of German cinema for the year 1933 published by Andor Kraszna-Krausz in *Close-Up*:

> two Germans who work on films have announced that they want to transpose phonetically with the photo cell the light reactions of plastics, and to compose them with their parallel visual impressions to obtain sound film accords.
>
> This extremity must have been suggested by the experiments of Oscar Fischinger whose compositions of dancing lines are the only kind of abstract film which can be found in the regular programme of the German cinemas, and which are well received by the public. Fischinger, who originally by synchronisation of his studies made real record pieces, has been trying recently—in order to obtain a more complete unity of picture and sound—to record decorative music in the *Lichtongerät* (light-sound) apparatus.
>
> Simpler, more thorough and practical seem to be the similar endeavours of Rudolf Pfeniger [*sic*], who after a long and difficult analysis, was successful in the calculation of sound writings, and also in drawing them with the hand.[59]

And indeed, upon closer examination of the manner in which each of these inventors *frames* his activities, it becomes clear that despite the superficial similarity they are each pursuing very different goals. Fischinger, as he himself is the first to admit, is basically interested in exploring the relationship between given graphic forms and their acoustic correlates, and how that isomorphism might allow one to make cultural-physiogonomic comparisons. When, for example, he suggests that "we should investigate the ornaments of primitive tribes in terms of their tonal character"[60] it is clear that his point of departure is the graphic mark. Besides this sociological interest, Fischinger also repeatedly argues that hand-drawn sound restores an artistic "sovereignty" to the filmmaker by once again giving him control over elements that the studio system had delegated to specialists. Invoking a rather hackneyed topos from romantic aesthetics, Fischinger insists that "real" art cannot tolerate such collective production because "this in the truest sense most refined and highest artistic activity comes to be only through, and directly out of, a singular personality, and the artwork that arises in this manner—for example works by Rembrandt, Bach or Michelangelo—are immediate creations of the highest power and profit *precisely from their handwritten*, irrational and personal qualities."[61] Despite the fact that, in what might well be an amusing tip of the hat to Pfenninger, the article from which this passage is taken is signed "Engineer Oskar Fischinger," it is clear that for Fischinger handwritten sound, indeed writing per se, is entirely in the service of a thoroughly anti-technological (irrational) artistic intention: "hand-made film renders possible pure artistic creation."[62]

Nothing could be further from the impulse behind engineer Pfenninger's fundamentally pragmatic and sober scientific investigations. Eschewing aesthetic discourse entirely, Pfenninger focused on the technological development of a new form of acoustic writing, a semio-pragmatics of sound whose function was to liberate composition from the constraints of both the extant musical instrumentarium and reigning notational conventions. Unlike Fischinger, who began with graphic forms and then explored what sort of sounds they produced, Pfenninger's primary focus was on the acoustic, in an attempt to establish what the precise wave form is that would allow one to re-produce a specific sound at will. Despite the potential visual appeal of their sine-wave forms, Pfenninger's curves are decidedly *not* ornaments but are rather, as numerous critics have rightly noted, "templates or print-types"[63]; that is, semiotic entities that can be combined to

Figure 3.17 Rudolf Pfenninger in his lab photographing sound strips for the optical sound track; film still from *"Tönende Handschrift"- das Wunder des gezeichneten Tones* (1931) (Filmmuseum Muenchen).

produce sounds in a linguistic—which is to say, thoroughly technical and rule-governed—manner. Unlike Fischinger's curves, which were continuous, Pfenninger's were discrete units. Indeed, in what is perhaps the most succinct manner of differentiating the two projects, while Pfenninger could (at least in theory) have used his method to re-produce every sound made by Fischinger's ornaments, the opposite is obviously not the case. Thus it is no surprise that from the start critics rightly insisted that Pfenninger's invention was not an ornamental practice as much as it was a new technique of acoustic notation, even going so far as to claim that he was in the process of "constructing a contrivance resembling a typewriter which, instead of letters, will set together sign waves in succession."[64]

Pfenninger's discovery was threatening not only because it challenged the hegemony of certain tonal systems (since graphic sound is both entirely free of overtones and entirely compatible with quarter-tone and other scale systems), but also because it represented a fundamental shift in the *status* of recorded sound. As pointed out most clearly in an anonymous review in the *Völkischer Beobachter* [*sic*], prior to Pfenninger, all recorded sound was always a recording *of* something—a voice, an instrument, a chance sound: "in this system, something audible can be recorded by the microphone only if it really exists; i.e., if it was produced somewhere *beforehand*. Rudolf Pfenninger, however, produces tones from out of nowhere."[65] If Pfenninger's synthetic generation of sound effectively *destroyed the logic of acoustic indexicality* that was the basis of all prior recorded sound, it also exposed the *residual iconic-indexicality* in Fischinger's only seemingly similar activities. Indeed, the experimental fascination with establishing the acoustic correlates *of* a profile or *of* a particular visual form at some level always also assumes that such sounds are sounds *of* something, even if that something is now simply a recognizable graphic trace. Thus, to the extent that Fischinger's work explores and uses the sounds made by various extant things (in his case, graphic forms), his work could be described as a sort of proleptic *musique concrète*, while Pfenninger's synthetic practice is closer to certain nonreferential, acoustically constitutive practices of electronic music. To the extent that Fischinger's ornaments function semiotically, they do so as "motivated" signs, whereas Pfenninger's curves depend, strictly speaking, on only the particular—and in the last analysis, arbitrary—properties of the selenium cell that is the basis of the particular optical cinema sound system he used to produce his sonic graphematics. And it is this crucial semiotic difference that ultimately explains why Paul Seligmann, a member of *Das neue Frankfurt* film club for whom Moholy-Nagy had screened works by both Fischinger and Pfenninger, credits only Pfenninger and not Fischinger with the invention of a functional system of acoustic writing: "It is in the end Pfenninger who discovered the path to acoustic writing. While Fischinger merely photographs sound as a process, Pfenninger captures it as individual images, which led him to develop templates by means of which particular sounds and sound groups can be repeated at will."[66] Indeed, in its rigorously systematic character, Pfenninger's research deserves to be compared closely not with Fischinger but rather with the very similar—and similarly analytic—investigations into synthetic sound undertaken at the same time in the Soviet Union by Nikolai Voinov and Aleksandr Ivanov, who cut out saw-toothed sound shapes from paper in the form of contoured combs, each representing a halftone, which could then be used repeatedly and in various combinations much like the basic formal vocabulary of visual animation; and by Evgenii Sholpo, who developed a very successful circular "disc" variation on Voinov and Ivanov's combs.[67]

Recorded Sound in the Age of Its Synthetic Simulatability

If Pfenninger's invention makes it possible to create sounds that—as he put it so wonderfully—come from out of nowhere, why is it, one might wonder, that the synthetic sound that accompanies the various films in the *Tönende Handschrift* series is so banally imitative of extant sonorities, even going so far as rendering Händel's Largo or the Barcarole from Offenbach's *Hoffmanns Erzählungen*? Is this yet another instance of a radically new technology for the generation of sound attempting to legitimate itself not by foregrounding its own unprecedented sonic capacities but by slavishly simulating well-known classical pieces—as was the case, for example, with the early performances that introduced the technological wonder of the Theremin?[68] Whatever the motivation might have been, and however trivial it might seem acoustically at first audition, the *effect* of hearing familiar repertoire emanating from a source that not only involved neither instruments nor musicians but consisted only in the systematic photographing of a graphematic vocabulary for an optical sound track, would have been deeply disturbing. And that discomfort stemmed not least from the fact that, while at this initial stage the sound could still be differentiated from the signature timbre of traditional instruments, it was—at least in theory—only a matter of time and technical refinement until it would no longer be possible to distinguish acoustically a sound generated synthetically from a sound produced by conventional means.

For some critics this immediately suggested that synthetic music would in the future render orchestras superfluous because, as the imaginative reviewer of the *Pester Lloyd* put it, "one could conjure up a phantom orchestra [*Geisterorchester*] which does not exist in reality but whose sounds are simply the result of an act of drawing."[69] Indeed, Fischinger himself effectively implied as much, rearticulating in some of his essays on hand-drawn sound Stuckenschmidt's earlier argument that music machines such as the gramophone would eliminate the necessity of the live performing musician as an intermediary between composition and realization. However, given the labor-intensive conditions of Pfenninger's synthetic sound techniques, it was hardly likely that synthetic sound would restage in an even more drastic fashion the all-too-recent labor-political drama that was the consequence of the advent of the gramophone and later the sound film (both of which eliminated in stages the need for full-time musical ensembles to accompany screenings).[70]

What the advent of synthetic sound *did* fundamentally change however was the ontological stability of all recorded sound. The introduction of optical film sound in the late-1920s had already made possible a previously unavailable degree of postproduction *editing*, thereby undermining the *temporal* integrity of acoustic recordings, which could now be patched together out of various takes at various times.[71] The invention of a functional means of generating synthetic sound, however, seemed likely to push this challenge to the so-called authenticity of sound recordings even further. Although it was unlikely that one would be able in the near future to create entire compositions by synthetic means *ex nihilo*—or perhaps better, *ex stylo*—what was decidedly possible was minimal and punctual interventions into the fabric of extant recordings. This is precisely what Humphries had done—incorporating just a few unnoticeable substitutions into the otherwise intact optical soundtrack of a film. But what made these changes so disturbing was precisely the fact that, while indistinguishable from the rest of the spoken words, Humphries's synthetic voice was just that—synthetic—and thus opened up a fundamental doubt about the status of *everything* on the soundtrack. Indeed, the extent of the critical reaction to his efforts was itself a good barometer of the threat they represented to a certain— indexical—ideology of recorded sound. For while the cut threatens the integrity of the recording as a continuous event, it does not in any way undermine the indexicality of the recording process as such, which continues to govern all of the now rearranged pieces just as much as it did before they were edited. Pfenninger's invention of synthetic sound, on the other hand, represents nothing less than the incursion into the acoustic domain of postproduction composite adjustments–often referred to as "corrections" or "improvements"—that are no longer of the order of the indexical. No longer the re-phenomenalized trace of a prior acoustic event, as tones from out of nowhere, they are no longer sounds *of* anything but are, instead, simply a set of *graphic* (i.e., non-acoustic) instructions.

Most of the reactions to the *Tönende Handschrift* simply registered the profound anxiety that the undermining of sonic indexicality provoked—without being able to articulate its sources: typical in this regard is the statement "the consequences of this discovery are so monstrous, so spooky, that at this moment we cannot fully grasp them."[72] One particularly astute critic was, however, able to identify exactly what was at stake:

> Just as a photographic plate can be retouched and beautified by the art of the photographer, in a similar manner one will be able to modify the spoken word, the sound and modulation of the human voice, to its utmost perfection. A wide domain of acoustic re-touching has here opened up for the film industry, and no singer will ever again run the risk of not having been able to hit the high C perfectly.[73]

Die Kurve des Lautes „n"

So kompliziert sieht ein „a" in der Tonschrift aus

„u"

Figure 3.18 "The curve of the sound 'n'; This is how complex the sound 'a' looks in sound writing."

Once off-pitch notes can be corrected, late entries adjusted, disturbing overtones eliminated, and unpleasant sonorities rendered more agreeable, then every recording of music can in theory be "perfect." Indeed, as Herbert Rosen insists, such tweaked productions might, in fact, sound better than the "originals": "Indeed, we will even go so far as to say that all these presentations will be significantly better, purer and more lacking in any blemishes than the authentic recordings! Since all the contingent possibilities on the one hand, and all the shortcomings that are characteristic of a number of musical instruments on the other, will now be eliminated by the sounding handwriting."[74] But this new quality in recorded music is, of course, bought at a price, since now one can no longer "know" what exactly the status is of the performance it registers. In other words, a technological doubt has been introduced into the indexical readability of recorded performance. At any point what one is hearing might be the product of a synthetic, Pfenningerian postproduction intervention that is unrecognizable as such. This is the beginning of a far-reaching undecidability—recorded sound in the era of its referential ambiguity—that, decades later and in the wake of a much expanded repertoire of studio interventions, would lead to the rise of "direct-to-disc" mastering and so-called live recordings as an (ultimately futile) attempt to restore the prelapsarian untroubled indexicality of recorded sound prior to the moment of its synthetic simulatability.

Coda: The Afterlife of Synthetic Film Sound

As it turns out, the five films in the *Tönende Handschrift* series—the first results of Pfenninger's experimentation with synthetic sound—were also his last. In a 1953 interview Pfenninger explained the lukewarm response to his invention in the early 1930s as follows: "The time was not ripe, my invention came twenty years too early."[75] Or perhaps too late: only a few years later Pfenninger's films would be designated "seelenlos und entartet" (soul-less and degenerate) by the Nazis,[76] and thus, not surprisingly, work in this domain effectively came to a halt. While Moholy-Nagy himself explored some of the challenges raised by Pfenninger's technique in 1933 in the form of a short experimental film entitled *Tönendes ABC* (Sounding ABC) whose optical sound track was rephotographed such that it could be projected on the image track simultaneously with the sound (allowing one to *see* the same forms that one was also *hearing*),[77] besides a brief mention of synthetic sound in W.L. Bagier's 1934 documentary *Der Tonfilm*, Germany would quickly cease to be the fertile ground for work on synthetic film sound that it had been for the previous few years.[78]

Elsewhere, however, especially in the wake of the extensive publicity surrounding the release and international distribution of the *Tönende Handschrift* series, "hand-drawn sound" quickly became something of an international sensation, albeit a very brief one. In America even commercial films such as Rouben Mamoulian's 1931 *Dr. Jekyll and Mr. Hyde* took advantage of the uncanny acoustic possibilities afforded by the new technique as a means to provide a sonic correlate to the transformation of the gentleman into the monster and vice versa. The result was, as one commentator described it, "a vivid, synthetically created sound track built from exaggerated heart beats mingled with the reverberations of gongs played backwards, bells heard through echo chambers and completely artificial sounds created by photographing light frequencies directly onto the soundtrack."[79] In France, following a few articles on *Die tönende Handschrift* in French and Belgian journals,[80] hand-drawn sound began to appear in film sound tracks, most notably in the work of Arthur Hoérée whose practice of *zaponage*, a technique that involves using a dark paint or stain called Zapon to touch up the optical sound track, was employed to great effect in Dimitri Kirsanoff's 1934 *Rapt*.[81] While I have not been able to establish the extent to which actual synthetic sound appeared in Italian film, the issues involved were at the very least known there; for example, the German music theorist Leonhard Fürst (who had written about Fischinger in *Melos*) gave a lecture on new techniques of film sound on May 2, 1933, at the International Music Conference which took place during the May Festival in Florence. Following this lecture, which included screenings of a reel of Fischinger's *Tönende Ornamente,* Eisenstein's *Romance Sentimentale* (France, 1930), and Pfenninger's *Tönende Handschrift,* explanatory essays began to appear on the subject in both technical and touristic journals.[82] In the Soviet Union the fruits of the wide-ranging local research into synthetic sound began appearing in the sound tracks of films such as *Plan velikikh rabot* (*Plan of Great Works,* 1931), *Kem bït* (*Who to be,* 1931), and *Gibel sensatsii* (*The end of a sensation,* 1931), and then somewhat later in *Symphony of the World* (Soviet Union, 1933), in the "Ivoston" group's 1934 *Prelude by Rachmaninoff,* and in Grigori Alexandrov's short collaboration with Sergei Eisenstein entitled *Romance Sentimentale* (France, 1930).

In the wake of Humphries's sensational 1931 breakthrough in London one might have expected, in turn, that quite a lot of new work would be done on synthetic sound in England. While this seems not to have been the case (although it is entirely possible that synthetic sound continued to be used discretely by the major studios when needed), the British interest in the subject did, however, lead to what is arguably the most significant *reception* of work in synthetic sound done elsewhere. Pfenninger's technical breakthrough was reported at length in numerous richly illustrated articles in British professional journals such as *Wireless World* and *Sight and Sound.*[83] At about the same time—and in addition to the already mentioned London Film Society screenings of synthetic sound films by Fischinger on May 21 and December 10, 1933 (the latter with Moholy-Nagy's *Tönendes ABC*), and a screening of *Die tönende Handschrift* on January 14, 1934 (described in the program notes as "the most elaborate attempt so far made to use synthetic sound for cinema purposes")—on January 13, 1935, the Film Society screened a double bill of films *about* sound: Bagier's *Der Tonfilm* and the British documentary *How Talkies Talk* (Dir. Donald Carter, 1934), which the program notes describe as follows:

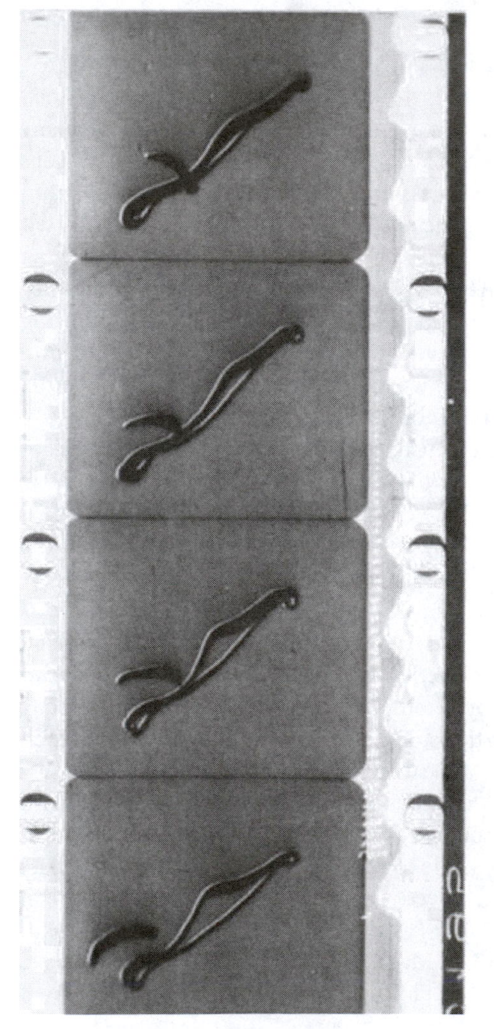

Figure 3.19 Four frames from "Loops" (Norman McLaren, 1940) with hand-drawn sound and image. Reprinted by permission of the National Film Board of Canada.

> Two films showing the different system of sound recording. Of special technical interest in the English film is the actual photography of the photographic trace of the light beam which was developed while it was being filmed. The process of recording is normally carried out in the dark, but, by choice of the right film stock for the motion picture camera, and by illuminating the scene with a light which does not affect positive film (on which the recording is done), a picture has been obtained of the actual recording sound waves. The sounds which are shown are synthetic, but they follow the mechanical wave form of sound.[84]

It was also in London, just over a year later, that a young Scottish art student was hired by John Grierson to work for the General Post Office (GPO)

Film Board. This student, Norman McLaren, would go on to become arguably the world's most well-known and prolific proponent of synthetic sound. In the years following the above mentioned screenings and articles, McLaren began his first experiments with synthetic sound, scratching directly onto the sound track in an improvised manner for *Book Bargain* (1937), to take just one example from his time with the GPO film unit, and in the abstract films *Allegro, Dots, Loops,* and *Rumba* (the last consisting of only a sound track without visuals), which he made in 1939 for the Guggenheim, then known as the Museum of Non-Objective Paintings, in New York. But when asked about the inspiration for the much more systematic technique of generating synthetic sound that he developed in the early 1940s, McLaren credits the Continental experimental films shown at the Glasgow School of Art, where he had studied from 1932 to 1936:

> Amongst them was a film called *Tonal Handwriting* made by a German engineer from Munich—Rudolph Phenninger [*sic*]. He had evolved a system. First of all, the film consisted of a documentary showing how he did it. He had a library of cards and a camera. He'd pull out a card, film a frame and so on, and then at the end of that he had a little cartoon. He had music with this, quite lively, not distinguished, but very lively. This is the basis on which I developed my card system.[85]

It was this neo-Pfenningerian method of "synthetic animated sound"—involving a library of one-by-twelve inch strips, each with from one to 120 iterations of a hand-drawn sound-wave pattern that could produce every semitone across a five-octave range—which McLaren used in later films with synthetic soundtracks such as the stereoscopic *Now Is the Time* (1951), *Two Bagatelles* (1952), the Oscar-winning *Neighbors* (1952), and *Blinkety Blank* (1955). McLaren detailed his method in a series of introductory and technical essays that would be instrumental in disseminating the technicalities of the procedure.[86] Using this technique, McLaren produced for the National Film

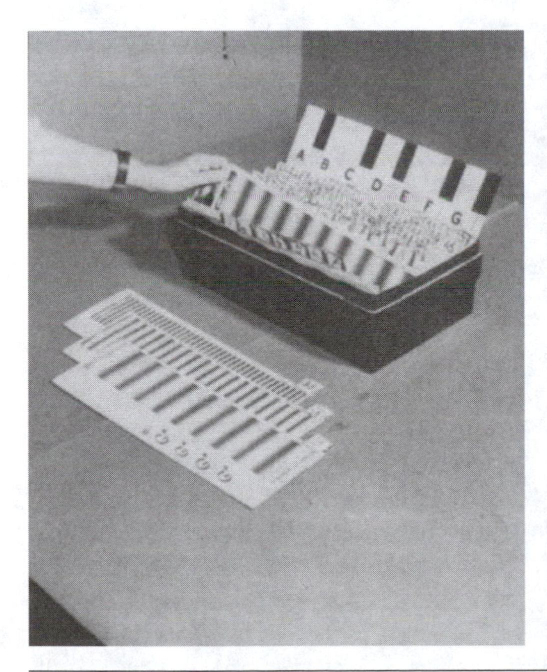

Figure 3.20 Left: Cards containing soundwave patterns which, when photographed, produce musical sounds. Right: McLaren "composing" a film score using cards containing sound-wave patterns. Both images reprinted by permission of the National Film Board of Canada.

Board of Canada what is arguably the magnum opus of the synthetic sound film, the seven-minute-long *Synchromy* (1971), in which one sees the abstract patterns that are at every moment creating the sounds that one is hearing. The result, as described by a contemporary critic, is "a fascinating exercise in the 'perception' of sound."[87]

The subsequent chapters in the rich and fascinating history of synthetic sound—which is, alas, far too extensive a subject to be dealt with here—unfold across a number of domains, ranging from avant-garde cinema (especially experimental animation) to the development of new notational systems and technologies for the production of sound. The former includes, to take just three examples, the work of the Americans John and James Whitney in the 1940s (who employed a pendulum device to generate an entirely synthetic optical sound track for the "audio-visual music" of their *Five Abstract Film Exercises* of 1943–1945), the experimental short *Versuch mit synthetischem Ton* (Test) by the Austrian underground filmmaker Kurt Kren in 1957 (with an entirely

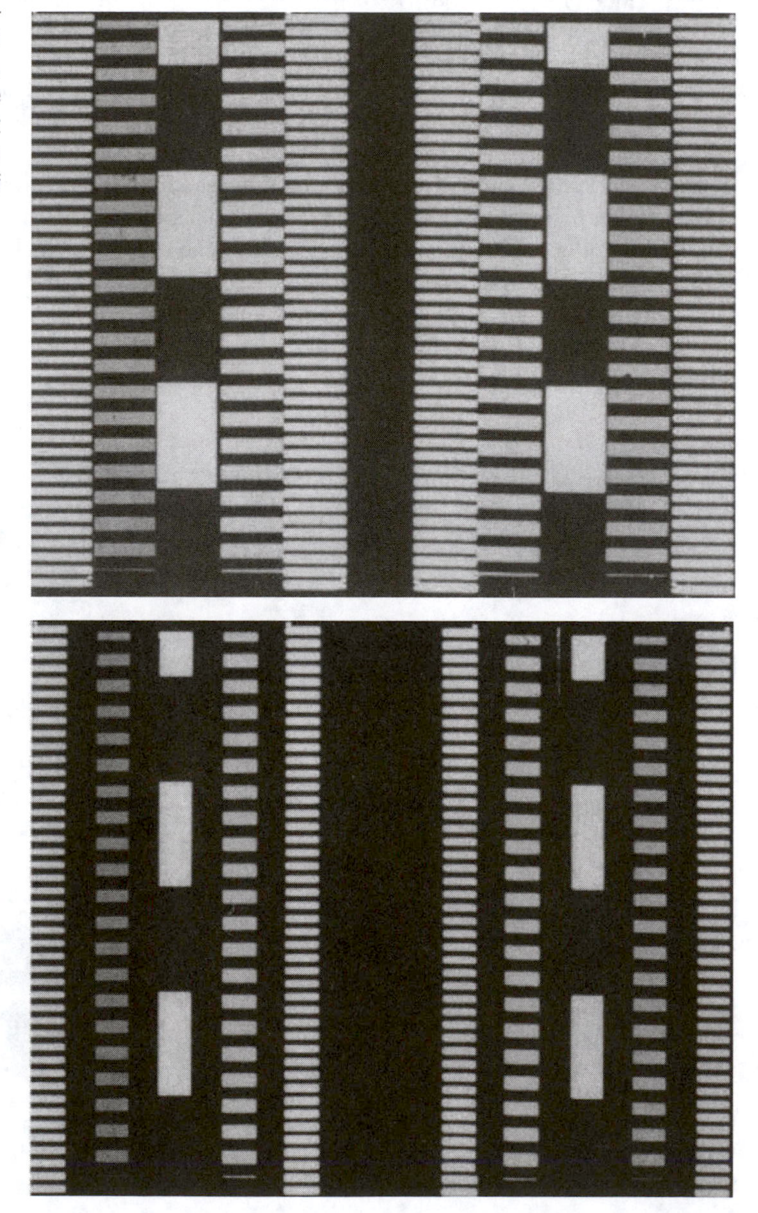

Figure 3.21 Film stills from *Synchromy* (Norman McLaren, 1971). Soundtrack shifted on to the image track; what you see is what you hear. Reprinted by permission of the National Film Board of Canada.

"scratched" optical sound track), and the films of Barry Spinello in the 1960s (whose synthetic sound tracks—for example in *Soundtrack* [1970]—were generated both by drawing and painting directly on the celluloid and by means of self-adhesive materials such as microtape and press-apply lettering).[88] The latter locates Pfenninger's method in the complex history of the invention of new recording media such as magnetic tape and of new synthetic sound technologies such as Harald Bode's 1947 melochord (which was used in the 1950s in Stockhausen's studio for electronic music in Cologne), Harry F. Olson's famous RCA Electronic Music Synthesizer (which was first introduced in 1955), Robert Moog's pathbreaking modular synthesizer built in 1964, and the later proliferation of MIDI interfaces that have rendered the experience of, and work on, music as a graphic material an almost quotidian affair.

But besides its genealogical importance, Pfenninger's *Tönende Handschrift* is also of great, thoroughly contemporary *theoretical* interest, offering as it does a remarkable proleptic parallel in the domain of the acoustic to the development that is at this moment transforming the status of much visual representation. If, as some have argued, the advent of digital imaging has thrown into question many of the referential assumptions that heretofore characterized the various fundamentally indexical nineteenth-century visual media such as photography and cinema, then the aesthetico-political consequences of this paradigmatic shift are of fundamental importance. Just as Pfenninger's technique of synthetic sound—especially when it operates as a simulacral "correction" of traditional sonic material within an otherwise indexical recording—fundamentally undermines the presumed *homogeneity* of the indexical field, opening it up to a doubt whose epistemologically contaminatory consequences cannot be contained, so too does the increasing prevalence of a similar semiotic hybridity in the visual domain—such as, to take an obvious example, the completely computer-rendered 3-D creatures that inhabit an otherwise live-action cinematic landscape in the Disney film *Dinosaur* (Dir. Eric Leighton, 2000)—throw the indexical status of the entire visual field into question. In light of Lev Manovich's suggestion that in media-historical terms the advent of the digital episteme can be described as a turn from an optical to a *graphic* mode of representation that in fact characterized the nineteenth-century media out of which cinema was developed,[89] the essentially graphematic nature of Pfenninger's synthetic sound technique in turn reveals a key dimension of this graphic turn of new media—its fundamental status not so much as drawing but qua inscription as a techno-logics of writing.

Notes

This essay is part of a much longer forthcoming study of the genealogy of synthetic sound whose initial phases were generously supported by the Pro Helvetia Stiftung and the Princeton University Committee on Research in the Humanities and Social Sciences. Further helpful input, access to materials, and suggestions were kindly provided by Jan-Christopher Horak (then at the Munich Film Archive), the late William Moritz, Roland Cosandey (Vevey), and Rebecca Gomperts (Amsterdam). I would especially like to thank Sigrid Weigel (then Director of the Einstein Forum, Potsdam) for the opportunity to first present this material at the 1999 Potsdam conference on the "Cultural and Media History of the Voice," and Daniela Peters for her helpful and intelligent editorial support. This essay first appeared in German as "'Töne aus dem Nichts'. Rudolf Pfenninger und die Archäologie des synthetischen Tons" in: Friedrich Kittler, Thomas Macho and Sigrid Weigel, Eds., *Zwischen Rauschen und Offenbarung: Zur Kultur- und Mediengeschichte der Stimme* (Berlin: Akademie Verlag, 2002), 313–355. It is dedicated to the accomplished film animator Marianne Pfenninger, who so graciously and unhesitatingly gave me access to the archive of her late father's pioneering work.

1. "Synthetic Speech Demonstrated in London: Engineer Creates Voice Which Never Existed," *New York Times*, 16 February 1931: 2.
2. On the institutional history of the development of these pioneering optical sound systems in Europe and the United States, see Douglas Gomery, "Tri-Ergon, Tobis-Klangfilm, and the Coming of Sound," *Cinema Journal* 16, no. 1 (Fall 1976): 50–61. On the early history of the Tri-Ergon unit at UFA, see Hans Vogt, *Die Erfindung des Tonfilms* (Erlau: Vogt, 1954), Wolfgang Mühl- Benninghaus, *Das Ringen um den Tonfilm: Strategien der Elektro- und derFilmindustrie in den 20er und 30er Jahren* (Düsseldorf: Droste, 1999), esp. 21–41, and Guido Bagier, "'Ton mehr aufdrehen—verstärken!'" in *Das Ufa-Buch*, ed. Hans-Michael Bock and Michael Tötenberg, 244–247 (Frankfurt/M.: Zweitausendeins, 1992). For details on the prehistory and early triumphs of the Vitaphone system, see Mary Lea Bandy, ed., *The Dawn of Sound* (New York: The Museum of Modern Art, 1989).

3. See Wolfgang von Kempelen, *Mechanismus der menschlichen Sprache nebst der Beschreibungen einer sprechenden Maschine* (1791; reprint, in facsimile form, Stuttgart-Bad Cannstatt: Friedrich Frommann Verlag, 1970): 183.
4. Friedrich Kittler, *Gramophone, Film, Typewriter*, trans. Geoffrey Winthrop-Young and Michael Wutz (Stanford: Stanford University Press, 1999): 83; trans. modified. First published as *Grammophon, Film, Typewriter* (Berlin: Brinkmann & Bose, 1986): 129.
5. For a fascinating discussion of the function of acoustic technologies in the nineteenth-century debates on language transcription and phonetics, see Wolfgang Scherer, "Klaviaturen, visible speech und Phonographie: Marginalien zur technischen Entstellung der Sinne im 19. Jahrhundert," in *Diskursanalysen I: Medien*, ed. Friedrich Kittler, Manfred Schneider, and Samuel Weber (Opladen: Westdeutscher Verlag, 1987): 37–54.
6. Scherer, "Klaviaturen, visible speech und Phonographie," 48. It was this very Henry Sweet who later served as the model for Professor Higgins in G. B. Shaw's *Pygmalion*.
7. His account of the event can be found in Count [Theodor A.L.] du Moncel, *Le Téléphone, le microphone et le phonographe* (Paris: Hachette, 1878). The anecdote is cited from the "Authorized translation with additions and corrections by the Author," published as *The Telephone, the Microphone and the Phonograph* (1879; reprint, New York: Arno Press, 1974): 244–245.
8. Kittler, 59. Indeed, as Kittler goes on to explain, Berliner's gramophone gained its worldwide popularity as a flat disc technology that delivered sound "pre-recorded" by the record industry while Edison's (cylinder) phonograph enabled individual users themselves to record and store the sounds of daily life but as a result remained a very private (and thus commercially much less lucrative) medium.
9. Alfred Mayer, "On Edison's Talking-Machine," *Popular Science Monthly*, 12 March 1878, 723.
10. H[arry] A. Potamkin, "'Playing with Sound,'" in *The Compound Cinema: The Film Writings of Harry Alan Potamkin*, ed. Lewis Jacobs (New York: Teacher's College Press, 1977): 87. First published in *Close Up* 7, no. 2 (August 1930). Reprinted as a facsimile in *Close Up: A Magazine Devoted to the Art of Films*, vol. 7: *July–December 1930*, ed. Kenneth Macpherson and Bryher (New York: Arno Press, 1971): 114.
11. This account of Dyer's testimony at the congressional hearings can be found in Lisa Gitelman, "Reading Music, Reading Records, Reading Race: Musical Copyright and the U.S. Copyright Act of 1909," *The Musical Quarterly* 81, no. 2 (Summer 1997): 275. Reprinted in an expanded version as Chapter 3, "Patent Instrument, Reading Machine" in Lisa Gitelman, *Scripts, Grooves, and Writing Machines: Representing Technology in the Edison Era* (Stanford: Stanford University Press, 1999): 97–147.
12. "Die Frage, ob die Eindrücke auf den Platten und Walzen als Schriftzeichen im Sinne des § 41 St. G.B. anzusehen seien, müsse bejaht werden. Die Laute der menschlichen Stimme würden beim Phonographen in gleicher Weise fixiert wie durch die Buchstabenschrift. Beide seien eine Verkörperung des Gedankeninhaltes, *und es mache keinen Unterschied, daß die Buchstabenschrift diesen Inhalt durch das Auge, der Phonograph ihn aber durch das Gehör vermittele*, denn auch die Blindenschrift, die ihn durch das Tastgefühl vermittele, sei eine Schrift im Sinne des § 41," "Die Sprechmaschinenschrift und die bestehenden Gesetze," *Phonographische Zeitschrift* 7, no. 9 (1906): 198. (Unless otherwise noted, all translations are author's own.)
13. Alexander Melville Bell, *Visible Speech: The Science of Universal Alphabetics; Or Self-Interpreting Physiological Letters for the Writing of All Languages in One Alphabet* (London: Simpkin, Marshall, 1867). See also Scherer, "Klaviaturen, visible speech und Phonographie," 49f. Interestingly, this difference in focus was already manifest at the turn of the century, as evidenced by the opening line of a 1903 editorial that reads: "Strangely enough, phonographic writing is generally of greater interest to linguists than it is to phonograph technicians." "Die phonographische Schift auf den Platten von Plattensprechmaschinen," *Phonographische Zeitschrift* 4, no. 42 (1903): 577.
14. Both L. Hermann in 1890 and Bevier in 1900, for example, used a mirror mounted on a fine tracing device such that it reflected a beam of light onto a film which thereby registered photographically the outlines of the phonographic squiggles, while a later apparatus developed by a certain appropriately named Edward Wheeler Scripture was able to generate a graphic translation of the contents of the phonographic grooves magnified over 300 times laterally and five times in length. These and numerous other devices are discussed at length in the remarkable early study by Dayton Clarence Miller, *The Science of Music Sounds* (1916; reprint, New York: MacMillan, 1934). A useful contemporary account of the race to produce a device that would successfully capture sound waves photographically can be found in the report on "Vorrichtung zur photographischen Aufnahme von Schallwellen," *Phonographische Zeitschrift* 1 (1900): 13. The later fascination of such optical translations of acoustic data for the field of musicology is evident in William Braid White, "Music Made Visible," *Proceedings of the Music Teachers' National Association* 53 (1929): 102–107.
15. Bernard Holland, "A Man Who Sees What Others Hear," *New York Times*, 19 November 1981, C28. According to this article, "Before an audience in the auditorium of Abington Hospital, near Philadelphia, two weeks ago, Stimson Carrow, professor of music theory at Temple University, handed Dr. Lintgen a succession of 20 long-playing records chosen by Mr. Carrow and 10 of his graduate students. All identifying labels and matrix numbers were covered over, but Dr. Lintgen, simply by taking the records in his hands and examining their groove patterns in a normal light, identified the piece and the composer in 20 cases out of 20." This story was also picked up by both *Time* ("Read Any Good Records Lately?" 4 January 1982, 82) and, a few years later, by the *Los Angeles Times* (Al Sekel, "The Man Who Could Read the Grooves," 19 October 1987, B3), the latter story probably provoked by very similar reports about a thirty-three-year-old Englishman named Tim Wilson who made the rounds of British and American talk shows demonstrating his ability to identify unlabeled records simply by "reading" the grooves. At least one contemporary visual artist, K.P. Brehmer, was so inspired by the story that he dedicated a work to him—"Komposition für Tim Wilson II" (1986). More recently, MIT professor Victor Zue, an expert in natural-language processing, amazed research colleagues in the field of speech recognition with his capacity to read what people were saying in spectrographs, the digital translation of a voiceprint.

16. The most recent chapter in this story is the development by a young Israeli computer scientist of software that is capable of visually "reading" a twelve-inch record (using a simple flatbed scanner) and then translating the patterns within the grooves into sounds. While the results of Ofer Springer's transcriptions are described as "barely recognizable as the original music" the purpose of the experiment was not high fidelity but, as Springer himself explains, "to show an audio signal could be visually recoverable from a record." (Leander Kahney, "Press 'Scan' to Play Old Albums?," *Wired News*, see http://www.wired.com/news/digiwood/0,1412,57769,00.html). Downloadable examples of the resulting sonic re-phenomenalizations as well as the souce code for the decoder are available at Springer's homepage www.cs.huji.ac.il/~springer.
17. Douglas Kahn and Gregory Whitehead, eds., *Wireless Imagination: Sound, Radio, and the Avant-Garde* (Cambridge: MIT Press, 1992): 11.
18. Ludwig Wittgenstein, *Tractatus logico-philosophicus* (1921; facing-page German and English edition, with an introduction by Bertrand Russell, London: Routledge & Kegan Paul/ New York: The Humanities Press, 1986): 39; translation modified.
19. Rainer Maria Rilke, "Ur-Geräusch," in *Sämtliche Werke*, vol. 6 (Frankfurt/M: Insel-Verlag, 1966): 1087. First published in *Das Inselschiff* 1, no. 1 (1919/1920): 14–20; translated as "Primal Sound," in Kittler (1999): 39 (translation modified).
20. Kittler (1999), 44. As Kittler puts it: "Before Rilke, nobody had ever suggested decoding a trace that nobody had encoded and that encoded nothing. Ever since the invention of the phonograph, there has been writing without a subject."
21. Kittler (1999), 45.
22. László Moholy-Nagy, "Production-Reproduction," in *Moholy-Nagy*, ed. Krisztina Passuth (London: Thames and Hudson, 1985): 289–290. First published in *De Stijl* 5 (1922) #7: 97–101.
23. László Moholy-Nagy, "Neue Gestaltung in der Musik: Möglichkeiten des Grammophons," in *Der Sturm* (Berlin) (July 1923) #14 (reprint, in facsimile form, Nendeln, Liechtenstein: Kraus Reprint, 1970): 102–106. In English as "New Form in Music: Potentialities of the Phonograph," in Passuth, *Moholy-Nagy*, 291–292.
24. This is the concluding line of Stuckenschmidt's highly controversial 1925 essay "Die Mechanisierung der Musik," *Pult und Taktstock* 2, no. 1 (1925): 1–8, a text which is reprinted (albeit without the symptomatic introductory disavowal by the journal's editors) in Hans Heinz Stuckenschmidt, *Die Musik eines halben Jahrhunderts: 1925–1975* (Munich and Zurich: Piper, 1976): 15. Compare also Stuckenschmidt's "Maschinenmusik," *Der Auftakt* 7, nos. 7–8 (1927): 152–157. For a representative essay available in English, see his "Mechanical Music," *The Weimar Republic Sourcebook*, ed. Anton Kaes, Martin Jay, and Edward Dimendberg, 597–600 (Berkeley and Los Angeles: University of California Press, 1994). First published as "Mechanische Musik, *Der Kreis* 3, no. 11 (November 1926): 506–508. While the curious development of Stuckenschmidt's critical positions itself merits further treatment than is possible here, it is interesting to note that by 1929, in "Moderne Musik auf der Grammophon-Platte," *Der Auftakt* 9, no. 1 (1929): 12–15, Stuckenschmidt had entirely abandoned his earlier arguments for gramophone-specific music and instead promoted the use of the new medium as a means of disseminating the works of contemporary composers such as Schoenberg.
25. Stuckenschmidt, "Die Mechanisierung der Musik," 15.
26. Stuckenschmidt, "Die Mechanisierung der Musik," 15.
27. For an intelligent treatment of the aesthetic and institutional context of this fascinating moment in twentieth-century music history, and for the rather remarkable story of the recovery of the actual discs with which Hindemith did his experiments in 1930, see Martin Elste, "Hindemiths Versuche 'grammophonplatten-eigener Stücke' im Kontext einer Ideengeschichte der mechanischen Musik im 20. Jahrhundert," *Hindemith-Jahrbuch* 25 (1996): 195–221. For more on the various compositions written especially for the Welte-Mignon rolls, see Werner König, "Über frühe Tonaufnahmen der Firma Welte und die Werke für das Welte-Mignon-Reproduktionsklavier," in *Jahrbuch des Staatlichen Instituts für Musikforschung Preußischer Kulturbesitz 1977* (Kassel: Verlag Merseburger Berlin, 1978): 31–44.
28. See Heinrich Burkard, "Anmerkungen zu den 'Lehrstücken' und zur Schallplattenmusik," *Melos* 9, nos. 5–6 (May–June 1930): 230; and Ernst Toch, "Über meine Kantate 'Das Wasser' und meine Grammophonmusik," *Melos* 9, nos. 5–6 (May–June 1930): 221–222. Compare also Toch's earlier essay, "Musik für mechanische Instrumente," *Musikblätter des Anbruch: Sonderheft Musik und Maschine* 8, nos. 8–9 (October–November 1926): 346–349.
29. See, for example, "Die tönende Handschrift," *Lichtbild-Bühne* (Berlin), 3 December 1932.
30. This passage can be found in a little-known reprint of Moholy-Nagy's 1923 *Der Sturm* essay "Neue Gestaltung in der Musik"—along with a new preface and postface—under the title "Musico- Mechanico, Mechanico-Optico" in a special "Musik und Maschine" issue of the *Musikblätter des Anbruch*, nos. 8–9 (October–November, 1926): 367.
31. Paul Hindemith, "Zur mechanischen Musik," *Die Musikantengilde* 5, nos. 6–7 (1927): 156. Also cited in Elste, "Hindemiths Versuche," 213.
32. László Moholy-Nagy, "Az új film problémái" (1928–1930), *Korunk* 5, no. 10 (1930): 712–719. In French as "Problèmes du nouveau film," *Cahiers d'art* 8, nos. 6–7 (1932): 277-280. In English as "Problems of the Modern Film," *New Cinema* no. 1 (1934), reprinted in *Telehor* (Brno) 2, nos. 1–2 (1936) and Passuth, *Moholy-Nagy*, 314; translation modified.
33. Moholy-Nagy, "Problems of the Modern Film," 314. The section of the citation following the ellipsis reappears, slightly modified, in Moholy-Nagy's later study *Vision in Motion* (Chicago: Paul Theobald, 1947): 277.
34. Moholy-Nagy, "New Film Experiments," in Passuth, *Moholy-Nagy*, 322; emphasis added. This text was first published as László Moholy-Nagy, "Új film-kísérletek," *Korunk* 8, no. 3 (1933): 231–237. In the *Vision in Motion* volume, Moholy even reproduces two photographs of Pfenninger working on the production of synthetic sound tracks, quite possibly the first publication of these images in book form: Moholy-Nagy, *Vision in Motion*, 276–277.
35. -f., "Film-Experimente: Zu einer Matinée im Frankfurter Gloria-Palast," *Rhein-Mainische Volkszeitung*, 5 December 1932. Besides Moholy-Nagy, the impressively cosmopolitan and theoretically sophisticated lecture series which the film club of "das neue Frankfurt" ran parallel with its screenings also featured speakers such as Rudolf Arnheim, Paul Seligmann, Joris Ivens, and Dziga Vertov. The 1932–1933 program of screenings and lectures is reprinted in Heinz

Hirdina, ed., *Neues Bauen/Neues Gestalten: Das neue Frankfurt/die neue Stadt. Eine Zeitschrift zwischen 1926 und 1933* (Dresden: Verlag der Kunst, 1984; reprint 1991): 341.

36. Typical, in this regard is the statement by the historian of animation Gianni Rondolino: "*Tönende handschrift*, realizzato da Pfenninger nel 1930–31, è forse il primo essempio di musica sintetica." Rondolino, *Storia del cinema d'animazione* (Torino: Einaudi, 1974): 141. A rare extended treatment of this curiously neglected acoustic moment in the history of avant-garde cinema can be found in Hans Schleugl and Ernst Schmidt Jr., *Eine Subgeschichte des Films. Lexikon des Avantgarde-, Experimental- und Undergroundfilms*, vol. 2 (Frankfurt/M: Suhrkamp, 1974): 938–940. On the other hand, in what is otherwise a well-informed and wide-ranging account of the subject, there is no mention of Pfenninger whatsoever in Robert Russett, "Experimenters in Animated Sound," in *Experimental Animation: Origins of a New Art*, rev. ed., ed. Robert Russett and Cecile Starr, 163–177 (New York: Da Capo, 1976), nor in the otherwise intelligent assessment of "Avant-garde early sound films" in Russell Lack, *Twenty Four Frames Under: A Buried History of Film Music* (London: Quartet Books, 1997): 104–111.
37. One of the few texts in English that focuses on this fascinating and largely unknown chapter, which I explore at some length in my forthcoming study, is Nikolai Izvolov, "The History of Drawn Sound in Soviet Russia," trans. James Mann, *Animation Journal* 7 (Spring 1998): 54–59. See also Hugh Davies, "Avraamov, Arseny Mikhaylovich," in *New Grove Dictionary of Musical Instruments* 1, ed. Stanley Sadie (London: MacMillan Press, 1984): 91; Davies, "Sholpo, Evgeny Aleksandrovich," in *New Grove Dictionary of Musical Instruments* 3, ed. Sadie, 377; and Davies, "Variaphon" [*sic*], in *New Grove Dictionary of Musical Instruments*, ed. Sadie, 716.
38. The most extensive and well-informed overview of Fischinger's work can be found in William Moritz, "The Films of Oskar Fischinger," *Film Culture* 58–60 (1974): 37–188. More recently, Moritz contributed a lengthy study, entitled "Oskar Fischinger," to the catalogue *Optische Poesie: Oskar Fischinger Leben und Werk* (Frankfurt/M: Deutsches Filmmuseum, 1993): 7–90. On the history of early abstract cinema, see also *Film as Film: Formal Experiment in Film, 1910–1975* (London: Hayward Gallery/Arts Council of Great Britain, 1979): 9–35; Standish Lawder, "The Abstract Film: Richter, Eggeling and Ruttmann," in Lawder, *The Cubist Cinema* (New York: New York University Press, 1975): 35–64; and Hans Richter, "Avant-Garde Film in Germany," in *Experiment in the Film*, ed. Roger Manvell, 219–233 (London: The Grey Walls Press, 1949).
39. As reported in numerous journalistic accounts of Diebold's lectures on "den gezeichneten Film" (hand-drawn film) in the Südwestdeutschen Rundfunk ("Bernhard Diebold vor dem Mikrophon," *Frankfurter Zeitung*, 18 June 1932) and in the cinema Kamera Unter den Linden (H.A., "Bernhard Diebold über Fischingerfilme: 'Das ästhetische Wunder,'" *Lichtbildbühne*, 1 June 1932).
40. This description is from the program note to a screening of Fischinger's "Early Experiments in Hand-Drawn Sound" at The Film Society in London on May 21, 1933. The note is reprinted in *The Film Society Programmes 1925–1939* (New York: Arno Press, 1972): 277. Although the program lists the date of these experiments as 1931, entry no. 28 in Moritz, "The Films of Oskar Fischinger," entitled "Synthetic Sound Experiments," is described as "several reels of sound experiments from 1932 [that] survive in marginal condition." On December 10, 1933, the Film Society screened another program of Fischinger films entitled "Experiments in Hand-Drawn Sound," dated 1933. The 1993 Fischinger filmography *Optische Poesie* includes only one entry, however, entitled *Tönende Ornamente* (Sounding Ornaments) and described as follows: "Experimente mit gezeichneter Lichttonspur (Experiments with hand-drawn optical soundtrack) s/w, Ton, 123 m, 5 min. (weiter 500 m auf Nitrat erhalten)" (106).
41. Moritz, "The Films of Oskar Fischinger," 51. According to another source, Fischinger made a similar discovery with the geometric motif that the ancient Egyptians used to designate the snake: when reproduced on the optical sound track, the acoustic result was exactly the hiss characteristic of this animal! See Georges Hacquard, *La Musique et le cinéma* (Paris: PUF, 1959): 34.
42. "Persönliche, nationale, charakteristische Eigentümlichkeiten werden sich naturgemäß auch im Ornament ausdrücken lassen. Der Deutsche bevorzugt bei seinem Stimmansatz einen heftigen Anschlag. Dies entspricht einer bestimmten heftigen Kurve, während des Franzosen weicher Ansatz sich auch entsprechend im Ornament anders gestaltet. Es gibt also auf diese Weise eine ebenso deutliche 'Mundschrift' wie es eine 'Handschrift' gibt." Oskar Fischinger, "Klingende Ornamente," *Kraft und Stoff* (suppl. to *Deutsche allgemeine Zeitung*) 30 (28 July 1932). This edition includes extensive illustrations. The text was reprinted, albeit with occasionally significant variations, as "Tönende Ornamente: Aus Oskar Fischingers neuer Arbeit," *Filmkurier*, 30 July 1932; and "Klingende Ornamente! Eine neue Basis der Kunst?" *Saarbrücker Landeszeitung*, 11 September 1932. A typescript in the Fischinger Papers in the Center for Visual Music Archive (Los Angeles) bears the title "Klingende Ornamente, absoluter Tonfilm" (Sounding Ornaments, Absolute Sound Film), and another textually identical typescript carries the title "Der Komponist der Zukunft und der absolute Tonfilm" (The Composer of the Future and the Absolute Sound Film). Interestingly, the passage cited here is dropped in the *Filmkurier* reprint.
43. Although in his 1974 study Moritz writes that this lecture took place a full year earlier in August 1931 (Moritz, "The Films of Oskar Fischinger," 52), in his 1993 essay on Fischinger the period of both the experiments in drawn sound and the Berlin lecture on the subject is identified as late summer 1932 (Moritz, "Oskar Fischinger," 33). The 1932 dating is also confirmed by the publications of Fischinger's own writings on sounding ornaments—for example, Oskar Fischinger, "Was ich mal sagen möchte," *Deutsche allgemeine Zeitung*, 23 July 1932; and the texts cited in n. 42 above—as well as the extensive journalistic accounts that only begin to appear in the late summer and fall of 1932. See, for example, Fritz Böhme, "Verborgene Musik im Lindenblatt: Die Bedeutung von Fischingers Entdeckung für den Tonfilm," *Deutsche allgemeine Zeitung*, 30 July 1932; M. Epstein, "Elektrische Musik: Neue Wege der Musikaufzeichnung," *Berliner Tageblatt*, 24 August 1932; Margot Epstein, "Gezeichnete Musik: Oskar Fischinger's 'Tönende Ornamente,'" *Allgemeine Musikzeitung*, 25 November 1932, 591; and Fritz Böhme, "Gezeichnete Musik: Betrachtungen zur Entdeckung Oskar Fischingers," *Deutsche Frauen-Kultur* 2 (February 1933): 31–33. The confusion as to the year might

be due to Fischinger's highly proprietary relationship to this subject matter and his desire to show that his experiments had preceded Pfenninger's (which were also first made public in the summer of 1932)—an imperative that often led to postfactum redating of works.

44. Gustav Hegi, *Illustrierte Flora von Mittel-Europa* (Munich: J.F. Lehmann, 1906–1931).
45. Pfenninger's early animations include *Largo* (1922), *Aus dem Leben eines Hemdes* (1926), *Sonnenersatz* (1926), and *Tintenbuben* (1929). For the history of the studio, see Petra Putz, *Waterloo in Geiselgasteig: Die Geschichte des Münchner Filmkonzerns Emelka (1919–1933) im Antagonismus zwischen Bayern und dem Reich* (Trier, Germany: WVT, 1996).
46. Given how long Pfenninger would have needed to translate his initial insight into the functioning system with which he would complete his first synthetic soundtrack in 1930, it seems reasonable to assume, as numerous music and media historians have done, that Pfenninger's discovery of synthetic sound effectively took place "perhaps as early as 1929" (Hugh Davies, "Drawn Sound," in *New Grove Dictionary of Musical Instruments*, vol. 1, ed. Sadie, 596–597). The 1929 date is also invoked by Peter Weibel, who states that "Rudolf Pfenninger invented 'sounding handwriting' in 1929 in Munich by taking sounds drawn onto paper strips and shooting them one by one directly with the camera, thereby incorporating them into the optical soundtrack" (Weibel, "Von der visuellen Musik zum Musikvideo," in *Clip, Klapp, Bum: Von der visuellen Musik zum Musikvideo*, ed. Veruschka Body and Peter Weibel [Cologne: Dumont, 1987]): 84.
47. Odo S. Matz, "Die tönende Handschrift," *Prager Deutsche Zeitung*, May 1931; reprinted in *Wahrisches Tageblatt*, 21 July 1931. See also "Töne aus dem Nichts: Die phantastische Erfindung eines Müncheners," *Telegrammzeitung*, September 1931; and "Töne aus dem Nichts: Eine Erfindung Rudolf Pfenningers," *Acht Uhr Abendblatt* (Berlin), 12 October 1931. The latter report clearly has taken Humphries's achievement as the obvious litmus test for Pfenninger's innovation when it insists that "the inventor is currently working on drawing from out of nowhere the sound curves of the human voice using a paint brush and drafting pen, based on precise physical experiments." Matz's nationalistic casting of Pfenninger's discovery is, moreover, by no means exceptional: under a widely reproduced photo of the inventor working with strips of sound waves in his studio, the *Bayerische Zeitung* of October 31, 1932, has a caption that reads, "Gemalter Tonfilm—eine bedeutsame *deutsche* Erfindung" (Painted Sound Film—A Significant *German* Invention); emphasis added.
48. Moritz, "Oskar Fischinger," 31. Moritz's account goes on to describe how "through the study of extant soundtracks he quickly mastered the calligraphy of traditional European music such that he was able to record 'Fox you have stolen the goose' and other simple melodies." In the context, this plausible but otherwise undocumented claim to have also mastered a technique that is clearly similar if not identical to Pfenninger's systematic procedure is most likely wishful apocrypha.
49. "Tonfilm ohne Tonaufnahme: Die Sensation der tönenden Schrift," *Film-Kurier*, 20 October 1932.
50. I have been able to track down reviews and/or announcements of screenings of some or all of the *Tönende Handschrift* series at the Capitol in Berlin; the Phoebus Lichtspiele and the University in Munich; the Capitol-Lichtspiele in Halberstadt; the Emelka-Theater in Münster; the Goethehaus, Imperator, and Universum-Lichtspiele in Hannover; the Kristall-Palast in Liegnitz; and the Brussels Filmweek. The film's distribution in Holland is confirmed by the existence of a print with Dutch text frames in the collection of the Nederlands Filmmuseum in Amsterdam.
51. What follows is merely a sampling of the more detailed reviews in the major European papers and film journals, leaving aside the vast number of reprints, local reviews of the various screenings, radio programs, interviews, academic lectures, and other public events promoting Pfenninger's work in the months that followed: Jury Rony, "Tonfilm ohne Tonaufnahmen," *Pressedienst der Bayerischen Film-Ges.m.b.H.* 13 (19 October 1932); Karl Kroll, "Musik aus Tinte," *Münchener Zeitung*, 19 October 1932; St., "Die tönende Handschrift," *Deutsche Filmzeitung* 43 (21 October 1932); Zz., "Tönende Handschrift," *Germania*, 21 October 1932; "Handgezeichnete Musik," *Vossische Zeitung*, 21 October 1932; "Tönende Handschrift," *Kinematograph* (Berlin), 21 October 1932; -n., "Die tönende Handschrift: Sondervorführung im Marmorhaus," *Film-Kurier*, 21 October 1932; "Pfenningers 'tönende Handschrift' im Marmorhaus," *Lichtbild-Bühne*, 21 October 1932; wkl, "Von Ruttmann bis Pfenninger: Zu der Sondervorführung in den Kammerlichtspielen," *Münchener Zeitung*, 22 October 1932; Dr. London, "Pfennigers [sic] 'tönende Handschrift,'" *Der Film*, 22 October 1932; -au-, "Gezeichnete Töne: Neue Wege für den Tonfilm," *Berliner Morgenpost*, 23 October 1932; -e., "Tönende Handschrift: Sondervorstellung im Marmorhaus," *Deutsche allgemeine Zeitung* (Beiblatt), 23 October 1932; W.P., "Der gezeichnete Tonfilm," *Frankfurter Zeitung*, 2 November 1932; "Erschließung einer unbekannten Welt: Gezeichnete Musik," *Tempo* (Berlin), 2 November 1932; and K.L., "Die tönende Handschrift," *Neue Züricher Zeitung*, 27 November 1932.
52. N., "Klänge aus dem Nichts: Rudolf Henningers [sic] 'Tönende Handschrift,'" *Film-Journal* (Berlin), 23 October 1932; K[urt] Pfister, "Die tönende Handschrift," *Breslauer Neueste Nachrichten*, 24 October 1932; -au-, "Gezeichnete Töne"; A. Huth, "Die tönende Handschrift: Eine sensationelle Erfindung," *Nordhäuser Zeitung u. Generalanzeiger*, n.d.
53. "Es klingt nach gedeckten Orgelpfeifen, nach gestopftem Horn, nach Harfe, nach Xylophon. Es klingt seltsam unwirklich." K. Wolter, "Gezeichnete Tonfilmmusik," *Filmtechnik*, 12 November 1932, 12–13. Not surprisingly, the reaction to Fischinger's experiments was much the same. According to Moritz's account, "When Fischinger picked up the first reels of *Sounding Ornaments* from the lab and had them screened there on the lab's projectors, the technicians were shocked by the strange sounds and feared that further reels with this noise could ruin their machines." "Oskar Fischinger," 33.
54. R. Prévot, "Musik aus dem Nichts: Rudolf Pfenningers 'Tönende Handschrift,'" *Münchener Neueste Nachrichten*, 20 October 1932; emphasis in original.
55. W.P., "Der gezeichnete Tonfilm."
56. wbf, "Handgezeichnete Musik," *Telegramm-Zeitung*, 19 October 1932.
57. Paul Bernhard, "Mechanik und Organik," *Der Auftakt* 10, no. 11 (1930): 239.
58. See, for example, Kroll, "Musik aus Tinte"; wkl, "Von Ruttmann bis Pfenninger"; and "'Tönende Handschrift': Neuartige Wirkungsmöglichkeiten des Tonfilms," *Kölnische Zeitung*, 25 October 1932.

59. A[ndor] Kraszna-Krausz, "Beginning of the Year in Germany," *Close-Up* 10 (March 1933): 74–76. The comparison is cast in somewhat different terms—artistic versus commercial—in the program notes for a May 21, 1933, screening of Fischinger's "Early Experiments in Hand-Drawn Sound" at the Film Society in London. In this text, which likely was either written by Fischinger or based on materials he provided, the audience is advised that what they are about to see is "not to be confused with the similar effects invented by Hans [sic] Pfenninger. The latter system has been developed commercially to form the musical accompaniment to puppet and cartoon films." *The Film Society Programmes 1925–1939,* 277.
60. "Die Ornamente primitiver Völker sind zu untersuchen auf ihren Klangcharakter." This phrase from a typescript of Fischinger's 1932 essay "Klingende Ornamente" in the Fischinger Papers in the Iota Foundation Archive, does not appear in any of the published versions I have found.
61. "Dieses im wahrsten Sinne reinste und höchste Kunstschaffen formt ausschließlich aus einer einzigen Persönlichkeit direkt, und diese Kunstproduktion, die so entsteht, etwa Werke von Rembrandt, Bach oder Michelangelo, sind unmittelbare Schöpfungen höchster Potenz und gewinnen *gerade durch das handschriftliche,* Irrationale und Persönliche." Ingenieur Oskar Fischinger, "Der absolute Tonfilm," *Dortmunder Zeitung* (1 January 1933); emphasis added. Reprinted as "Der absolute Tonfilm: Neue Möglichkeiten für den bildenden Künstler," *Der Mittag* (Düsseldorf), January 1933; also in *Schwabischer Merkur,* 23 January 1933. A typescript of this article in the archive of the Center for Visual Music with the title "Der Komponist der Zukunft und der absolute Film" bears a handwritten date of 1931–1932, which might well be an example of the retro-dating mentioned earlier.
62. Fischinger's fundamentally antitechnological stance is articulated unambiguously in an essay that explains hand-drawn music as part of his project of "absolute film," which it defines as "a film that only embraces technology to the extent that it really serves to realize artistic creations. It is thus a redemption of art and of the artistic personality in the context of film production." See Böhme, "Gezeichnete Musik."
63. "Schablonen oder Drucktypen," Max Lenz, "Der gezeichnete Tonfilm," *Die Umschau* 36, no. 49 (3 December 1932): 971–973.
64. "Soundless Film Recording," *New York Times,* 29 January 1933, sec. 9, 6. As far as I can tell, no such device was ever constructed.
65. "Nach diesem Verfahren kann nur etwas hörbares durchs Mikrophon aufgenommen werden, das wirklich vorhanden ist, d.h. das irgendwo *vorher* erzeugt wurde. Rudolf Pfenninger aber schafft Töne aus dem Nichts," "Tönende Handschrift," *Völkischer Beobachter,* 25 October 1932; emphasis added.
66. "Pfenninger, schließlich, hat den Weg zur Tonschrift gefunden. Während Fischinger nun den Ton fließend photographiert, nimmt er ihn bildweise auf, was zur Fertigung von Schablonen führt, mittels deren bestimmte Töne und Tongruppen immer wiederholt werden können," Paul Seligmann, "Filmsituation 1933" in: Hirdina, ed., *Neues Bauen/ Neues Gestalten*: 341; first published in *Die neue Stadt* (1932–1933), no. 10. Echoing Seligmann's conclusion, the intelligent critic of the *Münchener Zeitung* also recognized that Pfenninger's project is fundamentally different from those of his predecessors and that the credit for the invention of synthetic sound "thus belongs entirely and exclusively to him" ("*gehört also ihm ganz allein*"). His argument however—"Rudolf Pfenninger's 'sounding handwriting' *creates* sounds from out of nowhere whereas Walter Ruttmann's, Moholy-Nagy's and Oskar Fischinger's film studies *arise out of* the music or a musical-rhythmic experience" (emphasis added)—gets the different directions of the creative vectors exactly wrong: however musical the inspiration for their genesis might have been, Fischinger's curves are not *derived* from sound; they generate it, whereas Pfenninger's curves *are* in the last analysis derived *from* the sounds that they analytically re-create (Wkl, "Von Ruttmann bis Pfenninger").
67. An intelligent, illustrated, contemporary account in English can be found in Waldemar Kaempffert, "The Week in Science," *New York Times,* 11 August 1935, sec. 10, 6.
68. This is admirably documented in the wonderful film *Theremin: An Electronic Odyssey* (Dir. Steven M. Martin, 1993).
69. "Gemalte Musik" *Pester Lloyd* (Budapest), 26 November 1932.
70. This idea nevertheless remained in circulation for quite some time. In 1936, for example, Kurt London is clearly referring to the techniques of synthetic sound when he suggests that in the future "one might do without an orchestra and instruct a composer to put his music together in patterns upon paper, which would then be photographed and then produce a very strange and quite unusual sound." London, *Film Music* (London: Faber & Faber, 1936): 197.
71. A striking example of the dramatic new possibilities of acoustic montage afforded by the Tri-Ergon system is Walter Ruttmann's eleven-minute and ten second image-less "film" entitled "Weekend." Ruttmann's sonic collage, which took full advantage of the ability to edit sounds made possible by recording on the optical film sound track (instead of the gramophone), was broadcast on Berlin radio on 13 June 1930 and, after decades during which it was considered lost, was rediscovered in New York in 1978. It is readily available, together with a series of contemporary "remixes" by John Oswald, to rococo rot and others, on an Intermedium label CD (Rec 003).
72. Lac., "Erschließung einer unbekannten Welt: Gezeichnete Musik," *Tempo* (Berlin), 2 November 1932.
73. ky., "Die tönende Schrift: Eine Umwälzung auf dem Gebiete der Tonwiedergabe," *Kölner Tageblatt,* 18 November 1932; reprint in *Solinger Tageblatt,* 3 December 1932.
74. Herbert Rosen, "Die tönende Handschrift," *Die grüne Post* 24 (11 June 1933): 14.
75. "Die Zeit war noch nicht reif; meine Erfindung kam 20 Jahre zu früh," Hans Rolf Strobel, "Musik mit Bleistift und Tusche: Der Filmklub zeigt heute Rudolf Pfenningers Kurzfilme," *Münchener Abendzeitung,* 4 May 1953.
76. Despite this negative assessment, Pfenninger was able to stay in Germany both during and after the Third Reich, working as a production and set designer in the Geiselgasteig film studios outside Munich, where his credits included animation work on *Wasser für Canitoga* (Dir. Herbert Selpin, 1939), set design for *Das sündige Dorf* (Dir. Joe Stöckel, 1940) and for *Hauptsache Glücklich!* (Dir. Theo Lingen, 1941), production design for *Einmal der liebe Herrgott sein* (Dir. Hans H. Zerlett, 1942) and for *Orient Express* (Dir. Viktor Tourjansky, 1944), set design for *Der Brandner Kasper schaut ins Paradies* (Dir. Josef von Baky, 1949), *Das seltsame Leben des Herrn Bruggs* (Dir. Erich Engel, 1951), and *Nachts auf den Straßen* (Dir. Rudolf Jugert, 1952) as well as production design for *Aufruhr im Paradies* (Dir. Joe Stöckel, 1950).

77. Years later Moholy-Nagy recalled that the sound track for *Tönendes ABC* "used all types of signs, symbols, even the letters of the alphabet, and my own finger prints. Each visual pattern on the sound track produced a sound which had the character of whistling and other noises. I had especially good results with the profiles of persons" (Moholy-Nagy, *Vision in Motion* [Chicago: Paul Theobald, 1947], 277). Unfortunately, there is no way to examine more carefully what appears to be the rather surprisingly Fischingerian character of this film, because, while it was shown at the London Film Society on December 10, 1933, it is now considered lost.
78. Moritz documents in great detail the fascinating and, of course, ultimately futile attempts by Fischinger and his supporters well into the mid-1930s to convince the fundamentally antimodernist National Socialist regime of the great cultural value of the genre of "absolute film." Moritz, "Oskar Fischinger," 35–45. Fischinger even went so far as to write a letter to Goebbels in December 1935 demanding both respect and financial support for his films. However, thanks to Ernst Lubitsch, by February 1936 Fischinger and his films were on their way to Hollywood to take up a job at Paramount. It was here that he worked on the "Toccata and Fugue in D Major" sequence of Disney's *Fantasia* directed by Samuel Armstrong (1940).
79. Arthur Knight, *The Liveliest Art* (New York: MacMillan: 1957): 158.
80. See, for example, P.W., "Les sons synthétiques de l'ingénieur Pfenninger," *XXme Siecle* (Brussels), 16 December 1932; and "Curieuse expérience," *Musique et instruments,* 285 (1933): 265.
81. See Arthur Hoérée and Arthur Honegger, "Particularités sonores du film *Rapt,*" *La Revue musicale* 15 (1934): 90; and Arthur Hoérée, "Le Travail du film sonore," *La Revue musicale* 15 (1934): 72–73. For a more general context of the reception and development of synthetic sound in France, see Richard Schmidt James, *Expansion of Sound Resources in France, 1913–1940, and Its Relationship to Electronic Music* (Ph.D. diss., University of Michigan, 1981).
82. See, for example, the extended discussion of Pfenninger in a lengthy article by Luciano Bonacossa, "Disegni animati e musica sintetica," *La vie d'Italia* 40, no. 8 (August 1934): 571–582, esp. 578–582.
83. Herbert Rosen, "Synthetic Sound: Voices from Pencil Strokes," *Wireless World* (London), 3 February 1933, 101; and Paul Popper, "Synthetic Sound: How Sound Is Produced on the Drawing Board," *Sight and Sound* 2, no. 7 (Autumn 1933): 82–84. Only a few years later *Sight and Sound* published a more general account by V. Solev entitled "Absolute Music," *Sight and Sound* 5, no. 18 (Summer 1936): 48–50; an earlier version of this article had appeared as "Absolute Music by Designed Sound," *American Cinematographer* 17, no. 4 (April 1936): 146–148, 154–155.
84. Program note, *The Film Society* (London), 13 January 1935; reprinted in *The Film Society Programmes,* 310.
85. Interview with McLaren in Maynard Collins, *Norman McLaren* (Ottawa: Canadian Film Institute, 1976): 73–74.
86. Besides the early essay he authored with Robert E. Lewis entitled "Synthetic Sound on Film," *Journal of the Society of Motion Picture Engineers* 50, no. 3 (March 1948): 233–247, McLaren wrote "Animated Sound on Film," a pamphlet first published by the National Film Board of Canada in 1950 and then in a revised version as "Notes on Animated Sound," *Quarterly of Film, Radio and Television* 7, no. 3 (1953): 223–229. This text was subsequently reprinted in Roger Manvell and John Huntley, *The Technique of Film Music* (1959; revised and reprinted, New York: Focal Press, 1975): 185–193; excerpted in Russett and Starr, eds., *Experimental Animation,* 166–168; and reprinted almost without acknowledgment in Roy M. Prendergrast, *Film Music: A Neglected Art* (1977; reprint, New York: W.W. Norton, 1992): 186–193. See also Valliere T. Richard, *Norman McLaren, Manipulator of Movement. The National Film Board Years, 1947–1967* (Newark: University of Delaware Press, 1982).
87. David Wilson, "Synchromy," *Monthly Film Bulletin* 39, no. 466 (November 1972): 241.
88. See John Whitney and James Whitney, "Audio-Visual Music," in *Art in Cinema,* ed. Frank Stauffacher, 31–34 (San Francisco: SF-MOMA, 1947; reprint, New York: Arno Press, 1970); reprinted in P. Adams Sitney, ed., *The Avant-Garde Film: A Reader of Theory and Criticism* (New York: New York University Press, 1978): 83–86. John Whitney and James Whitney, "Notes on the 'Five Abstract Film Exercises,'" in *Art in Cinema,* ed. Stauffacher. John Whitney, "Bewegungsbilder und elektronische Musik," trans. Ruth Cardew, in *Die Reihe: Information über serielle Musik* 7 (1960): 62–73; in English as "Moving Pictures and Electronic Music," *Die Reihe* (English ed.) 7 (1965): 61; reprinted in Russett and Starr, eds., *Experimental Animation,* 171–173. On Kren, see Hans Scheugl, ed., *Ex Underground Kurt Kren* (Vienna: PVS, 1996), esp. 161. See also Barry Spinello, "Notes on 'Soundtrack,'" in *Experimental Animation,* ed. Russett and Starr, 175–176; first published in *Source: Music of the Avant Garde* 6 (1970).
89. Lev Manovich, "What Is Digital Cinema?" in *The Digital Dialectic: New Essays on New Media,* ed. Peter Lunenfeld, 173–192 (Cambridge: MIT Press, 1999): 175.

Part II

Archives

4

Memex Revisited

Vannevar Bush

An Austrian monk, Gregor Mendel, published a paper in 1865 which stated the essential bases of the modern theory of heredity. Thirty years later the paper was read by men who could understand and extend it. But for thirty years Mendel's work was lost because of the crudity with which information is transmitted between men.

This situation is not improving. The summation of human experience is being expanded at a prodigious rate, and the means we use for threading through the consequent maze to the momentarily important items are almost the same as in the days of square-rigged ships. We are being buried in our own product. Tons of printed material are dumped out every week. In this are thoughts, certainly not often as great as Mendel's, but important to our progress. Many of them become lost; many others are repeated over and over and over.

A revolution must be wrought in the ways in which we make, store, and consult the record of accomplishment. This need holds true in science, in the law, in medicine, in economics, and, for that matter, in the broadest subjects of human relations. It is not just a problem for the libraries, although that is important. Rather, the problem is how creative men think, and what can be done to help them think. It is a problem of how the great mass of material shall be handled so that the individual can draw from it what he needs—instantly, correctly, and with utter freedom. Compact storage of desired material and swift selective access to it are the two basic elements of the problem.

I began worrying over this matter more than a quarter century ago, and some twenty years ago published an essay about it called "As We May Think" (*The Atlantic Monthly*, 1945). Next in this present discussion I want to present some thoughts from that earlier paper. Then we will have a look at what has happened during the past two decades and try to see if we are any closer to the means of the needed revolution.

In that essay I proposed a machine for personal use rather than the enormous computers which serve whole companies. I suggested that it serve a man's daily thoughts directly, fitting in with his normal thought processes, rather than just do chores for him.

If it is to fit in with his normal thought processes, the heart of the matter is selection. Our present ineptitude in getting at the record is largely caused by the artificiality of systems of indexing. When data of any sort are placed in storage, they are filed alphabetically or numerically, and information is found (when it is) by being traced down from subclass to subclass. It can be in only one place, unless duplications are used; one has to have rules as to which path will locate it, and the rules are cumbersome. Having found one item, moreover, one has to emerge from the system, like a dog who has dug up a buried bone, and then re-enter the system on a new path. This is a serious handicap, even with the high-speed machinery just now beginning to be applied to the problem of the libraries.

The human mind does not work that way. It operates by association. With one item in its grasp, it snaps instantly to the next that is suggested by the association of ideas, in accordance with some intricate web of trails carried by the cells of the brain. The mind has other characteristics, of course: trails not frequently followed are apt to fade; few items are fully permanent; memory is transitory. Yet the speed of action, the intricacy of trails, the detail of mental pictures, is awe-inspiring beyond all else in nature.

Man cannot hope fully to duplicate this mental process artificially. But he can certainly learn from it; in minor ways he may even improve on it, for his records have relative permanency. But the prime idea to be learned concerns selection. Selection by association, rather than by indexing, may yet be mechanized. Although we cannot hope to equal the speed and flexibility with which the mind follows an associative trail, it should be possible to beat the mind decisively in the permanence and clarity of the items resurrected from storage.

To turn directly to that earlier discussion:

Consider a future device for individual use, which is a sort of mechanized private file and library. It needs a name. To coin one at random, "memex" will do. A memex is a device in which an individual stores all his books, records, and communications, and which is mechanized so that it may be consulted with exceeding speed and flexibility. It is an enlarged intimate supplement to his memory. What does it consist of? It consists of a desk. Presumably, it can be operated from a distance, but it is primarily a piece of furniture at which an individual works. On its top are slanting translucent screens, on which material can be projected for convenient reading. There is a keyboard, and sets of buttons and levers. Otherwise, memex looks like an ordinary desk. In one end is its stored reference material. The matter of bulk can be well taken care of even by present-day miniaturization. Only a small part of the interior of the memex is devoted to storage, the rest to mechanism. Yet if the user inserted 5,000 pages of material a day it would take a hundred years to fill the repository. So he can be profligate, and enter material freely. Most of the memex contents are purchased on tape ready for insertion. Books of all sorts, pictures, current periodicals, newspapers, are thus obtained and dropped into place. And there is provision for direct entry. On the top of the memex is a transparent platen. On this our user places longhand notes, photographs, memoranda, all sorts of things. When one is in place, the depression of a lever causes it to be recorded on a blank space in a section of the memex memory.

Memex has, of course, provision for consulting the record by the usual scheme of indexing. When the user wishes to consult a certain book, he taps its code on the keyboard, and the title page of the book promptly appears before him, projected onto one of his viewing positions. Frequently-used codes are mnemonic, so that he seldom consults his code book; but when he does, a tap of a key or two projects it for his use. Moreover, he has supplemental levers. By deflecting one of these levers to the right he runs through the book before him, each page in turn being projected at a speed which just allows a recognizing glance at each. If he deflects the lever further to the right he steps through the book 10 pages at a time; still further speeds scanning to 100 pages at a time. Deflection to the left gives him the same control backwards. A special button transfers the user immediately to the first page of the index. Any book of his library can thus be called up and consulted with far greater facility, comfort and convenience than if it were taken from a shelf. And his personal library is voluminous; if he had it present in paper it would fill his house or office solidly.

He has several projection positions; hence he can leave one item in position while he calls up another. He can add marginal notes and comments, for the nature of his stored record is such that he can add or erase, quite as readily as though he were adding notes to the page of a book.

So far, all this is conventional; a mere projection forward of present-day mechanisms and gadgetry. It affords an immediate step, however, to associative indexing, the basic idea of which is a provision whereby any item may be caused at will to select another, immediately and automatically. This is the essential feature of the memex; the process of tying items together to form trails is the heart of the matter.

When the user is building a trail, he names it, inserts the name in his code book, and taps it out on his keyboard. Before him, projected onto adjacent viewing positions, are the items to be joined. At the bottom of each there are a number of blank code spaces; a pointer is set to indicate one of these on each item. The user taps a single key, and the items are permanently joined. In each code space appears the code word. Out of view, but also in the code space, is automatically placed a set of dots as a designation; and on each item these dots by their positions designate the index number of the other.

Thereafter, at any time, when one of these items is in view, the other can be instantly recalled merely by tapping a button adjacent to the code space. Moreover, when numerous items have been thus joined together to form a trail, they can be reviewed in turn, rapidly or slowly, by deflecting a lever like that he used for turning the pages of a book. It is exactly as though the physical items had been gathered together from widely separated sources and bound together to form a new book. But it is more than this; for any item can be joined into numerous trails, the trails can bifurcate, and they can give birth to side trails.

To give you a simple example, the owner of the memex, let us say, is interested in the origin and properties of the bow and arrow. Specifically he is studying why the short Turkish bow was apparently superior to the English long bow in the skirmishes of the Crusades. He has dozens of possibly pertinent books and articles in his memex. First he runs through an encyclopedia, finds an interesting but sketchy article, and leaves it projected. Next, in a history, he finds another pertinent item; he ties the two together. Thus he goes, building a trail of many items. Occasionally he inserts a comment of his own either linking it into the main trail or joining it, by a side trail, to a particular item. When it becomes evident to him that the elastic properties of available materials had a great deal to do with the superiority of the Turkish bow, he branches off on a side trail which takes him through text books on elasticity and tables of physical constants. He inserts a page of longhand analysis of his own. Thus he builds a trail of interest through the maze of materials available to him.

His trails do not fade. Several years later, his talk with a friend turns to the queer ways in which a people resist innovations, even of vital interest. He has an example in the fact that Europeans, although outranged, still failed to adopt the Turkish bow. In fact he has a trail on it. A touch brings up the code book. Tapping a few keys projects the head of the trail. By lever, the user runs through it at will, stopping at interesting items, going off on side excursions. It is an interesting trail, pertinent to the discussion. So he sets a reproducer in action, records the whole trail, and passes the record to his friend for insertion in his own memex, there to be linked to a more general trail.

Now, is this all a dream? It certainly was, two decades ago. It is still a dream, but one that is now attainable. To create an actual memex will be expensive, and will demand initiative, ingenuity, patience, and engineering skill of the highest order. But it can be done. It can be done, given enough effort, because of the great advances which have been made in mechanization, the instruments which have already been built in great numbers to aid man's computations and his thoughts, the devices already used for storing and consulting masses of data, the ingenious elements of electric and magnetic circuits which have been developed during the last two decades.

New and powerful instrumentalities have come into use to help it on its way toward birth. Highly sensitive photocells capable of seeing things in a physical sense; magnetic tapes that instantly record with utter faithfulness music or vision; advanced photography which can record not only what is seen but also what is not; transistors capable of controlling potent forces under the guidance of less power than a mosquito uses to vibrate his wings; cathode ray tubes rendering visible an occurrence so brief that by comparison a microsecond is a long time; transistor combinations which will carry out involved sequences of operations more reliably than any human operator and thousands of times as fast; miniaturization of solid-state devices which will put the complex circuitry of a radio set into the volume of a pinhead; video tapes which put the moving episodes of a football game onto a little strip of film, and instantly reproduce it-there are plenty of mechanical aids with which now to effect a transformation.

So it can be done. Will it be done? Well, that is another question. The great digital machines of today have had their exciting proliferation because they could vitally aid business, because they could increase profits. The libraries still operate by horse-and-buggy methods, for there is no profit in libraries. Government spends billions on space since it has glamour and hence public appeal. There is no glamour about libraries, and the public do not understand that the welfare of their children depends far more upon effective libraries than it does on the collecting of a bucket of talcum powder from the moon. So it will not be done soon. But eventually it will.

To look forward to memex we will lean on what has already been done. Machines of today fall into two great divisions, first those that supplement man's muscles and his senses, and second those that aid his mind. We do not need to deal with the former, although they have made possible our modern civilization with all its benefits and its dangers. The latter are sometimes included under the general term of thinking machines, but this is an unfortunate expression, for they do not think, they merely aid man to do so. They are of two sorts, analytical machines and data-handling machines, and these are sometimes combined. The great example of the first sort is the digital machine. It is often called a computer, but this is a misnomer, for the machine does far more than to compute. A single large unit costs several million dollars. Our present business organizations could not operate without it. Properly instructed, it can do about anything a man can do using pencil and paper, and do it a million times as fast. The only things it cannot do are those which distinguish a man from a machine.

It is told what to do by the insertion of a coded tape, and the preparation of this tape is called programming, of which more will be said later. When the computer has completed its job, it delivers its results by rapidly operating a typewriter, or sometimes by drawing them on a screen. It works entirely by using numbers, although these may also represent letters or instructions, and these numbers are in the binary system, that is, to a base two instead of the usual base ten. It gets the numbers it works on from the input tape, or from its own memory, where great masses of data are stored. The tape, and subsidiary instructions stored in the memory, tell it how to manipulate numbers for all its purposes. Its main element is an elaborate network of electric circuits. These can manipulate numbers by addition, subtraction, multiplication, and division. Thus far it is indeed a computer. But it has, importantly, other circuits which can perform the operations of logic, and it is these which give the digital machine its great power. As a simple example, these can examine a set of numbers and pick out the largest. Or they can follow one set of instructions or another according to the results of the moment. The machine does all of these things very rapidly indeed, many million operations a second.

Another type of analytical machine is the analogue machine. These are nowhere near so precise as the digital machines, but they are far less expensive and are genuinely useful for exploratory purposes, especially in engineering. The principal form is the differential analyzer which appeared some thirty years ago. To use one of these in examining a problem, say the problem of how a suspension bridge will behave in a gusty wind, one assembles an electric circuit which follows the same physical laws as the bridge, though usually with a different time scale, and which then moves a point of light on a screen in just the way in which the bridge will swing in the wind. One has set up an electrical circuit which obeys the same differential equations as the physical system under study, and which hence behaves in the same way, and then one watches it perform, usually by the pattern it produces on an oscillograph.

There are also special-purpose analytical machines which do not belong to either of these classes. An early one of these is the tide-predicting machine. There are also machines for statistical analysis, evaluating correlation coefficients and the like, and for solving integral equations, or interpreting x-ray diffraction patterns of crystals. Some of these have been crowded out by the great success of the digital machine, but they include ideas which should not be forgotten. Everything that can be done on analogue or special machines can also be done on a digital machine, although often not so neatly or flexibly or inexpensively. Data-handling machines are also of various sorts, from the extremely simple card catalog up through the numerous ways of manipulating punched cards. The memory component of the digital machine is probably the most remarkable of the data-handling devices. Another should be mentioned as well. This is the rapid selector, which first appeared some twenty years ago. This would take a roll of photographic film containing 100,000 or so items in single frames, and select desired items from these in accordance with a code in the margin. It could do so while viewing the items at the rate of 1000 per second. And it printed out the selected items on a short piece of similar film.

Each item could consist of a page of print, drawings, or photographs. There are now a variety of modern forms of this device. Some of them combine the sorting and ordering facility of the punched-card equipment with rapid selection by code. The same sort of thing can of course be done with magnetic tape.

The evolution of data-handling equipment thus has involved two important features: compression, which allows great masses of data to be stored in a small space, and rapid access, by which a single piece of information can be located and reproduced in a very brief time. The development of detailed devices or elements did not alone make this whole range of equipment possible. There is another, and very important, general consideration which should be noted: Over three centuries ago Pascal constructed a calculating machine which embodied many of the essential features of recent keyboard devices, but it could not then come into use. The economics of the situation were against it; the labor involved in constructing it, before the days of mass production, exceeded the labor to be saved by use of it, since all it could accomplish could be duplicated by sufficient use of pencil and paper. Moreover, it would have been subject to frequent breakdown, so that it could not have been depended upon; for at that time and long after, complexity and unreliability were synonymous.

Only a century ago, Babbage, even with remarkably generous support, could not produce his great arithmetical machine. His idea was sound enough, but construction and maintenance costs were then too heavy. Inexpensive construction is a new thing. Had a Pharaoh been given detailed and explicit designs of an automobile, and the tools with which to work metal, and had he understood them completely, it would have taxed the resources of his kingdom to fashion the thousands of parts for a single car, and that car would have broken down on the first trip to Gaza. Machines with interchangeable parts can now be constructed with great economy of effort. In spite of much complexity, they perform reliably. It is this reliable complexity, attained at reasonable cost, produced by hard work and the rigors of competition over many years, together with the advance of basic science, and finally man's ingenuity, which has now made it possible to lighten the burden on man's mind, as earlier developments lifted the load from his muscles.

An excellent example of how the advance goes forward is the history of the thermionic tube and the transistor. The thermionic tube was, at its inception, largely a matter of ingenious tinkering, without much reliance on science. Edison, who was no scientist, noted a current from the filament of one of his electric lamps to a plate he put in, but he did nothing about it. De Forest, who probably knew still less science, added a grid between them, and the thermionic tube was born. For many years it was erratic in operation and likely to fail at any moment. Then engineers learned to make a really good vacuum and it became much better. Finally it became so reliable that it could be installed in an amplifier of a submarine cable at the bottom of the sea and expected to last for forty years. It became so rugged that, in the proximity fuzes of the war, it could be put into a shell, fired out of a gun, and still be expected to work as a sensitive electronic detector.

Then came the transistor, which has superseded the tube for most purposes. This certainly did come out of the application of science. A group of men, working on the theory of electric conduction in solids, soon saw how the phenomena they predicted, and checked in the laboratory, could be put to use. The transistor, which can be as small as the eye of a fly, requiring extreme precision of construction, rugged and long lived when once built, is perhaps the most versatile device man has yet produced. With the use of very little power, and in a small space, it will amplify, modulate, rectify, and do dozens of other things. It is one of a whole family of devices based on the use of semiconductors: photoelectric cells, rectifying valves, etc. When the transistor is combined with other elements, resistors and capacitor, sealed in a resin, an assemblage the size of a thimble will do all that used to be done by a radio receiver as large as a suitcase. More than this: by some very modern methods of depositing very thin layers of material in a vacuum, the whole thing can be reduced to a thin wafer the size of a flyspeck, and a thousand such can be produced identically in a single manufacturing operation.

A very great advance—possibly the greatest so far—as we look toward the future memex, is magnetic tape. We have known it for some time in dictating machines. It hit the market modestly soon after the war, and, around 1958, tapes appeared capable of carrying great detail, so much so that video tapes appeared carrying an entire television broadcast with its 70,000 or so complete pictures on a single reel. The idea is a simple one, and the tape is merely a plastic strip covered with magnetic material in finely powdered form. As it passes over an electromagnet, the voice, picked up by a microphone, causes the strength of that magnet to vary, and these variations become impressed on the tape in the form of its magnetization. A wavy form of air vibrations from the voice becomes an identically wavy form of magnetism on the tape. Then, when the tape is run in front of a coil, the voltages there produced can be amplified and fed into a loudspeaker, and a replica of the original voice appears as sound waves in the air.

All this is now ancient history. But it is relatively new to put on the tape the variation of light impinging on a photocell as an optical system sweeps its view over a scene, and to do this so that all the details of a complex scene are thus recorded in a small fraction of a second. And then to reverse the process and reproduce the scene to a viewer a thousand miles away.

In our living rooms, we watch a football game. A television camera is scanning the scene line by line, twenty-four pictures a second. The response of its photoelectric equipment, transmitted a thousand miles to our living room, conveys the intensity of an electron beam which sweeps over our TV screen and reproduces the play as it occurs. But the output of the camera also sweeps over a fairly broad magnetic tape, and magnetizes it. Thus, a few moments after a play occurs, the tape record can be re-scanned, and the result transmitted to our TV set, so that we can see the play over again. To accomplish this, using a reasonable amount of tape, has required a great compression of the magnetic record. But it has been done and is now accepted by television viewers as a commonplace. Another important feature of magnetic tape, for our future memex, is that it can be erased. Fortunately, this is easy. One merely sweeps a permanent magnet over the tape and the record is gone. When we take a photograph we are stuck with it; to make a change we must take another whole photograph. But with a magnetic tape which presents to us a picture one can cancel half a line, if he will, add a changed line, or put in a marginal note or code. The moving finger writes, but its record is not here irrevocable.

The advent of the laser may bring photography back into competition for memex storage. It can produce such a small spot of light that there is a factor of 100 or more on compactness compared to magnetic storage. The spot can be intense, so much so that it is used to bore small holes in diamonds, and this means a photographic record can be made in a very short interval, and read out equally rapidly for projection. The film used can be of such low sensitivity that daylight will not affect it appreciably, and the usual processes of development can be avoided, which means parts can be obliterated and additions made to the record. Beyond this the laser renders possible an exciting process called holography, which may render it possible to project the record so that it is three-dimensional. This is an utterly new form of three-dimensional projection, for it is as though the original scene or model were actually present, and one can move about and view it from various angles. There are many tough problems to be solved before the use of the laser for such purposes becomes practicable. But, for a long view ahead, it exhibits a wholly new field of versatility in which ingenuity will certainly produce results.

There is a point here worth pausing to consider for a moment. For the purposes of memex we need a readily alterable record, and we have it. But alteration of records has a sinister connotation. We watch a girl on the screen moving her mouth and someone else is doing the singing. One can put into a man's mouth for all to hear words he never spoke. The ingenuity which special-effects men use on television is often amusing, sometimes powerfully dramatic, sometimes annoying, as when a razor is seen to shave sandpaper. Advancing technology is making it easy to fool people. It would be well if technology also devoted itself to producing forms of records, photographic, printed, sound-recorded, which cannot be altered without detection, at least to the degree of a dollar bill. But it would be still more effective if the code of morals accepted generally rendered it a universally condemned sin to alter a record without notice that it is being done.

It is thus fairly clear that there is no serious problem today in assembling, editing, and correcting the record, or in compressing it into as small a volume as we may need for memex. If we wish it, a whole private library could be reduced to the volume of a matchbox; similarly, a library of a million volumes could be compressed into one end of a desk. If the human race has produced, since the invention of movable type, a total record in the form of magazines, newspapers, books, tracts, advertising blurbs, correspondence, having a volume corresponding to a billion books, the whole affair, assembled and compressed, could be lugged off in a wheelbarrow.

Compression is important not only to keep us from being swamped, but also when it comes to costs. The material for a microfilm private library might cost a nickel, and it could be mailed anywhere for a few cents. What would it cost to print a million copies? To print a sheet of newspaper, in a large edition, costs a small fraction of a cent. The entire material of a private library in reduced film form would go on ten eight-and-one-half-by-eleven-inch sheets. Once that was available, with the reproduction methods now available, duplicates in large quantities could probably be turned out for a few cents apiece beyond the cost of materials.

Mere compression, of course, is not enough; one needs not only to make and store a record, to add to it at will, and to erase, but also to consult it. As things are now, even the modern great library is not generally consulted; it is nibbled at by a few. How to consult the new compressed record is a major question in selective analysis. The great digital computers of today keep their extensive records in various ways. The records constitute their memory, which they consult as they proceed with computation. They use magnetic tapes or disks. But they also use great arrays of minute toroids of magnetic material, interlaced with fine wires. The reason for these latter is the necessity of rapid access. The fast access, in a computer, is fast indeed, and has to be. Times, for them, should be mentioned in nanoseconds, or billionths of a second. In a nanosecond light will move only about one foot. That is why it is important to keep the components of a computer small; its speed of operation is sometimes limited by the time necessary to get an electric pulse from one part to another. The storage in little toroids can respond in times like these.

No problem of speed of access need bother the future memex. Indeed, for memex we need only relatively slow access, as compared to that which the digital machines demand: a tenth of a second to bring forward any item from a vast storage will do nicely. For memex, the problem is not swift access, but selective access. The indication of a possible beginning here is to be found in the rapid selector mentioned earlier. When items on frames projected for viewing can readily have codes entered in their margins, by which they can automatically select other items, we have a significant step toward memex. But the access problem is by no means solved. The storage of memex will be huge, and all parts of it need to be promptly available.

Clearly, we need to study further how the human brain meets this puzzle. Its memory system consists of a three-dimensional array of cells, each cell very small compared to even the volume of magnetic tape used for a single impulse, and the magnetic tape is two-dimensional. We make three-dimensional storage, for example, by an array of toroids, but the units here are huge compared to a cell. Somehow the brain consults this full array and brings into consciousness, not just the state of one cell, but the related content of thousands, to recall to us a scene of a decade ago. We have very little idea as to how it is done. In fact we do not even know what we mean when we write "consciousness." If there is a roadblock in the path toward a useful memex, it lies in this problem of moderately rapid access to really large memory storage.

The heart of this problem, and of the personal machine we have here considered, is the task of selection. And here, in spite of great progress, we are still lame.

Selection, in the broad sense, is still a stone adze in the hands of a cabinetmaker. Yet, in a narrow sense and in other areas, something has already been done mechanically on selection. The personnel officer of a factory drops a stack of a few thousand employee cards into a selecting machine, sets a code in accordance with an established convention, and produces in a short time a list of all employees who are females, live in Trenton, and know Spanish. Even such devices are much too slow when it comes, for example, to matching a set of fingerprints with one of five million on file. Selection devices of this sort have now been speeded up from their previous rate of reviewing data at a few hundred a minute. The great computer will enter its active memory and select a desired item in a microsecond or less, if it is told just where to go for it, and in an interval which is still very brief if it has to hunt for it.

So much for the methods of storing record and of retrieving items from storage. But what about the making of the record? Is it possible that somewhere during this procedure we may find ways of anticipating the selective needs to be encountered later when one wishes to consult that record? Our record-making system of today should remind us of the covered wagon; we are bound to have to improve it and in doing so we must have an eye to the possibilities of coding, cross-linking, and all else that will be requisite to selective access.

Today, to make a record, we still push a pencil or tap a typewriter. Then comes the business of digestion and correction, followed by an intricate—and largely cockeyed—process of typesetting, printing, and distribution. To consider the first stage of procedure, will the author of the future cease writing by hand or typewriter and talk directly to the record? He does so (indirectly) even now, of course, by talking to a stenographer or into a dictating machine. And there is also the stenotype, that somewhat disconcerting device encountered in court or at public meetings. The operator strokes its keys languidly and looks about the room and sometimes at the speaker with a disquieting gaze. From the machine emerges a typed strip which records, in a phonetically simplified language, what the speaker is supposed to have said. Later this strip is retyped into ordinary language, for in its nascent form it is intelligible only to the initiated. It would be fairly easy to rig a device to operate a stenotype as one talked. In short, if anyone wishes to have his talk directly produce a typed record, all the elements are here. All he needs to do is to take advantage of existing mechanisms—and alter his language.

Our present languages are not well adapted to mechanization. True, digital machines can be made to translate languages, Russian into English, for example. As with their writing of poetry or composition of music, one wonders, not that they do it badly, but that they do it at all. So far, machine translation has not become really useful. But it is improving, and the study that is being devoted to the problem is showing us much about the nature of languages themselves. It is strange that the inventors of universal languages, none of which have ever caught on, have not seized upon the idea of producing one which better fits the technique for transmitting, recording, and modifying speech. The business of communication between men and machines thus is a complex affair. Men's language has grown without reference to machine use, and now, if we try to talk directly to a machine, it will not understand us. Even if we write or type our material, we have to be careful to put it in form that the machine can grasp.

We see a simple example of this in the numbers put on bank checks with magnetic ink, so that machines can sort them. They have a faint resemblance to figures as we ordinarily write them, but to the machine their altered form is entirely clear.

A better example occurs with the digital computers. These can do extraordinary things, but only if they are given explicit and detailed instructions on how to do them. The process of instruction, programming, uses a special language, incomprehensible to the layman, learned by a human operator only after careful study and experience, but lucid and unambiguous to the machine. There are several new languages under development for this purpose of telling digital computers what to do and how to do it. They are in terms of binary numbers, when they enter the machine, for that is the natural language of the computer.

We will not expect our personal machine of the future, our memex, to do the job of the great computers. But we can expect it to do clever things for us in the handling of the mass of data we insert into it. We particularly expect it to learn from its own experience and to refine its own trails. So our means of communication with it merits careful consideration. Usually we will tell it what to do by pushing a button or moving a lever. Pushing just one button will often call up a fairly complicated internally stored set of instructions. This will serve for ordinary use. But it would be nice, and easily arranged for, if the machine would respond also to simple remarks. If Fido will respond to "lie down," the machine ought to respond readily to such a remark as "hold it." This matter of a memex learning from its own experience merits some discussion. A digital machine can now be caused thus to learn. Such machines, for example, can be set up to play checkers with a human opponent. Chess is too much for them, because of its complication, which merely means that it calls for an excessive amount of storage and time, but they do very well at checkers. In fact, they can learn to beat a good player. In the digital machine's memory is stored a large number of positions that may occur in a game, and possible following moves to be used. But positions and moves are rated in accordance with assumed values. Confronted with a position, the machine consults its memory and chooses the best-rated move to use. But now comes the real point. It continually alters the rating of the moves in accordance with its success or failure. If a move results in a more highly valued position, its rating goes up, and if it results in catastrophe, it goes down. In this way the machine learns. Playing at first a very poor game, it finally becomes expert. A memex can be constructed to do similar things. Let's say its master is a mechanical engineer, and that he has a trail which he uses very frequently on the whole subject of heat transfer. The memex notices (we have to use such terms; there are no others) that nearly every time he pursues the trail there are a series of items on which he hardly pauses. It takes them out of the main trail and appends them as a side trail. It also notices that when he comes to a certain item he usually goes off on a side trail, so it proceeds to incorporate this in the main trail.

It can do more than this; it can build trails for its master. Say he suddenly becomes interested in the diffusion of hydrogen through steel at high temperatures, and he has no trail on it. Memex can work when he is not there. So he gives it instructions to search, furnishing the trail codes likely to have pertinent material. All night memex plods on, at ten or more pages a second. Whenever it finds the words "hydrogen" and "diffusion" in the same item, it links that item into a new trail. In the morning its master reviews the new trail, discarding most of the items, and joining the new trail to a pertinent position.

Does this sort of thing sound bizarre or far-fetched? Machines are doing more surprising things than this today. Much needs to occur between the collection of data and observations, the extraction of parallel material from the existing record, and the final insertion of new material into the general body of the common record. For mature thought there is no mechanical substitute. But creative thought and essentially repetitive thought are very different things. For the latter there are already powerful mechanical aids. We shall need still more.

In particular we have delved far enough into the chemical processes by which the human body operates to grasp the fact that we shall never come to full understanding in this enormously complex field until our processes of reasoning have been greatly refined, and divested of all the clutter of repetitive acts which now take up most of the time that we consider we are devoting to thought. For this reason there will come more machines to handle advanced mathematics and manipulation of data for the scientist. Some of them will be sufficiently bizarre to suit the most fastidious connoisseur of the present artifacts of civilization.

The scientist, however, is not the only person who manipulates data and examines the world about him by the use of logical processes, though he sometimes preserves this appearance by adopting into the fold anyone who becomes logical, much in the manner in which a British labor leader is elevated to knighthood. Whenever logical processes of thought are employed—that is, whenever thought for a time runs along an accepted groove—there is an opportunity for the machine. In fact a machine which will manipulate premises in accordance with formal logic has already been constructed. Put a set of premises into such a device and turn the crank; it will readily pass out conclusion after conclusion, all in accordance with logical law, and with no more slips than would be expected of a keyboard adding machine.

Logic can become enormously difficult, and it would undoubtedly be well to produce more assurance in the use of it. The machines for higher analysis have usually been equation solvers. But we now have equation transformers, which will rearrange the relationship expressed by an equation in accordance with strict and rather advanced logic. Progress here is a bit inhibited by the exceedingly crude way in which mathematicians express their relationships. They employ a symbolism which grew like Topsy and has little consistency; a strange fact in that most logical field.

What might be the consequences of the developments we have been discussing? Assuredly they would not be limited to the men of science. It could be hoped that the writing of history and biography, for example, would improve, not just in accuracy, but in art, as the writer is able to turn the drudge part of his task over to a tireless assistant, always willing to work when he is, and never at a loss to divine what he wishes to remember. Wholly new forms of encyclopedias will appear, ready-made with a mesh of associative trails running through them, ready to be dropped into the memex and there amplified. The lawyer will have at his touch the associated opinions and decisions of his whole experience, and of the experience of friends and authorities. The patent attorney will have on call the millions of issued patents, with familiar trails to every point of his client's interest. The physician, puzzled by a patient's reactions, will strike the trail established in studying an earlier similar case, and run rapidly through analogous case histories, with side references to the classics for the pertinent anatomy and histology.

Another area in which new machine accomplishments are needed is organic chemistry. These accomplishments are just beginning to appear. There are millions of organic compounds that have been studied, and an unlimited number of possible ones, many of them no doubt useful. The organic chemist is in a tough spot. His memory is severely taxed, and much of his time is consumed in labor that does not call on his true skills. He ought to be able to turn to a machine with a specification of a compound, in terms of either its form or its properties, and have it immediately before him with all that is known about it. Moreover, if he then proposes a chemical manipulation on such a compound, the machine should tell him, within the limits of knowledge at the time, just what will happen. It would do so by using the known laws of chemistry, and the chemist should turn to experiment in the laboratory only for confirmation, or when entering unexplored territory. We are a long way today from such a situation. But machines can certainly do this, if we build them intelligently and then tell them what to do.

The historian, of whom I have spoken above, with his vast chronological account of a people, can parallel this with a skip-trail which stops only on the salient items; he can follow at any time contemporary trails which lead him all over civilization at a particular epoch. There will be a new profession of trailblazers, those who find delight in the task of establishing useful trails through the enormous mass of the common record. The inheritance from the master will become, not only his additions to the world's record, but for his disciples the entire scaffolding by which they were erected. Each generation will receive from its predecessor, not a conglomerate mass of discrete facts and theories, but an interconnected web which covers all that the race has thus far attained.

When the first article on memex was written, the personal machine, the memex, appeared to be far in the future. It still appears to be in the future, but not so far. Great progress, as we have seen, has been made in the last twenty years on all the elements necessary. Storage has been reduced in size, access has become more rapid. Transistors, video tape, television, high-speed electric circuits, have revolutionized the conditions under which we approach the problem. Except for the one factor of better access to large memories, all we need to do is to put the proper elements together—at reasonable expense—and we will have a memex.

Will we soon have a personal machine for our use? Unfortunately not. First we will no doubt see the mechanization of our libraries, and this itself will take years. Then we will see the group machine, specialized, used by many. This will be especially valuable in medicine, in order that those who minister to our ills may do so in the light of the broad experience of their fellows. Finally, a long time from now, I fear, will come the personal machine. It will be delayed in coming principally by costs, and we know that costs will go down, how much and how rapidly none can tell.

It is worth striving for. Adequately equipped with machines which leave him free to use his primary attribute as a human being—the ability to think creatively and wisely, unencumbered by unworthy tasks—man can face an increasingly complex existence with hope, even with confidence. Presumably man's spirit should be elevated if he can better review his shady past and analyze more completely and objectively his present problems. He has built a civilization so complex that he needs to mechanize his records more fully if he is to push his experiment in its proper paths and not become bogged down when partway home by having overtaxed his limited memory. His excursions may be more enjoyable if he can reacquire the privilege of forgetting the manifold things he does not need to have immediately at hand, with some assurance that he can find them again if they prove important.

The applications of science have built man a well supplied house, and are teaching him to live healthily in it. They have also enabled him to throw masses of people against one another with cruel weapons. They may yet allow him truly to encompass the great record and to grow in the wisdom of race experience. He may perish in conflict before he learns to wield that record for his true good. Yet, in the application of science to the needs and desires of man, this would seem to be a singularly unfortunate stage at which to terminate the process, or to lose hope as to the outcome.

5

Out of File, Out of Mind

Cornelia Vismann

It is well known that almost every administration in the West has been haunted by the professional furor to take notes of everything spoken within its office walls. This practice of total documentation leaves not a single spoken word without a written equivalent. The officials working in these administrations record in order to act, and act only by recording. Administrative execution has, in other words, always meant execution on paper. The phantasmal belief that files can and are meant to record all governmental proceedings and happenings in their entirety has fueled the categorical imperative of Western administrations to make records and keep files. This belief has been fundamental to the administrative practice of recording and filing for at least the last two centuries. Max Weber, the German bureaucracy-expert of the 19th century, transformed this practice into a principle. "The management of the modern office," he wrote, "is based upon written documents (the 'files'), which are preserved in their original or draft form, and upon a staff of subaltern officials and scribes of all sorts."[1] Weber formulated this principle at the very moment when another medium of communication was emerging: the telephone. This new non-script based means of communication threatened the existence of files insofar as it had the potential to usurp extra- and intra-administrative communications from the documentary universe of the written word. To prevent this from happening, record keeping was implemented as a bureaucratic principle. From then on files began to pile up all over—files which historians, far from complaining about the masses of paper, would eventually take as their preferred source. The administrative workers, however, have since been drowning in files. For them records are the monsters they have to do battle with every day.

The imperative of administrations to record every action as an execution on paper causes all kinds of problems. A person, after all, can be held responsible for something on file, something which, according to who is doing the looking, in retrospect should not have been recorded at all. Recently, troubles of that kind became quite acute in Germany when, during the change of government in 1998, state officials of the defeated party sought to evade accountability. They tried to reverse the logic of producing facts by, as in the proverb *quod non est in actis non est in mundo*,[2] making the files, along with their damaging content, physically non-existent. Their hope was that a written document once removed from a file would disappear from memory altogether. Out of file, out of mind, to revise an age old saying. In mistaking the materiality of files for their content, German officials created a void in administrative documentation which did not, of course, go without a political aftermath.

Like those German officials, a farmer in one of the late 19th century Swabian writer Johann Peter Hebel's famous calendar anecdotes called *Prozeß ohne Gesetz* ("Trial without Law") also mistook the physicality of the law for its content. The man was looking for legal advice in the city. After he had told his advocate about a quarrel he was involved in, he learned he had no chance whatsoever; the lawyer pointed to a certain paragraph in one of his law books, which stood against his case. But the farmer did not give in easily and was determined to win his case. When the lawyer was not looking, the farmer tore out the page from the law book on which the law in question was printed. Afterwards, he bribed the lawyer to take up his case, despite the seeming impossibility of winning. But then, quite unexpectedly, he received a favorable verdict. Although this was due to some formal reason (the other side had not appeared in court), the farmer praised himself for having been so smart as to tear out the page with the disturbing law on it—"otherwise he would not have won the case."[3]

The confusion between the material and the hermeneutics of law made Hebel's provincial character a happy man. Urban people, however, are usually not so happy even though they too often misunderstand the logic of disappearence. In the case of the German officials, things seemed more difficult when files instead of legal codes were at stake. As physical entities and as recorded data, files display a rather complicated duality. The murderers of Julius Caesar, for instance, tried to eliminate files in the real in order to extinguish unwanted data within the symbolic. Their aim was not only to have a dead Caesar, but a completely forgotten one. So they were concerned with the question of how to extinguish Caesar's consulate *ex tunc*, that is, how to undo his political deeds from his rise to power up to his death. They tried to achieve this goal by burning his consulate files. How ineffective or only symbolic this case of practiced *condamnatio memoriae* was can easily be measured by the numerous biographies of Caesar which have appeared ever since. Hence despite the file-burning act, the consul was not forgotten in history. But even if the ineffectiveness of destroyed or removed records might prove Caesar's murderers wrong, it does not follow from this that the simple equation between files and the world, between the physicality of storage and the existence of data in the order of signs, is wrong altogether. On the contrary: what is not in the records in the first place can hardly be remembered. So if one does not want an action in the real to become significant, it should certainly not be recorded. Consequently, only harmless data—data that cannot be used against the record keeper him- or herself—will make its way into the files. This kind of screening undermines the Weberian administrative rule that no official action can be performed off the record.

The documentary power of files—their ability, more exactly, to hold someone responsible for his or her action—leads to "the effect that everybody who wants to put something on record will think ahead of time about what and how to formulate it."[4] The formal legal tone of this quotation sounds as if it might be from Max Weber; it was actually uttered by a contemporary lawyer. But the phenomenon of manipulative and selective documentation or self-censorship was already known to Weber. His definition of *bürokratische Verwaltung* reveals an insight into the fatal dialectics between record keeping and data-exclusion: "Bureaucratic administration always tend to exclude the public, to hide its knowledge and action from criticism as well as it can."[5] And the best way to hide knowledge and action is, one could add, not to put it on record in the first place. The "critique" Weber mentions alludes to the period of enlightenment when people were outspokenly critical about the exclusionary tendencies of dynastic record keeping. Apart from their claims to freedom, equality, and fraternity, they also wanted all files open to the public. When these demands were eventually met and archival files were made more or less accessible, however, government secrets did not vanish. The *arcana imperii* changed only their status. A political secret was no longer what was kept in the files, locked away in chanceries or hidden in obscure archives; a secret was exactly that which was off the record.

The shift in the definition of political secrets from a well protected arcanum in the files to total concealment happened around 1900. This is a relatively recent development compared to the far older tradition where writing down was connected not with producing facts (by taking notes), but with erasure. In ancient and medieval times the act of writing was at once production and elimination. If a sign is drawn into a wax tablet, the surface is destroyed and by that act of destruction a readable trace is produced. The object that performs this trace is called a *stilus* in Latin. One end of this pencil is sharp in order to draw or better tear lines into the wax tablets, the other end is round in order to erase or rather cancel those lines. This object for a concerted writing-canceling operation, literally the act of *stilum vertere*, led eventually to the chancery, an institution designed to do nothing but produce signs and cancel them.[6] Here cancellation has to be taken in its most literal sense, a crossing out of writing. The task of the chancery workers was to copy and then cross out drafts of official papers, thereby rendering the original useless while indicating that a copy had been made.

Those erasures did not effect total disappearance or purgation of what was written, but left traces. These traces of a performed erasure were exactly what medieval chanceries were so keen on. As agencies of deconstruction avant la lettre, the chanceries did not only accept the trace of erasure as an unavoidable side-effect within the production of official paperwork, they established it as an indispensable step of the whole procedure: No document was to leave the chancery without the draft having been made unreadable by a gridwise deletion. The act of canceling, implemented to prevent unauthorized copying, was so prominent that the chancery even got its name from this act, the word being etymologically derived from "cancel" or "cancellation."

This connection between cancellation and chanceries faded eventually. It was already forgotten when the Grimms defined that institution in their famous dictionary. Writing and its storage effects dominated the once powerful corresponding act of canceling. Hence a chancery is defined in Grimm's dictionary merely as the place for "execution in writing."[7] This definition echoes the practice of what Weber called *Aktenmäßigkeit*, the principle that management is based upon files. At the beginning of the 19th century, writing, storage and legibility had already formed a stable relationship in which cancellation had no place, showing that cancellation was at that time simply no longer an integral cultural practice. The only reminder today of those grid-like cancellation marks is a rather unspectacular sign without any operative function, the rhomb or cancel character (#), that is used for "number." Starting in 1928, IBM reserved a place for this sign on its punch cards. When these punch cards were then used as a basis for the ASCII-code, this symbol became incorporated into the modern computer key system.

The decline of the once-powerful institution of cancellation was initiated by a seemingly minor but nevertheless far-reaching change in the procedure of making chancery letters. Starting around 1500, drafts were no longer cancelled, but stored in a legible state. When cancelled letters subsequently appeared, they were handled with deep suspicion, a possible hint of forgery. Cancellations then became an object of meticulous regulation. A statute of the chancery of Maximilian I from 1497 required, for example, that erasures should only be made if a high-ranking secretary had allowed them. The allowance had of course to be documented, otherwise the authenticity of a document with traces of a cancellation would have remained questionable. Script cancellation thus changed from being a sign of truth's guarantee to being its own worst enemy. At the beginning of the 20th century, when a Berlin-Frankfurt rubber factory baptized one of its products, an eraser, "veritas," it revealed a good sense of historical irony.

By canceling the act of cancellation, the institution which gained its power through this act likewise lost its importance to Western administrations. In its place arose the governmental obsession with total documentation and with it the increasing importance of the archive. This preoccupation with the archive reached its peak around 1800, when the archive was considered to be the "soul" or the "memory" of the Prussian State. From then on it was only a question of time until those storage-institutions would cause severe problems in the real, as the physics of files took on monstrous proportions. Symbolic actions of elimination were not able to cope with the masses of paper in the real. Considering the domination of storage over cancellation, it sounds like a desperate appeal when the expert on administration and system theorist, Niklas Luhmann, emphasizes the obliterative power of files in contrast to their storage function: "By the phenomenology of files one could get the impression that the system could hardly be moved under the weight of its history. But one would fail to notice that files not only 'organize' memory but also the act of forgetting."[8] Even if files would implement collective forgetting, however, the gravitational forces of the colossal masses of paper still clearly pull towards a total recording for the purposes of memory.

The dogma of complete documentation and the tendency towards more and more detailed reports led to the well-known proliferation of files. Archivists are not, after all, c(h)ancellors. Even when the storage weight became unbearable and the data became outdated before it could be used, archivists were not even trained as file-eliminators. The only way to liberate the world from the crushing weight of files became their material destruction. A virtual third institution, after chancery and archive, emerged: that of wastepaper. A Prussian statute dated 1876 accordingly instructs the record-keeping agencies to divide their files into three sections: those in use for the current administration; those to go to the archives, because they are historically valuable; and finally, those records of no further use.

Files designated as wastepaper, it seems, slip through all categories and definitions. Whereas chanceries clearly operate in the symbolic, and archives could be perceived as borderline-institutions of the imaginary, wastepaper dealers perform their work of destruction neither truly in the imaginary realm nor truly in the symbolic. The act, which operates in the shadow of the symbolic order, crosses instead the threshold into the real. There is no language for what is expelled from the symbolic. The deed that follows after the act of cassation at the interface between thesaurus and trash, between archive and garbage, falls into the void of the symbolic order. It is neither integrated into the current administration, nor does it belong to the order of the archive. Consequently there exists no order, rule, or instruction for the destruction of files. After the order to destroy, the mode of destruction itself is never prescribed. Whereas there exist statutes over statutes within Germany for the handling of files in the use of administrations and in the archive, the handling of files which are categorized "out of use" is not itself regulated. One does simply not find any how-to instructions for the destruction of files.

Because of their non-discursive state, files separated out from the rest are destined to disappear. However, as long as they have not been shredded, they can be reintegrated into the symbolic order. The return of the repressed, so to speak, can be taken literally here. Abandoned files introduced into the order of the symbolic process information again. For that reason, the already mentioned Prussian statute was so providential to require an official certificate, quite exemplary for contemporary data protection, to confirm the complete elimination of the files designated to be turned to pulp. The measure was taken in order to prevent all kinds of misuse—if not "pulp fictions"—with the files.

More recently a new problem with files has emerged: the problem of uncontrolled record-deletions. This is a problem that can logically only arise after rules for storage and deletion have been put in place, which can be broken, as soon as a person wants to get rid of unpleasant information stored in the files. In this circumstance, the old question of what can disappear from the world once it is on record becomes current again. When the Ministry for Official Security (*Ministerium für Staatssicherheit*, MfS) of the former GDR was confronted with the question of how to let thousands of files disappear without attracting any attention, it chose—among others—the most simple of all possible methods: the method of purloined letters. The MfS put the potentially damaging files in a place where surely nobody would take notice of them—out in the garbage on the streets.

But files which are removed from secret offices to public streets are not automatically out of the world, as people often tend to think. The "illusion of disappearance" that Michel Cahn wants to shatter by the model of ecological recycling is destroyed when the waste is processed into information again.[9] As the manager for a company that offers "File destruction in a van on the spot" has remarked, "As you know, data scandals, where confidential files are found in the streets, are [...]

almost usual." This quote can be found printed out on a full-length page advertisement that not accidentally adorns the back page of the official 1992 German zip-code directory published by the German postal service. At that time, the reorganization of zip-codes after the German reunification had made formulas, questionnaires and printed stationary invalid. Document-destruction in a van came in handy when trying to prevent the misuse of this material.

A door-to-door paper shredding service is a rather late development in the modes of paper-destruction. The shift from manual to mechanized destruction around 1900 made the wastepaper dealer, the emblematic figure of Baudelaire's *chiffoniers* or Benjamin's *Lumpensammler*, obsolete. It is interesting to note that paper shredders neither are mentioned in manuals about office-equipment nor is their origin recorded in histories of the office or bureaucracy. This lacuna only affirms its fall out of the symbolic.

Paper shredders probably did not come into German public consciousness until after the re-unification process. Starting in November 1989, the so-called *Reißwölfe* (literally: tearing wolves) went into full speed on Normannenstraße in Berlin where the MfS was located. It was reported that the shredders soon became overtaxed and were therefore hastily replaced by models from West Berlin. After a short period of time the Western "tearing wolves" were also overtaxed. The newspapers reported that they had "overeaten" themselves, as if the files had been a fat prey and shredders wild beasts. Indeed the animal ancestors of the mechanical wolves received their name from their undomesticated tendency to tear apart all kinds of edible beings. Wolves derive their name from the German verb *wolfen* which means "to tear." They are, as an etymological dictionary defines them, "tearing animals." If it is true that *wolfen* and *reißen* both mean tear, then *Reißwölfe* seems to be a mere doubling that emphasizes the act of tearing through repetition. An explanation for this redundancy could be that it serves as a differentiation-device: wolves (*Wölfe*) as tearing animals can thus be differentiated from wolves as tearing machines (*Reißwölfe*).

The English notion for *Reißwölfe* underlines their tearing forces quite well. They are called "devils." According to Webster's *Third New International Dictionary*, they are understood to be "any of various machines, appliances, or devices: as [...] a machine for tearing or shredding something or for grinding material into bits (as stock for papermaking, woollen for shoddy, or fur for felt)." The task of the first devils was not to tear paper, but to break down cotton and lambswool to produce yarn at the end of the 19th century. The devilish machines of the industrial age had to grind the material down and tear it up into a single fiber.[10] The fibers were torn up six times and unified again six times, until six to the power of six fibers were produced. In contrast to these rather complex machines for constructing and destroying, the modern file shredders were just designed to destroy. From their deconstructive mechanical ancestors they adopted only their destructive side. Paper shredders tear and do nothing else. It therefore follows that the machine employed under the German "verbrannte Erde" (scorched earth) policy in 1944 and 1945 would be named *Schienenreißwolf*, "tearing wolf for rails." It seemed to function like a normal train, but in its wake left a trail of ripped-up rails, devastating the terrain.

Whether shredded by machine or by hand, the elimination of files is dirty work. In contrast to digitalized data storage, paper files cannot be eliminated by clean delete orders. By comparing the advantages and disadvantages of hardware and software deletion, the staff of the chancellor's office in Germany became media experts almost overnight, when they became aware of the differences between the clumsy materiality of paper files and the seemingly trouble-free destructibility of digitalized data. When confronted with the question of how to erase government information, especially data on the political unification process before the new government was to take over, officials of the chancellor's office discovered files in quite a different light. As with all shifts in power, the election of the Social Democrats as the majority party in the fall of 1998 was a delicate moment. At issue was what should be transferred to the next legislation and what should, for political and strategic reasons, best be withheld.

As the integral force for continuity and tradition in power, files had for centuries played key roles when rulers have wanted to cut dynastic threads. There have been times, like those of the medieval German kings, when regimes tried to avert a succession by stealing files and taking them hostage; there have been times when files were burned in order to interrupt the continuity of power or when files were simply stolen in order to wage war against another country. It was not until the 19th century that legislation for storage and keeping records tried to prevent those modes of bureaucratic sabotage. But despite those regulations for keeping governmental files,[11] hardly a single byte produced and stored in the *Bundeskanzleramt*, the German chancellor's headquarters, was transferred to the newly elected government in 1998. And that was quite a significant loss, since the missing data dealt with the agency of government management (*Geschäftsführung der Leitung*). This agency for coordinating the government can be compared to a search engine for the entire state file archive.[12] So when the agency's complete files were deleted, material spared elsewhere from the orgy of destruction was as useful as library books in the right place on the shelf but without an index system.

Almost two years later, when the new government attempted to call up files from the former chancellor as part of the investigations into the above mentioned CDU Party financial affair (in the context of the reunification process), the missing files became evident. The cleaning house of the federal chancellor's office subsequently turned into the subject of an investigation. A committee was given the task of determining what actually happened during those "days of the tearing wolves."[13] That this was a political scandal is probably more obvious within the American legal tradition, where public access to government files has been granted in the Freedom of Information Act, than it was at first in the German one. Behind this right stands the idea that files belong to the public, an idea still foreign to German jurisdiction and legislation. No statute comparable to the Freedom of Information Act exists in the German law system, although there are signs that the dogma of administrative and governmental files are not subject to any kind of public interest is going to be changed. Statutes for the accessibility of state files are in preparation. But presently government and administrative files are still regarded by German law as neither private nor public property. They simply "belong" to the State.

The practice of understanding files as state property began in the Roman Empire, 78 A.D. At that time, Rome had expropriated the file keepers such as magistrates. With that coup, the Imperium Romanum had laid the groundwork for the filing-monopoly of the state and, along with it, the monopoly of the State itself. But the de-privatisation or rather expropriation of files only opens another battle over the private/public distinction. Ministers and other public service employees take official notes and also private ones—a fact of which the ministers of Helmut Kohl's cabinet, charged with withholding official files, took advantage. They claimed that the files in question were in their personal belonging, designed for private use only. To complicate the distinction between private and public even more, they invoked a category sometimes applied by archivists: the "private-official," das *Privatdienstliche*, something which can also be taken as an interesting contribution to the debate over the public/private sphere. Even the beloved German institution of *geistiges Eigentum*, intellectual property, was mentioned by the ministers as an argument in defense of their total autonomous power over the files against the charge of having breached paragraph 133 of the German Criminal Code (*Verwahrungsbruch*) according to which they are legally responsible for safeguarding the files.

Another line of defense was also tried, this time one taking advantage of existing uncertainties over handling non-paper files. The officials accused with repressing official documents claimed that paper-files and computer files have to be handled differently, arguing that the rules for keeping conventional records are not applicable to digital ones. With that argument a sensible gap in legal regulations was found and dramatized. It is true that there are hardly any explicit norms for electronic record-keeping that respect its peculiarities while recognizing its similarities with "classical" record-keeping. The government is therefore at present preparing to reform the whole field.

According to current law, most of the administrative regulations on how to handle and keep files[14] have to be made applicable to digitalized data via analogies.

The chairman's report of the committee supposed to search for the missing files of the former chancellor, which was released in July 2000, gives insight into the differences between paper files and computer files with respect to data elimination. The staff of the chancellor's office, for example, refrained from destroying physical files because it would, in its own words, attract too much attention. The concrete existence of files is a fact which cannot disappear from the world as easily as digitalized ones. The report also shed light on the startling fact that everything, even insignificant information, was destroyed. As the German official in charge of classified materials (*Geheimschutzbeauftragte*) affirmed, the previous government did not select incriminating data in particular. He explained bluntly that he had agreed in those days of decision and destruction with the chief of the central administration of the chancellor's department that the files should vanish indiscriminately. He is quoted as saying, "in order to search pointedly for incriminating material, too many members of the staff would have had to be employed, so that in the end [the task] would not have been manageable at all. Therefore general deletion was the only solution."[15]

However, the findings of the committee's investigation of the missing files in 1998 were not as negative as one might expect. Quite a lot of the lost information could be retrieved. First, this was made possible by finding the voids, which was not an easy task. A lost file, after all, can only be discovered if there exists a hint that something is missing. In order to detect a gap in a stack of records, it is necessary to combine the real and the symbolic. A retrieval system for files such as index cards or a registration of some kind serves as a reference between the two universes. So even if files are destroyed, the signifier of the destruction still exists and reveals the loss—unless it is destroyed itself. In an administration becoming more and more interlinked, that kind of total elimination seems less and less likely. At least records kept in parallel files point at what is missing. In order to prevent this from happening, the chancellor's headquarters reduced computer links from the beginning. Like all powerful institutions, it worked on the basis of total asymmetry: nobody had access to the headquarters computer system, but all governmental computer systems ran through the headquarters system. By that structure, the chancellor's headquarters gained optimal control over file links within the different offices and retained autonomy in cancelling the information stored only in its own system .

The findings of the filing search committee reported missing files and index-card systems, yet in most cases, copies existed so the allegedly lost information could be recovered. In terms of computer files, nearly all data of the period in question were deleted. But the committee found 99 backup files in the chancellor's office with almost 1 million specific data files. The committee differentiated between lost paper files and lost computer files, thereby unwillingly affirming the line of defense of the "suspects"[16] mentioned above. But the difference they made had no strategic purpose. It was instead tied to the logic of the respective recovery possibilities of conventional and computer files. One could almost make a correlation between the simplicity of elimination and the chances of retrieval: the less manual work involved in the act of elimination, the higher the chances of restoring the information and vice versa. The classical paper files, which exist in shredded form, are usually restored by applying complicated archaeological practices. For the recovery of "Stasi-files," special methods were developed which adopted and refined the archaeological work of combining the puzzles of ancient broken pottery. The edges of torn paper documents, for example, were scanned horizontally and vertically using a computer so that the two corresponding sides, the positive and negative, could be found and attached again, as in the use of a symbolon,[17] which is taken literally here.

With respect to deleted electronic data, the recovery process is less tedious, at least if they are deleted in a so-called salvage mode, known to ordinary computer users as the delete key. In that case, the cancelled information can be made readable again with some technical effort—this was the case with the first deletions in the chancellor's headquarters. Later data was purified in saver mode, the so-called purge mode. With that mode of deletion, the information is irretrievably lost. Although the virtual bonfires in the purge mode are far from a medieval purgatorio—they do not leave ashes and smoke behind—the activation of this mode seems, even in modern rational times, to be a prerequisite for salvation. Deliverance means—in a bureaucratic as well as in a religious context—forgetting or deleting what has been recorded in the files or accordingly in the book of life. In other and more profane words: out of file, out of the world contains a promise of salvation, however unfullfilable it will be in the end.

Notes

1. Max Weber, *Economy and Society. An Outline of Interpretative Sociology*, vol. 2 (Berkeley / Los Angeles / London: University of California Press, 1978): 957.
2. For details on this proverb as well as on files and filing-systems in general see Cornelia Vismann, *Akten. Medientechnik und Recht* (Frankfurt a. Main: Fischer Verlag, 2000) forthcoming as *Files: Mediatechnique and the Law* from Stanford University Press.
3. See Cornelia Vismann, "Von der Poesie im Recht oder vom Recht in der Dichtung–Franz Kafka und Johann Peter Hebel." In: *Fremdheit und Vertrautheit. Hermeneutik im europäischen Kontext*, edited by Rainer Enskat and Hendrik J. Adriaanse (Leuven: Peeters, 2000): 275–282
4. Thomas-Michael Seibert, "Aktenanalysen. Zur Schriftform juristischer Deutungen." In: *Kommunikation und Institution*, vol. 3 (Tübingen: Narr, 1980): 34.
5. Weber, *Economy and Society*, 992.
6. For this and the following paragraph with further reference see Cornelia Vismann, "Cancels: On the Making of Law in Chanceries." In: *Law and Critique*, 7:2 (1996): 131–151.
7. Jacob und Wilhelm Grimm, "Kanzlei." In: *Deutsches Wörterbuch*, Bd.11 (München: Deutscher Taschenbuch Verlag, 1984 [Reprint of the Edition Leipzig 1873].
8. Niklas Luhmann, *Organisation und Entscheidung* (Opladen: Westdeutscher Verlag, 2000): 160.
9. Michael Cahn, "Das Schwanken zwischen Abfall und Wert. Zur kulturellen Hermeneutik des Sammlers." In: *Merkur* 45 (1991): 674–690.
10. For a good description of "Reißwolf" see *Meyers Großes Konversations-Lexikon*, 6th ed (Wien / Leipzig: Bibliographisches Institut, 1909), vol. 16.
11. *Gemeinsame Geschäftsordnung der Bundesregierung* (Common order for the management of the federal government), annex: *Registratur-Richtlinie* (Registration-guideline); *Bundesarchivgesetz* (Federal archive law for storage of closed records).
12. Thomas Kleine-Brockhoff and Bruno Schirra, "Operation Löschtaste." In: *Die Zeit*, 20 July 2000, 3.
13. Their task was limited to the investigation according to the federal disciplinary statutes (Bundesdisziplinarordnung). Since this law applies only for lower ranking staff, the investigation committee had no authority to question the chancellor himself or other high ranking officials who actually ordered the elimination of the material. It became thus an investigation into the shadowy world of elimination at work behind the scenes of political responsibility.
14. Kleine-Brockhoff and Schirra, "Operation Löschtaste."
15. Confidential annex to the "Hirsch-Report," witness hearing minutes, quoted in Thomas Kleine-Brockhoff and Bruno Schirra, "Operation Löschtaste."
16. They are not suspects in the strict legal sense, because the committee did not do the work of a state attorney.
17. See Michel Foucault, "La vérité et les formes juridiques." In: *Dits et écrits*, vol. 2 (1970–1975) (Paris: Gallimard, 1994): 558–623.

6

Dis/continuities

Does the Archive Become Metaphorical in Multi-Media Space?

Wolfgang Ernst

In this paper, I address (multi) media archaeology in two parts: first, an epistemological reflection on the term "media archaeology" and second, literal case studies. But, before I begin (*arché*), I want to reflect on the term "archaeology of multi-media" itself. Having been trained as a historian, a classicist and an archaeologist (in the disciplinary sense), I have always felt uneasy with the predominance of narrative as the uni-medium of processing our knowledge of the past. Theoretically, works like Michel Foucault's *L'Archéologie du Savoir*[1] and Hayden White's seminal *Metahistory*[2] have helped me express this unease with the rhetoric of historical imagination. It took, however, a new infrastructure of communicating realities—the impact of digital media—to put this critique of historical discourse into media-archaeological terms and practice. But caution: Even when we claim to perform media-archaeological analysis, we easily slip back into telling media stories.

The archaeology of knowledge, as we have learned from Foucault, deals with discontinuities, gaps and absences, silence and ruptures, in opposition to historical discourse, which privileges the notion of continuity in order to re-affirm the possibility of subjectivity. "Archives are less concerned with memory than with the necessity to discard, erase, eliminate."[3] Whereas historiography is founded on teleology and narrative closure, the archive is discontinuous, ruptured. Like all kinds of data banks, "it forms relationships not on the basis of causes and effects, but through networks"; instead of being a medium of cathartic memory, "the archive is traumatic, testimony not to a successful encounter with the past but to what Jacques Lacan has referred to as the 'missed encounter with the real'"[4]—that is, an allegory of the impossible bridging of a gap.

Archaeology, as used by Foucault in a somewhat playful, delusory way, is a term that does *not* imply the search for a beginning; it does *not* relate analysis to a kind of geological excavation. Thus it differs substantially from what the *Oxford English Dictionary* defines as archaeology: "indicating the material or substance of which anything is made or consists." So what happens if we apply this Foucauldian term to the genealogy of media?

Part I: An Epistemological Reflection on the Term "Media Archaeology" Pre-Histories of the Computer?

So, how does media *archaeology* differ from media *history*?

To answer with an anecdote: Hewlett-Packard has now acquired the garage on which the company based its advertising campaign *The Garage Principle*. This garage is the primal hut of Silicon Valley where, in 1939, Bill Hewlett and David Packard began constructing technical apparatuses, out of which emerged the Eldorado of microchips. This garage is now listed, under the number 976, as a monument of American heritage (inventories count memory, rather than narrate it). The tragedy of this media monument is that, while the garage has survived, the first technical instruments produced by these pioneers have not.[5] That is, the empty frame remains, but the more tricky technological artifacts, which are always just temporary configurations and not tightly coupled things, are lost.[6] This difficulty culminates in the fragile endurance of computer programs, which only recently have become the objects of archives.[7] Media archaeology describes the non-discursive practices specified in the elements of the techno-cultural archive, without simply reducing the archive to its technical apparatuses. Media archaeology is confronted with Cartesian objects, which are mathematizable things,[8] and let us not forget that Alan Turing conceived the computer in 1937 basically as a *paper machine* (the most classical archival carrier), not necessarily dependent on its electronic implementation (this is a question of speed in calculating).

The so-called *8-Bit Museum*, the homepage for 8-bit computers and video games, is an example of the computer-based Internet developing an archive of its own genealogy (an unbroken lineage so far), reminding us of the wonderful archaeological époque of the 8-bit computer when "computer" did not automatically equal "Windows-PC":

> In this mythical time before the MByte had been invented, interaction with the computer was somewhat different from today. Valiant users fought through endless listings to glean a few tricks from others, one wrestled mercilessly for every single byte, programs were relentlessly optimized until they could be run even on a 1MHz chip, tragedies unfolded when a cassette with important data stubbornly signalled ?LOAD ERROR, and in general, fighting the computer was not always easy.[9]

Media archaeology is not only about re-discovering the losers in media history for a kind of Benjaminian messianic redemption. Media archaeology is driven by something like a certain German obsession with approaching media in terms of their logical structure (informatics) on the one hand and their hardware (physics) on the other, as opposed to British and U.S. cultural studies, which analyze the subjective effects of media, such as the patriarchal obsession with world-wide order and hierarchies in current hypertext programming languages as opposed to digital options of—female?—fluidity.[10]

"There are no archives for computer games."[11] The real multi-media archive is the *arché* of its source codes; multi-media archaeology is storage and re-reading and re-writing of such programs.[12] As opposed to the copyright on software programs, which extends for 75 years in the U.S., software piracy successively creates a kind of anarchical archive, an *anarchive* of otherwise abandoned software as cultural evidence. Media *history* is not the appropriate medium to confront such an archive and to perform such a re-reading and re-writing. Media history seeks to privilege continuities instead of counting with discontinuities, since any implicit narrative, which is always a linguistic operation, permanently produces connections between heterogeneous parts.

Consider, for example, two examples in current media research: *Renaissance Computers*, edited by Neil Rhodes and Jonathan Sawday, and a Frankfurt *Literaturhaus* conference called *Book Machines*. *Renaissance Computers* expressly draws a parallel between the media revolution from manuscripts to printing in Europe enabled by Johann Gutenberg in 1455 and Martin Luther's use of printed text for the distribution of Protestant messages (*theses*) in 1517, and the present digital technology era. The symbolic machines of the sixteenth-century "methodizer" Peter Ramus (Pierre de la Ramée) are presented as a *pendant* to the computer of today, and they claim there exists "an indisputable resemblance between the effects of the printing press and those of the computer... in the increased volume of information."[13] This claim still thinks media from the vantage point of alphabetical texts, but audio-visual data banks make all the difference. The authors want to "explore the technology of the early printed text to reveal how many of the functions and effects of the modern computer were imagined, anticipated, or even sought after long before the invention of modern digital computing technology,"[14] but computing is not about imagination and texts, but rather the alliance of engineering and mathematics. Here, a well known historiographic trope (*synekdoché*) lurks around the corner: the desire of occidental man to privilege continuity against the experience of ruptures, thus saving the possibility of an unbroken biographical experience. Against such analogies, however, media archaeology insists on differences. In this context, this means highlighting the fact that the Renaissance *ars combinatorial*, unlike the universal discrete machine named the computer, was not able to calculate on its own, even less store data in random access memories or registers. The coupling of machine and mathematics that enables computers occurs as a mathematization of machine, not as a mechanization of mathematics. While the book has, for half a millennium, been the dominant medium of storing and transmitting knowledge, the computer is able, for the first time, to *process* data as well. What separates technological chance in the fifteenth century from the digital époque is the computer's genesis in World War II, driven by the need for fast number crunching; the difference is between the symbolic (in Lacan's sense: writing, letters) and the mathematical real (computing).

In 1999, Frankfurt *Literaturhaus* organized a conference on *Book machines* (a term coined by Thomas Hettche). On this occasion, the media archaeologist Friedrich Kittler pointed out the differences rather than the continuities between memory media: he argued that analogue broadcast media, which are linear-sequential and base their storage on the principle of the tape, should be afraid, for they will be swallowed by the Internet. According Kittler, books, however, share with the computer "the deep quality of being discrete media." Both are combinatoric machines; the only difference is that books are resident memories, while the computer can automatically read and write.[15] On the Internet texts are, for a while, not falling silent, which is why "Internet archaeology" is necessary (Denis Scheck). But who is responsible for this kind of documentation? Classical archives and libraries do this kind of documentation only exceptionally; for the new kind of memory there are not fixed *lieux de mémoire* any more, not in the sense of institutions, but rather rhizomes within the net itself. While the stability of memory and tradition was formerly guaranteed by the printed text, dynamic hypertexts—the textual form of the Internet—will turn memory itself into an ephemeral, *passing* drama.

A Forerunner of the Internet?

The historian of science Rolf Sachsse describes Wilhelm Ostwald and his "organisation of organisers" (*Die Brücke* in Munich between 1911 and 1914) as a "multi-mediatic" forerunner of the Internet.[16] So too does Jonathan Sawday, when he asks if our contemporary "idea" of the "net" or "web" was "foreshadowed in the Renaissance, at least as a conception."[17] Does this imply a history of ideas instead of media archaeology? But how can media of the past be addressed? Narratively or by discrete alphanumeric ciphering, such as signatures of documents and objects? These questions are tricky because the answers themselves depend on the very agencies being thematized: the archive, the library and the technical museum. Whatever will be said has already passed a process of selection, transport, inventorization and storage according to classification, a signal processing circle best described in terms of cybernetics and information theory. Significantly, the archaeologist of knowledge itself, Michel Foucault, made the signal-to-noise ratio—the relation between message and noise—the subject of a talk in 1966, reminding us that "Freud a fait des énoncés verbaux des malades, consideérés jusque'là comme bruit, quelque chose qui devait être traité comme un message."[18] Some of Foucault's own talks have been recorded on tape. In this audio-archive, the signal-to-noise ratio enters the memory of Foucault itself—a kind of techno-corpse with Foucault's recorded voice, which conveys both message and noise because of material corruption. We are dealing with what history calls *tradition* in the sense of transmission of signals, which the media archaeologist sometimes can decipher from noise only when technical filters are applied. At this point, media archaeology replaces philology as the art of deciphering texts.

Sven Spieker (University of California, Santa Barbara) recalls the link between the media archives of the early 20th century avant-garde and its contemporary, the emerging science of psychoanalysis, a connection theorized by Benjamin in his conception of "the optical unconscious." In Benjamin's conception, imaging media are archaeologists of images that could otherwise never be seen by the human eye (ranging from telescope to radiological scans). The unconscious archive, though, is rather close to the computer, as defined by Jacques Lacan ("ça compte," rather than "raconte"):

> The Freudian unconscious . . . must (also) be understood as a media theory whose centerpiece, the "psychical apparatus," belongs in the same context as other storage media, such as the camera (to which Freud often compared the psyche) or cybernetics (Lacan). Significantly, the Freudian archive-unconscious is capable of storage only to the extent that it crosses out or makes illegible the signatures on other objects stored in its archive, which means that the unconscious is not a machine for remembering but, rather, a machine that continuously erases previous entries in order to replenish its storage capacity.[19]

Multi-Media?

When using the term *multi-media*, we have to remember that we are already victim to a discourse inaugurated by the Microsoft Corporation when it started to release its *Windows* aesthetics. Multi-media describes the way or method of production, the forms of its transport, not its object or content.[20] While a printed letter can only carry the meaning of one phonetic unit, one byte can encode 256 different textual, acoustic or visual options.[21] The term multi-media is thus an interfacial betrayal on the computer screen: in digital space: the difference between the aesthetic regimes only exists for the human user, simulating the audio-visual human senses *under one surface*. A close reading of the computer as medium, though, reveals that there is no multi-media in virtual space, only one medium, which basically calculates images, words, sounds indifferently, since it is able to *emulate* all other media. The term multi-media is a delusion. By flattening the difference between print, sound and image and technically sending them in one standard channel only, such as the telephone line (a sequential operation that separates this procedure from spatial bundling), the computer makes these data accessible almost instantaneously. It effaces the resistance to access characteristic of the traditional archive thus far, though in practice there is still delay, caused by a multi-medial multiplication of data transfer resulting in traffic jams. With RealVideo and RealAudio, for example, delayed transfer, which is "tradition" (in Jack Goody's terms) in the age of print, is substituted by the Asynchronous Transfer Mode (ATM), media-archaeological discontinuity in its most technical sense. While we see one part the video on screen, the next part is already loaded in the background—a coupling of storage and transfer in realtime, a flooding of the World Wide Web by the archive itself.

How can the notion of multi-media be applied to the cultural technology of archiving? As in traditional culture, multi-media first requires archival space, a large storage space like an optical disk for audiovisual data to be kept for processing.[22] But multi-media is not just the extension of the textual archive; hyperlinkability, the very virtuality of multi-media as defined by Ted Nelson, involves the interconnectivity of different media. This option is blurred by the notion of hyper*text*, which just extends what every academic text already does by connecting the textual flow with the apparatus of footnotes. HTML as a protocol means more than just texts.[23] As Nelson says of Vannevar Bush's 1945 design of an associative, micro-film based memory machine, the famous *Memory Extender (MEMEX)*: "Bush rejected indexing and discussed instead new forms of interwoven documents."[24]

Importantly, Nelson coined the term *docuverse*, which in a way is responsible for the iconic desktop metaphor of current *Windows* interfaces, and which rather than instigating a genuinely media-archaeological thinking of the computer, prolongs the metaphor of archival spatial order. The German media scientist Hartmut Winkler made Nelson's term the basis of his computer-archaeological book *Docuverse*, which took for granted the language-based structure of the Internet. He wrote this a few years before the *pictorial turn* in the Internet took place, a turn made technologically possible by data compression algorithms and broadband-transmission of real audio and real video (streaming). Significantly, downloaded images generated by web-cams are no longer called an *archive* (a term which belongs to paper-based memory), but a *gallery* (the visual realm). That is why the U.S. visionary of digital architectures, David Gelernter, points towards the data flow (*lifestream*) as a future alternative to the current desktop-metaphor of present interfaces that still carry, with file-like icons, an anachronistic archivism dating from old-European times of secretaries and offices, instead of rethinking digital storage space in its own terms. Temporal dynamics will thus replace spatial metaphors and catachrestic uses of terms from architecture. A media archaeology of the file has recently been written by Cornelia Vismann:[25]

> This archaeology of law is at one end framed by predecessors of files like the administrative lists in Babylon, at the other end by file-like text administrating systems in computer programs. There it becomes evident that filing technologies have always been the prehistory of the computer as well, which with its *stacks*, *files* and *registers* inherits diverse occidental administration practices.[26]

Emphatic memory (on hard-disks) in Gelernter's scenario is being replaced by a future of the computer as a place of intermediary, passing storage: "The Lifestreams system treats your own private computer as a mere temporary holding tank for data, not as a permanent file cabinet."[27] Future, present and past are but segments, functions of marking differences within a data stream which is time-based rather than space-based.

Fahrenheit 451

An interruption to remind you of another utopia, a film classic which has been probably prematurely classified as science fiction, François Truffaut's *Fahrenheit 451*. In it, a new medium—film and its techno-allegorical other, TV—takes the burning of its mediatic predecessor, the book, as its object. And indeed, the light points of digital signals on the screen literally efface the classic book format as the dominant storage medium.

Another key element defining multi-media, namely interaction, is an aspect Bertholt Brecht highlighted in the 1920s for the emerging medium radio, insisting that it can be used bi-directionally, rather than only being broadcast unilaterally.[28] The unidirectional communication of books still dominated the user experience. The computer, through its possibilities for interactivity, "play" and the creativity of hypertext, is now rapidly undoing that idealization of stability, and returning us to a kind of textuality that may have more in common with the pre-print era. Thus, Vincent Gillespie has argued that the contemporary user's experience of hypertext "... seems ... to be similar to a medieval reader's experience of illuminated, illustrated and glossed manuscripts containing different hierarchies of material that can be accessed in various ways."[29] With different *hierarchives*, a network is not a text any more, rather an archi(ve)-tecture. As long as the keyboard of computers is alphabet-based like a typewriter, the paradigm of printing remains dominant; progressively, though, the mouse-click is replacing the key-stroke as the means of directing the monitor, and the orientation is shifting to visually-perceived information landscapes.

The fundamental difference, though, between a classical print-based archive and multi-media storage is interaction—which at the same time increases the memory capacities of the user, in contrast to just reading or looking at things and commemorating them. The traditional archive has, so far, been a *read only memory*—printed texts reproduced through inscription, not rewritten by reading (a concept still maintained by the CD-ROM). In multi-media space, however, the act of reading, that is the act of re-activating the archive, can be dynamically coupled with feedback.[30]

In multi-media space, sound and images can be shifted, cut, stored, and re-loaded as in word-processing software. Thus, the archival regime is being extended from text to audiovisual data. At the same time, however, and as a kind of revenge by audiovisual data for being subjected to texts, this extension changes and dissolves the very nature of the archival regime. Consider, for instance, the necessity of compressing digital video streams in order to make them storable and transmittable. While in occidental tradition every letter counts in the transmission of an archived text—which is the lot of a whole discipline called philology—by compressing and decompressing digital images subtle amounts of data are lost. This might be almost undetectable to the weak human eye, an organ that has been deceived in its perception since the origin of time-based media like film, but in the world of military target calculations this one bit of absence or difference might lead to fatal errors. Multi-media, then, is for human eyes only.

The Relation between Print and Multi-Media

The usual vantage point from which we talk about the archive—at least from a European cultural point of view—is still the notion of the print-based, paper-formatted archive. The media-archaeological task, then, is to re-think archival terminology in order to embrace a multi-media concept of the archive. The book belongs to the first external memory devices through which culture as memory-based has been made possible,[31] but the book now has lost its privilege as the dominant external memory of alphabetic knowledge. Europa is still book-, that is library- and archive-base-fixated; in contrast, the media cultures in the U.S. have already developed a culture of permanently recycling data, rather than eternally fixed memories.

While traditionally the archive has institutionally, and even legally, sealed off a data bank from immediate access, "there is no ending online. There's no closure, no linear basis. It's about bringing it in, checking it out, constantly evaluating."[32] Thus, the archival media memory is de-monumentalized, just as Erasmus perceived when he put together his *Adages*: "I could add things even during the printing, if anything came to hand which should not be left out"—mobile letters. But then, Sawday's comment falls back on a media-historical analogy, which is inherently teleological or rather symbolic rather than allegorical: "What Erasmus had was the new technology of print. What he already knew he needed was a computer." This anachronism corresponds with what even Rhodes and Sawday must finally admit is a difference between the effects of Renaissance print and contemporary computer technologies: "Print culture tended to produce a concept of the text as a relatively fixed and stable entity: the book. The great, multi-volume, 'standard' editions... stand as monuments... and... are also monuments to a belief in the stability of the printed word, and the possibility of freezing, for all time, that which has been thought and said."[33] This freezing is opposed to the constant dynamic flow of information in cyberspace. So, if archaeology deals with monuments—is it still the right method for analyzing digital topologies?

Of course, there is a constant and permanent movement between the media-archaeological layers of writing. This text of mine has been written and processed on computer, then evidently printed out on paper. This printing gave it, for a moment, the aura of a "final version," and an archival stability and authority against constant re-writing. On the way to Brown University, where

I first gave this paper as a talk, I added a lot of handwritten notes which re-turned it, in part, to a manuscript. The following steps to this publication, the editorial practice, confirmed the recognition that "there is no last word in textual matters."[34] Media-archaeology replaces the concept of a historical development from writing to printing to digital data processing through a concept of mediatic short-circuits; the discreteness of digital data, for example, has started with the ancient Greek alphabet already providing a model of elementary analysis of both speech and writing.[35] Of course, multi-media computing makes the medieval chart re-processible in its multi-media semiotics, no longer reducing it to its literal information by printing the document. In the Renaissance, the media format *book*—and multi-media archaeology is about formats—in contrast to the sequential reading of rolls (*volumen*), offered new options of data retrieval by supplementary tables of contents and indexes,[36] since for the first time, numerical data (page numbers) were combined with discrete text units (the single page), which facilitated rapid alphabetical search (as a classificatory system). In digital space, however, every bit can be addressed on a multi-media level (text - image - sound). Addressing is no longer limited to sentences, words, letters. Images could never be directly addressed by a book retrieval, unless indexed by words. Image- or sound-based retrieval of pictures and music would lead to a genuinely multi-media search engine culture. Maybe, in North America, American Indian culture and the ideological opposition of the first immigrant generations to old literate Europe has preserved a sense of orality which has made it easy for the second-order orality of gramophone, telephone, radio and TV broadcast to spread rapidly. Marshall McLuhan's media utopia of the wired global village could originate only in America, while Europe's book-oriented media culture stays on the side of writing.[37]

For the longest time in cultural history, storage of data and the means of operating them have been kept separately. The symbol-processing machine (the computer in its von-Neumann-architecture) though does not separate data and programs any more; rather both are deposited equally in the working memory of the machine, to be differentiated only in the actual moment of data processing. Suddenly, a psychoanalytic insight becomes technically true—the dialectic of archive and transference: "I think the challenge is to think the two as convergent: as two interdependent and inseparable moments perhaps in a single process."[38] The difference, though, between all old media like the book and the computer lies in the simple evidence that books cannot be (re-)programmed once printed. Thus the computer cannot easily be made compatible with a (media) history; it rather has an *arché*, a (archeo-)logics of its own.

The Silence of the Archive

The invention of printing distances the reader from the text, beholder from the image, creating a kind of "silence of the archive" through the silent reading situation. This situation corresponds to the media-archaeological insistence on confronting absences and silences, as opposed to the multi-media phantasy of a "talking" archive (Leah Marcus). Today, another desire for historical continuity over all discontinuities emerges: "the computer bridges the gap between manuscript and print" again.[39]

An inscription above the entrance to the Vatican Library in Rome demands without ambivalence: *Silentium*. "We associate libraries, collections of knowledge, and systems for memory retrieval with silence and hence with permanence."[40] It is exactly this kind of silence, which the archaeology of knowledge learns to confront while resisting the temptation of turning silence apotropaically into the discourse of historic talk. In ancient and medieval times, reading was performed aloud. The printing press silenced the voice, which returned as an inner hallucination again and again. Milton, for example, "thought of the perusal of printed volumes not as a purely visual activity but as a form of displaced orality—a conversation with kindred spirits who were long dead or at great distance."[41] This corresponds to the archival phantasm of history as a function of printing.

Our attitude towards phonographically-recorded sound sources[42] matches the situation of every historian: both strive to make an archive of (in the broadest sense) scriptural bodies (texts, partitures, wax cylinders) resonate. Activation of the archive in the pre-media age meant an energetic charging by *re-enactment* (Collingwood): Jules Michelet, historian of the French Revolution, believed he heard hallicunatorily the murmuring of the dead in the archive, as if documents were already the logocentric derivate of a gramophone. By his writings, he himself became a resonant body, a *medium* for the voices of the dead. Instead of apparatuses, it was historical discourse that functioned as a drogue of imagination, helping him to this kind of self-perception: "Dans les galeries solitaires des Archives où j'errai vingt années, dans ce profond silence, des murmures cependant venaient à mon oreille."[43] Is this now being replaced by the multi-mediatic interface illusion of the computer? "In recent years, the computer is no longer silent"; audio-visual perception supplements the traditional "reading" of texts—an "assimilation via the ear as well as the eye. Such a multi-leveled"—that is, *multi-media*—"'talking' archive would do more than make a significant number of early books conveniently available for downloading.... It would allow us to begin to reenter a mind set that was endemic to the early modern era, even though it has long been lost to us in the era of silent libraries."[44] That means (multi-)media archaeology, no longer "literally," but synesthetically.

Global Memories

While the term "archive" seems to describe all sorts of data banks in the World Wide Web almost universally, it also blurs the dis/similarities between old (print) and new (digital) archives. It is exactly the "multi" of multi-media that separates old from new archives. In contrast to two thousand years of basically written history, the advent of audio-visual recording media has led to genuinely multi-media "global memory" projects like the music-ethnological Berlin gramophone archive (E. M. v. Hornbostel) around 1900 and the film *Archive de la planète* of world cultures (A. Kahn) around 1930, resulting in the *Encyclopaedia Cinematographica* of moving nature (Institute for Scientific Film in Göttingen after WWII), which turns the archive into a discrete matrix of life itself. *Encyclopaedia cinematographica* has been the name of a film project of the German Institute of Scientific Film (Göttingen) which, under the guidance of the behavior studies scholar Konrad Lorenz, attempted to fix the world of moving beings on celluloid (up to 4,000 films). Like the medical films produced at the Berlin hospital Charité between 1900 and 1990 that the media artist Christoph Keller has secured from being thrown away as trash, this visual encyclopedia forms an archive that gains its coherence not from the internal but the external criteria of classification.[45]

As opposed to multi-media aesthetics, digital archaeology tries to get beyond sight and sound, since behind the images and noises we are confronted with "practices in which visual images no longer have any reference to an observer in a 'real,' optically perceived world," but rather refer to electronic mathematical data where abstract visual and linguistic elements coincide and are circulated.[46] Finally, the Human Genome Project reminds us that the apparent multi-media images and sounds of life are being replaced by a strictly numerical archive calculating rather than narrating life; if German a pun may be allowed: *zählen* (counting) instead of *erzählen* (narrating).

Part II: Case Studies in Media-Archaeology: The Virtual Reactivation of a Lost Sound Storage Medium: Hornbostel's *Phonogramm-Archiv*

Occidental phonocentrism has always been striving to find the means to store the human voice in the memory apparatus, be it the "dialogical" hallucinations of speaking with the dead in historical imagination. New technical means since late nineteenth century make it possible to inscribe traces of the human voice both literally in the already established archival institutions of cultural memory and in the epistemological "archive" (Foucault) as dispositive of cultural (re)cognition.

The notion of the archive is in transition, moving towards the audio-visual. As long as there have been archives, the phantasm of recording the acoustically real (i.e., the non-writable) has generated rhetorical, symbolic and scriptural forms of memorizing sound in supplementary ways. Despite the emergence of the phonograph, this new type of record was still subject to forms of inventorization and administration developed in the context of paper-based archives. (Multi-)Media archaeology seeks to reconstruct phantasms of memorizing sound in a pre-technical age and point out the discontinuities which arose with the invasion of audiovisual records into traditional archives, libraries and museums in the twentieth century. It culminates in a plea for rethinking the options of retrieval under new media conditions—transcending the notion of the archive itself.

In Germany, the invasion of the Edison phonograph into the Gutenberg-galaxy of cultural memory inaugurated a century that, for the first time, was also endowed with an audiovisual memory. In the same year that Sigmund Freud fixed his psychoanalytic interpretation of dreams,[47] the psychologist of acoustic phenomena, Carl Stumpf, and in his steps the music ethnologist Erich Moritz von Hornbostel, founded at the Berlin University a world-wide phonographic archive of wax-cylinder recordings of people threatened with extinction.[48] What appears rather unique, even idiosyncratic in the case of Hornbostel´s ethno-phonographical archive, should be read as part of an overriding multi-media practice of global classification, data processing and information storage, leading to early twentieth century efforts to create a universal science of cultural documentation (like Paul Otlet's *Mundaneum* in Brussels for meta-bibliography). As an example of a cinematographic *global memory*-project striving to make the memory of the world (later UNESCO's obsession) audio-visually recyclable, consider the Parisian banker Albert Kahn's project (died 1940), which from 1910 sent cameramen around the world to register images that might soon vanish.[49] Today, after two World Wars have effaced a lot of these objects, this collection is being preserved in Boulogne-Billancourt as an *Archive de la planète*. This memory, currently being made accessible on digital video disk, addresses a past from which no material archaeological relic has survived.

At the end of the twentieth century, the destiny of von Hornbostel's phonographic archive has been reversed, returning the collection to dissemination once again, and it is difficult to re-assemble this archive scattered by World War II.[50] Frozen voices, banished to analogue and long forgotten storage media, wait for their (digital) de-freezing.[51] At this moment, the fact that technical memory is "audio-visual" for human ears and eyes only manifests itself; the digital processing of such data equalizes the sensoric notion of multi-mediality itself. The Berlin Society for the Enhancement of Applied Informatics has developed a procedure to regain audio signals from the negative tracks in galvanized Edison wax cylinders by opto-analytic deviation: endoscopic recording devices "read" the sound traces graphically, re-translating them into audible sound by algorithmically transforming visual data into sound. Digital memory ignores the aesthetic differences between audio- and visual data and makes one interface (to human ears and eyes) emulate another. For the computer, the difference between sound and image and text, if they counted, would count only as the difference between data *formats*.[52]

When the ethnologist M. Selenka visited the American Indian Wedda tribe in 1907, she made the natives speak or sing into a phonograph, which she instantly played back to the speakers' joyful recognition.[53] With the media mystery of physically real recordings of sound and images, humans receive a multi-media mirror effect (in the Lacanian sense) that sublates the clear-cut difference between presence and absence, present and past. Strangely enough, we can to listen to this play-back today in exactly the same quality as the American Indians could in 1907: an example of the above mentioned opto-electronic archaeology of sound can be appropriately experienced via the World Wide Web.[54] Message or noise? Only the media-archaeological operation of reading the inscribed traces opto-digitally makes the otherwise inaccessible sound recording audible again. Synesthetically, we can *see* a spectrographic image of sound memory—a look straight into the archive.[55] The opto-digital *close reading* of sound as image, though, dissolves any semantically meaningful unit into discrete blocks of signals. Instead of musicological hermeneutics, the media-archaeological gaze is required here—a reminder of light-based sound inscription in early film.

Retrograd—*Excavating an Archive of Medical Films*

The term "archive" is frequently assumed to cover all activities of storing. "*Yet archives are not . . . collections, and their media-archaeological specificity and reproductive (mnemonic) strategies have to be carefully evaluated.*"[56] What, then, *is* an archive?

An archive is not an arbitrary quantity; not any collection of things can be an archive. The archival regime of memory is not an idiosyncratic choice, but a rule-governed, administratively-programmed operation of inclusions and exclusions that can be reformulated cybernetically, or even digitally.[57] Still, an intended archive can be subject to deformation, as illustrated by the collection of medical films produced at the Berlin hospital Charité from 1900–1990. Once intended as a film archive of general medicine,[58] the lot was not re-assembled and published multi-medially until recently. As a result of German reunification, the film institute of the Charité was closed within three days. Some material was lost, the rest was packed into sacks and placed in the Charité attic. Here, a filmmaker's camera searched for the last piece of evidence of what was once there.[59] Media archaeology, unlike media history, deals with absence. When looking at these films, it becomes apparent that images are weak, since they dissolve into nothing without archival authority.[60] Thus, a Foucauldian archaeological gaze is needed; that is, an active regime of ordering. Buried in analogue media, these images remain irretrievable for the moment; only the website performs this act of memory as media archaeology: we digitally (re-)*move* the cinematographic stills.[61]

This example demonstrates that the archaeology of multi-media no longer takes place in ground archives, but rather in virtual space. Without a fundamental, material support, however, it is no longer *arché*-ology in the classical sense, but rather cybernetic archaeo*logistics*. Consider more closely the QuickTime *movie* of a surgical operation on a shank in the Berlin Charité clinic from 1903.[62] Here the camera gaze allies itself with its object: it doubles the chirurgical gaze;[63] the anatomy of the body corresponds to the discrete, jumping images of early film. The ultimate media-archaeological gaze is opto-technical. The surgical amputation and the filmic cut coincide. Surprisingly, at the end of this short film, the surgeon Professor Bergmann looks and bows at the camera (whose camera-man was Oskar Meßter, later founder of the German UFA film industry) as though he was addressing a theatrical audience.[64] This gesture recalls the arena-like situation of the anatomical theatre established since the Renaissance. Keller's archival time-cut reveals a media-archaeology of medical films, thereby generating a parallel memory not of recordings of past reality, but of the ways images are consciously and apparatively constructed.[65] In the multi-media archive, code and culture coincide.[66]

An off spring of this medical film archive, the secret Nazi medical film project between 1941 and 1945 at the Charité, was later thrown by the SS into lake Stößensee near Berlin when the Red Army approached. There was literally a media-archaeological moment when divers detected these films in 1993 and rescued them; just three of several hundred film rolls could be deciphered at all, one of them showing (on heavily damaged film material) a naked man who performs several movements, apparently directed by outside orders. Correspondingly, a film by the Greek director Angelopoulos called *Ulysses' Gaze* is about a filmmaker who wanders through the Balkans in search of three reels of film from the early 1900s that were never developed. The final scene takes place in the ruins of Sarajevo where the reels of film are magically developed by an archivist, barricaded underground. When the filmmaker finally gets hold of the undeveloped film reels and they are developed, nothing can be seen on them any more, just blank frames.[67]

Media-archaeology deals with gaps and confronts absences. Of course, every film is always already itself an archive of movements, conserving modes of motion. Nevertheless the pioneer of film montage in Russia, Wsewolod Illarionowitsch Pudowkin, who with the Leningrad behaviorist Pawlow did a film in 1928 with the title *Functions of the Brain* insisted, that each object which is being recorded and projected by film is dead, even if it once moved in front of the camera.

Between Reading and Scanning

The computer does not literally read texts any more, but scans them, thus perceiving writing as an image, a cluster of signals (whether or not they may be finally re-composed to the form of a text page or an image). Signal processing replaces pure reading. The computer reduces signals to the smallest possible alphabet; still "the two most important directing signals which link the central processing unit of the computer to external memory are being called READ and WRITE."[68]

The media artist Angela Bulloch uses a key visual, a sequence from Michelangelo Antonioni's film *Blow Up* (1966): the protagonist, a photographer, hiding behind a tree takes photos to discover a murder; but the closer the camera looks in order to identify the spot (of the murder?), the less the photo serves as evidence for an apparent murder. As the German critic Karl Kraus once argued, "the closer one looks at a word, the further it looks back."[69] The artist extends this process of identification by yet another magnification, enlarging the digital scan of this scene in great blocks of single pixels and thus exploding the image within a sequential modular system of her so-called *pixel boxes*, in which one pixel is represented in a 50 × 50 cm monitor, attached to complex RGB lighting systems and which can be generated and programmed with any digital information.[70] This disillusion of the image's betrayal of the human eye reveals the media-archaeological scanner-gaze of the computer looking at a different kind of archive, no longer looking for just letters. The pixel modules also point to the fact that digital images are composed hyper-indexically by pure information, unlike referential images like those of classical photography, which still suggest a pre-discursive real. These modules developed by Angela Bulloch and Holger Friese reveal that multi-media archaeology requires technical skills. A pixel, which is the smallest conceivable picture element, only makes sense semantically when it appears within a group. To discern an image, the distance between the viewer and the group of pixels must be large if the light square made by a single pixel is 50 × 50 cm. In this situation, close reading can be performed only by the computer, and the computer is thus the true media archaeologist.

"I want control over every pixel" (Andreas Menn):

> In digital space the elements of files are discrete states. For digital images this means: There is nothing between one and its adjacent pixel. Discrete states though are unperceivable by human senses; the physiology of human perception and body are being characterized by the analogue, the continually floating. The digital thus arrives with the disappearance of the body therein.[71]

But, at the other end of this expulsion, the body *re-enters*. While interrogating the materiality of the pixel, the media artist Menn media-archaeologically decides to produce each pixel manually with his own body: "I work with my body in front of a digital camera; my appearance in the visual field equals 'one,' my disappearance equals 'zero.' I am being scanned by the camera."[72] From a distance, the writing, performed by pixels based on images of his body, reads: "I only want to work digitally."

What looks like an image on the computer monitor is nothing but a specific actualization of data (*imaging*). The computer thus renders data visible in a time-based way; the static notion of the image is being replaced by a dynamic one:

> This variability marks a fundamental chance of imagery. As opposed to classical image media like photography and film in the case of the computer-generated image the visual recording is not fixed invariably on a physical carrier, the negative, but always "fluid".... At any point of time digitally stored "images" can be manipulated, thus making the notion of the "original" state redundant.[73]

Visual Archiving: Sorting and Storing Images[74]

Cultural memory of images has traditionally linked images to texts, terms and verbal indexes. Confronted with the transfer of images into digital storage, non-verbal methods of classification are gradually gaining importance. Rather than the archival question, the search methods used to find pictorial information pose a problem to video memory, for they are still limited to models developed for retreiving text. What new kind of knowledge will exist exclusively in the form of images? What part of traditional knowledge can be transformed into images and what part might just vanish? Techno-image archaeology[75] seeks to rethink the notion of images, considering the process of archiving as organizing all that can be visually accessed as knowledge. In terms of technology, an archive is a coupling of storage media, data format (content) and address structure. Methododically this implies leaving behind the description of single objects in favour of an investigation of data sets.

In his 1766 essay "Laocoön," G. E. Lessing discusses the aesthetic conflict between the logic of language and the logic of images in terms of a genuinely multi-media semiotics: *pictura* is no longer—as declared by Horace—*ut poiesis*; time-based media (like dramatic speech and linear narratives) differ from space-based media (like simultaneous pictures).[76] Walter Benjamin, from a different perspective, reiterates that history appears in sudden images rather than narrative stories. Jules-Étienne Marey and Eadweard Muybridge chrono-photographically transformed an otherwise temporally experienced sequence (movement) into a spatial series (of discrete moments), close to the present aesthetics of the mouse-click. The digitization of images today provides a technical basis of inquiry into this conflict (i.e., the rather simultaneous aesthetics of websites as opposed to the moving image on the TV screen), so that the computer medium can ground that investigation. It would not make sense to retell a teleological story of image processing that finally reaches its aim in digitization; on the contrary, this history of images needs to be revised from the digital point of view. For example, how can archives be related to algorithms of image processing, of pattern recognition and computer graphics?

In sharp contrast to hermeneutics, the media-archaeological investigation of image archives does not take images as carriers of experiences and meanings. The relation between vision and image cannot be taken as the guideline for investigation, since image processing by computers can no longer be re-enacted using the anthropological semantics of the human eye. The methodological starting point is rather an archaeology of multi-media based on Claude Shannon's mathematical theory of communication, as well as the practices and concepts of data-structure oriented programming, amidst the insisting ruins of the Gutenberg galaxy. The *artes memoriae* have been visual techniques of memorization from the rhetorics of Antiquity to the Renaissance. Museums—collections, images of picture galleries, catalogues—since have always dealt with programming material image banks. The struggle for visual knowledge in (literally) the age of enlightenment in the eighteenth century led to visual encyclopedias and their visualizations (like the *planches*, i.e., the visual supplement of the big French *Encyclopédie* edited by Diderot and d'Alambert). Photography then has been the switching medium from perception to technology, creating the first technical image archives, and movies themselves have been archives (Hollywood and the rules of image sequences).

When it comes to (re-)programming image-oriented structures in the digital databases of given image archives, priority has been given to the development of a visually-addressable image archive.

By combining Multiresolutional Image Representation with simple Octree structures, a variable archive module might be applied. This would allow us to test algorithms by creating different visual sequences and neighborhoods. Most operators of image processing and pattern recognition such as filters and invariant transformations can be integrated in the structure of a database in order to make accessible a cluster of images. The next step might be the development of an interactive and visual agent capable of "intelligent" retrieval of images by graphical sketches.

Archival terminology, however, still carries grammato-centristic notions of data storage, but image and sound memories should no longer be subjected to uni-media, text-based retrieval. Usually, a subject index refers to categories that themselves refer to a register that, just like a conventional book library, assigns film titles a catalogue number. The catalogue number in turn refers to an actual film at one particular spot within the corridors of the storeroom, or in virtual space: a *link* refers to an actual website. But the alphabet as guiding indexical order of image and sound inventories today is being replaced by the algorithm—a kind of writing which is not just written language.[77]

It was writing that enabled cultural memory by storing remembrance outside man; at the same time, though, it reduced tradition to one channel of communication. Is this still true for the seemingly polyphonic multi-media age, when audio and visual data can be transmitted without scriptural meta-data? In digital space, when not only every film, but every still in every film, or even more—every pixel in every film frame—can be discretely addressed, titles no longer subject images to words, but alphanumerical numbers refer to alphanumerical numbers. Thus, the archive transforms into a mathematically defined space; instead of being a passive container for memorizable data, the techno-archive (as dispositive) actively defines the memory of images. Digital space is no longer an anthropological prosthesis to man, but a genuinely medially generated form. Whereas kinematographic forms of narrative still conform to human ways of perception by translating themselves into technical operations as instrumental extensions of human senses (eyes and ears), electronics directs images according to its own rules, only remotely connected to human perception.[78] The montage of images is being replaced by invasive digital intervention into the image itself, replacing narrative with calculation. Thus a genuinely image-based image retrieval is possible—an archive beyond icono*logical* semantics, based on computing algorithms which perform similarity-based image sorting. On a new technical level this brings us back to the visual administration of knowledge in the age of similarity (the Renaissance, the Baroque) which in the meantime had been replaced by the age of classification (Enlightenment, Neo-Classicism) as described by Foucault in *Les Mots et les Choses*.

> Clearly, . . . there is a tension between a system in which bite-sized pieces of information could be manipulated and rearranged and that sense of the "order of things" (the structure of correspondence), which underpinned the world views given a new lease on life by the medium of print. Here again there is a strange resemblance to modern conditions . . . The early modern version of field theory and chaos theory is Montaigne's observation that "toutes choses se tiennent par quelque similitude" (similitude binds everything together) and this is where poetry . . . enters the realm of the Renaissance Computers.[79]

The Renaissance and Baroque curiosity cabinets performed an aesthetics of pre-multi-media collecting, which leads Claire Preston to draw "an analogy between electronic search operations and the methods of the *curiosi* of early modern science and antiquarianism"[80]—with *analogy* itself being a figure of resemblance, as opposed to the Cartesian notion of difference which can be (mathematically) calculated. Collectors in the seventeenth century "imposed structure on the apparent disarray of the phenomenal world by searching for 'matches' . . . amongst the otherwise jumbled elements of their study." Systems of resemblance—visual patterns that may appear to us entirely fortuitous—were expressed by "horizontal or vertical contiguity" in the cabinets and illustrations. These efforts were driven by the belief that creation was coherent, and that the task of the scholar was to uncover and display this lost coherence—a kind of theological archaeology of knowledge, based on the assumption that what looks contingent to men, is a hidden coherence, a kind of pattern recognition in God's eye:

> In a world which seemed to present itself as a wilderness of forms, a variety of analogous or synonymous systems could provide the equivalent of a visual search-engine, much as we search a modern electronic database by finding an exact alphabetic or ASCII match for a flagged semantic item.... Dominique du Cange, the sixteenth-century French philologist, suggested (incorrectly) that the words 'musaeum' and 'mosaic' were cognate.... What all the cabinets and their encyclopaedias share is a syntax of resemblance or identity which is nearly always signaturist in its insistence on occluded and idiosyncratically selected likeness; their patterns are to be read as comparative contingencies or juxtapositions, as a system of potential *matches*.[81]

Is the notion of the printed encyclopedia as an alphabetical order of things still useful or is it a hindrance to thinking the cultural image banks of the future? Similarity-based image-retrieval belongs much more to a "senseless formal principle, which is exactly because of its dullness as useful as the alphabet is in a lexicon."[82] The Italian art historian Giovanni Morelli praised such a senseless method of comparing images as scientific, since it was objective; that is why a current image retrieval program is named after him: "Its salient feature is that it matches, sorts and classifies pictures exclusively on their visual characteristics."[83] The characteristics that it uses are derived directly from the process of digitization, and here the system differs from the historical Morelli method: "The automated "Morelli" system is not concerned with establishing authorship. It is concerned with providing an objective means of describing and identifying pictorial characteristics, such as form, configuration, motif, tonality and (ultimately...) colour."[84] Since the comparison of images here is of a simple overlay kind, and points of similarity and difference are recorded during the process of comparison, the central criterion is a simple matching process—a visual equivalent of the well known word search that is a standard feature of every word-processing and database computer software. This process of similarity-based image retrieval is possible only because the digitized image is an image that is stored as a set of quantifiable elements.[85]

René Descartes once criticized the category of resemblance as the fundamental experience and primary form of knowledge, denouncing it as a confused mixture that must be analyzed in terms of identity, difference, measurement, and order. Likewise, the data transfer compression program MPEG-7 tries to establish standards of *content-based* audiovisual retrieval: "The goal of MPEG-7 is to provide novel solutions for audio-visual content description."[86] A multi-media content description interface, though, is no longer a print-based archive. Media-archaeology thus means rethinking the notion of the *archive* subversively, hyper-literally, even at the risk that it might be more useful to replace it media-culturally in favor of agencies of dynamical *transfer*.

The multi-media archive deals with truly *time-based media* (which are images and sound), with every image, every sound only existing for a discrete moment in time. Freezing an electronic image means freezing its refresh-circle. Already, the temporal order of film is an effect of a ranging of discrete, in themselves statical (photographical) series of images one after another, unlike their correlative digital images, which are not simultaneous spatial entities but in themselves already composed by lines, which are refreshed permanently, that is time-based. In both cases, human perception is cognitively betrayed; the better knowledge, though, is on the side of the apparatus. As with the Williams-tube in early computing, where images were used for data storage since the picture elements died with a certain temporal deferral, the effect of an electronic "image" for humans is based on the minimal after-image intermediary memory—turning the image into a slow memory function.

Archival Phantasms (the Internet)

The emergence of multi-media archives has confused the clear-cut distinction between the (stored) past and (the illusion of) presence and thus is more than just an extension or re-mapping of well-known archival practices. The archival phantasms in cyberspace are an ideological deflection of the sudden erasure of archives (both hard- and software) in the digital world. "The twentieth century, the first in history to be exhaustively documented by audio-visual archives, found itself under the spell of what a contemporary philosopher has called 'archive fever,' a fever that, given the World Wide Web's digital storage capacities, is not likely to cool any time soon."[87]

Does the *archive* become metaphorical in multi-media space? This is a plea for archiving the term *archive* itself for the description of multi-media storage processes. Digital archaeology, though, is not a case for future generations, but has to be performed in the present already. In the age of digitalizability, that is, the option of storing all kinds of information, a paradoxical phenomenon appears: Cyberspace has no memory.[88]

Cyberspace is not even a space, but rather a topo-logical configuration. That is why the metaphorical application of the Renaissance *ars memoriae* to Internet memory is a mis-application. There are no *lieux de memoire*, rather there are addresses. In the Internet, the address structure of communication and the address structure of archival holdings merge into one. From place to pure address: Traditionally, "only what has been stored can be located"—and *vice versa*.[89] Today, on the contrary, the Internet generates a "new culture of memory, in which memory is no longer located in specific sites or accessible according to traditional mnemonics, and is no longer a stock to which it is necessary to gain access, with all the hierarchical controls that this entails"[90] (called "archontic" by Derrida).

A necessary precondition for any data retrieval is addressibility, the necessity of being provided with an external—or even internal—address. In Plato's dialogue *Meno* "it appears as if the matter of memory is but an effect of the application of techniques of recall"[91]—*is there no memory*? Is the World Wide Web simply a technique of retrieval from a global archive, or does it mark the beginnings of a literally *inventive* relationship to knowledge, a media-archaeology of knowledge that is dissolving the hierarchy traditionally associated with the archive?

As a machinic net of finite automata, the Internet has no organized memory and no central agency, being defined rather by the circulation of discrete states. If there is memory, it operates as a radical constructivism: always just situationally built, with no enduring storage. This invokes the early notion of *museum* as a cognitive and empty, rather than architectural or institutional space: "*Museaeum* was an epistemological structure."[92] Similarly in neurophysiology, memory operates like the imaginary in the formation of mental images: since there is no fixed place for images in the mind (at least not locatable), mental images are generated like images on an electric screen which have to be constantly refreshed. Oswald Wiener asks whether it makes sense at all to speak of mental *images*, if they have to be physiologically scanned in a time-based process, i.e., as a set of discrete (light-) moments in time[93]—in Lessing's sense a shift from visual to temporal indexicality (and *vice versa*, according to Benjamin). Can the Internet itself be separated from the notion of an<->archive at all? If an archive is a hallucination of a comprehensive lot, is then the Internet an archive? The Internet is no archive indeed, but a collection.[94] The function of archives exceeds by far mere storage and conservation of data. Instead of just collecting passively, archives actively define what is at all archivable. In so far as they determine as well what is allowed to be forgotten, since "the archival operation first of all consists of separating the documents. The question is to know what to keep and what to abandon."[95] Such is the difference between a paper-based (state-)archive in the strict, memory-institutional sense, and the Internet: The archive is a given, well-defined lot; the Internet, on the contrary, is not just a collection of unforeseen texts, but of sound and images as well, an anarchive of sensory data for which no genuine archival culture has been developed so far in the occident. I am talking about a truly multi-media archive, which stores images on an image-based method and sound in its own medium (no longer subject to verbal, i.e., semantical indexing).[96] And finally, for the first time in media history, one can archive a technological dispositive in its own medium.[97]

Dis/order

What separates the Internet from the classical archive is that its mnemonic logic is more dynamic than cultural memory in the printed archive. Although the Internet still orders knowledge apparently without providing it with irreversible hierarchies (on the visible surface), the authoritative archive of protocols is more rigid than any traditional archive has ever been. Traffic overload in the computer networks led the Clinton administration to build a new, separate system—the Internet II, restricted to scientific (and military) communications. Thus the remaining Internet somewhat adopts the so-called *chaotic storage* method in economy: "The World Wide Web and the rest of the Internet constitute a gigantic storehouse of raw information and analysis, the database of all databases. . . . The more serious, longer-range obstacle is that much of the information on the Internet is quirky, transient and chaotically 'shelved'"[98]—leading to archival phantasms of disorder. At the same time, memory in cyberspace is subject to an economy of memory not generous to gaps and absences.

Data transfer is incapable of transmitting non-information, while "in face-to-face interaction, much of what is most valuable is the absence of information, the silence and pauses between words and phrases."[99] Cyberspace is based on the assumption that unused space is economic waste—a result of the scarcity of storage capacity in early computing. Is the Internet really a medium through which self-organization produces the first comprehensive cultural memory?

This anarcho-archive is rather a fluid intermediary Random Access Memory. Who then archives the Internet? "Abandonware Community Triumph" is the name of such an initiative, which archives software and keeps it accessible. However, this quickly leads to a conflict with copyright, as exemplified by the current discussion over access to the most important of all archives: the files of the Human Genome Project. With the print-fixation of the traditional archival terminology, we run the risk of overlooking the fact that a different kind of archive is being built in non-public, proprietary ways by entrepreneurs like Bill Gates with his *Corbis* image bank, which holds the *digital* copyright of a lot of European historical imagery. This image bank, opposed to copyright law and the "legalistic infrastructure"[100] so well developed for textual authorship (the institution of the *dépôt légal* (national libraries), is based on different digital copyrights.[101] Probably two kinds of memories will remain—a radical rupture: Like in Ray Bradbury's *Fahrenheit 451* a new memory burns an old one.[102] This nostalgia is of course a phantasm surviving from the age of print. The alternative is a media culture dealing with the virtual an-archive of multi-media in a way beyond the conservative desire of reducing it to classificatory order again. Data trash is, positively, the future ground for media-anarchaeological excavations.[103]

Notes

1. Michel Foucault, *L'Archéologie du Savoir* (Paris: Gallimard, 1969).
2. Hayden White, *Metahistory* (Baltimore: Johns Hopkins UP, 1973).
3. Sven Spieker, proposal for a research project: *"Archive fever": Storage, Memory, Media* (draft 2000), n.p.
4. Ibid.
5. News reported by Detlef Borcher in his column "Online." In Die Zeit, no. 43, 19th October 2000, 46; on related subjects, see Bruce Sterling's "*Dead Media Project* in the WWW.
6. In a media-genealogical sense, there are no proper media, but rather a constant coming-into-being of media; see Joseph Vogl, *Medien-Werden: Galileis Fernrohr*. In *Archiv für Mediengeschichte* Vol. 1 (Weimar: Universitätsverlag, 2001), pp. 115–123, esp. p. 120.
7. On a proposed "museum" of software programs, see Wolfgang Hagen, "Der Stil der Sourcen. Anmerkungen zur Theorie und Geschichte der Programmiersprachen," In *HyperKult: Geschichte, Theorie und Kontext digitaler Medien*, edited by M. Martin Warnke et al. (Basel / Frankfurt a. M.: Stroemfeld, 1997), 33–68.

8. See Friedrich Kittler, *Eine Kulturgeschichte der Kulturwissenschaft* (München: Fink, 2000), 12th lecture, 228–249; further Richard Dienst, *Still Life in Real Time. Theory after Television* (Durham / London: Duke UP, 1994).
9. http://www.zock.com; last modified: 30th October 2000.
10. On media hardware in Germany, see: Deutsches Technikmuseum Berlin (http://www.dtmb.de/Rundgang/indexJS.html); and Heinz-Nixdorf-Museum (http://www.hnf.de/museum/comp4all.html); in a more global context: Apple History (http://www.apple-history.com), and the rubric "Classic Computing" (http://listings.ebay.de/aw/listings/list/category8086/index.html).
11. Konrad Lischka, "Verlassene Kunst. Softwarepiraten retten das digitale Erbe," *Die Zeit* No. 4, 18 January 2001, 34, referring to the (up to now) only exhibition of computer games archaeology in the Maryland Science Museum, organized by the group Electronic Conservancy (http://www.videotopia.com).
12. Some links to that kind of software (thanks to Tilman Baumgärtel, Berlin): The Abandonware Community Triumph (http://tact.tuol.org); the Altair: http://www.nostalgia.itgo.com/Computer/Altair/Altair8800en.html; the Altair Emulator: http://www.threedee.com/jcm/emu8080/index.html.
13. Neil Rhodes/Jonathan Sawday, "Introduction: Paperworlds. Imagining the Renaissance Computer." In *The Renaissance Computer: Knowledge Technology in the First Age of Print*, edited by Rhodes / Sawday (London / New York: Routledge, 2000): 1–17 (10 and 12).
14. Neil Rhodes/Jonathan Sawday, "Introduction: Paperworlds. Imagining the Renaissance Computer," 13.
15. Sebastian Domsch, "Diskretion ist Maschinensache," *Frankfurter Allgemeine Zeitung* Nr. 259 v. 7. (November 2000): 52.
16. See Rolf Sachsse, "Das Gehirn der Welt: 1912. Die Organisation der Organisatoren durch die Brücke. Ein vergessenes Kapitel Mediengeschichte," *Mitteilungen der Wilhelm-Ostwald-Gesellschaft zu Großbothen e.V.*, 5. Jg., issue 1 (2000): 38–57; the text can—in the best tradition of Ostwald's "global brain" project—be addressed in the Internet—as subject and object of this theme, under: http://www.heise.de/tp/deutsch/inhalt/co/2481/1.html. He calls it multi-mediatc, since it encompasses visual cultural artefacts of communication as well.
17. Jonathan Sawday, "Towards the Renaissance computer." In *The Renaissance computer*: 35.
18. Michel Foucault, "Message ou bruit?" In: *Concours médical*, 88e année (22 octobre 1966): 6285f (Colloque sur la nature de la pensée médicale); reprinted in: M. F., *Dits et Écrits I* (Paris: Gallimard, 1994): 557–560 (559).
19. Spieker, proposal, n. p.
20. Volker Kahl, "Interrelation und Disparität. Probleme eines Archivs der Künste." In *Archivistica docet: Beiträge zur Archivwissenschaft und ihres interdisziplinären Umfelds*, edited by Friedrich Beck (Potsdam: Verl. f. Berlin-Brandenburg, 1999): 245–258 (252).
21. See Friedrich Kittler, "Von der Letter zum Bit." In Horst Wenzel (ed.), *Gutenberg und die Neue Welt* (Munich: Fink, 1994), pp. 105–117.
22. The computing part of the computer, however, is not storage, but dynamic calculation.
23. See the research project "Hypertext. Theorie und Geschichte," University of Kassel (http://www.uni-kassel.de/fb3/psych/pers/meyer/wz2/httg.htm).
24. Theodor H. Nelson, "As We Will Think." In *From Memex to Hypertext: Vannevar Bush and the Mind´s Machine*, edited by James M. Nyce and Paul Kahn (San Diego/London: Academic Press, 1991), 259 (245 and 253).
25. Though somewhat reducing this claim back to history is the very sequential unfolding of arguments following the format of the book.
26. Cornelia Vismann, *Akten. Medientechnik und Recht* (Frankfurt/M.: Fischer Verlag, 2000) (book jacket text).
27. Our candidate for replacing the desktop is called 'Lifestreams.'" David Gelernter, *Machine Beauty. Elegance and the Heart of Technology* (New York: Basic Books, 1997), 106.
28. Bertolt Brecht, *Der Rundfunk als Kommunikationsapparat. Rede über die Funktion des Radios* (1932). In *Ausgewählte Werke* vol. 6, (Frankfurt/Main: Suhrkamp, 1997), 146–151.
29. Rhodes/Sawday, *The Renaissance Computer*, 12, referring to: Vincent Gillespie, "Medieval Hypertext: Image and Text from York Minste." In *Of the Making of Books: Medieval Manuscripts, Their Scribes and Readers. Essays Presented to M. B. Parkes, Aldershot*, edited by P. R. Robinson and Rivkah Zim (Aldershot, England: Scolar Press, 1997): 208f.
30. We are familiar with this from the technical options of the re-writable magneto-optical disc.
31. "Informationen, die nicht im externen oder internen Speicher einer Datenverarbeitungsanlage abgelegt sind, können überhaupt nicht als Daten verwendet werden. Der Grad der Öffentlichkeit ergibt sich dann aus den Zugriffsmöglichkeiten auf den Speicher." Michael Giesecke, "Als die alten Medien neu waren. Medienrevolutionen in der Geschichte." In Rüdiger Weingarten (ed.), *Information ohne Kommunikation? Die Loslösung der Sprache vom Sprecher* (Frankfurt/Main: Fischer, 1990), 75–98, here: 86.
32. Mark U. Edwards, Jr., *Printing, Propaganda, and Martin Luther* (Berkeley/Los Angeles/London: University of California Press, 1994), 163; see also Neil Rhodes/Jonathan Sawday, *The Renaissance computer*, 12.
33. Rhodes/Sawday, "Introduction: Paperworlds. Imagining the Renaissance Computer," 11.
34. Rhodes/Sawday, "Introduction: Paperworlds. Imagining the Renaissance Computer," 11.
35. See Derrick de Kerckhove, "Die atomare Kommunikation." *Schriftgeburten. Vom Alphabet zum Computer* (München: Fink, 1995), 143–158.
36. See Rhodes/Sawday, *The Renaissance computer*: 7.
37. Heinz Schlaffer, Introduction to: Jack Goody/Ian Watt/Kathleen Gough, *Entstehung und Folgen der Schriftkultur*, transl. Friedhelm Herborth (Frankfurt/M.: Suhrkamp, 1991): 7–23 (7).
38. Samuel Weber, e-mail to the author, 14 October 1998.
39. Leah S. Marcus, "The Silence of the Archive and the Noise of Cyberspace." In Sawday/Rhodes, *The Renaissance computer*, 18.

40. Marcus, "The Silence of the Archive and the Noise of Cyberspace," 24.
41. Marcus, "The Silence of the Archive and the Noise of Cyberspace," 25.
42. For example Erich Moritz von Hornbostel's music-ethnological Berlin *Phonogramm-Archiv* after 1900; see Sebastian Klotz (ed.), *"Vom tönenden Wirbel menschlichen Tuns": Erich M. von Hornbostel als Gestaltpsychologe, Archivar und Musikwissenschaftler* (Berlin / Milow: Schibri, 1998).
43. Jules Michelet, "Histoire de France, preface of 1869." In *Oeuvres Complètes IV*, edited by Paul Viallaneix (Paris: Flammarion, 1974), 24.
44. Leah S. Marcus, "The Silence of the Archive and the Noise of Cyberspace," 27–28.
45. Exhibition at Kunstbank Berlin, May 2000.
46. Jonathan Crary, *Techniques of the Oberserver* (Cambridge, MA: MIT Press, 1990).
47. Sigmund Freud, *Die Traumdeutung (*Leipzig/Wien: F. Deuticke, 1900).
48. See W. E., "Hornbostels Klangarchiv: Gedächtnis als Funktion von Dokumentationstechnik." In Klotz (ed.), *Vom tönenden Wirbel*, 116–131; furthermore the catalogue no. VI (*Wissen*) of the exhibition *7 Hügel. Bilderund Zeichen des 21. Jahrhunderts* in the Berlin Martin-Gropius-Bau, Berlin 2000 (esp. the hypermedia installation "MusikWeltKarte"). See as well the preceedings of the Annual Conference of the International Association of Sound and Audiovisual Archives—IASA—, Vienna, 18–23 September, 1999: *A Century of Sound Archiving.*
49. See Sabine Lenk, "Die *Autochrone-* und Filmsammlung des Albert Kahn." In *Früher Film in Deutschland [= KINtop 1. Jahrbuch zur Erforschung des frühen Films]* (Basel / Frankfurt/M.: Stroemfeld / Roter Stern, 1992), 120–122. Carl Stumpf actually deplored the absence of optical memory in the Berlin Phonogram Archive; see Otto Abraham/Erich Moritz von Hornbostel, "Über die Bedeutung des Phonographen für vergleichende Musikwissenschaft" *Zeitschrift für Ethnologie* 36 (1904): 222–236.
50. Susanne Ziegler, "Das ehemalige Berliner Phonogrammarchiv." In Annegrit Laubenthal (ed.), *Studien zur Musikgeschichte. Eine Festschrift für Ludwig Finscher* (Kassel et al.: Bärenreiter, 1995), 766–772, here: 771.
51. On phonographic metaphors of writing in medieval times *avant la lettre* see Horst Wenzel, "Die 'fließende' Rede und der 'gefrorene' Text. Metaphern der Medialität." In *Poststrukturalismus. Herausforderung an die Literaturwissenschaft*, edited by Gerhard Neumann (Stuttgart/Weimar: Metzler, 1997).
52. See Artur Simon (ed.), *Das Berliner Phonogramm-Archiv 1900–2000. Sammlungen der traditionellen Musik der Welt*, (Berlin: VWB, 2000): 209–215.
53. Quoted in Max Wertheimer, "Musik der Wedda," *Sammelbände der Internationalen Musikgesellschaft* Vol XI, No 2 (January-March 1910), 300–309 (300).
54. http://www.gfai.de/projekte/spubito/index.htm.
55. See the spectrogram of a recontructed recording of Wedda chants in Ceylon 1907 on the SpuBiTo web page.
56. Spieker, proposal, n. p.
57. For the German case, see Heinz Hoffmann, *Behördliche Schriftgutverwaltung. Ein Handbuch für das Ordnen, Registrieren, Aussondern und Archivieren von Akten der Behörden*, 2. Aufl. (München: Oldenbourg, 2000).
58. See the article by C. Thomalla, "Ein medizinisches Filmarchiv [A medical film archive]," *Berliner Klinische Wochenschrift* No. 44 (1918)
59. Christoph Keller, "Lost/Unfound: Archives As Objects As Monuments." In catalogue *ars viva 00/01 - Kunst und Wissenschaft* for the exhibition by prizewinning artists of Kulturkreis der deutschen Wirtschaft im Bundesverband der Deutschen Industrie e. V (Berlin: M8 Labor für Gestaltung, 2000), n. p.
60. Stefan Heidenreich, "Die Wirklichkeit mag keine Bilder, über die Ausstellung der *Encyclopedia Cinematographica* des Medienkünstlers Christoph Keller in der Kunstbank Berlin, Mai/Juni 1999," *Frankfurter Allgemeine Zeitung* Nr. 126 (Berliner Ausgabe Berlin edition) 31 May 2000, BS8.
61. http://www.medfilm.de.
62. http://www.khm.de/~chrk/medfilm/03.unters.html.
63. Michel Foucault did not only write about the *Birth of the Clinic* as an "archaeology of the medical gaze," but himself defers, by making it a subject of research, his dis/continuity with the three-generation-long tradition of the Foucault family as surgeons.
64. Thanks to Thomas Elsaesser for this precise look at the Quick-Time movie. See as well Christoph Keller, "Lost / Unfound," on the 1900 movie.
65. Presstext – Christoph Keller: "encyclopaedia cinematografica" kunstbank Berlin Mai 2000.
66. Comparatively, see the records on the film-based so-called Institute for Cultural Research (Institut für Kulturforschung/Kulturfilm-Institut), dating from 2nd June 1933 until 4th July 1935, in the Archive for the History of the Max-Planck-Institute, Berlin-Dahlem, I. Abteilung: Kaiser-Wilhelm-Gesellschaft, Repositur 1A: Generalverwaltung der KWG, no. 1041.
67. Thanks to Lisa Parks (University of California, Santa Barbara) for reminding me of this corresponding scene.
68. Friedrich Kittler, "Computeranalphabetismus." In *Literatur im Informations-Zeitalter* (Frankfurt/M/New York: Campus, 1996), 237–251 (239).
69. "Je näher man ein Wort ansieht, desto ferner sieht es zurück": Karl Kraus, *Pro domo et mundo* (Munich: Langen, 1912), 164.
70. Such is the installation *BLOW_UP T.V.* of Angela Bulloch in the gallery Schipper & Krome, Berlin, September to November 2000.
71. Andreas Menn, textual supplement (Cologne, July 2000) to his digital video *Workout* (1999).
72. Ibid.
73. Claudia Reiche, "Pixel. Erfahrungen mit den Bildelementen." *Frauen in der Literaturwissenschaft* Rundbrief 48 (August 1996), 59–64 (59).
74. For co-authoring of this chapter, thanks to Stefan Heidenreich and Peter Geimer (both Berlin).

75. On technical images and the notion of the techno-imaginary, see: Vilem Flusser, *Kommunikologie*, edited by Stefan Bollmann/Edith Flusser (Frankfurt/M.: Fischer, 1998).
76. See Hans Ulrich Reck, "Metamorphosen der Archive/Probleme digitaler Erinnerung." In *Metamorphosen. Gedächtnismedien im Computerzeitalter*, edited by Götz-Lothar Darsow (Stuttgart-Bad Cannstatt: frommann-holzboog, 2000), 195–237 (223).
77. Hartmut Winkler, abstract for his lecture "Theorie und Geschichte der Schrift" at the University of Paderborn, winter term 2000/2001.
78. Knut Hickethier, *Film- und Fernsehanalyse* (Stuttgart: Metzler, 1993), 158.
79. Rhodes / Sawday, *The Renaissance computer*: 13, referring to: Michel de Montaigne, *Oeuvres complètes*, edited by Albert Thibudet and Maurice Rat (Paris: Gallimard, 1962), 1047, and to N. Katherine Hayles, *The Cosmic Web: Scientific Field Models and Literary Strategies in the Twentieth Century* (Ithaca, NY: Cornell UP, 1984).
80. Claire Preston, "In the Wilderness of Forms: Ideas and Things in Thomas Browne's Cabinets of Curiosity." In Rhodes/ Sawday, *The Renaissance computer*: 170–183 <abstract, 170>.
81. Claire Preston, "In the Wilderness of Forms: Ideas and Things in Thomas Browne's Cabinets of Curiosity," 174f.
82. "*Image retrieval* und visuelles Wissen," Lecture by Stefan Heidenreich at EVA conference in Berlin, 13 November 1997.
83. William Vaughan (Birkbeck College, University of London), "Automated Picture Referencing: A Further Look at 'Morelli,'" *Computers and the History of Art* Vol. 2 (1992): 7–18 (7).
84. Vaughan, "Automated Picture Referencing: A Further Look at 'Morelli'": 8.
85. Vaughan, "Automated Picture Referencing: A Further Look at 'Morelli'": 9.
86. http://www.ircam.fr/produits/techno/multi-media/Cuidad/mpeg7_info.html.
87. Draft Sven Spieker (n. p.), referrring to: Jacques Derrida, *Mal d'archive. Une impression freudienne* (Paris: Galilée, 1995).
88. Christoph Drösser, "Ein verhängnisvolles Erbe," *Die Zeit* (23 June 1995): 66.
89. Harriet Bradley, "The seductions of the archive: voices lost and found," *History of the Human Sciences* Vol. 12 No. 2 (1999): 107–122 (113).
90. Howard Caygill, "Meno and the Internet: between memory and the archive" *History of the Human Sciences* Vol. 12 No. 2 (1999): 1–11 (10).
91. Caygill, "Meno and the Internet: between memory and the archive": 2.
92. Paula Findlen, "The Museum: its classical etymology and renaissance geneaology" *Journal of the History of Collections* 1, no. 1 (1989): 59–78, *abstract*.
93. See the Documentary film by Matthias Brunner and Philipp Pape (Berlin), *Am Anfang war die Machine* [At the beginning, there was the machine], Germany 1999, in program of International Video Festival X, Bochum, May 2000.
94. See, though, the *Open Archive* Movement and http://www.archive.org.
95. Arlette Farge, *Le goût de l'archive* (Paris: Seuil, 1989), 87.
96. See, for example, Adaweb (http://adaweb.walkerart.org).
97. For these somewhat auto-poetic links in the Internet thanks to Tilman Baumgärtel once more: http://www.0100101110101101.ORG; http://www.dejavu.org, announcing: "Experience the history of the web! Go to the emulator to re-live an era in the history of the web! Or go to the timeline to read about the old times!" See as well: http://www.w3history.org.
98. Editorial: "The Internet. Bringing order from chaos," *Scientific American* vol. 276 no. 3 (March 1997), 494 (49).
99. Steven G. Jones, "Understanding Community in the Information Age." *Cybersociety. Computer-mediated Communication and Community*, edited by Steven G. Jones (Thousand Oaks 7 London/New Delhi: 1995), 10–35 (28).
100. Rhodes / Sawday 2000, "Introduction: Paperworlds. Imagining the Renaissance Computer," 12.
101. See Thomas Hoeren, "Urheberrecht in der Informationsgesellschaft" *GRUR* issue 12/1997. On the technical regulations of cyberspace beyond legalism see Lawrence Lessig, *Code and other laws of cyberspace* (New York: Basic Books, 1999).
102. Sebastian Handke, "Die neue Flüchtigkeit. Wer archiviert das Internet? Archivwissenschaftler und Medienarchäologen diskutierten in der Mikrolounge des WMF über die Zukunft der Erinnerung," *taz* <Berlin> no. 6264 (7 October 2000), 26.
103. Links to recycling: the Redundant Technology Initiative (http://www.lowtech.org) and Mark Napier's www.potatoland.org. On the term "anarchaeology," see Siegfried Zielinski, *Archäologie der Medien* (Reinbek: Rowohlt, 2002).

7

Breaking Down

Godard's Histories

Richard Dienst

And we: spectators, always, everywhere,
turned toward everything around us and never beyond!
We are filled up. We make order. It breaks down.
We make order again and break down ourselves.

—Rilke, *Duino Elegies*

In 1995, when Pierre Bourdieu wanted to attack the institutions of television, he decided to use television to deliver his message. In two televised lectures, later published as *On Television*, Bourdieu presents a careful sociological critique of the mechanisms that constitute authority and privilege in the media, along with a strident denunciation of the market pressures, themselves transmitted by television, which have wrecked the hard-won accomplishments and dissolved the self-generated autonomy of artistic and intellectual cultures. But Bourdieu seems to recognize that even the most clear, most reasonable critique will not strike television where it lives. At the beginning of his analysis, he suggests in passing that what is *really* needed is "a true critique of images through images—of the sort you find in some of Jean-Luc Godard's films."[1] As an example, strangely enough, Bourdieu cites only *Letter to Jane*, one of Godard's great didactic works of the early 1970s. It consists largely of a lengthy critical commentary spoken over a single magazine image of Jane Fonda in Hanoi. It might be considered a high point in the "essay-film," but today we would have to say that "the critique of images by images" cannot proceed one by one, one film for one photo. Just as, on one hand, it is never enough to offer a "critique" of television that would be content to pick out the better programs or the more adventurous programmers, it is always too much, on the other hand, to put all images under suspicion when insisting upon the irreducibly spectacular dimension of contemporary culture.

In his current work, Godard does something other than critique. He tries to redefine the power of the image for our historical moment, to make images that would enable the stubborn work of remembrance and imagination, and thereby—perhaps—change the scope of the life we might be able to live, in spite of everything. It is a project that beggars belief. How on earth, after having conceded the exercise of images for so long to the ruling powers, can we ever again use them to tell, to show, to make history?

To show what an answer to that question might look like, we should address Godard's most massive work— the *Histoire(s) du cinéma* [Histories or Stories of cinema]. It is hard to know what to call it without begging all the questions it raises from the very first look: its media, genres, modes, and substances cannot be declared in advance, let alone the question of whether it belongs to "history" or "cinema." For Godard, the very idea of a history of cinema becomes possible only because it has already broken down, and with it, the irreplaceable bonds between art and politics that once sustained the territories and the worlds that cinema promised to make habitable. Godard's work situates itself in the technological and political order that is already replacing all of that, unraveling everything that cinema might have held together. But his "return journey" to cinema is neither nostalgic over its loss nor hopeful for its recovery. Instead, he sets out solely to look for (hi)stories that have been neither seen nor heard, in the hopes of finding whatever might remain unrealized amidst a history that has already realized too much.

To begin, a few points of orientation. Upon its completion in 1998, the *Histoire(s)* comprised a work in eight parts, delivered in three different media: video, printed book, and audio CD. It is crucial to distinguish between them. In its video form, the work was released in installments from 1988 to 1998, all of them appearing on French and European television, as well as being screened at various festivals and museums. The total running time is 4 hours and 22 minutes. To coincide with the appearance of the final episodes, the publishing house Gallimard released a four-volume print edition of the *Histoire(s)*. These books are in no way a mere souvenir of the video-film: they present a different "variation" of the material, calling for a different kind of attention. Here, still images are arranged across the pages alongside a printed text of most (but not all) of the words spoken in the video. Some phrases, invisible or inaudible, have been added. The words are laid out in blocks that do not consistently reflect the pacing of the soundtrack or the interplay of speakers, taking advantage of the printed page to rearrange the relationship between phrases, and the relationship of phrase to image. Because Godard's video technique has always relied upon superimposition, onscreen captioning, and an arrhythmic conception of mixing, the presentation of "stills" poses a real challenge: the translation into the printed form offered Godard a chance to reconceive the selection of visual elements, to seize instants where the composition reaches a heightened intensity, to treat verso and recto as units of montage, and to redraw the balance of what can and cannot be discerned. Stripped of the luminosity of the screen, the still images require the eye to work with a different intensity, peering and scanning across the page at its own pace. In short, the production of the books presented an opportunity to remake the videos as a certain kind of artwork, as if finally fulfilling, somewhere between the flux of transmission and the binding of print, the frustrated convergence of painting, poetry, and philosophy so often attempted in Godard's films.

Next, the complete soundtrack of the *Histoire(s)* was released on CD in 1999, packaged within a different set of books that include an accompanying script printed in French, German, and English (which transcribes the verbal track more closely, but without attempting to describe the action or designate the images themselves). With the CDs the isolation of one layer of images—the sound-images—again permits a shift in the balance between elements. The play of different voices, the variable foreground and background of music and words, the startling changes of speed, the recurrent crack of cloudbursts and whirr of film-editing machines, and above all, the presence and possibility of silence: all of this registers the matter of the *Histoire(s)* in a rather distinct way, posing difficulties unsuspected and easily avoided while sitting in front of the video screen. It begins to seem as though the sounds have been entertaining relationships among themselves all along, behind the back of the screen. Perhaps there is no obvious reason to prefer one version of the work over the other two. The most scrupulous engagement would face the imposing task of working through each episode in each material form, looking not so much for a synthetic experience, much less the overall message, but rather for those features that go unregistered, unremarked in the others. Only then would it become possible to encounter the spark of montage, which is also the spark of history, in the midst of all the slipping, stuttering and looping—as if it takes such a vast and meticulous assemblage to generate even the chance of such a spark.

Whatever it tells us about the interplay of media, this complex of material makes a startling point: the history of cinema can be told, it seems, everywhere but in cinema. Yet, as the initial episodes insist, the history of cinema is the only history that needs to be told, because only cinema has been capable of telling the story of its time. But it failed, and that is the real story. Only cinema could have constructed the linkages between technology and life required by the modern world; only cinema offered a way to show one life to another without threatening both. Montage—understood first as the production of connections, comparisons, constellations and other kinds of relationality—is the only form, the only technique, cinema has to offer to help us live historically. All the rest—its plots, its clichés, its obedience to ruling ideas and awful prejudices—would be precisely what montage could undo by cutting open. But, again, cinema got caught up in everything but pursuing its only real task. The "beautiful care" [beau souci] of montage turned out to be a burden heavier than any film, or any filmmaker could bear: it is nothing less than the obligation to make history out of every image, to know how to slip dreams into reality, when to splice memory into the flow of forgetfulness, and moments of beauty into the unfolding catastrophes of modern life. Cinema let us know that all these images are somehow *there*, adjacent to each other, if not to us. We do not lack for images; if anything, the images lack a "we" who could bring them together.

Given the density and complexity of this work, its aleatory drift, its idiosyncratic mix, one might begin to wonder in what possible sense it could be called a history. Surely there have been histories of cinema, and even the most elaborate theory of cinema—Deleuze's—unfolds around a fairly clear-cut historical framework, punctuated by the rupture of 1945. But even here, history seems to come from the outside—we still want to know if there can there be a history of images *through* images. What if the stockpile of the imaginary, once stirred, can no longer sit still, so that every image, at any moment, might swing around to demand its moment in the light? Where is the narrative in that? What kind of history is it when, on one hand, the most solid visual documents—testimonies of tremendous work and suffering—can suddenly be turned into flickering sketches of color? And on the other hand, when the great scripted fictions and glorious spectacles can reappear, stripped of ideology, as images of some lasting truth?

To outline one path through these questions, I want to make a few propositions that follow a few clues laid out in the *Histoire(s)*. Moreover these propositions, like the *Histoire(s)* itself, do not really concern Godard's work or even cinema itself, but rather turn outward toward images in general—"all of them that ever were, or ever will be"—if indeed we can still imagine facing history from such an impossible angle.

The first proposition is simple: *Images are what remain to be seen*. Let me insist upon this statement to the letter: images remain, and they are seen. Their remaining is what lets them be seen, it is what constitutes their visibility. So, as soon as we see something that looks like an image, we ought to ask about the preservation, persistence, even the survival of what is there to be seen.

"Remaining" is a matter of recording and transmission, certainly, but also of marking time. Images can "remain" in ways or in places that words and sounds cannot; which is not to say that images are always and everywhere more permanent or decisive, but rather that they make a certain kind of impression, imprinted all at once by the reliefs of time in a way that makes it possible for us to handle the stuff-in-flux we call life and history. Such a process is "technological" in the most broad sense of the term: that through which living groups reproduce and extend themselves. Images belong to the general organization of matter that makes things happen: they are constitutive components of any practical environment, they help to teach us what has been done and can be done. Even in a setting where images are constantly destroyed or used up, they belong to a common economy where durability and efficacy are set to work: images are never more or less "ideal" or "transcendental" than tools or dwellings. Images do not wait, suspended in the ether, until a body comes along to give them weight; rather, an image makes its appearance by weighing upon the earth, even when it looks as if it is waiting for someone to treat it as something ethereal. Images can't help it if they slip in and out of visibility: they depend on us to set them up and knock them down. That is why, in the anthropological framework proposed by Andre Leroi-Gourhan, images and language belong to the same plane of exteriorization and differentiation that is said to inaugurate a specifically human history. (It may turn out that this turn to prehistory is a misleading way to tell the story, begging all the crucial questions, given that images and language have led us hither-and-yon about what "humanity" and its "history" might be, but let's leave that question aside for now.)[2]

By starting with the reminder that images must somehow remain to be seen, we underline the fact that images only ever exist because they belong to a given and inherited technical system, even when they have been transferred from one system to another. They can be considered instruments of capital in a basic sense: that with which historical beings lay claim to the workings of time through the products of their labor. Being able to see an image as an image requires an enormous investment of collective and cumulative effort over many millennia. The hand and the eye, in working together, learn how to help each other, and thereby each gives to the other the means to realize their common capabilities. As Godard puts it in the *Histoire(s)*: "It is said that some think and others act, but the true human condition is thinking with one's hands."[3] Petroglyphs and photographs alike are created by preserving something for view: it is the gesture and skill of preserving itself that is being affirmed, whether or not we can recognize something else—the arc of a running animal, the familiar expression of a stranger—being preserved as well. By the same token, a world of disposable images can teach us to see nothing other than a world of disposable landscapes and disposable populations: carelessly made and carelessly kept, all of them can now disappear in an instant.

What distinguishes today's digital images from ancient bone inscriptions or cuneiform tablets is not a greater degree of imaginariness, but rather the different kinds of investment of the material world by human practices, carried out in each case by the whole ensemble of technical mediations required to present such images as images. It is not the images themselves that are more or less abstract, but rather the relationships in which they are set to work. Somehow each image is made to remain—whether as incision, as representation, as reproduction, as transmission, or as recombination—and in distinguishing between such different kinds of image, we begin to set forth a general anatomy of social relationships. When critics explore how particular sets of images act upon different kinds of spectators, they are investigating not simply the properties of a given text, but the division of labor and the cultivation of skills that are engendered and enforced by acts of seeing. In fact there is no reason to assume that images are ever made to be seen "in general," by anybody and everybody: such a global democracy of vision has never been attempted, let alone brought into being, by any regime of image-making (although Hollywood, now backed by the WTO, will finally make a stab at it under the fiction of a free market).

All of which is to say: We should not take the "presence" or the "operation" of images for granted. Whenever they appear, they deserve to be treated as a strictly circumstantial accomplishment. If there is an image, it must have survived the moment in which it was made, traversed the vicissitudes of transmission, outlasted the perils of objecthood, and taken a place in a particular practical context ruled by a veritable economy of visibility. Every image is both a historical thing, some mark made to order vision as soon as it is made, *and* a historical event, a spark of vision that organizes a world of relationships around it. Before (and in spite of) anything else, an image makes visible its own persistence and passage as an image, testifying to a system of social and technical supports that necessarily operate more-or-less unseen as such. Images "remain" according to the temporalities of a given technical system, rather than the temporalities of the visible world it projects. An image of a horse running or a bird flying testifies not so much to the speeds of horses or birds, but to the speeds of technologies that now traverse the earth and the air. So when Godard shows us a time-loop of a Muybridge cat trotting along a white wall, followed by a kinescope gull superimposed on a girl's pensive face, we remember all over again that the nuances of motion and emotion have become so visible only by passing through the flickering rhythms of a machine, which has itself become part of an enormously profitable production system. In cinema, each image, each frame, is made to remain only until the next one. What subjects experience as "persistence of vision" or "movement" on the screen is the material trace of a technical logic of multiplicity and reproducibility, which has finally achieved the economies of scale and speed that allow modern subjects to imagine themselves calibrated to their mode of production.

Godard understands this dynamic well: "*there is not an image, there are images ... [As] soon as there are two, there are three*."[4] This principle of multiplicity or plurality holds true on the level of perception as well as the level of montage, because it suggests that the two operations are really the same: images take shape only in the midst of an action, never as inert discrete objects "out there" and never as completed blocs of memory "in here." Once an image is dissolved into the fits and flows of sensation, or, on the contrary, resolved into clear idea (where it would be equivalent to a caption or a concept), it loses its character as an image. Which is to say: images can be defined only as an exception to bodily experience or to pure reason. But alongside that formal and psychological principle is an ethical one: one image is never enough, either to serve as the basis of subjectivity or to hold the place of the other. Images "remain" only as long as they "rely upon" other images, but this reliance is less a matter of lacking of self-sufficiency than of drawing upon the multitude of forces that criss-cross between them.

We might then speak of a material arithmetic of images: for example, the punctum described by Barthes is a third thing "added on" to the encounter of photograph and spectator; conversely, the "image of movement" described by Eisenstein (in his "Laocoön") is clearly "subtracted" not only from the sequence of images but from the perception of the spectator.[5] And there is already a long line of Lacanian bookkeepers, multiplying and dividing images by each other to calculate the resultant psychic forces. But because images remain, and because they are never alone, such equations remain unfinished: to insist on the materiality of each image, no matter what kind of text is being examined, is to shift the emphasis from the meaning of images to the efficacy of their powers. And if there is never just one image, there is never just one "power of images" or "imaginary power."

Nevertheless, one often hears about the "power of images"—indeed, everybody believes in the power of images, if they believe in anything at all. Evidently it is hard to speak about images without invoking several kinds of power in the same breath. But it is also hard to talk about *power* except through the images deployed in its name. Every idea of social or psychic power will seem vague, its meanings stretched between the flutters of bodily sensation and the constant pressures of social control, unless the exercise of power is grasped through the instance of images. If the notion of *power* alerts us to recognize every kind of task that images can do, the notion of the *image* prevents us from treating power as something inaccessible, ineffable and unanswerable. Images, we might say, are the earthly remains of a worldly clash: every image sets forth a particular disposition of powers, from the drawing powers of desire to the staying powers of stubbornness.

Walter Benjamin might have been offering an ancestral motto for Godard's project when he wrote: "History breaks down into images, not stories."[6] And we can now see that the inverse must also be true: History is built up out of images, not stories; at least the history Benjamin wants to tell in the *Passagen-Werk* will be composed that way, for a very special reason. It is a story of how capitalist modernity seizes command of the present, steering it along the course of "progress," by mobilizing dream-images of archaic pasts and fabulous futures. Both the object and the technique of Benjamin's history involve the construction and deconstruction of images in the most precise sense: he wants to develop an optics of historical imagination—exposing the open secrets of commodity fetishism and the official stagecraft of historical development—to reveal alternate temporalities already at work, ready to interrupt the scripted schedule of civilization. For him, looking back on the nineteenth century's unfinished legacy of catastrophes, the images left by history were literally "remains," ruins and ghosts of lost time. By taking the side of images in opposition to the isolated exercise of memory and the forgetful drive of capitalist accumulation, Benjamin acted as if the preservation of those remains might harbor the awakening of life in them, if only for a moment, long enough to remind us that bourgeois society cannot impose its destiny upon the visible world. More importantly, it reminds us that what is "given" to us as the inheritance of history and the accumulated accomplishments of social life are, indeed, imaginary products, suffused with wishful thinking in the best and worst senses.

Again: images remain, and in remaining they sometimes appear to signal the moment of death. We have learned to see how the advent of photography altered the temporal horizons of modern social life by putting the instant in touch with the eternal, no longer through the unpredictable stroke of the aesthetic but through the impersonal intercession of complex mechanical and chemical processes.[7] For Godard, the moment of death is fixed when the act of reflection upon life and death becomes unreflectingly material: he tells the story of Lot's daughters fleeing Sodom and Gomorrah, who are turned to pillars of salt, as a way of recalling that film begins with "silver salts" to capture the passage of light, so that taking a picture is also a fatal "look back" at the world.[8] The need to record what happens to us is here expressed in the name of a "humanity" whose sufferings are regularly forgotten, ignored and hidden: it is Godard's most impassioned political passage in the whole work, where he accuses the "governments of Europe" (but not just Europe) of outright barbarism for failing to stop, failing to see, and failing to name as crimes the atrocities in Bosnia and Chechnya. (And alongside, we see etchings of torture, a quick shot of Mitterand, the Mostar Bridge, a photo of refugees, a painting of finely-dressed women in an opera box, the logo of Universal Pictures...) Our moment in history can readily be characterized by an crippling inability to see what is around us: perhaps, no matter how many images we have of Bosnia, or how few we have of the starving and dying victims of the U.S. assault on Iraq, we still need an etching by Goya or a blast of Mozart to see something there. That is not due to some basic flaw in images, and it would be mistaken to treat the air of sadness about mortality as a quality of the images themselves. Images remain, but the work of remembering and mourning is ours alone, precisely because it is labor that can only be done by us in the present. Through images we call upon ourselves to mourn what remains without being seen: all of the ways we live out the passage of history, through the obligations imposed by the past, through the relays of responsibility that constitute the present moment as something other than a spectacle for our consumption, and finally, through the collective inheritances that define the world as a balance-sheet. (For some there is fertile wealth and for everybody else there are interminable debts). In that sense, every image, insofar as it preserves what is lost and keeps it in view, remains entirely on the side of critique and operates only within the reign of the present. But insofar as images can "remain" by preserving nothing but themselves, they outlast every defeat and forget all deaths. The sense of fatality conveyed by images is our problem, not theirs.

The issues of loss, inheritance, and obligation flow throughout Godard's *Histoire(s)*. Like Benjamin's, Godard's work is a rescue mission: he wants to save images from the breakdown of cinema and the closure of its era. In principle, every kind of image ought to be saved: not just those made by cinema, but artworks, news photos, snapshots, indeed, everything that cinema itself once seemed to threaten. These are the elements that call the history back to life, even when that life is unbearable: "for nearly fifty years/ in the blackness/ the people of the darkened halls/ burned the imaginary/ in breathe warmth into the real/ which now takes its revenge/ and wants real tears and blood."[9] This moment of retribution is indeed the turning point: as for the relationship of cinema to the monstrous crime of the camps, we hear a moment of praise for the documentary impulse: "a single rectangle / thirty-five millimetres wide/ saves the honor of all reality." But we also hear about an unexpected kind of redemption: "if George Stevens hadn't been the first to use/ the first sixteen millimetre colour film at Auschwitz and Ravensbrück/ there's no doubt that Elizabeth Taylor's happiness would never have found a place in the sun."[10] The terrible colors of the real are transformed into luminous black and white. When, in the course of this commentary, the screen flashes the word "Dasein" followed by news photos of Nazi rallies, disfigured corpses, segments of Lang's *Metropolis*, Rossellini's *Germany Year Zero*, and Fassbinder's *Lili Marleen*, the lines of complicity seem to point everywhere without really reaching their marks. This passage remains unresolved; its montage must honor both its historical reference, no matter how unrepresentable, as well as its impulse toward art, no matter how banal, forgetful, or guilty. *In light of this diffused and broken-down history, every image is provisional, but all images are final.* If there is a materialist brand of hope, here it is, and it is all the more remarkable because Godard wrests it from his melancholy angel.

There are two postures and two kinds of politics available in this moment of breakdown.

First, there is Godard as the defiant and apocalyptic artist—"Saint Godard" as Philippe Sollers calls him—who, in his withdrawal from cinema as a public and collective endeavor, claims the right to call down curses upon it. This would be an exemplary kind of "autonomy," insistently dramatized as solitude. (It also draws upon the imaginary map of Europe in which a Swiss village is both outside and somehow in the center of the whole continent.) It is a position in which he can make everything he touches his own, and a bit Lear-like, he stakes all of it on the sovereign love he bestows and withholds at will. This is the figure who seems to inherit both of the traditions in twentieth-century European radical aesthetics—the one who can still wield the rhetoric of engagement (perhaps more Malraux than Sartre), but who pursues the strategies of "autonomous art" prescribed by Adorno, extrapolated into unexpected formal somersaults. (Here it would be best to talk about Godard's music, especially the way his sampling of modernist music precipitates new kinds of negations of it.) If the *Histoire(s)* obey the logic of neither history nor cinema, it is because Godard continues to operate as if his own work proves that the old logics are no longer binding.

As opposed to this hermetic figure, it is possible to discern a rather different kind of position, for which the idea of an "intellectual" might still fit. Instead of exemplary withdrawal, this Godard in fact pursues a stunningly direct intervention into public life, using television in exactly the way it is always used: to address governments, nations, peoples, and the world, all without enforcing the usual boundaries between them. In answer to Bourdieu, it could be seen as a move beyond critique, toward an affirmation of images in the face of their programmatic degradation. From this perspective, the *Histoire(s)* must be seen as an insistent demonstration of the connections between spectatorship and citizenship, castigating the political system for poverty, incivility, and war, while insisting that everybody, all of us, are responsible for what has been put on view. Traveling a kind of magic circle, Godard lays out the most idiosyncratic and opaque paths of thinking to arrive at a view of this historical conjuncture *sub specie aeternitas*. Set against the world-historical dead-end of cinema and television, the *Histoire(s)* shake images free and set them loose. That may be both the best and the worst solution. If Godard's work is neither "autonomous" nor "universal," it is because this profusion of images can figure both loss and creation, mourning and expectation.

And so, a final proposition, which is just a slight alteration of the first one: *images are what remains to be seen*. As long as an image remains in sight, it has not been seen: to be visible means that it awaits another look. When Godard declares: "the image is firstly a form of redemption, and listen—I mean redemption of the real,"[11] his insistence seems impatient, as though we are forgetting the most fundamental point: that we look at images only when we are not done with them, nor they with us. What images "save" is not a physical Real, nor even a psychoanalytic one: instead this Real must be something else.

Near the end of the final episode we hear Godard speaking:

> Yes, the image is happiness
> but nearby, nothingness dwells,
> and all of the power of the image
> can only be expressed by calling out to it.
> Perhaps it is still necessary to add

the image capable of negating nothingness
is also the look of nothingness on us.[12]

Even as we look forward to images, images look back on us: this lesson, repeated so often by artists and philosophers (this passage echoes Blanchot) has never allowed us to settle down in the ebb and flow of the visible, where anticipation and aspiration meet the judgments of history. This much might be learned from Godard's work: that no matter how weak or fleeting its charge of hope may be, every image carries one, simply by being seen. Every throwaway advertisement, every difficult art-film, every news report, every .jpg stockpiled in the hard drive bears its own measure of potentiality, capable of overcoming both the actual and the virtual disposition of things. But if we piece together the whole story of the *Histoire(s)*, we see that images call out to not one but two kinds of nothingness which are far apart: there is the blank page of Mallarmé, where anything might yet be written, and that of Hegel, who tells us that periods of happiness have no place in history, and are blank pages there. No matter how fast the images come—twenty-four or thirty frames a second, or just once, slowly gathered over a lifetime—it remains our task to look for what might be real, or true, or somehow worth saving about them, as long as they remain to be seen.

Notes

1. Pierre Bourdieu, *On Television*, trans. Priscilla Parkhurst Ferguson (New York: The New Press, 1998), 10–11.
2. For an elaboration of the broad conception of technology and its deconstructive reckoning with the notion of "humanity," see Bernard Stiegler, *Technics and Time, 1: The Fault of Epimetheus*, trans. Richard Beardsworth and George Collins (Stanford: Stanford University Press, 1998), esp. chapter 2, "Technology and Anthropology." For Leroi-Gourhan, whose work inspires both Steigler and the recent "mediology" of Régis Debray, see *Gesture and Speech*, trans. Anna Bostock Berger (Cambridge: MIT Press, 1993).
3. Jean-Luc Godard, *Histoires du cinéma(s)*, (Paris: Gallimard, 1998), volume 4, 45. Henceforth I will cite this edition by volume and page number. The CD and book edition was produced by ECM Records (Munich, 1999).
4. Jean-Luc Godard, *Jean-Luc Godard par Jean-Luc Godard, tome 2: 1984–1998*, ed. Alain Bergala (Paris: Cahiers du cinéma, 1998), 430. Henceforth cited in the text as "GPG2."
5. GPG2, 192
6. Walter Benjamin, *Das Passagen-Werk*, volume 1 (Frankfurt: Suhrkamp, 1982), 596. This phrase is extracted from note [N11, 4].
7. See especially Eduardo Cadava, *Words of Light: Theses on the Photography of History* (Princeton: Princeton University Press, 1997), esp. page 128.
8. *Histoires du cinéma(s)*, 3: 31.
9. Ibid., 1: 111.
10. Ibid., 1: 86, 131–3.
11. Ibid., 3: 149.
12. Ibid., 4: 299–300.

8

Ordering Law, Judging History

Deliberations on Court TV

Lynne Joyrich

To illustrate the proper use of the word "archive" in its definition of the term, *Webster's Revised Unabridged Dictionary* offers the following sentence, taken from Richard Allestree's seventeenth-century tract on correct speech and judgment entitled *The Government of the Tongue*: "Our words . . . become records in God's court, and are laid up in his archives as witnesses."[1] Bringing together notions of recording, witnessing, and judging in his guide to ethical life through speech, Allestree conjures an image of a heavenly court that also serves as a sort of evidence file of enunciations, a realm of both law and the chronicling of discursive details, which—if only we could see these court records—might finally provide us with the means to know and to value appropriate codes for living. Yet what are the relations that exist between discursive, judicial, archival, and viewing systems? What does it mean to come to know life through court recording, and how is that knowledge produced, held, and disseminated, whether through immediate divine law or our own mediated versions? How might we visualize this court and this archive, and what effects—discursive, epistemological, ethical—might therefore emerge? In this essay, I attempt to approach such lofty questions through what might appear to be a somewhat lowly route, for I take this image of court proceedings which one might witness and record, not as just metaphorically instructive, but quite literally: as the image that I witness on my television set when I tune in to Court TV.

A network devoted to recording and televising trials, from the mundane to the dramatic, the Courtroom Television Network (better known as Court TV) would seem, in many ways, to embody Allestree's vision of judgment time at the ultimate bench, particularly since its own vision (involving, as I will elaborate, its own time of judgment) is presented as coming from a rather godlike perspective, the camera never revealed and so never attributed to a particular source. It might also appear to strive, in some respects, to his image of the archive: in order to provide it with material for further television productions, the network maintains a video library of everything it's covered, carefully logged so that all of the cases, lawyers, judges, and defendants who have appeared on the network are catalogued and cross-referenced in so many "records laid up as witnesses"—though viewers' access to this archive seems to exist at the mercy of the powers-that-be at the network headquarters, granted by the discretion of one of its executives based on the persuasiveness of speech of the supplicants.[2] Of course, for those who cannot gain access to Court TV records in this matter, the network sponsors an easily accessible Web site (CourtTV.com), with two other affiliated sites as well (TheSmokingGun.com and CrimeLibrary.com), on which one might find, among other things, case files and summaries; copies of pieces of evidence entered into trial records; a sampling of court documents; reports on legal and political news; crime statistics; a directory of "mob bosses," "outlaws," and "unique gang organizations" as well as directories of serial killers and "terrorists, spies, and assassins"; a compilation of wills of famous people; mug shots of the rich and (in)famous; a list of "stupid crimes and misdemeanors"; a glossary of forensic and investigation terms; compilations of clues for "cold cases" not yet solved or brought to trial; a verdict and judgment archive; an audio and video archive; message boards and chat transcripts; and, of course, programming and scheduling information—a whole constellation of data and details about our judicial system and the television network devoted to covering it.[3]

However, such actual (or virtual) archives, investigations, and judgments are not my primary interest here. Rather, I am interested in a more Foucauldian notion of the archive—conceptualized as a system that governs the range of possibilities for and dissemination of statements, the ordering of a discursive network—and in the knowledges that *this* archive makes possible.[4] Yet in discussing Court TV, one might find a double meaning to these terms. Court TV is certainly, in institutional media terms, a network; but, beyond just indicating its position within the television industry, the word "network" suggests the way in which media outlets such as television channels function as sites for discursive and epistemological production. How the law is ordered on television (on this and other networks), how its statements are enounced (and, by this, I, along with Foucault, do not simply mean legal statements) determines the judgments we reach.

The cases presented on Court TV do literally reach judgment (at least most of them do), but "judgment" here too has a double meaning. A judgment is, of course, a judicial decision, the legal determination of a court of law; but the word also refers, in a more general way, to a critical faculty, the capacity for reason, an operation of knowledge. It might be reasonable to assume that an analysis of the Courtroom Television Network would privilege the former definition, the juridical meaning of the term "judgment," over the latter, the less governed use. Yet, in this essay, I am actually more concerned with the other sense of the word—with the way it alludes to the very deployment of sense: judgment as an epistemological procedure, one that might be activated in our engagement with television and the law. The deliberations I'd like to consider are thus not just those of the juries who hear specific Court TV cases; I find the "case" of Court TV itself suggestive for considering the nature and order of epistemological deliberation—that is, for thinking through the ways in which, through various cultural and media forums (including, importantly, television networks), processes of knowing are offered or refused.

The texts on or about the Courtroom Television Network provide examples of such forums, and, to recognize this, I've broken my discussion of its "telepistemology" into sections marked by the titles of current or past Court TV programs. This is not because I want to analyze these programs as discrete works set off from the flow of the channel or from TV as a whole (quite the contrary); it is merely to note how the medium itself categorizes and names some of the strategies of knowing that traverse and organize its flow, to acknowledge the way in which television differentiates and/or interweaves its discourses and, in so doing, constructs processes of knowledge from which we, as educators and cultural critics, might also learn.

1. "In Practice": Televisual and Intellectual Histories

My interest in mass cultural deliberations began years ago when I was completing a book on television, gender, and postmodern culture, investigating how discourses on the media figure television in particular terms that guide how we discuss and "know" it and how television, in turn, provides us with particular figurations that help to organize our understanding of our culture. During that time, I found myself more and more drawn to the problem of how media and mass cultural forms impede—and even more intriguingly, how they also create—specific modes of knowledge: not only what might be considered archives of cultural citations but epistemologies and ways of seeing for audiences who become involved in the texts and cultural forms. I thus began my own deliberations on questions of popular knowledges and pleasures, the personal and pedagogical, mass culture's "instruction" of our will to know and its ordering of the thoughts we're then led to entertain.

Most members of the academic world make a number of assumptions about knowledge: what it is, who has it, how to accumulate it, and how to impart it. It is the creation, organization, and dissemination of knowledge that, in fact, defines our sphere of labor, and scholarly practitioners thus tend to claim the product of this labor, knowledge, for themselves. Historians of discursive regimes as well as theorists and educators working within the realm of social epistemology have questioned the boundaries drawn within and around this scholarly expertise, and my work is indebted to their investigations of the ways in which knowledge is objectified and disciplined—subject to particular ruling paradigms, selection processes, and technical divisions that guide research and training.[5] These investigations emphasize how professionally transmitted knowledge is situated and formed; in a reverse but hopefully complementary movement, I'd like to consider how forms other than those professionally recognized as educational or epistemological ones also transmit knowledge—how what we might call "cultural epistemologies" operate out- or along-side of the scholarly situation.

Struggles over knowledge certainly define and divide communities within the academic arena, but what occurs for other communities, in other discursive arenas? How might knowledge be defined differently and work in different cultural and subcultural formations? Do mediated and/or popular texts have distinct rules for knowledge production, compilation, and dissemination? What possibilities then exist for media instructors and workers in teaching such forms, and in what ways do such texts instruct us? As these questions suggest, I hope in this essay, and in the larger intellectual project of which it is a part, to interrogate the relationship between different constructions of institutionalized (and non-institutionalized) knowledge, including those of the media as well as those of the academy itself.[6] Rather than assuming that knowledge is somehow the "property" of the academic world—that it properly belongs within academia (and academia alone)—I am interested in exploring how socially and culturally significant knowledge might be activated and deployed within a number of spheres, even those that have been considered trivial or illusionary, too "mass"-oriented or conversely not enough, merely "personal" or overtly "political."

Though our popular "escapisms" are typically counterposed to cultural enlightenment, the knowledge generated through what escapes our knowing scholarly judgments may most reveal the stakes in our epistemological desires.[7] For that reason, the popular texts that I am interested in addressing are ones that I feel personally invested in—not simply texts that I've enjoyed but those that I also feel have taught me something through, rather than in spite of, this enjoyment (texts that I believe have educated my desire to know as they've encouraged me to entertain a specific set of ideas, or at least ways of forming ideas). Having been an enthused viewer of the channel, I would include the Courtroom Television Network in that category. In fact, as a network that quite openly engages with the notion of "knowing judgments"—yet does so, I suggest, in a distinctive way through its specific televisual conventions—Court TV provides an ideal site for analyzing the formation of knowledge on television and for considering both how that formation might intersect with and/or differ from scholarly conceptions of knowledge and how it produces a particular epistemic and viewing appeal. I cannot judge the value of the legal education I've received (or not) from Court TV, but I can certainly testify to my fascination with the phenomenon of "trial television" and the way in which it incites my own epistemophilia, my desire to know.

Indeed, there have been times in which I've been drawn to my TV set as if literally summoned by the Court TV anchors. Given the huge growth in the cable network's ratings over the past few years, apparently many others have been called as well.[8] The channel now reaches 75.8 million homes with (at the time of writing) a Nielsen rating of .9, taking as its chief competitors in the cable industry channels such as A&E, the SciFi channel, TVLand, BET, Bravo, and MSNBC.[9] Remarking on the network's place within the U.S. media landscape, one Court TV executive with whom I spoke suggested that, together with CNN and MTV, Court TV has had the largest impact on the shape of today's television of the array of cable networks[10]—an assessment with which I am tempted to agree, not only because of Court TV's focus on law and investigation (so key both to our litigious society and to the narrative themes of U.S. television) but also, as I elaborate in the essay, because of its construction of time and space, narrative and discourse, and, through those, its construction of knowledge. Despite the network's (relatively) small size, the influence of a channel such as Court TV thus need not be so surprising given its particular televisual expression of the discourses of a legalistic culture—a culture in which, as former Chief Justice Warren Burger argued, the courts are expected "to fill the void created by the decline of church, family, and neighborhood unit," one in which the law (not to mention television itself) hence holds great significance, authority, and fascination.[11] It is Court TV's deployment of this legal and epistemological fascination that I would like to examine so as to better understand the operations of its particular televisual "courting."

Let me then move into the details of my analysis of the Courtroom Television Network with a brief account of my involvement with Court TV. While it made its debut in 1991, I really began to get interested in the station in the summer of 1993 (the beginning of a significant period, not only for my description of some specific cases, which largely draws from that moment, but for the network itself, since this was when the Courtroom Television Network was first gaining prominence and popularity, thus making an analysis of the terms through which it promoted itself and treated its topics at that time particularly telling). I had occasionally tuned in before that period of rising network popularity to check out the then-new cable channel, but had always found it somewhat dull: nothing much ever seemed to happen. Yet that summer I found myself waking up each morning with anticipation, looking forward to watching my favorite show: "Menendez." What had changed? Was it just because the Menendez brothers' trial was great drama? Well, in part, yes, it was. But this wasn't the only reason, nor was it the only case on Court TV that I followed. Over the length of time in which the Menendez trial was gearing up and going on,[12] I also watched (and became very intrigued by) the murder trial that resulted from the Detroit Police beating of Malice Green, and later, the trial commonly referred to as the "Reginald Denny case," that of some of those involved in the L.A. uprising following the beating of Rodney King.[13] In other words, I became involved with many narratives on Court TV—with its multiple narrative structure as a whole. Even if somewhat repetitive, it no longer seemed boring to me. Its daily constancy was part of its appeal, and instead of leading to monotony, this seemed as if it opened up a whole network of interlocking discourses and stories: the many cases, the many conflicting stories within each case, the stories of the lawyers themselves, and even the stories of the Court TV reporters (some of whom moved onwards and upwards to what, in an epistemic hierarchy, were seen as "better" media outlets).

Because I've been a long-time soap opera fan, Court TV's narrative form was quite satisfying to me; I was comfortably familiar with its pluralized and serial storytelling, articulated largely within the times and the terms of the everyday (statements enunciated in "legalese" notwithstanding), and, during that period, my Court TV viewing was very much like that of my soap opera viewing. Many of those who've appeared on the channel (the Menendez brothers, Lorena and John Wayne Bobbitt, William Kennedy Smith, and so on) can be compared to a soap opera's cast of characters; indeed, they often were in the popular press, though more as a means to belittle viewer interest than to consider how the daytime serial constitutes a particular way of seeing, correlating, and understanding. But more interesting and relevant to my argument is that the very narrative patterns typical on soaps and Court TV corroborate each other, even down to the predictable but no less troubling fact that stories of sex, greed, and familial dysfunction—the overarching themes of the Menendez trial—were granted more screen time and narrative priority than stories of race and political disenfranchisement—the themes of the Detroit and L.A. trials.

Strangely (or maybe not so strange, as I hope to elaborate), the case that came to rivet all of America—the O.J. Simpson case (which was covered from the summer of 1994 through the fall of 1995, not only by cable's Court TV but, to a large degree, by many mainstream broadcast channels as well)—didn't really grab me. Given my love of Court TV, this may seem a bit hard to explain. Perhaps it was that my own identification as a soap opera viewer came into conflict with my identification as a trial viewer: annoyed by the way in which news of O.J. constantly interrupted and impeded my soap viewing, I joined in the chorus (or one might say "jury") of those viewers who angrily demanded that O.J. coverage be confined "over there"—on the Courtroom Television Network, in a particular category of televisual discourse—rather than let loose on all of the network's daytime schedules.[14]

A common remark volleyed back to such complaints was that viewers who were kept from their daytime dramas didn't really miss anything: all of the scandal and family chaos that one might expect from one's soap was available in the Simpson case itself (a case that, as many commentators noted, brought together the themes of sexual and racial tension that I mentioned in my schematization of previous cases a moment ago).[15] But I would argue that this characterization of the knowledge and pleasures of O.J.T.V. isn't actually true—and it's not true in a way that goes beyond the obvious point that, while scandalous and crisis-ridden, the Simpson trial did not involve the particular family scandals and crises in which regular soap viewers were invested at the time. For rather than maintaining the soap opera-like structure of Court TV that I mentioned before, coverage of the O.J. trial actually disrupted that format, arresting the movement between multiple narratives by the more totalized attention given to this one trial. (It was still not completely totalized; adhering to network founder and CEO Steven Brill's commitment to providing a mix of legal hearings,[16] Court TV continued to cover other trials, but the Simpson case heavily outweighed them in the daily schedule and the station's narrative and discursive emphases. It was also not long after this case that the network began, more broadly, to change its profile, redirecting some of its energy away from its daytime trial coverage, with the multiplicity of court-related activity that involved, to a more focused attention on singular stories, particularly as narrated through prime-time documentaries and syndicated detection and crime series).

In other words, Court TV's typical "balance" slightly shifted with (and after) the Simpson case. I'm not just referring to a balance of diverse legal perspectives nor even to that of diverse statements and cases per se, but more importantly, to its balance of boredom and excitement, public and private captivation. For Court TV (at least pre-Simpson Court TV) made a double offer of pleasure-in-knowledge. On the one hand, enacting cable television's strategy of "narrowcasting," it promised to grant a somewhat obscure knowledge of court cases (creating my delight in knowing so many little details about the Menendez brothers, not to mention their lawyers and their psychiatrist, that no one else around me knew, making me a kind of private, living repository of Menendez miscellanea). On the other hand, asserting television's traditional power to construct "media events," Court TV provided participation in a fully shared (almost official) fascination. As the Simpson trial became a national obsession (as opposed to an aficionado specialty, subcultural interest, or dissenting TV judgment), it incited a different type of engagement, activating that more totalized conception of knowledge encapsulated in the term "TV coverage." No longer so invested in the oscillation of private and public interest, secret and shared knowing, esoteric or exhaustive archive, Simpson-era Court TV ascribed to a slightly altered epistemological structure.

The ensuing visibility of the channel in the more openly lawful domain of civic duty may have elevated it in the court of public opinion (not to mention that of ratings and advertising revenue), but this also altered Court TV's usual format, leading to what was, in effect, a generic and epistemic shift in the network's textual composition. This took the form of two interrelated changes: a move from the temporal and narrative structure of the continuing serial to something more closely resembling a self-contained mini-series (albeit, a very prolonged one), and simultaneously, a reconstruction of the processes of knowledge formation it elicits from what I would call, after Claude Levi-Strauss, "bricolage" (a piecemeal collection) to those of a more overarching "science" (an abstracted plan).[17] While less to my personal liking, it was this shift, and then, post-Simpson, the return (at least in the daytime Court TV schedule) to something close to its previous epistemic mode, that led me to reflect on the ways in which Court TV's trial coverage generally operates to construct narrative, time, and, through them, knowledge and pleasure—the elements I turn to in the next sections.

2. "Trial Story": Narrative on Court TV

In some ways, the deployment of narrative on the Courtroom Television Network seems to be much like the deployment of narrative in the law itself: while each depicts itself in terms of the ordering of facts in the pursuit of the truth, they are actually more about producing stories than excavating a buried reality and a hidden *veritas*.[18] This, indeed, is what troubles many critics: justice abstractly defined as "right" vs. "wrong," order vs. chaos (with truth and order, of course, on the side of the right) becomes, in effect, a materially determined tale-telling contest. The law maps out rules for constructing, assembling, and presenting these tales while court cases pit competing narratives against one another, leaving juries and judges to decide which stories are more credible and/or elegant. To some, this might not seem all that different from TV's ratings wars, for television also involves a number of intersecting and competing narratives, drawing from a range of genres and discursive forms in order to encode multiple (and sometimes conflicting) knowledges and modes of address.

This parallel (the mutual strategy of creating knowledge through interlocking yet divergent narratives) suggests that law is an ideal forum for television—a suggestion more than borne out by the extensive history of television law shows beginning in the late 1940s and early 1950s with such shows as *Court of Current Issues* (1948–51), *Your Witness* (1949–50), *Famous Jury Trials* (1949–52), and *They Stand Accused* (1949–54).[19] This "trial" shot of courts on TV continued into the following decade and beyond, as demonstrated, for instance, by the long running series *Perry Mason* and *Divorce Court* (which aired in the '50s and '60s, for nine and twelve years respectively, before being revived again in various versions in the '70s, '80s, and '90s).[20] An interesting example that points to the interpenetration of television and the law—and to its sometimes contradictory effects—can be seen in the 1954 "real-life" murder trial of Dr. Samuel Sheppard, who was accused of bludgeoning his pregnant wife but claimed to have fought off an intruder who entered the house and committed the crime. Not only did this case generate enormous attention to the relationship between media and law (Sheppard, defended by attorney F. Lee Bailey, eventually won an appeal in 1966 on the grounds that the media coverage had impeded his ability to receive a fair trial—a strategy also attempted with some of the more recent cases I've mentioned), but it definitively extended the media/law relationship as well: the story supposedly gave rise to the popular TV series *The Fugitive* (1963–67), the final episode of which garnered a larger audience than any other single episode of a regular series up until that time (and not surpassed again until the revelation of "who shot J.R." on *Dallas*, thirteen years later).[21]

A similar sort of media/law interpenetration might be seen in more recent television history which again boasts a number of (otherwise very different) court shows. To name just a few, one might consider *Ally McBeal*, *Family Law*, *The Guardian*, *JAG*, *Judging Amy*, *L.A. Law*, *Night Court*, *The Practice*, and, of course, the various *Law and Order*s, the original of which was modeled on the earlier text *Arrest and Trial* (1963–64) and itself provided the model for Court TV's own program *The System*, once described by Steven Brill as a "nonfiction *Law and Order*."[22] *The People's Court* was one of the best known television programs to sell itself on just that type of "nonfiction" status noted by Brill: an early precursor to today's spate of reality TV shows, on that program "real" plaintiffs and defendants agreed to bring their small-claims cases before the retired Judge Joseph Wapner, in exchange not only for the privilege of appearing on television but of having the producers reimburse them for the outcome.[23] Other programs (both before and after *The People's Court*) have exploited the mix of "real" and "dramatic," "factual" and "fictional" elements as well. (For instance, on the previously mentioned *Divorce Court* and on the 1958–65 program *Day in Court*, actors portrayed litigants in fictional or "real-life" stories, but the cases were argued by actual lawyers in front of law professors or retired judges).

The popularity of such "actual law" shows seemed to ensure the success of not only many new examples of reality programming, but of Court TV itself, whose own success in turn arguably helped to promulgate a number of other (though frequently short-lived) programs of both the narrative and non-narrative variety (such as *Courthouse*, *The Home Court*, *Jones and Jury*, *Judge for Yourself*, *Judge Greg Mathis*, *Judge Hatchett*, *Judge Joe Brown*, *Judge Judy*, *Style Court*, and more). It also established connections between Court TV and other television outlets, with, for example, Court TV providing news footage for various networks, developing syndicated programming (such as *Inside America's Courts*, a show that was produced by—but not aired on—the Courtroom Television Network), and entering into joint ventures (such as Court TV and NBC's plan to co-produce the program *Trial By Fire*, initially considered for the 2003 season); and as I'll elaborate a bit later, explicit historical and textual links were created between Court TV and Steven Bochco's critically acclaimed drama *Murder One*.

While the network and syndicated examples that I've given might all be described as "court shows," these programs run the gamut of formats: from prime-time to daytime and late-night programming, including soap operas, police stories, military dramas, sit-coms, "dramadies," current affairs, talk shows, makeover programs, and "real case" situations. Court TV itself has drawn on many of these televisual traditions: not only does it bring us into the real courtroom in its continuous, live coverage of actual trials,[24] but (even before turning to prime-time airings of various syndicated programs) it has explored numerous other genres as well. Over its history, it has included its own current affairs and talk show programs (*Washington Watch*, *In Context*, and what were called its "Open Line" segments); aired small-claims disputes à la *People's Court* (*Instant Justice*); and produced narrativized summaries of some of its most dramatic trials (*Trial Story*, from which I take this section's title); it once even considered developing a Saturday morning block of informational programming for youth (dubbed "Court TV Kids").[25] That last plan was scrapped for a program that provides educational videos directly to the classroom, but Court TV still airs a variety of types of texts: discussion programs (*Catherine Crier Live*); investigation series (*Extreme Evidence*; *Forensic Files*; *I, Detective*); criminal biographies (*Mugshots*); what might be labeled legal gossip shows (*Dominick Dunne's Power, Privilege and Justice*; *Hollywood at Large*; *Hollywood Justice*); even interrogation/detection game shows (*Fake Out*; *House of Clues*). It thus brings together not only TV's many narrative forms but television's fictional and non-fictional modes, its entertainment and educational roles.

Exploiting its similarities to yet differences from the surrounding televisual landscape, Court TV has promoted itself in typical marketing fashion through patterns of product recognizability and distinction, comparing and contrasting itself with other (fictional and non-fictional) media offerings. This was particularly important to the network when it was first gaining national recognition though not yet moving into serious prime-time competition—in other words, when Court TV was attempting to sell the public on the value of its steady daytime diet of live court coverage (a diet that the network had to promise, through intertextual reference, wouldn't be too bland). For instance, one early station advertisement cited press reviews of the channel, including this typical quote from the *Boston Herald*: "Watching Court TV is better than watching *L.A. Law*"—a recommendation that operated by assuring us that non-dramatized television is indeed more dramatic than its fictional cousins. Another ad—also declaring Court TV's heightened television status precisely, and paradoxically, by proclaiming its very lack of televisuality—stated, through a voice-over echoed by on-screen titles [the words emphasized below]:

> A powerful new TV network is captivating America. There's *no script. No actors.* The lighting and sound can be *downright lousy.* But the human issues and emotions are more vital and riveting than most Hollywood movies. *Court TV. Real trials. Real people. Different television.*

The simultaneously self-depreciating and self-congratulatory comment about the network's "lousy" picture further resonates with debates over Court TV's visual style. These too center on the relationship of its camerawork, or lack thereof, to the dual roles I've mentioned (dramatic license or lawful accounting of the real). Differently read in terms its adherence to and/or abdication of narrative, Court TV's minimalist aesthetic has been lauded from those on both sides of the "entertainment"/"education" spectrum. Just to give two examples, an admiring Steven Bochco identified its camera's "dispassionate" gaze as the key to Court TV's success; it is this, he implies, that allows viewers to become passionately involved in the dramatic legal scene.[26] Ironically echoing Bochco's valuation, although here emphasizing the camera's ability to capture not viewers but truth (a pedagogical rather than popular triumph), Yale law professor John Langbein describes the courtroom camera as a "godsend" able to correct the misperceptions derived from watching *Perry Mason* and Bochco's own *L.A. Law*.[27] Extending the long take beyond even our wildest Bazanian dreams, Court TV appears to be both revelatory and engrossing, equally orchestrating contemplation and consumption through its pans, dips, and zooms.[28]

My point in noting these visual and narrative features, though, is not to debate if Court TV, in recalling the "golden days" of live television or in referencing more recent public service channels (such as C-Span), really does educate us, judiciously informing us of "the whole truth" of our legal system; or if, alternatively, as critics state of other "actuality media" (such as webcams and TV's reality shows, both of which also promise the real but, of course, involve complex staging), it instead lures and manipulates us through an unwarranted use of theatrical devices and sensationalizing effects.[29] Not only is there no such division (Court TV is simultaneously seeking more dramatic, high-profile cases *and* continuing to develop various educational school products, informational online services, and classroom CD-ROMs), but I would caution against making such demarcations without carefully examining the connotations of the terms, the epistemic ranking they presume of "realism" over "sensationalism." That is to say, I am less interested in mapping a distinction between "real" and "dramatic" television, fair and unfair coverage, legitimate or illegitimate learning, than in exploring the forms of knowledge and pleasure that our fascination itself might yield or evade, the kinds of ideas that our very enjoyment allows (or inhibits) us to entertain. Given what I've said about the many narrative forms that Court TV employs, these pleasures might seem as if they need no further discussion; as I've described it, Court TV sounds exciting and diverse, pulling together a colorful collection of segments and stories. But as previously noted, the experience of watching it is often quite the reverse: particularly for the uninitiated, it can seem rather monotonous and tiresome. For all of the time that it can take, little seems to happen; transmission may be live (actually on a 10-second delay), but it still seems to many to be deathly dull.

3. "Instant Justice": The Time(s) of Court TV

What is this structure of temporality in which time ceaselessly goes forward in coverage that keeps us ever up-to-date, even though it feels as if it barely moves at all (a feeling that was particularly strong in the early years of Court TV when prime-time programming simply consisted of a repetition of the day's events, though still incited today by the network's relentless coverage of its cases and discussion of these same cases in its news reports and talk shows)? If anything, it is this temporality of what one might call flowing suspension (or, conversely, suspended flow) that seems to distinguish Court TV. Though the Courtroom Television Network is often described in terms of its premonition of and participation in "the 500-channel, interactive, instant-information television of the future," by relying on familiar conventions from news, sports, talk TV, and television drama, Court TV may only be "new" in its structure of time.[30] For Court TV seems to construct a dual (if not triple) sense of time, split between instantaneity and protraction, teleological advancement and spiraling reverberation.[31]

In some ways, particularly in its moments of punctuation—its opening credits and station identifications—Court TV heightens television's illusion of urgency and immediacy. Network promotions feature dramatic music and fast-paced editing, juxtaposing, to give one example from an early station promo, quick images of people running up courthouse steps, lawyers pulling guns and other explosive evidence out of plain paper bags, suspects being fingerprinted, a judge's gavel slamming down: images that lend the iconography of a detective show and the excitement of a police chase to the otherwise slow progression of a court hearing (a strategy even further heightened in Court TV's recent "the investigation channel" teasers). As with other "real-time" televisual offerings, the hearings themselves are enlivened through the use of a game or sportscasting model: commentators engage in a play-by-play analysis of each "team's" strategies, bolstering the network's production of temporal urgency by implying that the tide can turn at any instant, that any subtle shift in tactics might be the decisive moment in the case. This notion of time has indeed been graphically expressed. The logo that was used for Court TV's *Prime Time Justice* (a nightly wrap-up program that reviewed the day's trials) resembled a clock face that filled up the screen as the "hands" revolved in full circle until they "caught up" with themselves, justice empowered to close crime and detection's, law and order's, temporal gaps.[32] Such sounds and graphics suggest an insistence, almost anxiety, of time: the urge to push forward, the need to keep up with the future while putting events in their present place.

As might be expected of a channel that so carefully records and keeps track of its time, this temporal advancement doesn't just point toward the future; it also carries the archival weight of the past, reminding us of a revered legacy. Many of the network promos gave Court TV an honorable (if melodramatic) heritage of televisual and civic duty. In one, familiar images of "great old moments" captured on film or videotape (John Kennedy's funeral, a spaceship launch, a response to a terrorist attack) flashed on the screen as a narrator claimed:

> For almost three decades, nothing has touched our lives more dramatically than the events we've watched on live television. Today only one network is dedicated to bringing you the kind of real-life drama that can only be found on live television. *Before you hear it on the evening news; before you read it in the morning paper; watch it happen live. Court TV: be there live.*

While underscoring the immediacy of its coverage (repeating variants of "live" six times), this ad validated the promise of such presence by references to important moments (not quite) gone by, placing the network in a zone that, curiously, seems to exist both in and outside the ephemeral passage of time. In fact, one might argue that, in its temporal claims and practices, Court TV adheres to Michel Foucault's description of the "particular level" of the archive as that of neither pure tradition nor "the welcoming oblivion that opens up to all new speech the operational field of its freedom"; rather, "between tradition and oblivion, it reveals the rules of a practice that enables statements both to survive and to undergo regular modification."[33] It is this "regular modification," a mix of constancy and change which suggests both an immediate history and an historical immediacy, that seems to grant Court TV, if not a legal, then a certain discursive and, I would argue, epistemological authority—one that the network uses for promotion but one that might also operate for viewers, as I elaborate below, in quite different ways.

Further suggesting the paradoxical notion of immediate tradition, another network ad made such a conflation even more emphatic, interweaving televisual and national history in order to authorize Court TV's own jurisdiction.

> When the Founding Fathers wrote the Constitution, they made sure that trials in the bold new experiment called America would be public; so courtrooms were built with large audience galleries and people watched the debates that transfixed the community.... [Here, there's a brief visual and aural insert of a dramatic moment from a trial concerning life and death, illness and bodily decay: "She may not die; she may end up a vegetable," testifies one teary witness. The voice over then continues.] Courts haven't changed and neither has our desire to witness these events first-hand. Court TV: an idea as old as America.

Combining an emotional reminder of the rush of passing time that live TV can capture with an appeal to the solid traditions of our nation's past, these station promotions both mark Court TV's place within a trajectory of temporal progression, an advancing march of time that viewers are invited to eyewitness, even as they also interrupt—and so retard and impede—that very sense of linear progression.[34]

In other words, while I've been discussing the channel's production of a sense of urgency as time goes by (despite, or perhaps because, of its equal claims to weighty tradition), as viewers might have guessed from even how long it takes for the Court TV "clock" to apprehend its own temporal passage, the sense of time insistently proceeding onward is not the network's only mode of temporality. In between the moments of dramatic punctuation which suggest a stirring temporal advancement are the actual proceedings: the "real time" coverage of its cases that to many viewers seems tedious at best. As previously mentioned, such slow unfolding of daily events across multiple narrative registers resembles a soap opera more than a fleeting tactical match (Court TV's nod to sportscasting and game show conventions notwithstanding). This proliferating, serialized deployment of story and time itself has important discursive and epistemological effects: the ensuing inter- and intra-case flow creates suggestive resonances and reverberations that may then feed back into our understanding of the various trials, the law as a whole, or perhaps most intriguingly, our culture's "laws" of gender, sexuality, generation, race, and class. It is thus not simply the second-by-second amassing of informative details that produces Court TV's effects of knowledge; this operates alongside an extended deployment of interwoven stories and identifications that provide both a context and a counterpoint for such details, yielding an epistemological network that is, at the same time, a social and cultural map.[35]

Let me give an example to illustrate my point. During the trial of the white Detroit Police officers Walter Budzyn and Larry Nevers who beat and then killed an unemployed African American steelworker named Malice Green, not only did I become very involved with the stories of racial prejudice and class anxiety told from the witness stand, but also in the intersecting tales (equally telling in terms of race and class) of the lawyers on the case. These focused on Kym Worthy, an African American district attorney who had worked largely to prosecute black defendants but began making a name for herself as really tough on racist cops, and Michael Bachelor, Budzyn's African American defense lawyer who once worked with Worthy in the D.A.'s office before moving into private practice where he largely defended black clients; mid-trial, he began receiving death threats for his new work on behalf of the cops, and the Court TV reporters noted that he grew increasingly haggard and depressed throughout the case. The insights gleaned from Worthy's and Bachelor's "private" stories of professional pressure and advancement televisually testified to the social tensions and political problems of urban life as much as (if less chillingly than) the violent beating of Malice Green itself did, adding a "personal touch" that made Court TV's coverage all the more affecting (indeed, even these lawyers' names sound like TV star material).

At the same time as the layered dramas of race and class placement were being enacted in the Budzyn and Nevers case, Court TV was also covering the first Menendez trial—a trial in which class ambitions and ethnic heritage (or perhaps class complacency and ethnic disavowal) clearly played important roles even though the Court TV reporters and commentators rarely mentioned these aspects as they directed attention instead to questions of dysfunctional relationships, childhood abuse, and legal culpability. As with the Detroit coverage, these familial and sexual tales were complemented by arresting images and stories involving the lawyers or others tied to the case: attorney Leslie Abramson's habit of patting her client Erik Menendez on the head and back like a small child until the judge put an end to such visible displays of maternal affection; Abramson's adoption of her own child that briefly interrupted the court proceedings; and finally, the long convoluted tale of adultery and sexual abuse that affected the brothers' psychiatrist, his marriage counselor wife, and his maybe-lover/maybe-prisoner who first broke the news to the police about the Menendez brothers' murder confession after eavesdropping on a session. More than just demonstrating the intersection of legal and therapeutic discourses, the interlocking narrations of personal detail and social division—ranging over issues of sexuality and violence, gender and generation, class and ethnicity, privilege and power—echoed against one another, creating layers of intertwined meaning through the passing of time, an epistemological network activated by a temporal entanglement.

While not as drawn out and involving smaller casts of characters, similar descriptions could be given of other Court TV examples. In an essay in *The Nation*, Lewis Cole writes of his interest in (indeed education through) Court TV's treatment of John Wayne and Lorena Bobbitt's cases in 1993–94 (the charges brought against him in a marital sexual assault trial, which Court TV reported on but did not televise, and those brought against her for severing her husband's penis in the "malicious wounding" trial, which Court TV showed in full).[36] Much more than simply the material for a collection of off-color anecdotes, what emerged for Cole across the two cases—or, more accurately, in the amount of testimony and coverage in one as compared to the paucity in the other, the concern and attention over his scars, not hers—was a slow revelation of domestic violence, an understanding of the state's (witting or unwitting) complicity in such violence, and even a sense of the way in which Court TV itself participates in this complicity by failing to challenge certain state tactics and terms of the law, not to mention those of TV itself. It was thus precisely what many critics have attacked as Court TV's "sensationalist" treatment—meaning not only its sometimes headliner cases but, as with the textual construction that defines the genre of the "sensation novel," also its spiraling and intersecting temporal and narrative paths, its meandering and/or abrupt shifts from one case to another—that resulted in the network's exposure of the limitations of the law and the media as well.

Cole describes the effect of this as a "living drama," "a lesson in social reality," a lesson, I would argue, initiated through Court TV's management of time.[37] Brought together, the various examples I've given of the network's cases and commercials suggest that, through its production of flow and segmentation, Court TV offers a dual register of temporality (with corresponding spatial coordinates as well). In the movement across moments of punctuation and continuity, compression and extension, time is constructed as both something to be kept up with (the urgent time of public consequence) and as something to be mulled over, even savored (the amplified time of private revelation). What is caught between this double dimension of time and space is precisely, I would argue, the temporality of history itself—not only the personal histories of those involved in the cases nor U.S. legal history in the making, but a social and cultural history: a map and a chronicle of the changing discursive categories (legalistic and popular) by which citizens are identified, called to the stand, and held accountable. In his discussion of the archeology of knowledge, Foucault laments the difficulty of apprehending, and so interrogating, what he calls our "historical *a priori*"—the historicity of the archive that is our own discursive regime, which can emerge for critique, he writes, only in "fragments" since it exists at the thresholds and discontinuities of statements, at "the border of time that surrounds our presence, which overhangs it, and which indicates it in its otherness."[38] Yet despite the difficulty of seeing and so learning from this archive, networks like Court TV, through its own operations at the "borders of time," may (even if unwittingly) give us a glimpse.

4. "The System"/"In Context": Summoning Knowledge Through Court TV

How then does Court TV deploy or displace historical knowledge? As I've described it, the flow of the channel shuttles the viewer back and forth between the public and the private, the urgent and the aimless, detail and overview, update and record, rapid change and steady continuation. Never directly addressing how these dimensions come together (how the prolonged formation of personal identity is itself an unstable political process), Court TV maintains our fascination in historical narration without revealing how its narratives are themselves historically constructed. While, as Cole's example demonstrates, viewers are certainly not prevented from (and might even be motivated toward) considering Court TV's discursive choices (its adherence to various televisual, legal, and political conventions), the network itself does not subject its own construction of stories and timelines to nearly the same scrutiny as those of the lawyers' who appear on the air; its particular location between and across distinct temporal and narrative registers is effective without being avowed. It is almost as if its two modalities of time act as continual alibis for one another. We are given the evidence of history but no conclusive verdict; it is witnessed without being held up for trial.

It is perhaps because of this absent causality that viewers often attempt to enter into the course of events themselves—to impact the outcomes that Court TV vows merely to report by doing such things as offering themselves as witnesses or calling the lawyers seen on the air with advice on the cases, attempting to connect their personal viewing histories with the public proceedings of TV and the law. In this way, viewers may strive to bridge the gaps between televisual and historical time, private and civil space, personalized and depersonalized knowledge. While such attempts directly to determine legal and televisual action can be naive, this nonetheless points to the way in which Court TV uniquely solicits an historical and epistemic engagement, inviting us (even if inadvertently) to entertain a self- and social-consciousness. Typically, today's popular representations have us know history in one of two opposite, yet paradoxically complementary, ways. It may be portrayed, on the one hand, in purely personal terms—what we might call the *Forrest Gump* version of social change (after the well-known film where, thanks to cinematic effects, our innocent and somewhat unaware hero winds up at the center of almost all the pivotal events of the twentieth century) in which history is simply the sum total of individual experiences—or, on the other hand, as episodes that are totally outside of the self, passing pictures we just sit back and watch (the other side of the *Gump* phenomenon in which history is something for which the individual, at least the ethical individual, is present, but in which he is not consciously involved nor actively participating). A comparable example demonstrating these two modes of conceptualizing history (as total yet depersonalized, or as personal yet decontextualized) might be drawn from Court TV itself, embodied by the differences between the station's aforementioned library (a video archive that is complete and permanent yet inaccessible to viewers) and its website (a selection of links that is easily accessible yet fragmentary, dispersed, and constantly changing). Yet because of the textual construction that I've attempted to detail, Court TV may still provoke viewers to try to corroborate these historical modalities and knowledges—to investigate the ways in which ephemeral and dispersed "private" affairs (be these violent family secrets or merely TV viewing habits) become the ordered and recorded "public" events of media and the law.

In other words, while station promos have proclaimed that the network "let[s] justice speak for itself," what truly speaks to the audience may be that which lies between the things that are actually told. In this way, Court TV alludes to the operations of historical and discursive construction less through its "direct examination" than through the very procedures of knowledge formation that its narrative and temporal strategies incite in the viewers. From its inception, the Courtroom Television Network has been involved in a debate about its relationship to knowledge: while it touts its purely informative function (this idea is largely what has opened courtrooms to video cameras since the 1970s), others argue that it can only adversely affect the sum of knowledge possessed by the viewers (always seen as potential jurors), either by letting them know too much (so that they'd have to pretend lack of knowledge if called for a trial), or by deluding them into a false sense of expertise when in fact they don't know much at all.[39]

Despite claims from the network, it is thus not that Court TV plays a simple educational and public service role; such claims have been rebutted by critics who cite its "sensational" appeal, its personalization of the issues, its capitalization on commercial television's interruptive form. At most, they suggest, it yields scandalous knowledge: not just a knowledge of the scandals of the rich and famous (and, I'd add, the poor and historically erased) but a scandalous spreading of knowledge itself: the profusion and diffusion of procedures of knowing. Moving from moment to moment, segment to segment, case to case, Court TV fails to give any one coherent account of the law, let alone a singular definition of Justice and Truth.[40] Yet this failure of total information (or knowledge understood as totalized information) is precisely what I find instructive in it.[41] It's too easy, I believe, to condemn television for the way in which it impedes knowledge (in this case, how Court TV might obscure, rather than clarify, our knowledge of history, society, media, and law); rather, television gives us particular ways of knowing, specific means of engagement with our times and our world.

My engagement, and, it would seem, that of other Court TV viewers as well, lies in the interstices—in what might be glimpsed between the different stories, the dual temporalities, the various precepts (civic and social) deployed in the coverage. Through these spaces we might enter others, extrapolating (even if "unlawfully") from the cases to our social formation, creating indeed a new network and archive of personal and political recognitions, entertaining and educational pleasures, public and private knowledges. For those of us interested in conceiving such knowledge, whether through reading media texts or producing them, this is an important lesson in itself. It is not by disavowing or simply indicting the tools of the media (in this case, television's narrative, spatial, and temporal possibilities) that we reformulate their laws; it is only in giving them a hearing that we might recodify the historical, discursive, and epistemological systems by which our culture legislates what and how we know.

5. "Body of Evidence": Serving a Citation Through "Law TV"

I'd like to draw my discussion of the deployment of knowledge in and through Court TV toward its conclusion by noting an instance of what elsewhere I've termed television's "self-receptivity," the way that TV texts figure possibilities for their own (or other televisual texts') reception: here the way that Court TV has been used to figure possibilities of televisual knowledge.[42] As I mentioned in my previous comments on narrative and law, there are certainly overlaps between the Courtroom Television Network and other televisual traditions and even a number of specific interfaces between the Court TV flow and other stations' texts (its references to a variety of programs as well as return references to Court TV). For example, one dramatic encounter in the first Menendez trial became commonly known as "the *Perry Mason* moment" of that trial—a moment at which the prosecutors, just like the famed lawyer of that designated television show, manipulated Erik Menendez into revealing a lie he had made on the stand. Conversely, coverage of the Menendez case provided fodder for other television texts, serving as the basis for, among other things, several TV movies and an episode of the series *Law and Order*. Of course, dramatization of notable trials is not particularly surprising. Another, perhaps less expected example of a competing television forum recalling Court TV (particularly amusing in the light of Court TV's claim to be "different television," despite its borrowings from other genres, such as sportscasting) comes from an ad for ESPN's coverage of women's tennis: that cable network, the ad asserts, "gives a whole new meaning to Court TV."[43]

But the case that I find most telling is one in which Court TV is more than just another intertextual reference for a media-savvy audience, for at least one TV program has centrally included the Courtroom Television Network both in its production history and in its diegesis, where, interestingly, it is distinctly marked as a significant site of knowledge. I'm referring to the drama *Murder One* (which, despite only running from 1995–1997, was well respected in the world of TV, influencing, through its treatment of its topic, not only the genre of the law show, but through its temporal construction, other types of series, such as *24*, as well) and, specifically, to *Murder One*'s incorporation of a barely disguised Court TV in the form of a fictional cable station entitled "Law TV." Actually, there is very little need for even this disguise: *Murder One* creator Steven Bochco was explicit in his acknowledgement of Court TV, while Court TV's Steven Brill gave *Murder One* his stamp of approval (describing this drama as "a natural progression" from his own attempts at television entertainment/legal education). In several interviews, Bochco identified the Courtroom Television Network as an inspiration for the program (motivated, he claimed, by his fascination with the coverage of the Menendez case and the O.J. Simpson pre-trial hearing): "'I'm a real Court TV junkie,'" Bochco stated in one such interview, "'I mean it's the best show on TV. I'll admit we have absolutely no shame that in 'Murder One' we have, some might say, ripped off Court TV. I say we've made them a little homage.'"[44] To further underscore the imbrications of televisual and legal narratives in the Court TV/*Murder One* nexus, Bochco originally wanted Howard Weitzman, the man he had chosen to act as legal consultant for *Murder One*, to act as legal consultant *on Murder One*'s Court TV-clone "Law TV"—in other words, to play himself in the show within the show. That became impossible when Weitzman accepted an executive position at MCA-Universal, a job even more central to the entertainment world than that of either *Murder One*'s behind- or its on-the-screen consultant.[45]

However twisted this tangle of legal and media representation became, Bochco devised the idea of such show-within-the-show commentators to serve a very practical purpose: keeping viewers up-to-date on the one murder story that propelled *Murder One*. In this way, the use of the notion of "progression" to describe the relationship between Court TV and *Murder One* might be seen as more than just part of a mutual back-patting session between two auteurs (the two Stevens). Shifting Court TV's singular focus from the level of network to narrative (a narrowcasting of story rather than spectators), *Murder One* progressed through only a single case for a whole season, requiring a familiarity with prior diegetic events for viewers who wished to follow the developing plot. Bochco thus refigured Court TV as his program's "Law TV" in order to preserve and recap information, attempting to harness precisely the narrative, temporal, and epistemological drive that I've described so as to provide the knowledge required for *Murder One*'s own construction of narrative and time.

Though reviewers labeled *Murder One*'s temporal format and narration of one murder case a "TV experiment" and called attention to its "new method of storytelling," this textual construction recalls not only nonfiction extended trial coverage but, as fictional precedents, PBS-type multi-part drama (its continuity, focus, and "quality" status) and, beyond that, nineteenth-century serial novels (indeed, *Murder One*'s episodes were titled as chapters).[46] While then hardly new in narrative form, one might nonetheless claim that *Murder One*'s method of storytelling was distinct from the one that Bochco had previously introduced to prime-time television through innovative programs like *Hill Street Blues* and *L.A. Law*: the (now dominant) dramatic use of interwoven storylines that overlap within and between episodic time.[47] In contrast to such multiple plotting with intersecting, commonplace details, *Murder One* returned to a singular, almost archetypal, epic battle of good vs. evil, focused on the struggle between high-profile yet honorable defense attorney Ted Hoffman and smooth yet slimy millionaire Richard Cross (the villain at the center of all the misdeeds, yet not the one actually charged with the crime). Signaled by the program's elegant visuals and a baroque harpsichord theme, the traditional, even mannered nature of this conflict thus departed from the dispersal of narrative and viewer identification so prevalent in today's prime-time drama (not to mention, of course, in daytime TV).[48]

It may have been this precise composition (on both the textual and ideological level) that led Betsy Frank, executive vice president and director for strategic media resources of Zenith Media Services, to write in her widely read annual preview of each fall's television season that *Murder One* is "simply too good for viewers, especially men, to ignore."[49] This singling out of male viewers may seem odd, especially in light of Court TV's own female fan base.[50] (Such an audience appeal also appears to have been shortsighted, as coverage of the O.J. Simpson case surpassed *Murder One* in the ratings).[51] Yet I believe that this address indicates something about the way in which *Murder One* departed from Court TV's narrative, temporal, and epistemological structures as much as it attempted to lean on their support. Whatever sensational excesses surrounded its central story, *Murder One* largely strove to order the law, to understand law and order through binary terms, and to place this understanding back into legally authorized hands. While both *Murder One* and Court TV owe something to the continuing narrative form of soap opera, their constructions of knowledge, discourse, and truth are thus quite distinct.

Given this distinction, it is interesting to consider the effects of *Murder One*'s references to Court TV, to examine what exactly is summoned through these "citations." The diegetic station Law TV may have been created in an attempt to secure viewers and provide stable information for *Murder One*, but is this how it actually operated, or did it construct a different type of archive, provide a different kind of knowledge? As I indicated, Law TV was acknowledged as an "educational" reference site throughout *Murder One* (in the simple sense that it was a means by which viewers, in and of the narrative, were given reports on the trial). There was a moment in the on-going story, however, when the nature of this knowledge and its implications became dramatized within the plot. This occurs when Annie Hoffman, seemingly loving wife of defense attorney Ted Hoffman, watches Law TV while her husband is in court. What she sees—Ted tearing a young girl to shreds on the stand by delving into sordid details from her past—has enormous consequences in and for the text. Both recognizing the absolute typicality of the event that she witnesses and failing to recognize the husband that she thinks she knows in the lawyer that she sees, Annie is eventually led to announce first her refusal to allow her daughter to testify in another case coming up for trial, and secondly, her totally unforeseen desire for a separation from her husband. I say that she's eventually led to these unexpected decisions because at first Annie is ashamed to admit that she possesses the knowledge that she indeed has; she lies about the fact that she's been watching Law TV. It is only later, in the course of an argument, that she blurts out this information—information that is thus marked as scandalous not merely for its content (knowledge of the testifying girl's previously scandalous lifestyle) but for its origin (knowledge from a seemingly illicit viewing choice). It may be labeled "Law TV," but by provoking Annie to question the gaps between and across public and private realms, work and family, momentary statements and long-lasting habits—in other words, by doing for her what I argue Court TV does for us—this text undermines the Hoffmans' marital security, actually upsetting and disordering the lawful television family.

This upset and disorder extended into the textual conventions of *Murder One* itself. Prior to this narrative development, *Murder One*'s "chapters" had consistently closed with intimate scenes between Ted and Annie, displaying a familial haven in the heartless world of criminal behavior and defense (a strategy reminiscent of the one used with Frank Furillo and Joyce Davenport in Bochco's *Hill Street Blues*). Annie's revised understanding of their relationship (or of the relationship between Ted's professional and personal spheres) necessitated the development of other means of (temporary) closure for the program's weekly "chapters"; after Annie's Law TV encounter, the moments of domestic resolution that once ended each episode were replaced by scenes of familial conflict and marital rift. For instance, the episode following the one of Annie's television viewing and the subsequent separation closed on a poignant scene of Ted, still at the office now that he has been removed from the home, on the phone to his daughter Lizzie as he tries to contain his emotions over the domestic upheaval by acting as if everything is fine. Quizzing Lizzie on the states' capital cities for an upcoming school test, Ted tells her that the memorization may be difficult but that he knows she can accumulate the facts. "See, you know more than you think you do," he says in a tone of pride mixed with longing, after she gets a number of the answers right. But the moment is poignant precisely because we, as viewers, know that more is going on than Lizzie herself knows—that knowledge (of scholastic subjects, of legal discourses, of family dynamics) involves more than simply communicating the attainment of the facts.[52]

Thus, while Law TV may have been designed simply to fulfill the purpose of factual collection and informational communication for the show, *Murder One* ironically pointed toward the impossibility of this very project. Indeed, Law TV's appearance in the text tended to signal a crisis of knowledge rather than its clarity: when shocking events occurred that jarred the characters out of a sense of epistemological complacency, the image often cut to that of a frame within a frame, our view suddenly represented as mediated through the Law TV camera. In this way, Law TV marked the discursive construction of the truth, our (and the characters') access to knowledge always "screened" (mediated, not necessarily "incorrect"). This is one way, I would argue, in which *Murder One* was itself much more "truthful" than most other television shows: unlike almost all dramas on TV, *Murder One* showed people doing something that we know, in fact, they do: watching television.[53] And when they watched TV—Law TV in particular—they were not simply educated and enlightened; neither were they simply misled or confused. Rather, they were disturbed (for instance, when one of the lawyers saw the woman he cared for announce a surprise marriage on the stand), defensive (when they watched the coverage of their performances before the jury or the press), amused (when noting funny little details that caught their attention), or even just distracted (when, in a running gag, people in the office tuned into the trial only to debate the attractiveness of a specific red-haired Law TV anchor—possibly a televisual in-joke on the then-growing fan appeal of some of Court TV's own anchors and/or of the fading fan appeal of *N.Y.P.D. Blue*'s David Caruso, the then-out-of-favor actor from Bochco's other program that aired that same year). As I would argue is the case with Court TV, Law TV then didn't so much just elucidate or obscure knowledge, retrieve or hide truth, as it redirected it, suggesting other examinations and inciting new (sometimes significant, sometimes trivial) deliberations.

6. "Closing Arguments": Knowing the Law/Laws of Knowing

In an anthology on feminist social epistemology, Naomi Scheman frames her interrogation of the values that determine epistemic judgment with a comparison between what she sees as feminist knowing and what she considers to be the mainstream media's construction and representation of knowing. She cites two examples from what we might term the "law and order" genre of film and TV: the "just the facts, ma'am" demand for direct disclosure (taken from the famous *Dragnet* tagline), and the "who wants to know?" challenge commonly posed to investigators seeking truth (typical across TV's law and detection genre).[54] In the first instance, knowledge is neutral and deemed value-free; in the second, it has a calculable value by virtue of its objective existence as a thing to be possessed. Neither image, claims Scheman, acknowledges the historical constitution of knowledge and the contingency of its judgments—features that must be taken into account in order to challenge the epistemological verdicts that delimit means of knowing. I do not want to take issue with Scheman's larger argument nor repudiate the social epistemologists' and/or discursive critics' goals. Indeed, as previously noted, Foucault's work demonstrates that archives of knowledge, while forming our "historical *a priori*" must, of course, themselves be historicized—that, as both Foucault and Scheman assert, we cannot take popular discourses of knowledge for granted as grounds for uninterrogated truth claims about our culture and ourselves. However, I believe that it's essential to examine media texts carefully before we make *our* claims and verdicts, to see the limitations *and* the possibilities of their particular "cultural epistemologies" as we try to devise epistemologies of our own.

The television texts that I've discussed are clearly ones defined by an ideological problematic of "law and order" (whether fictional or not), and there is certainly much that we might object to in these forms. Yet even those television texts that strive to order their (and our) laws may nonetheless allow—in fact, given their deployment of particular conventions of time, space, narrative, and discourse, sometimes even solicit—an epistemological engagement that is not quite so legally contained.[55] That is, the "laws of knowing" may operate according to different tenets than the laws portrayed, and as I hope I've demonstrated, this disparity may direct and redirect our attention toward the cultural laws and archives that legislate our social and epistemic formations. Thus, though Court TV is hardly just a matter of educational, public service television, and *Murder One* was clearly not just a reflection on this form, such texts may still yield formations of knowledge (or an acknowledgments of such formations) that could prove instructive—ways of knowing on which, so to speak, the jury is still out. Through their extended and/or instantaneous temporal structures, their overlapping and/or discrete narrative forms, they may produce modes of deliberation with which we might retry social, televisual, and epistemic norms.

Notes

1. *Webster's Revised Unabridged Dictionary*, 1913 ed., s.v. "archive."
2. Thanks to Meg Bogdan, vice-president for business and legal affairs at Court TV, for this and many other useful bits of information about the network; interview with the author, November 2003.
3. Available at http://www.courttv.com; http://www.thesmokinggun.com; http://www.crimelibrary.com.
4. See Michel Foucault, *The Archeology of Knowledge and The Discourse on Language*, trans. A.M. Sheridan Smith (New York: Pantheon Books, 1972), especially the section "The Statement and the Archive," 77–131.
5. See, for example, Steve Fuller, *Social Epistemology* (Bloomington: Indiana University Press, 1988), and the essays in Joan E. Hartman and Ellen Messer-Davidow, eds., *(En)gendering Knowledge: Feminists in Academe* (Knoxville: University of Tennessee Press, 1991). Michel Foucault's theorization of "disciplinarity" is crucial to this work; in addition to the previously cited *Archeology of Knowledge*, see, for example, *Discipline and Punish: The Birth of the Prison*, trans. Alan Sheridan (New York: Vintage, 1979) and *Power/Knowledge: Selected Interviews and Other Writings, 1972–1977*, ed. Colin Gordon, trans. Colin Gordon et al. (New York: Pantheon, 1980).
6. Discussing some of the difficulties that he has faced in teaching television studies, John Caughie writes: "There is a real risk in the theorizing and, particularly, in the teaching of television of opening up a gap between the television which is taught and theorized and the television which is experienced. Teaching seeks out the ordering regularities of theory. A television is constructed which is teachable, but may not be recognizable." John Caughie, "Playing at Being American: Games and Tactics," *Logics of Television: Essays in Cultural Criticism*, ed. Patricia Mellencamp (Bloomington: Indiana University Press, 1990), 50. It is precisely because of some of the problems that Caughie maps out in this insightful essay that I am attempting almost to reverse the typical procedure of TV studies that he describes here: that is, I am attempting to "recognize" the way in which television itself teaches us, even if initially the "theory" that TV produces is not necessarily recognizable as such.
7. One might also think of these "epistemological desires" in terms of what Jacques Derrida terms "archive fever." Although I do not explicitly discuss this text in the essay, Derrida's comments the relationship between archives, knowledge, and the law, and, more specifically, his comments on the tensions and paradoxes of the archive—between, for example, stasis and discovery, order and dispersion, preservation and disappearance, the public and the private, the consequential and the mundane—are certainly relevant to my discussion of the tensions and paradoxes of Court TV and to my analysis of the larger epistemological paradoxes that these both demonstrate and yield. See Jacques Derrida, *Archive Fever: A Freudian Impression*, trans. Eric Prenowitz (Chicago: University of Chicago Press, 1995).
8. The Courtroom Television Network was launched on July 1, 1991 but did not begin reporting its ratings until 1998, so it is difficult to judge the network's popularity in its early years. Initially, Court TV's focus was solely on trials, providing live coverage of courtrooms in the daytime (when the network, at that point, had its greatest audience) and then "shadowing," or replaying, that footage in the evening. Through some high-profile trials, the station gained attention and audiences; for example, according to estimates made in the fall of 1995 (just after the time of the particular trials that I discuss in the essay), Court TV reached between 18 to 25 million homes. Once the network began developing a separate prime-time line-up, largely consisting of crime stories (treated both through original non-fiction series and documentaries and through re-runs of fictional crime programs), it experienced another marked growth in popularity, also shifting its focus from daytime to prime-time appeal. For instance, it reported a 200% increase in prime-time

ratings from 1999 to 2000; by 2001, it had broken into the top ten rated ad-supported basic cable networks; and by 2002, it reached the Nielsen watershed mark of reaching over 70 million homes. Information from: Michael Burgi, "Cases Dismissed: Networks Maneuver to Expand Coverage of Trials Despite Camera Bans," *MEDIAWEEK* 5, no. 43 (November 13, 1995): 16; Arthur Unger, "Steven Brill of Court TV: Lights... Camera... Justice?," *Television Quarterly* 27, no. 3 (Winter 1995): 49; "Adlink and Time Warner Citycable to sell Court TV locally in country's top two markets," at http://www.courttv.com/archive/press/locala.html; "Court TV 2001 Prime Time Averages .7, up +17% over 2000 marking highest-rated/most watched year," at http://www.courttv.com/archive/press/best_year.html; "Court TV now tops 70 million homes, ends record February with a .82 Prime Time Rating," at http://www.courttv.com/archive/press/ctv_70.html; Jim Lyons, Vice-President and News Editor at Court TV, interview with the author, November 7, 2003; and Bogdan, interview with the author.

9. Nielsen Media Research defines its ratings in this way: "A household Rating is the estimate of the size of a television audience relative to the total universe, expressed as a percentage. As of September 1, 2003, there are an estimated 108.4 million television households in the U.S. A single national household ratings point represents 1%, or 1,084,000 households." Court TV's .9 rating thus means that, according to Nielsen, 975,600 households tune in to the network. (Unfortunately, here is not the place to remark on Nielsen's definition of its ratings as proportional and relevant to "the total universe"; needless to say, it indicates the importance that Nielsen grants both to U.S. television and, of course, to its own ratings system). For further information on its ratings, see the Nielsen Media Research Web site at http://www.nielsenmedia.com/ratings/cable_programs.html.
10. Lyons, interview with the author.
11. Warren Burger, "Isn't There a Better Way?" *American Bar Association Journal* 68 (1982), quoted in Helle Porsdam, "Law as Soap Opera and Game Show: The Case of The People's Court," *Journal of Popular Culture* 28, no. 1 (1994): 2.
12. Here, I 'm referring specifically to the first Menendez trial, which Court TV covered in full (the judge having allowed a single television camera in the courtroom). The brothers were tried together, though with two separate juries; opening statements began on July 20, 1993, and the trials ended with hung juries for both brothers (with mistrials declared on January 13, 1994 for Erik Menendez and January 25 for Lyle Menendez). The brothers were then retried in the fall of 1995, though this time, there was only one jury, and, significantly, there were no cameras in the courtroom: presiding Judge Stanley Weissberg, who also oversaw the first trial (in addition to having overseen, in 1991–1992, the criminal case against Laurence Powell, Stacey Koon, Theodore Briseno, and Timothy Wind for the beating of Rodney King) ruled that this second trial would not be televised because that would "increase the risk that jurors would be exposed to information and commentary about the case outside of the courtroom." After a much shorter trial (and, reportedly, a much less dramatic one—perhaps not unrelated to the absence of television cameras), the jury returned on March 20 with a verdict, for each brother, of two counts of first degree murder with special circumstances as well as conspiracy to commit murder. For details of the case, rulings, and trials (from one of Court TV's affiliated websites), see http://www.crimelibrary.com/notorious_murders/famous/menendez.
13. The cases against Detroit Police officers Walter Budzyn and Larry Nevers for the beating death of Malice Green began in June 1993 and ended with convictions of second degree murder for both in August of that year. However, when defense attorneys discovered that, during a break in deliberations, the juries had watched, among other videos, Spike Lee's film *Malcolm X*, which opens with shots of the Rodney King beating, they immediately called for mistrials. The convictions were upheld by the Michigan Court of Appeals in 1995, but later overturned (in separate actions and by separate higher courts) with the argument that Budzyn and Nevers were denied fair trials because of the jury's "prejudicial" video viewing. The cases of Damian Williams, Henry Watson, Antoine Miller, and Gary Williams, in connection to the beating of Reginald Denny, were covered by Court TV from their arrest in May 1992 through their sentencing in December 1993. In regard to the latter case, Court TV was involved in a legal battle of its own, involving its use of video footage of the Denny beating in promotional "teasers" for the network's coverage. In its decision on this matter (giving Court TV the right to air the video extracts), the U.S. Court of Appeals for the Ninth Circuit wrote: "In this age of television and news, it is frequently the image accompanying the story that leaves an event seared into the viewership's collective memory. The riots that shook Los Angeles in April 1992 are book-ended by two such images: the footage of police officers beating motorist Rodney King, which led to the trial and verdict that sparked the rioting, and the footage of rioters beating truck driver Reginald Denny, which through television synecdoche has come to symbolize in a few moments the multiple days of violence that swept over the city." For details, see Michael I. Rudell, ESQ. "Court TV's Use of Videos of Reginald Denny Beating Held To Be Fair Us [sic]" at http://www.fwrv.com/articles/artrud32.htm; and the Web site http://www.ca9.uscourts.gov/ca9/newopinions.nsf/F2C17F67DE4A8E3888256C94005C4F75/$file/0056470.pdf?openelement. While I do not explicitly discuss the various and fascinating ways in which television/video and law intersect in all of the cases that I give as examples, these details certainly point to the imbrication of the media and our court system which I attempt to describe in the essay above.
14. While I concentrate on the construction of temporality on Court TV in this essay, it would also be interesting further to consider the implications of the spacialization of the network involved in statements such as these (i.e., confining things "over there" on Court TV rather than "letting them loose" on other networks).
15. As just one example, see John Fiske, "Prologue: 'The Juice is Loose,'" *Media Matters* (Minneapolis: University of Minnesota Press, 1994), xiii–xxviii.
16. Brill discusses this in his interview with Unger, "Steven Brill of Court TV," 42.
17. Claude Levi-Strauss, *The Savage Mind*, trans. George Weidenfeld and Nicholson Ltd. (Chicago: University of Chicago Press, 1966).
18. In an interesting paper on Court TV's coverage of the Menendez Trial, Joan McGettigan and Margaret Montalbano discuss the network's narrativization of history; my inquiry runs along similar lines to the insightful analysis they develop. Joan McGettigan and Margaret Montalbano "Framed: The Menendez Brothers on Court TV," unpublished manuscript.

19. *Your Witness* (ABC, 1949–1950) involved court case reenactments; *Famous Jury Trials* (DuMont, 1949–1952) dramatized cases in a courtroom setting and also used flashbacks to fill out the stories; *They Stand Accused* (first called *Cross Question* and seen locally in Chicago in 1948 and then nationally on CBS in 1949; retitled as *They Stand Accused* while running on the DuMont network from 1949–1952 and again in 1954) was an anthology series that reenacted actual trials but with a jury drawn from the studio audience; *Court of Current Issues* (DuMont, 1948–1951) was not a law show *per se* but a public affairs program featuring debates on topical issues. Other early "court shows" include: *The Court of Last Resort* (NBC, 1957–1958; ABC, 1959–1960), a program that dramatized the work of a group of criminal law experts who aided defendants believed to be unjustly convicted; *Traffic Court* (first seen locally in Los Angeles in 1957 and then nationally on ABC from 1958–1959) which reenacted traffic cases; *The Verdict is Yours* (CBS, 1957–1962), a program with fictional yet unscripted cases, which used actual attorneys as the show's lawyers and judges, with jurors drawn from the studio audience; *Day in Court* (ABC, 1958–1965), a daytime series based on actual trials, with professional actors portraying the litigants and witnesses but, again, real attorneys as the lawyers and current or former law professors as the judges; *Morning Court* (ABC, 1960–1961), a spin-off of *Day in Court*; *The Court of Human Relations* (NBC, 1959), a personal advice show; *Courtroom U.S.A.* (syndicated, 1960), another program dramatizing actual court cases; and *Arrest and Trial* (ABC, 1963–1964) which initiated the formula used today on *Law and Order*. For information on these programs, see: Tim Brooks and Earle Marsh, *The Complete Directory to Prime Time Network and Cable TV Shows, 1946–Present*, 6th ed. (New York: Ballantine Books, 1995); Richard M. Grace, "Reminiscing: Courtroom Simulations Were Featured on Early Television," *Metropolitan News-Enterprise*, March 27, 2003, also available online at http://www.metnews.com/articles/reminiscing032703.htm; Richard M. Grace, "Reminiscing: TV Courtroom Shows Proliferate in the Late 1950s," *Metropolitan News-Enterprise*, May 8, 2003, also available online at http://www.metnews.com/articles/reminiscing050803.htm; and Alex McNeil, *Total Television: A Comprehensive Guide to Programming From 1948 to the Present*, Fourth Edition (New Yotk: Penguin Books, 1996).
20. *Perry Mason* aired on CBS from 1957 to 1966 and was revived as *The New Perry Mason* in 1973–1974 (also on CBS) and then again, in the form of a TV movie, in 1985; other *Perry Mason* TV movies regularly followed until star Raymond Burr died in 1993. The syndicated *Divorce Court* originally aired from 1957 to 1969; it reemerged from 1986 to 1991 and then again in 1999, airing until the present date.
21. The relationship between the Sheppard case and *The Fugitive* is mentioned in Charles S. Clark, "Courts and the Media: Can Pretrial Publicity Jeopardize Justice?," *CQ Researcher* 4, no. 35 (September 23, 1994): 819–836. The information about audience size comes from Brooks and Marsh, *The Complete Directory*, 381.
22. Steven Brill quoted in Massimo Calabresi, "Swaying the Home Jury," *Time Magazine*, January 10, 1994, 56.
23. The cases aired on *The People's Court* (syndicated, 1981–1993), presided over by the retired Judge Wapner, were all within the small claims court's $2,000 limit; the producers would pay the judgment up to that limit if the plaintiff won, giving an additional $50 to the defendant. If the defendant won, both litigants split the $500 fund set aside for such cases. These figures come from Porsdam's interesting analysis of the program, "Law as Soap Opera," 5. Slightly different figures (of a $1,500 limit with $25 going to the defendant if the plaintiff won) are given in McNeil, *Total Television*, 650. *The People's Court* was revived again in 1997, continuing in syndication to the present, presided over by such judges as Ed Koch (former New York City mayor), Jerry Sheindlin (who, in an interesting intertextual connection, is married to the star of another television court show, Judge Judy), and Marilyn Milian (the first Latina to preside on any U.S. reality court show).
24. Although I discuss Court TV's narrative strategies more than its "non-narrative" ones, it also, of course, employs strategies drawn from a variety of TV's "reality genres" (news, sports, public service television, talk shows, etc.).
25. Michael Burgi, "Law School for Kids: Court TV plans Saturday-morning educational block for children," *MEDIAWEEK* 5, no. 38 (October 9, 1995): 12.
26. Quoted by Andy Meisler, "Bochco Tests America's New Legal Savvy," *New York Times*, October 1, 1995, sec. 2.
27. Quoted by Burgi, "Cases Dismissed," 16.
28. Jim Castonguay, private correspondence with the author.
29. Similar debates about the educational vs. voyeuristic value of TV law emerged with *The People's Court*. According to its creator, Stu Billett, the program was conceived as a "combination of soap opera and game show," and it was criticized by some people for being too showy and misleading. Others, however, lauded the way in which *The People's Court* increased popular awareness of the court system (with even Sol Wachtler, former Chief Judge of the State Court of Appeals in New York, stating that it usefully "act[ed] as a primer."). Billett and Wachtler quoted in Porsdam, "Law as Soap Opera," 4, 8.
30. This particular example comes from Lewis Cole, "Court TV," *The Nation* 258, no. 7 (February 21, 1994): 243. Another essay that discusses Court TV in the light of "the promise of interactive media, information 'highways,' and no less than 500 TV channels to choose from" is Elayne Rapping, "Cable's Silver Lining," *The Progressive* 58, no. 9 (September 1994): 36.
31. It is in regard to this temporality, too, that I would argue that network executive Jim Lyons' comment on the importance and influence of Court TV, MTV, and CNN is particularly noteworthy. See note 10.
32. This notion of a temporal gap has been augmented by discussions with Carol Clover, private correspondence with the author.
33. Foucault, *Archeology of Knowledge*, 130.
34. This vision of Court TV may be shared by media commentators as well. For example, in a review of the network, Elayne Rapping calls up an image very similar to the ones promoted by Court TV's own references to the "founding fathers" and its attempts to compare and contrast its offerings to those of fictional TV. Rapping writes: "Thomas Jefferson, who believed trials should be public events, said as much in the early, more idealistic days of this nation. And he would, most certainly, have approved of cameras in the courtroom. Why shield the lawyers and judges from the scrutiny they deserve? Why let such anachronistic paragons as Ben Matlock and Perry Mason stand in for the less virtuous real

thing in this key area of public life?" Going on to make a point related to my later argument about the ways in which Court TV reveals our culture's "laws" of social and sexual division, Rapping continues: "Jefferson might be shocked at the content of the trials we are now watching with so much fascination. But that would be because he—an aristocratic white male of the Eighteenth Century—was ignorant of and insensitive to matters of race, class, and gender." Elayne Rapping, "Gavel-to-Gavel Coverage," *The Progressive* 56 (March 1992): 35.

35. In this regard, the construction of knowledge on Court TV is quite different from that of something like the television quiz show, which tends to define knowledge as merely the recall of isolated facts, outside of any context save that of consumerism itself. Despite, then, the pleasure that viewers may find in amassing informative details from Court TV's coverage (as I acknowledge in the text above in relation to my fascination with Menendez miscellanea), this is not the only pleasure that Court TV provides nor does the network deploy knowledge in purely an "informatic" mode.
36. Cole, "Court TV," 243–245.
37. Cole, "Court TV," 243.
38. Foucault, *Archeology of Knowledge*, 130.
39. For a discussion of this debate (and the larger context of formulations that pit a free press against a fair trial), see Clark, "Courts and the Media."
40. There are many such critiques of Court TV. Just one example can be found in John Leo, "Watching 'As the Jury Turns,'" *U.S. News and World Report*, February 14, 1994, 17. Even the very title of this article suggests that the "soapy" aspects of Court TV contribute to a warping of justice as attention turns toward image and emotion rather than abstract cognition.
41. Borrowing terminology from Kathryn Pyne Addelson and Elizabeth Potter, one might say that it is this refusal—or probably more accurately, failure—of totalized information that allows Court TV to move beyond a singular epistemology of "propositional knowledge" ("knowing that") to a different kind of epistemological structure ("knowing how" or "knowing a person or place"). See Kathryn Pyne Addelson and Elizabeth Potter, "Making Knowledge," in Hartman and Messer-Davidow, *(En)gendering Knowledge*, 262–264.
42. I discuss the notion of "self-receptivity"—the ways in which television inscribes models of reception in its texts—in Lynne Joyrich, *Re-Viewing Reception: Television, Gender, and Postmodern Culture* (Bloomington: Indiana University Press, 1996).
43. *TV Guide*, March 16–22, 1996, 71.
44. Quoted in Meisler, "Bochco Tests," 33. Bochco makes similar statements to Mary Murphy in "Plotting Murder," *TV Guide*, September 30–October 6, 1995, 12, and to Cheryl Heuton, "Bochco Show Gets '95 Go (Steven Bochco's 'Murder One' TV Series on ABC)," *MEDIAWEEK* 4, no. 36 (September 19, 1994): 5.
45. Meisler "Bochco Tests," 33.
46. The description of *Murder One* as a "TV experiment" comes from Rick Marin in a review of the program; *Newsweek*, October 2, 1995, 98. Likewise, Meisler describes *Murder One* as employing a "new method of storytelling"; Meisler, "Bochco Tests," 33.
47. *Hill Street Blues* aired from 1981–1987, and *L.A. Law* aired from 1986–194, both on NBC. Both also received a considerable amount of attention (as did creator Steven Bochco) for introducing some of the strategies of daytime soap operas (such as continuing narrative, multiple plotlines, and dispersed identification) into prime-time drama.
48. In an interesting essay on prime-time television programs, Charles McGrath cites precisely these features as the reason why he preferred other dramas over *Murder One*. See Charles McGrath, "The Triumph of the Prime-Time Novel," *New York Times Magazine*, October 22, 1995, 86. As noted in the text above, Court TV's Steven Brill stated that he liked *Murder One*; yet in one interview, he explained that it was nonetheless just the sort of polarized representation of heroes and villains that I (and other critics) locate in that program that generally troubled him about TV's fictional portrayals of the law. Quoted in Unger, "Steven Brill of Court TV," 52.
49. Quoted in Meisler, "Bochco Tests," 33.
50. As previously noted, Court TV did not at first report its ratings. Yet, in its early years (before the network began heavily promoting its prime-time schedule), then-CEO Brill noted that Court TV had its largest audience in the daytime, and he conceptualized the network's primary competition as soap operas and talk shows. Brill thus described the viewers as "people who watch daytime television," which "skews a little bit female...the typical COURT TV viewer is a relatively educated female." Quoted in Unger, "Steven Brill of Court TV," 48.
51. Bill Carter, "O.J. Simpson Trial Bests 'Murder One' in Ratings," *New York Times*, September 28, 1995, section B. *Murder One*'s low ratings, however, did not prevent the program from being judged, for what it's worth, as 1995's best new dramatic series according to both *TV Guide* and *The People's Choice Awards*.
52. Again relevant here is the distinction, addressed in note 35, between the construction of knowledge in and through Court TV (or, in *Murder One*, "Law TV") and the construction of knowledge on television quiz shows.
53. In his article on prime-time drama, McGrath writes: "It almost goes without saying that neither 'E.R.' nor 'N.Y.P.D. Blue,' for all their daring in other ways...has dramatized one of the most basic and elemental acts of private life in America—namely, TV watching itself.... The only way TV makes its presence known in these prime-time dramas is in the form of newspeople pushing their way into the station-house lobby or clamoring, vulturelike, outside the emergency-room entrance; in all of these confrontations, the camera is always seen as an antagonist, a disrupter of business and a falsifier of truth." McGrath, "The Triumph of the Prime-Time Novel," 86. As I've indicated in the text above, this is not the case with *Murder One* (although this does not prevent McGrath from comparing it unfavorably with these other programs). While our access to "Law TV" often comes at moments of scandalous knowledge on *Murder One*, in that program's diegesis, television is nonetheless accepted as an unremarkable part of the everyday world, as a particular kind of business and a particular (though certainly not infallible) means toward truth. While hardly a bearer of absolute knowledge, television is thus not condemned as the "falsifier" that McGrath identifies in other programs.

Rather, in *Murder One*, the institutions of television production and consumption are seen as existing in a complex relationship with the production of knowledge.

54. Naomi Scheman, “Who Wants to Know? The Epistemological Value of Values,” in Hartman and Messer-Davidow, *(En)gendering Knowledge*, 179.
55. In making a distinction between a text’s “ideological problematic” and the “epistemological engagement” it solicits in viewers, I am citing, extending, and playing on the distinction between a text’s “ideological problematic” and “mode of address” made by, among others, Charlotte Brunsdon and David Morley, *Everyday Television: “Nationwide”* (London: British Film Institute, 1978), and David Morley, *The “Nationwide” Audience: Structure and Decoding, BFI Television Monographs 11 (London: British Film Institute, 1980).*

Part III

Power-Code

9

The Style of Sources
Remarks on the Theory and History of Programming Languages

Wolfgang Hagen
Translated by Peter Krapp

"Le style c'est l'homme même"

—Buffon

"... the procedure ... whose mastery exerts a decisive influence over the style and the quality of the work of a programmer."

—Wirth

"The principles of style, however, are applicable in all languages"

—Kerningham

"A style of programming is based on an idea (possibly speculative) of a 'calculator' ... which is to work off the program"

—Stoyan

"Can we be liberated from the von-Neumann-style?"

—Backus

Let us begin with a thought experiment.

1. The Library of Modern Sources

Imagine a large library called "The Library of Modern Sources." What would the blueprint for such a source museum look like? We might arrange departments and divide them along the large groups of programming languages: procedural (FORTRAN, ALGOL, PASCAL and C), functional (LISP, ML or MIRANDA), declarative (LOGO or PROLOG), and a new department for the object-oriented (SMALLTALK, EIFFEL, C++); parallel and neuronal languages would be in development. However, as a minimal condition, all sources ever written must be available as code, plus the descriptions and sources of all compiler-, interpreter-, and assembler-codes that belong to each system, including all those texts, blueprints, tables and diagrams describing the machines that run those codes. We would collect everything belonging to the symbolic register of our project: everything written, all knowledge on each code ever put into signs and sketches. Would this be sufficient? Does the history of source code include only what has been registered symbolically? I am afraid our library would in the end also have to include the real machines themselves, plus running versions of all operating systems and development platforms. Otherwise the bulk of older code would remain incomprehensible. But does anyone have even a minimal number of computers that ever ran at their disposal? No. Our collection would at best document those codes that never actually ran on a machine.

With regret our thought-museum would have to declare at the entrance that the history of all computer source code, their "historia rerum gestarum" as the Roman historians put it, coincides with the "res gestae," with the events themselves. Thus our thought experiment is a logical impossibility.

2. An Archive of Source Codes

Not to speak of the problem of procuring the code, the source of the sources. Where are they kept? Of course I have to trust the archives of Big Blue and MIT, Xerox, Apple, Microsoft, of the U.S. Navy and U.S. Air Force, but I doubt whether they actually kept the source codes of the UNIVAC or the BINAC, of the DUAL and the SHACO machines at Los Alamos, or of the IBM 701 or the 704, or of the prototypes of these machines. We recognize a grave and fundamental problem of the archive. Between an abstract computer representing every artificial code and a real machine to be controlled there is at least one generational process of another machine. Some German computing manuals call that generic machine which compiles the source code into machine code the "translator." Concerning the language of abstract computers, the "source," the operation of the translator is like a crossing of that subterranean river that the Greeks identify with the realm of death. The transition from symbolic program-text to real machine-code kills the language which sets it in motion. Often enough, the transition is one from being (software) to nothingness (hardware). To try to describe the running machine, and we have to do this often enough, we resort to another symbolic machine, the "assembler." Its "dis-ensembling" debugging—a process one can set up even if the source is written in "higher" languages—discursifies the running program for us into another, new language which has little to do with the source code. This path offers only a literally ideal and deceptive continuity. In reality, there is a "breakpoint" between symbolic machines and their real runtime. We call that breakpoint the halt where the machine is still running and where we place our symbolic vocabulary in between—but we do not actually displace anything: a masked interrupt (i.e., a piece of software which is inherent in the machine itself) is tracing and debugging at this location of the last communication with the machine. Hardware description language (HDL) may show in a diagram or in a temporal logic design how this interrupt works. If such breakpoints do not clarify what happens in the pipelines of machine hardware, there is still one last remedy left: the "post mortem dump" or post mortem debugging. In the end, a conceptual last judgment. "But," as Hegel says in the *Phenomenology of Spirit*, "the life of Spirit is not the life that shrinks from death and keeps itself untouched by devastation, but rather the life that endures it and maintains itself in it."[1]

3. "There Is No Software"

At this stage, I want to point out a figure of secret idealism nesting in the thought of these symbolic connections, in the transitions and transformations of abstract machines of computer programs; a secret idealism that is so seductively obvious. You may find the same figure in a number of American philosophers of "electric language," such as Michael Heim or Jay David Bolter. They say computer programs and systems are a "demanding collection of programmed texts that interact with each other."[2] But in real computing machines, texts do not interact: electron diffusion and quantum mechanic tunnel effects run over all chips, n-million transistor cells in n^2 correlations. If you want to call this occurrences "interaction," then you have to face the current technology production treating such interactions as systemic barriers, as physical side-effects, distortions, etc.[3] Symbolic software programs and the real runtime actions of computing machines are not joined in the interference of a continued universality, as the Gutenberg Galaxy has taught us to expect, but differ discontinuously, often almost grotesquely. If this were not the case, we would not see the recurring waves of painful software crises on micro- and macro-levels. "If programming was a strictly deterministic process following firm rules," the Swiss computing scholar Niklaus Wirth writes laconically, "it would have been automatized long ago."[4] Or else there would be, as Kittler provocatively put it, "no software."

4. Genealogy of Computer Media

Let us return to the early days of the computer, when one could easily assert, "there is no software." We do not have to discuss the fact that computers are based on the mathematical model of the Turing machine, and that this model was widely known to American scientists since the late 1930s. But Turing did not write about "software" in 1937—he offered a negative proof showing that a general algorithm for the general solution of general mathematical problems cannot be proffered. It follows that whatever can be addressed in a finite description by an algorithm is positively calculable. Oswald Wiener's idea of the entire world as a universe of folded Turing machines has less to do with Turing himself than with the impact the massive development of digital storage media has made on us.[5] We may suspect that the number of bytes on all computers in the world has already surpassed the number of letters in all the books in the world. This entropic digression of storage media sends us to another mathematical model, equally responsible for the current media sea-change: namely Shannon's mathematical theory of communication developed in the 1940s. Were we to situate the computer medium as it appears to us today in an evolutionary model—as we will try for the sake of the argument—then this model would project the development of three overlapping evolutionary complexes onto a time axis, neither interchangeable nor initially corresponding. They are:

- the mathematical model of calculability
- the engineering technique of storage development and addressing
- the mathematics and physics of communications technology.

I hasten to add that the strangely nontransparent "terminus a quo" of our problem, namely the question of the origin of programming languages, could also be articulated in three sections:

1. a first approach of programming languages, in the early 1950s, which follows symbolic contiguities, but no mathematical mode
2. in the late 1950s, the counter-movement of mathematically oriented functional and declarative languages that must idealize the machine from which they abstract

3. in the early 1970s and again in the late 1980s, the incision of simulation, marking the entry of earlier media (writing, image, sound) in the "computer"—making it a medium in itself. This last step leads to the object-oriented languages that are defined as neither procedural nor functional.

This background for a sketch of the three large groups of programming styles cannot be reconstructed otherwise. Thus each theory of programming languages, like any media theory, observes one *a priori*, namely its own: it can only be written up as historical theory. I want to reconstruct the first of the three steps I recognize in the development of programming languages, and thus in the end show a bit more of what one may call style.

5. "Stilus" or Metonymic Style

There are few poetological and linguistic concepts that describe a fundamental property of language and are simultaneously an effect of that same property. "Style" is one of them. Latin etymology indicates that "stilus" originally denoted a sharp stake, used to break up soil in agriculture, in wartime in traps. Later it became the name for a tool for making marks in wax, made of wood, horn, or metal, one end sharp, the other flat to smoothen the slab.[6] "Stilus," therefore, is a tool to write and to erase the written as well, a writing/erasing tool. Hans Ulrich Gumbrecht noted the metonymy that this binary tool of contradiction commences in the first century BC. For as ancient Rome turned into an imperial monarchy and with the development of books and libraries the written document began to replace the politics of forensic orators, some hastened to invert the facts of the visible world on their writing pads. So "stilus" means not only pen, but also "the use of the pen ... the practice of writing, the manner of the writer," and even the "language of the writer" itself, the "stilus artifex." This "feedback"-inversion of the "stilus" became the primary stylistic act of language. Gumbrecht quotes Cicero: "Vertit stilum in tabulis suis, quo facto causam omnem evertit suam."[7] "He inverts in his writing how he acts to destroy all his things." The reversion of the pen, "stilum vertere," now means to distinguish the written by writing. This is metonymy and thus is the thing—to use the "style" is the style, and vice versa. Can the written, in software, make the written unrecognizable?

Cicero's style relies on omission, setting up the economy of erasure for the greatest effect in speech. A differentiated conceptual history follows throughout all Latin, Scholastic and Renaissance rhetoric and poetics until the Enlightenment offers this motto: "style is the man himself." Style is supposed to be the bourgeois subject of speech and writing, the producing author liberating the free, but paid genius, from the formal prescriptions of past centuries. There are after all some few books in computer science that warn against such unfettered subjective style exercises, but this is merely stupidity on both sides. In their Kafka studies, both Friedrich Kittler and Bernhard Siegert demonstrate how in the end style is always derived from media effects, which with the rise of technical media exert their effect also on literature.[8]

In the first century BC, when "stilus" became metonymical and in the concept of style a scriptural tool became its own effect, the next media transition, a new media *a priori*, sets its shift in motion. For the arms and hands of Cicero, Sallust, Terence or Caesar rest less and less on papyrus and more and more on the *caudex*, the bound book of wax slabs, as well as the *codex* of vellum. Both *caudex* and *codex* complement the y-axis (of a script roll and continuity of text) with an x-axis of the page, and a z-axis, the number of pages. This brings writing into a three-dimensional order for the first time in history, and thus makes it accessible as script/text with the aid of numeric and other symbols: page numbers, paragraphs, indents, marginalia, sections, chapters, comma, hyphen, colon and semicolon. Now one can know of a writing that one does not see, but which is referred to as if it were visible, a *virtualizing* effect of the codex and the birth of style. With the rise of the printing press in the 15th century, all those codex-symbols, never incorporated in the standard alphabet, were incorporated into another code—that of mathematics.[9] And so we arrive at the symbolic basis of the style of programming.

6. The Signs of the First Programs

What was the first computer program and what were its first symbols? This is like asking: what were the first computers in history? It is well known that there are no satisfying answers. The MARK computers by Howard Aiken, Konrad Zuse's Z3, Turing's COLOSSUS and ACE-machine, and Mauchly-Eckert's ENIAC participate in the first developments. There is a ubiquity of beginnings, a dissipation of the mechanic start up of the computer.

6.1. Dissipative Evolution

The historical facts about the huge engineering boom in America in building calculators and computing machines after 1941 are well known enough. That technological evolution pivoted upon the need to calculate a growing number of "firing tables" for all possible flying objects and projectiles, and later the famous ENIAC addressed the so-called hydrodynamic Los Alamos Problem. The numeric calculation of shock wave equations of an H-bomb implosion was carried out on the ENIAC during three long months, beginning in October 1945.[10]

6.2. Zuse

The loner Zuse, in his sparse rooms in the German Experimental Institute for Aeronautics around 1940, cut off from the world of scientists and lost in the chaos of the National Socialist research administration, nevertheless deserves a good credit for important parts of the evolution of the computer—particularly with respect to the deep structure of developing programmable calculators, which cannot be restricted to the efforts of Turing or von Neumann, but go back to the mathematical problems of Hilbert, Ackermann and Frege. Zuse responds directly to them, as Turing and von Neumann did implicitly. Therefore we can assume that even without Turing, the debates about the axiomatics and foundations of mathematics in the 1920s could have led to the computer. Given what we know today, the difference between German and American computer development is not caused by a scientific gap (with mathematical and logical bases), but a due to the huge differences in military and industrial support. America and England had grown their military-industrial-academic complex continuously, arguing with what they thought the Germans were going to develop before all else and reinforcing their efforts by the massive investment in Los Alamos. In 1942, the "Manhattan Project" united 2,500 scientists in America and England, Goldstine, Mauchly, Teller and von Neumann, as well as Turing. Konrad Zuse, in contrast, worked partly on his own and partly for an organization pulled apart by the competing interests of the Wehrmacht, the SS, the Navy and the Airforce, misjudged and not institutionally recognized or supported. However in the end, his influence after the war would even extend to the conferences of ALGOL 58 and ALGOL 60, if only because the Z4 and Zuse's own programs were saved, and some of his developments were known to Rutishauser, who brought the knowledge and Zuse's name to the U.S.[11]

6.3. ENIAC

One always says very generally that John von Neumann introduced elementary concepts of the Turing machine into the computer boom dominated by the Americans. But we can specify the historical time and place, namely the ENIAC team between spring and fall 1944. Here von Neumann is on a team with Brainerd, Goldstine, Eckert, Mauchly, and Burks, and these team-mates share the claim to have invented sequentially stored programming, although it is usually associated only with von Neumann—all the more since this was an engineering advance that inevitably followed from the architecture of the ENIAC.[12]

6.3.1. Research Inputs With its 18,000 tubes, the ENIAC was the first computer of such size—the model for the admired post-war "electronic brains" and the first proof in the history of technology that switches of this magnitude were possible. Contemporaries also knew which major engineering trends helped develop this computer. Contributing to the hardware were:

- the electronics industry which had reached a peak boom in receivers and transmitters in its radio days
- the mechanical and electromechanical industry active in arms production
- robust 100KHz tubes from radar technology.

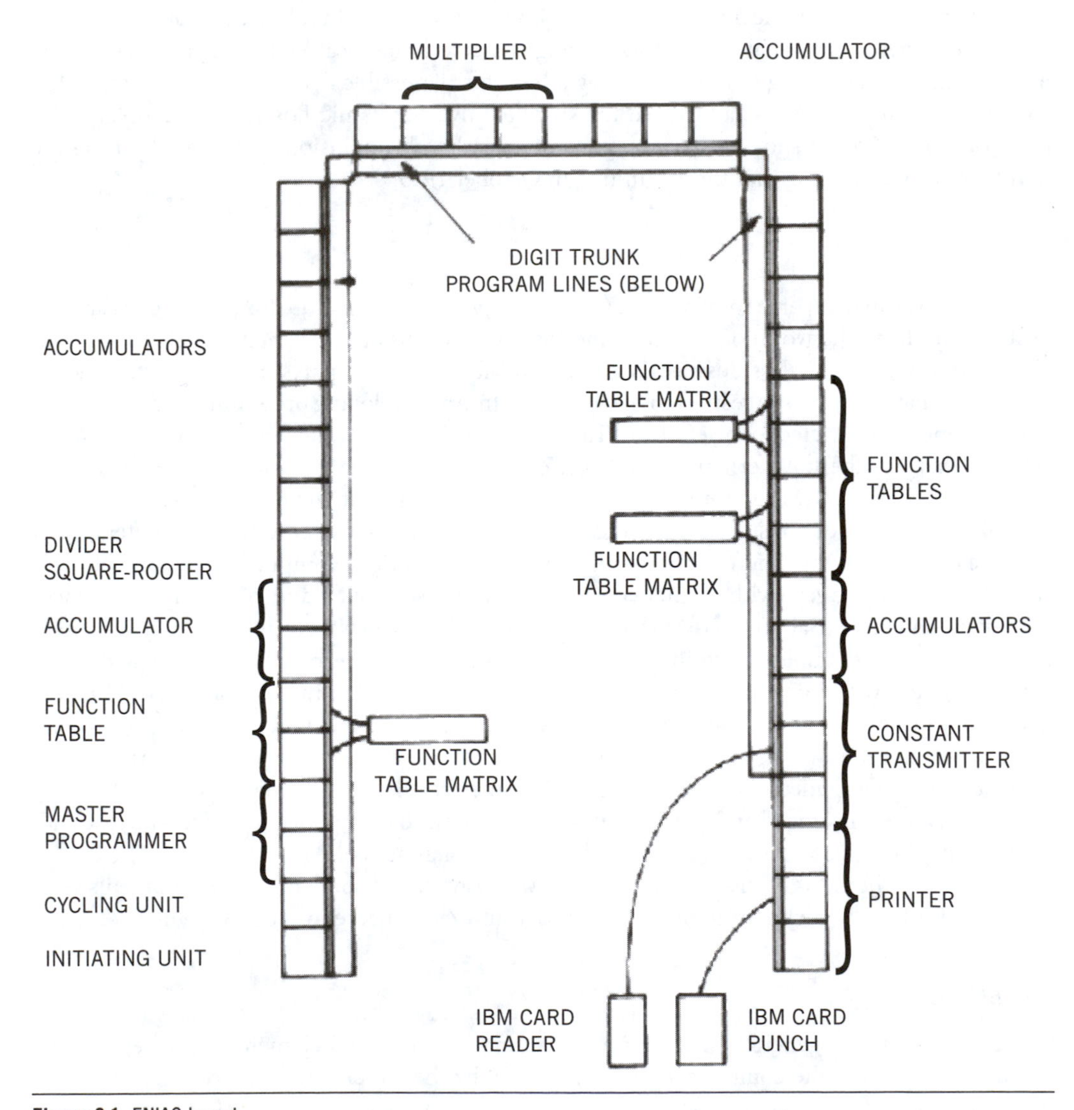

Figure 9.1 ENIAC layout.

Contributing to the computer architecture were:

- Vannevar Bush's differential analyzer as a model for machine organization,
- IBM switchboards for control, and experiences from Bell Labs and Aiken's MARK I.

All of these were the research input for ENIAC, which led inevitably to the concept of stored programs. ENIAC has two successor generations, the IAS, the WHIRLWIND (we will come back to this) to the IBM 700, which more or less directly leads into the present and to the more academically oriented product lines EDVAC, EDSAC and UNIVAC, which we will also encounter again.

6.3.2. Extension and Construction The ENIAC consists of four accumulators; one square rooter unit; one multiplication unit; three complex switchboards for matrix calculation; on the bottom left, the three control units and on the right, the punch-card input and output.[13] Each of these units had to be "programmed" first, that is wired together.

6.3.3. Programming The programming of ENIAC fell into two separate areas, numerical programming and, what the ENIAC-team called "programming proper."

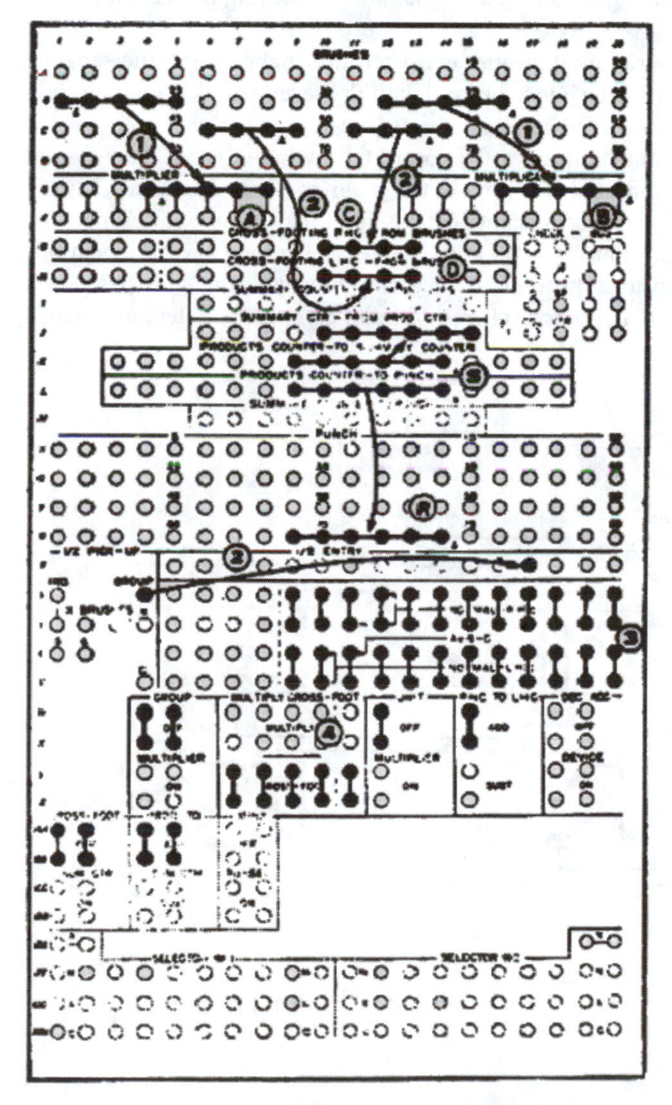

Figure 9.2 Programming of an accumulator unit. Wiring diagram for IBM 601 plugboard.

6.3.3.1. Numerical Programming The illustration shows the "programming" of an accumulator unit as used in the ENIAC. On top you see the set calculation "a*b+c+d." Working with these interfaces was called numerical programming, as Arthur Burks reports, and depending on the formula calculated, it had to be done for all accumulator, square root, or function units separately. For the most part, this specialized labor was carried out by the "ENIAC girls." In order to keep track of the sequence of such numerical programming, block diagrams such as the one shown in Figure 9.3 were used.

The diagrams determined what had to happen in the various units of the ENIAC, which accumulator would calculate which part of the formula, etc. One might call it numerical programming assignment, a first kind of addressing, for the accumulators were nothing but the intelligent storage cells of the calculator. There was no generalized symbolic notation for their coding, i.e., a plan for their connection. In such process diagrams, we find the predecessors of the first programming routines, as well as one of the oldest representations of calculation. For the Greeks drew their geometric figures and their derivations in a "diagramma"—and this was also the name of a scale, since musical sequences were understood as derived from cosmic geometry. For the ENIAC, program sequences are noted in diagrams, as the graphic interface between the mathematics of the formula to be calculated and the electronic plan for their numeric solution. The concrete implementation of such diagrams was laconically called "programming proper"—as if nothing was easier.

6.3.3.2. Programming Proper Programming proper consisted in synchronizing the separate units with the digit trunk or data bus on the one hand, and with the seven parallel program lines on the other hand. To this end, there were further diagrams as shown in Figure 9.4.

Transmit and receive in the accumulator had to be connected by hand; the three program controls of each accumulator were connected to the bus of the program lines. This is basically the archetypal innovation of the ENIAC. The electromagnetic differential analyzer and Zuse's machine still had a central, motor-operated unit controlling the speed of the calculations, like the inventions of Schickart or Babbage. ENIAC replaced those mechanical impulses with the completely new concept of ten bus cables upon which a cycling unit transferred a complex parallel pulse which

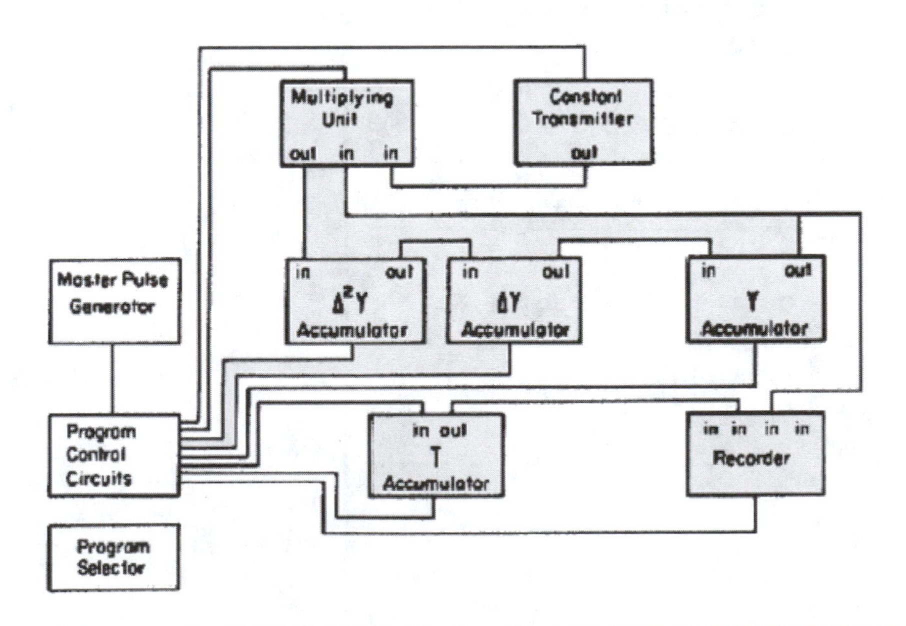

Figure 9.3 Block Diagram. Block diagram illustrating method of solution of $d^2 y/dt^2 = ky$ or $\Delta^2 y = k(\Delta t)^2 y$.

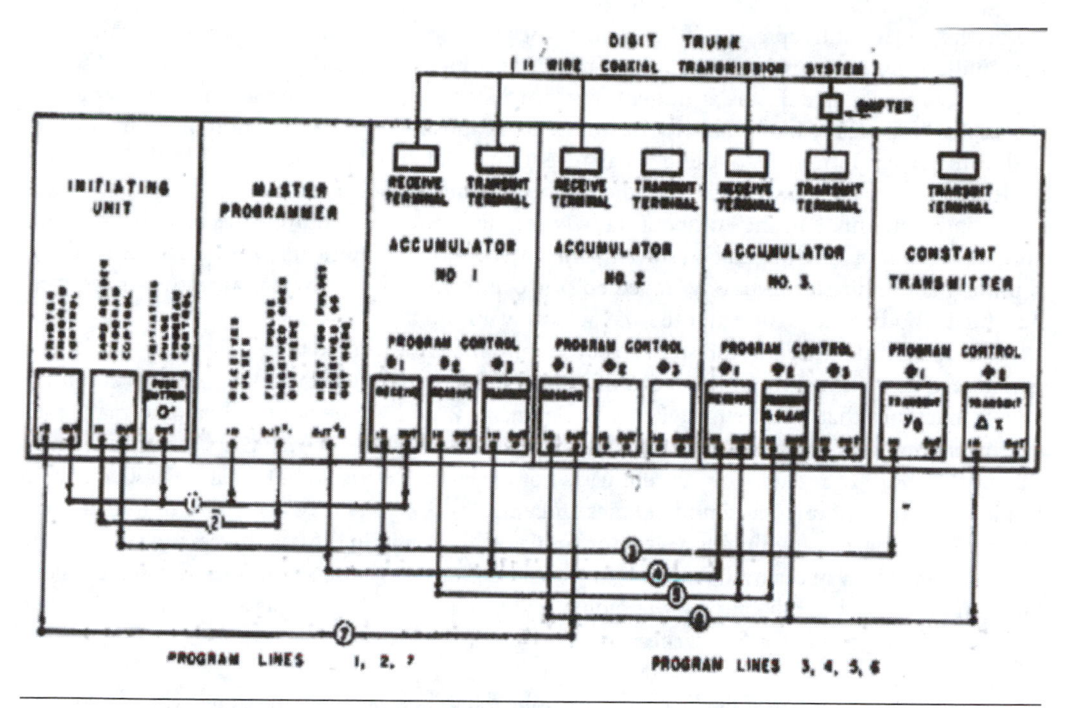

Figure 9.4 Simplified ENIAC program diagram (from Burke: 1980).

could be as fast as 3 micro-seconds, thus allowing switching at a veritable 3.3 KHz. The control of such a fast pulse had become possible with advances in radar technology. Once a program was set up, its running was static, with no recursions, inductions and conditional branching.[14]

6.3.4. Mercury Delay Lines It is obvious that the concept of stored programming developed from the tubes of the ENIAC and not from electromechanical calculators. The ENIAC provided a fast bus and its technology was robust enough to trigger impulse control-frequencies at 3.3 KHz running through a system the size of half a ballroom. This made it possible to approach mechanically impossible storage problems, by inventing all sorts of impulse-controlled delays. In the case of the ENIAC the team used the so-called "mercury delay lines" that Turing in his work on the English ACE-computer had used also. Pres Eckert had developed them in the spring of 1944 for radar units—another input from radar research into the computer.[15]

The "mercury delay line" was a slim tube filled with mercury, here two meters long, that could delay a pulsating ultrasound signal up to a millisecond. A deft switch allowed only ten tubes to refresh 1000 impulse bits in the space of a microsecond. The result was a phased electronic 1K-Bit storage unit for the cost of a few liters of mercury and ten tubes for ten dollars each. The mercury line was a revolutionary step: it reduced both the cost and the space required for memory at once by a factor of 100 to 1.

Between March and June 1945, the team built 32K of memory for data and program instructions for a new type of computer we now call the "von Neumann machine" by cascading 256 such mercury lines. The storage blocks required a phased bus architecture, the bus address logic condensed the distributed accumulators, multiplying units and square rooters onto one unit—the central processing unit (CPU)—and the whole system thus demanded a central control. Far from Turing's logic machines (which—as far as we know—was mentally not present in the ENIAC-Team and was definitely nothing von Neumann argued with), the synchronization of the described electronics architecture alone led to the legendary discrete sequentiality, giving up the generous parallel set-up the ENIAC still represented. Henceforth the guideline for the most successful machine type ever built by man was, in Burks' laconic words: "One thing at a time, down to the last bit!"[16] Single instruction, single data. In the arguments over the patent, John von Neumann would later admit that it was "practically impossible to list who was the apostle."[17] The innovation of binary storage and stored programming is anything but an invention or a defined patent. Moreover, the computer itself was not the end goal, neither for the engineers Mauchly and Eckert nor for the mathematician von Neumann.[18] In the competition between different architectures, the goal was to build another calculator that was able to solve nonlinear equations numerically, with great demands on rounding and induction. Nobody "invented" the computer as we know it now in the strict sense, nor did anybody want to invent it the way we know it now.

Von Neumann himself would soon go beyond this machine limited to mathematics that Turing had indeed described sufficiently, dedicating himself for another decade to the theory of automata and of machines that tolerate mistakes; von Neumann would write about cellular automata, matrix inversions, and neuronal learning. The scientific paradigm of the war years that give birth to the most important medium of the century is not the computer itself, but what we associate with Claude Shannon, namely thermodynamics and entropy as elementary principles of information theory. Read what Claude Shannon wrote on John von Neumann in the late sixties, and you see how information theory and automata research are interconnected.[19] For Shannon and von Neumann knew the problematic reliability of the "von Neumann machine" as well as any persons who ever had to deal with it. As Shannon writes:

> individual components must be built to extreme reliability, each wire must be properly connected, and each order in a program must be correct. A single error in components, wiring, or programming will typically lead to complete gibberish in the output. [...] The problem is analogous to that in communication theory where one wishes to construct codes for transmission of information for which the reliability of the entire code is high even though the reliability for the transmission of individual symbols is poor.[20]

Von Neumann's work on this topic concludes that reliable systems consisting of unreliable parts are possible if either the redundancy of parallel but similar components or their redundant connections are fantastically high.[21]

7. Von Neumann's Scores

After the question of the evolution of the "von Neumann machine," our interest turns to the question of how von Neumann programmed it. With the aid of Donald Knuth and Luis Pardo, I want to offer a provisional answer. Let us proceed with a rather useless little program written in ALGOL60 so as to not overly complicate the matter.

```
ALGOL 60:
begin INTEGER i; REAL ARRAY a[0:10];

REAL PROCEDURE F(t); REAL t; VALUE t;
   f:= sqrt(abs(t)) + 5 + t3 ;

   for i := 0 step 1 until 10 do read(a[i]);
   for i := 10 step -1 until 0 do
        begin
      y := f(a[i]);
```

```
    if y > 400 then write(i,"value too large")
      else write(i,y);
    end;
end.
```

The mere point of this program is that it contains the eleven-fold iteration of a formula that calculates the value of an eleven-digit array of REAL numbers in the inverse order of their input; if the result is > 400, the program issues a warning; if the result is smaller, it yields the numeric value. In von Neumann's notation, this program would look like Figure 9.5.

In three long pamphlets entitled "Planning and Coding Problems for an Electronic Computing Instrument" for the Army Research and Development Department, von Neumann explained this technique of programming. This is the stuff of advanced mathematics classes and I cannot go into it here. But I follow our colleague Jörg Pflüger in calling what you see here not a programming language but "planning"—on the one hand, a systematic planning diagram where the parts to be coded are simply entered into well defined boxes, and on the other hand a very open and repeatedly revised code notation which you do not see here, by which the people coding would execute what the different boxes of the diagram describe. The diagram itself is a loop of loops leading from i to e. I already referred you to the ancient Greek *diagramma* which also denotes a tonal system. Thus we may read this diagram more like a musical score than like a written notation of language.

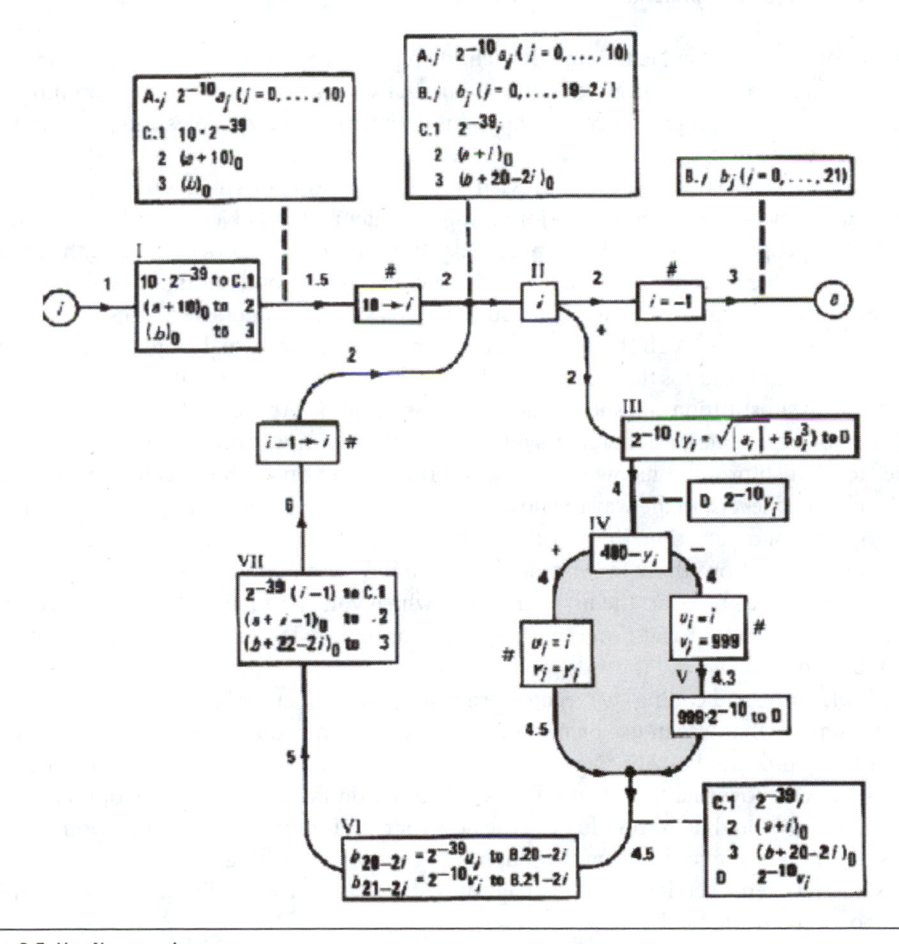

Figure 9.5 Von Neumann's score.

The rectangular boxes represent a mathematical orchestration of the work, while the parts and instrumentation are to a large extent free.

I have to limit myself to a few explanatory remarks. First you see that von Neumann leaves the scaling of the memory, the bitmapping of number types to the programmer. One had to get used to this. In the second box from the top on the left side for instance you read: "10*2 to the power of minus 39 to C.1"—which means 10 units of 40 bits each: set 10 40-bit words in storage area C.1. There are four types of boxes in this diagram:

1. "Operation boxes," which have a roman numeral, as in the box III, where after the 10th bit of the 40-bit word the arithmetic unit inserts the result of the formula "root of absolut a plus 5 a to the power of 3" with the order to shift the value to point D in the storage.
2. "Alternative boxes," also with roman numerals, which branch out depending on their prefix of the box value. For instance, in the straight line from i to e you see a box II inscribed with "i" which then branches to the lower right if i >+0, and to the left, in this case straight, if i is negative (smaller than zero).
3. "Substitution boxes," marked by a cross outside and an arrow inside. They do not signify code instructions but tell the reader of the diagram only how the value of a diagram variable has to be addressed.
4. "Assertion boxes," which simply give the value of variables, such as the last box on the straight line to e, i =–1, which means: the iteration is over.

Thus we can say that von Neumann's flow charts represent a graphic programming language. They show certain validation properties and commutative elucidations that have nothing to do with coding, but tell the programmer what happens to the operative variables and memory values when they re-enter a coded block.

Let us leave it at these hints. I have addressed only the most obvious attributes of the flow chart diagram, but it may suffice to remember Jörg Pflüger's statement: von Neumann had nothing like a concept of language in his mind. Even more, "planning and coding" does not raise the question whether it is possible to address a computer with a coherent symbolic notation. I show you this complicated flow chart to reveal how von Neumann suggested computers be addressed, namely by non-language means. Although they go together in even the most formal language, the planning score, which we see here keeps the four essential mechanisms separate: operation from alternation, alternation from substitution and substitution from assertion. There are no generative symbols, no substitutions of sign values, no literal or syntactic assertion, simply because the flow chart never even tries to furnish proof of its own correctness. This dissection of the structure of language is due to the changed levels, to the way the flow chart alternates concrete descriptions of operations with descriptions of description. In this way von Neumann prevents mathematical nonsense from creeping into his notation (think for instance of the simple mathematical nonsense in the C-command "x=x+1"). Let us be clear: the first manner in which von Neumann addresses the computer implements the conviction that programming does not require language. An egregious error by von Neumann, or shall we call it prudence?

Behind "planning and coding" we recognize a neat division of labor, if you will. The flow charts are reminiscent of what was in use before the von Neumann machine, and they can be read as a kind of perfection of the ENIAC diagrams. In this respect they also rely on programmers' habits. And since we know that the accumulators and function tables were largely operated by the "ENIAC girls" only and that the box-logic of the flow charts is nothing but a permutation of ENIAC diagrams, then planning and coding, although not a language, still has a lot to do with men as architects and women as coders—with a diagrammatic chasm, if you will, that is so self-evident that it is too easily overlooked.

8. Priesthood and Revolution

Perhaps we should add that the flow chart technique of programming as developed by von Neumann and Goldstine has remained rather theoretical. It was not implemented regularly on any calculator but rather served as conceptual crutch. Since the published and much discussed concepts of the ENIAC team—the technology of binary stored programming—there have been engineering solutions: immense sums were invested all over the U.S. to build computers, but there was no clear concept of how they should be addressed. What, therefore, was programming in the early fifties? According to John Backus:

> Programming in the early 1950s was really fun. Much of its pleasure resulted from the absurd difficulties that "automatic calculators" created for their would-be users and the challenge this presented. The programmer had to be a resourceful inventor to adapt his problem to the idiosyncrasies of the computer: He had to fit his program and data into a tiny store, and overcome bizarre difficulties in getting information in and out of it, all while using a limited and often peculiar set of instructions. He had to employ every trick he could think of to make a program run at a speed that would justify the large cost of running it. And he had to do all this by his own ingenuity, for the only information he had was a problem and a machine manual. Virtually the only knowledge about general techniques was the notion of a subroutine and its calling sequence. [...] Programming in the early 1950s was a black art, a private arcane matter involving only a programmer, a problem, a computer, and perhaps a small library of subroutines and a primitive assembly program. Existing programs for similar problems were unreadable and hence could not be adapted to new uses. General programming principles were largely nonexistent. Thus each problem required a unique beginning at square one, and the success of a program depended primarily on the programmer's private techniques and invention. [...]
>
> Just as freewheeling westerners developed a chauvinistic pride in their frontiersmanship and a corresponding conservatism, so many programmers of the freewheeling 1950s began to regard themselves as members of a priesthood guarding skills and mysteries far too complex for ordinary mortals. [...] This attitude cooled the impetus for sophisticated programming aids. The priesthood wanted and got simple mechanical aids for the clerical drudgery which burdened them, but they regarded with hostility and derision more ambitious plans to make programming accessible to a larger population. To them, it was obviously a foolish and arrogant dream to imagine that any mechanical process could possibly perform the mysterious feats of invention required to write an efficient program. Only the priests could do that. They were thus unalterably opposed to those mad revolutionaries who wanted to make programming so easy that anyone could do it.[23]

You may know this passage from Backus' report on the founders' years of FORTRAN well enough so I do not have to comment at length. But it leads us in a few sentences, and this is my intention in quoting it, to another stage, namely computers as media as well as machines. This stage would take another two or three decades, yet it was already discernible: something has started to produce priesthoods, mysteries and revolutionaries, something of considerable factual and economic impact as well as of immense technical importance, and it was expressing itself already without having a language. Somehow, we may understand from Backus, an imperative is indicated in the vehemently growing world of computers in the fifties, an imperative that makes priests, pioneers and revolutionaries dance around a golden calf, and the author of all these metaphors is none other than John Backus, who wrote probably the most influential programming language in computer history, namely FORTRAN. He concludes his *"confiteor"* with the sentence, "I am the culprit."

Perhaps to the uninitiated, to those who are not computer scientists, this mystery which is no mystery continues. So I insert a clarifying remark: John Backus wrote this in 1980, after his third important intervention in the history of programming languages, namely the Turing lecture of 1979 entitled "Can we be liberated from the von Neumann style?" Thus he discredited all the attempts to define languages in the way he and others had done before and aimed in the direction of a mathematically founded functionality. But even these attempts didn't solve the problem he describes in his retrospection. For at the birthof his language—the language he was going to specify and whose style would dominate the medium for many decades, perhaps until today—there way no language yet, but only a linguistic climate, as it were, an imperative, a denial as well as a demand.

```
SHORTCODE (sugg. John W. Mauchly, by William F. Schmitt, 1950)
Memory "Variable" i = W0, t = T0, y = Y0
11 inputs are addressed to the following words: U0, T9, T8,..., T0
constant: Z0 = 000000000000
          ZI = 010000000051
          Z2 = 010000000052
          Z3 = 040000000053
          Z4 = $$$TOO$LARGE
          Z5 = 050000000051 [5.0]

"Equation number recall information" [Labels]: 0 = line 01, 1 = line 06, 2 = line 07
 Short Code:
  Equations                          Coded representation
-----------------------------------------------------------------
  00        i = 10                   00 00 00 W0 03 Z2
  01 0:     y = (sqrt abs t) + 5 cube t   T0 02 07 ZS 11 T0
  02                                 00 Y0 03 09 20 06
  03        y 400 if<=to 1           00 00 00 Y0 Z3 41
  04        i print, 'TOO LARGE'
            print-and-return         00 00 Z4 59 W0 58
  05        0 0 if=to 2              00 00 00 Z0 Z0 72
  06 1:     i print, y print-and-return   00 00 Y0 59 W0 58
  07 2:     T0 U0 shift              00 00 00 T0 U0 99
  08        i = i – 1                00 W0 03 W0 01 Zl
  09        0 i if<=to 0             00 00 00 Z0 W0 40
  10        stop                     00 00 00 00 ZZ 08

Code-Equivalents:

01 - 06 abs value ln (n+2)nd power     59 print and return carriage
02 ) 07 + 2n (n+2)nd root              7n if= to n
03 = 08 pause 4n ifsto n               99 cyclic shift of memory
04 / 09 (                              58 print and tab
                    Sn, Tn,..., Zn quantities
```

The first implemented programming concept, as claimed since 1977 without contradiction, was Short Code. It is a language of imperatives, which simply follows an imperative. This concept also stems from the efforts of the ENIAC team—not from the mathematical corner around Goldstine and von Neumann, but from the technical pragmatism of John W. Mauchly. In 1949 he suggested this simple algebraic interpreter language, which you see here implementing our stupid algorithm. You will observe that it is "readable"—line 00 initializes i=10, line 01 the label 0, and behind it the square root formula. The interpreter jumps back there until line 08 assumes the value "–1." Already quite classical and very "spaghetti"-like.

I do not want to detain you with details, although some aspects might be rather interesting. Short Code was implemented on a UNIVAC in 1950 and already offered the programmer an "electronic dictionary," as the *ACM* magazine put it. Each arithmetic operation had a short code, hence the name. You see some of these short codes in the lower section. Short code did not know arrays, yet it shifted stored words cyclically, for instance in line 07. The simplifications are obvious: a limited dictionary of computer operations is born describing immediately what is to happen to the operands. Evidently the seeds of a language.

"With Short Code," the Remington Rand Corp. announced, "every mathematical equation can be evaluated by the mere means of notation. There is a simple symbolic transformation of equations into code [. . .] the necessity for special programming is eliminated. In our comparisons of computing time we observed a speed increase of at least 50:1 in comparison to manual programming." After the hundredfold gain in memory by the von Neumann machine, it was now a matter of gaining human storage as to become constitutive of a language—that is to say, of gaining time. Lacan would be happy getting this: programming history defines *language* as an instance of gaining mere storage time. Further: "Short Code will demonstrate its power as a tool in mathematical research and as a checking device for some large-scale problems," which means that it will also be checking priesthoods in Backus' sense, by means of a "simple tool," which will test mathematical arcana.[24]

9. FORTAN and Mariner I

Short Code is mentioned only because of its being the first putative interpreter language in computer history. It stems from the ENIAC team, just as von Neumann's concepts did. Otherwise it could not plead for much historical significance, like so many developments from the early years, although it represents quite well what everybody was looking for some years later, namely an algebraically oriented programming language. Short Code disappeared from the scene because the Remington Rand UNIVAC was used only by a small number of scientists. The pivotal condition for the scientific solution of complex problems was still missing, namely communication. Look at Jean Sammet, Grace Hopper, John Backus or anybody else—the late forties and early fifties in America are determined by the absence of what could be called a discourse on the linguistic speech-based control over the new digital computers. There was no national or international exchange, hardly a conference, no periodical publication until 1954. This further supports the thesis that the computer as the von Neumann machine was not a consciously assigned goal of the research organizations and of the scientific world in the U.S.

This is why the path to FORTRAN, which we will briefly outline, had to follow from another large-scale military project of the now icecold Cold War, the WHIRLWIND computer built at MIT in seven(!) years between 1945 and 1952. Since 1951 it had become the backbone for the "Semiautomatic Ground Environment Air Defense System" (SAGE), the first large computer project of the Air Force and the Navy. One must never forget that this very first net of computers ever built in the history of mankind is to be seen at the same time as the great grandfather of today's Internet, even though it had yet to experience the "EMP" shock of the hydrogen bomb.[25] The WHIRLWIND-computer (commanding the SAGE-net) not only connected regular radar oscilloscopes (or television screens) with a computer for the first time on a regular basis, it not only featured such puzzling devices as "light guns" touching the goals on the screen directly (later, in the '70, disarmed to "cursors" and "mouses" by the PARC-kids in Palo Alto, getting from there right into the first Mac's surfaces . . .) and probably the first keyboard ever, but it also attracted a huge crowd of scientists interested in programming, because the military organization of research worked as well as the scientific one did not.

```
1              v|N = {input},
2              i = 0,
3    1         j = i + 1,
4              a|i = v|j,
5              i = j,
6              e = i – 10.5,
7              CP 1,
8              i = 10,
9    2         y = F¹(F¹¹(a|i)) + 5(a|i)³,
10             e = y – 400,
11             CP 3,
12             z = 999,
13             PRINT i, z.
14             SP 4,
15   3         PRINT i, y.
16   4         i = i – 1,
17             e = –0.5 – i,
18             CP 2,
19             STOP
```

The language "style" which we see here was called "Laning/Zierler Algebraic Compiler" and was developed on and for the WHIRLWIND. The participants of the 1954 Office of Naval Research's first computer language conference were enthusiastic, although it does not live up to what is nowadays expected from a compiler.

A quick glance at how it works: the variables v and a are each indicated or subscribed by a vertical line to form an array, CP 1 or 3 in line 7 or 11 means a conditional jump order in the assembler style, namely, if the previous instruction yielded a negative result, jump to label 1 or 3. F to the power of 1 is the square root instruction; F to the power of 11 means an absolute instruction; lines 3 through 8 iterate in order to read the input v in to the variable a; lines 9 through 18 represent the core loop. After a few minutes, most of you will be able to read this Laning/Zierler Compiler well.

Now I want to quote what John Backus, the revolutionist, wrote in 1954. The Laning/Zierler Compiler immediately instigated the development of FORTRAN at IBM right after the conference in 1954 taking 18 man-years including the compiler. Backus writes, "a programmer may not be considered unreasonable if all he wants is formulas for the numerical solution of his problem, and perhaps a plan that shows him how his data are shifted from one storage hierarchy into another . . . No doubt, if he was to pursue this vigorously next week, he would be a psychiatric case, but perhaps next year he would be taken more seriously."[26]

Here is the result, less than a year later, in November 1954: the IBM FORmula TRANslation System, FORTRAN 0, developed under the auspices of John Backus:

```
1              DIMENSION A(11)
2              READ A
3    2         DO 3,8,11 J=1,11
4    3         I=11-J
5              Y = SQRT(ABS(A(I+1))) + 5*A(I+1)**3
6              IF (400. >=Y) 8,4
7    4         PRINT I,999.
8              GO TO 2
9    8         PRINT I,Y
10  11         STOP
```

> We think that FORTRAN offers a useful language for the formulation of problems which are fed into a machine solution... After one hour of instruction in FORTRAN, the average programmer will have completely understood the steps of writing a procedure in FORTRAN, and this without any further comments.[27]

I will not say anything new about FORTRAN; the DIMENSION declaration in line 1 of course demands an eleven-digit array of REAL numbers; line 2 reads them; line 3 shows the legendary DO statement in FORTRAN which in its 0-version denominates the start and end label, so in other words line 3 says: iterate from line with label 3 to line with label 8 until the variable J takes the value 11, then go to label 11 and end. The IF-statement was only binary then, but otherwise it is simple FORTRAN.

Here, in the language of the "lazy character," as Backus occasionally called himself, we also find the legendary programming error which made the Mariner I miss Venus by far (July 1962), and left its imprint on the software crisis of the 1970s: Instead of "DO 3 I = 1,3" the program read "DO 3 I = 1.3."[28] This mistake could not be recognized by the 18 man-year compiler project of IBM-FORTRAN whose speed still made Backus proud in 1980, because in the definition of FORTRAN spaces are not significant, owing to the fact that FORTRAN relied on punch cards where space has no significance. Consequently, because of a stop instead of a comma between 1 and 3, the result was interpreted as implicit declaration of the variable DO3I with a real value of 1.3, whereas it would have been correct to initiate a threefold iteration of all the program lines that we do not see here until the mark CONTINUE and the label 3. Needless to say, this is not just computer folklore.

10. Epilogue

The genesis of programming language styles up to and including FORTRAN is the genesis, as we know, of procedural and imperative languages. The fundamental weakness of these languages is something I do not have to impress upon computer scientists. And it is clear that declarative and functional languages are at least logically superior to the ones shown here, so I agree with Jörg Pflüger's thesis that the newer generation of object oriented languages represent an interesting synthesis of procedural structurability and functional logic design.

This paper presented a first draft discourse analysis containing a hypothesis on the development of programming languages. Strangely they do not stem from the recourse to a logical model, although the machines they control are explicitly based upon such a model. This remains a contradiction still to be resolved. The same goes for the question why those who developed the design of the machine, with full knowledge of the basic logical model, did not recognize that there is a demand for an effective and well-defined language in order to address that machine. My thesis is that for decades, the *arché*-structure of the von Neumann machine did not reveal that this machine would be more than a new calculator, more than a mighty tool for mental labor, namely a new communications medium. The development of FORTRAN demonstrates all too clearly how the communication-imperative was called on the machine from all sides. That imperative call obviously could not be detected in the *arché*-structure of the machine itself. It grew out of the Cold War, out of the economy, out of the organization of labor, perhaps out of the primitive numeric seduction the machines exerted, out of the numbers game, out of a game with digits, placeholders, *fort/da* mechanisms, and the whole quasi-linguistic *quid pro quo* of the interior structure of all these sources.

At any rate, communications media always have the structure of language, as we know since Freud. This side of and beyond explicitly spoken languages, they are characterized by the insistence of their inherent signifier, that is to say by contiguities and substitutions whose effects and traces can be visualized in graphs and diagrams, not in logical but in probabilistic and still unpredictable rules of generation.

Notes

1. Georg Wilhelm Friedrich Hegel, "Phänomenologie des Geistes" (1807), *Werkausgabe* vol. 3. (Frankfurt: Suhrkamp, 1970), 36. Georg Wilhelm Friedrich Hegel, *Phenomenology of Spirit.* (Oxford: Oxford University Press, 1977), 19.
2. Jay David Bolter, *Writing Space. The Computer, Hypertext and the History of Writing* (New Jersey: Lawrence Erlbaum, 1991), 9. See also Michael Heim, *Electric Language. A Philosophical Study of Word Processing* (New Haven: Yale 1987).
3. See Friedrich Kittler, "There is no software," *Draculas Vermächnis. Technische Schriften* (Leipzig: Reclam, 1993), 224–242, here: 242.
4. Niklaus Wirth, *Algorithmen und Datenstrukturen, Pascal* (Stuttgart: Teubner, 1983), 120.
5. Oswald Wiener, *Probleme der Künstlichen Intelligenz* (Berlin: Merve, 1988).
6. Alois Walde, *Lateinisches Etymologisches Wörterbuch* (Heidelberg: C. Winter, 1910), 738 and Hermann Menge ed., *Enzyklopädisches Wörterbuch der lateinischen und deutschen Sprache*, (Berlin: Langenscheidt, 1911), 716.
7. Hans Ulrich Gumbrecht, "Schwindende Stabilität der Wirklichkeit. Eine Geschichte des Stilbegriffs," edited by Hans Ulrich Gumbrecht and K.L. Pfeiffer, *Stil. Geschichten und Funktionen eines kulturwissenschaftlichen Diskurselements* (Frankfurt: Suhrkamp, 1986), 731.
8. Friedrich A. Kittler, "Im Telegrammstil," in Gumbrecht and K.L. Pfeiffer eds., *Stil.*, 358–370; Bernhard Siegert, *Relays. Literature as an Epoch of the Postal System* (Palo Alto: Stanford, 1999).
9. Jan Tschichold, *Was Jedermann vom Buchdruck wissen sollte. Ein Leitfaden für Drucksachen-Besteller* (Basel: Birkhäuser, 1949), and Florian Cajori, *A History of Mathematical Notations*, (Chicago: Open Court Publishing Company, 1928).
10. Arthur W. Burks, "From Eniac to the Stored-Program Computer," in *A History of Computing in the Twentieth Century*, edited by Nicholas Metropolis (New York: Academic Press, 1980), 311–344, here: 334. In the effort to avoid mistakes, the calculations took up immense amounts of memory. - "I have often been asked, 'How big was the ENIAC storage?' The answer is, Infinite. The punched-card output was not fast, but it was as big as you wished. Every card punched out could be read into the input again, and indeed that is how Metropolis and Frankel managed to handle cycle after cycle of the big problem from Los Alamos." Kathleen R. Mauchly, "John Mauchly's Early Years." *Annals Of The History Of Computing* 6 (1984), 547.
11. Nicholas Metropolis ed., *A History of Computing in the Twentieth Century* (New York: Academic Press, 1980), 522.
12. Court proceedings concerning the patent debate mark the end of the ENIAC team's collaboration. See Herman H. Goldstine, *The Computer from Pascal to Neumann* (Princeton: Princeton University Press, 1972).
13. As part of a virtual memory concept, stacks of punch-card inputs and outputs for the calculation of the Los Alamos problem, as Markly reports, would be fed by hand into the IBM card reader.
14. Arthur W. Burks allows in his description which I borrow here that the ENIAC was not simple and easy to program. "Programming proper" appeared more obscure and complicated to contemporaries than the central programming of the similarly built units of electromagnetic differential analyzers. For their firmly wired wheels and switches on the control panels one had advanced ideas of punch card control, also in Zuse, which would set the control panels by corresponding relays. The hundreds of inserts of the ENIAC which produced the wild cable trees visible on all photographs seemed less practicable for central or even stored programming.
15. Metropolis ed., *A History of Computing in the Twentieth Century*, 531.
16. Arthur W. Burks, "From Eniac to the Stored-Program Computer," in: *A History of Computing in the Twentieth Century*, edited by Nicholas Metropolis, ed., 311–344, here: 338.
17. T. Legendy and T. Szentivanyi eds., *Leben und Werk von John v. Neumann* (Mannheim-Wien-Zürich: Bibliographisches Institut, 1979), 57.
18. Friedrich Wilhelm Hagemeyer, a theoretical physicist from the old East Germany who earned his doctoral degree in sociology in the late 1970s with Wolf Lepenies, offers an exemplary idea of the complications that mark the genesis of scientific developments in the 1940s in his work on Claude Shannon's mathematical theory of communication. These complications in industrial research and development, civil technology, war economy and research, theoretical physics and mathematics result in a factual, autopoetic and planned research organism which he presents in his sociology of knowledge as an "evolutionary model of technological development" (Friedrich-Wilhelm Hagemeyer, *Die Entstehung von Informationskonzepten in der Nachrichtentechnik. Eine Fallstudie zur Theoriebildung in der Industrie- und Kriegsforschung.* Ph.D. thesis, Free University Berlin 1979, 8ff). In the case of Shannon's revolutionary work, it has the advantage of coming together in a relatively concise terminus ad quem.
19. H. Ulam, H.W. Kuhn, A.W. Tucker, and Claude E. Shannon: "John von Neumann," in *The Intellectual Migration: Europe to America 1930–1960,* edited by Donald Fleming and Bernard Bailyn (Cambridge: Harvard/Belknap, 1969), 235–269, here: 259ff.
20. H. Ulam, H.W. Kuhn, A.W. Tucker, and Claude E. Shannon: "John von Neumann," here: 256.
21. John von Neumann, "Probabilistic Logic and the Synthesis of Reliable Organisms from Unreliable Components," in *Automata Studies*, edited by Claude E. Shannon and John McCarthy, eds. (Princeton: Princeton University Press, 1956), 43–98.
23. John Backus, "Programming in America in the 1950s – Some Personal Impressions," in *A history of computing in the twentieth century,* edited by Metropolis, 125–135, here: 125–128.

24. Donald E. Knuth and Luis Trabb Pardo, "Early Development of Programming Languages," *Encyclopedia of Computer Science and Technology* (New York, Publisher, 1977): vol. 7, 419–493, here: 434. See C.A.R. Hoare, "Der neue Turmbau zu Babel. Rede zur Verleihung des Turing-Preises der Association for Computing Machinery, 1980," *Kursbuch* 75 (1980), 57–73, here: 61.
25. Wulf Halbach, "Virtualität und Ereignisse," in *Medien und Öffentlichkeit*, edited by Rudolf Maresch (München: Boer, 1996), 165–187.
26. Knuth and Pardo, "Early Development of Programming Languages," here: 456.
27. Knuth and Pardo, "Early Development of Programming Languages," here: 460.
28. G. J. Myers, "Software Reliability: Principles and Practices," John Wiley, 1976, here: 275.

10

Science as Open Source Process

Friedrich Kittler
Translated by Peter Krapp

Ladies and Gentlemen,

If I had begun my fifteen minute intervention not in the European style with "Ladies and Gentlemen" but instead addressed you in C style with "hello world," then I would presumably not be able to count on my liberty for too long. I would have had the pleasure of spending my time under a compiler and an assembler instead of entering silly ASCII keystrokes, but who knows whether this wonderful ASCII sequence, *print f ("Hello world");* will not soon be protected by source copyright, even here in Europe.

So I have two concerns. First, I worry about carrying coals to Newcastle, because I could not be here earlier. And secondly, I worry that academic freedom will stand or fall with the freedom of source code. In referring to academia, I mean above all the university, without wishing to talk self-servingly *pro domo*. For it is crucial for the university, since the Athenians, that the knowledge generated and passed on by it must be able to circulate without the protection of patents and copyrights, unlike in closed or even secret research organizations and industries. I would like to elucidate this history briefly with an eye to the dangers that imperil academia today.

The European universities I refer to were a creation of the High Middle Ages, and as far as I can see without any models or predecessors in any other cultures. This uniqueness is based on a media-technological reason: knowledge proliferated not only in oral explications or lectures from docents to students who in turn might become teachers, but in contrast to the ancients Greeks, at this university one had to work, not just chat. With the introduction of paper, which was cheaper and lighter than papyrus, universities ran *scriptoria*, where lectures were transcribed and handwritten notes copied, as well as libraries, which archived these processed data. Thirdly, to make the parallel to the contemporary global net obvious, universities also had their own medium of transmission: as incredible as this sounds, they had their own courier services.

In early modernity, the universities lost this wondrous hardware of processing, storage and transmission which defines every computer. The developing territorial states and later nation states disallowed or swallowed the mail services of butchers, monasteries and universities, and deregulation, as you know, only happened recently. After Gutenberg's invention of the printing press, the lion's share of the knowledge produced in the university fell to the system of books and publishing houses. Suddenly the universities ceased to write books and merely stored the books printed by others in libraries as well as in the heads of their students.

I leave open how Gutenberg's media revolution changed knowledge, modern universities, academies and labs. I am interested in this revolution above all in the context of open source and free software, as a rather precise model of what is going on these days. Without exaggerating too much, one may perhaps say that computers, at least on the software side, are also a creation of the university. If hardware, on its long march from tubes to transistors to integrated circuits, largely comes from military technology, the universal Turing machine as a concept (as software) stems from an academic dissertation, answering certain unsolved questions of the mathematical institute in Goettingen. Turing told Hilbert, if you like. Accordingly, the still dominant von-Neumann-architecture was developed by someone who made it from mathematics professor at Goettingen to chief consultant of the Pentagon. On its way to power, the knowledge contained in the computer and its algorithms has once again experienced the closure that threatened universities in their take-over by nation states.

As far as I know, the fastest and best algorithm testing prime numbers remains a trade secret of the Pentagon. The hope for pure, which is to say academic, mathematicians that pure math was never going to be abused for earthly ends, as Hardy still wrote during World War II, has been dramatically disappointed. But the parallel I draw between early modernity and the present only comes out fully in the so-called PC revolution. It was no accident that the garages and tinkerers' rooms that laid the groundwork for global firms like Intel and Apple were and are located next to or even on the grounds of institutions like the Rand Corporation or the Leland Stanford Junior University. The computer industry does what Gutenberg's printing press did when it took over and industrialized the calligraphy of the medieval university. The headhunters of Microsoft lurk around Stanford and at other doors of computer science departments, catch new programming serfs with new algorithms and squeeze them for five years, until the algorithms become proprietary and the coders, with their stock options, are dismissed into early retirement.

The worst aspect of this scandal seems to me that nobody talks about it. An American common law whose reach extends from the European Commission to the People's Republic of China has made an impossible concept of intellectual property as ubiquitous as it is unquestioned. Machines that, according to Turing's proof, are able to be not only all other machines but equally all human calculation, are now supposed to legitimate patents and copyrights more profoundly than ever. Machines that, according to the most recent results, run fastest and most efficiently when they were not programmed by programmers but by themselves, are supposed to belong to humans as private property—perhaps by way of euphemism for the capital corporate interests. Humanism, one might say, is today as in early modernity nothing but a fig leaf.

You all know better than me that the critique of this system can only be a practical one. Theoretical or historical remarks like mine can at best help not to lose one's overview among all the upgrades and benchmarks. However, it was practical when some programmers at the MIT resisted venality and when a computer science student at the University of Helsinki overcame the widespread fear of assemblers and cold starts. That is how immediately open source and free software are connected to the university. Look how much "edu" is in the sources of the Linux kernel. That is also how directly the future of the university depends on these free sources.

When the printing press and the nation state swallowed the media technologies of the medieval university, knowledge was left pretty much untouched on the content-level. The storage and transmission were privatized or nationalized, but data processing proper was still conducted in that beautiful old feedback circuit of eyes, ears, and writing hands. That is what changed with the computer revolution. Universal Turing machines make especially this data processing technologically reproducible. They see to it that the differences between the knowledge about technology, natural sciences, and humanities progressively disappear. This revolution, in other words, concerns all the faculties of the university, only to level their old distinctions that grew from media technologies.

The conflict of the faculties as Kant had described it may be solved peacefully, simply because it is no longer a matter of books against labs against counsel in the different faculties, but because all knowledge, including cultural knowledge, is processed in computers. This, it seems to me, grants an essential part of their chances to open source and free operating systems.

As usual, mathematicians may have been the first to grasp what this freedom delivers. Worldwide, two academic publishers distribute the mathematical journals of record. It should be obvious to catapult these journals from the Gutenberg galaxy into the Turing galaxy, especially after Adobe & Co successfully pirated almost the entire set of lead fonts from Garamond to Zapf. All mathematical knowledge would move to fully electronic publications, and their price as well as their copyright would be under the control of the said two global publishers. It is possible that this calculation goes awry, though: weeks or months before the essays or dissertations that advance mathematics land on an editor's desk in Heidelberg or Amsterdam, they are already on the computer server of a mathematical institute. This kind of bypass operation is more obvious than commercial distribution. The university can put its innovations online.

If this example, which I did not make up, is imitated, then the outlook for the commercialization of software is not rosy. The only way remaining to make knowledge proprietary would be to embed it in hardware. Once something is burned into a chip, it belongs to the firm who invested millions into the design and billions into their mass fabrication. No university can compete with that, regardless of whether it still depends on the financial support of a nation state or (more likely) already drifted off into medieval independence.

It seems as significant to me as it is sad that our congress deals with open source and free software, but does not even begin to discuss the possibility of open hardware. Since Gutenberg, constellations where, as in the middle ages, the hardware of knowledge resides with knowledge seem unthinkable. It is my impression that there are only two hopes left for hardware. Either academic freedom, while not building its own CPUs, can still produce a critique that would make faulty chips, or such stupid command sets as the complex instruction set of Intel, impossible. The division bug on the first two steps of the Pentium processor—A and B I believe—was discovered by a university, which forced Intel to conduct a recall that cost millions.

Or else, this chip- and hardware-production may act as its only possible critique. For if the prices for design and production of a machine that can be all other machines may climb to astronomical levels, they may also drop to zero. The first practical success of Turing's machine was that the Wehrmacht had forgotten an elementary fact: anything encoded by a machine can also be decoded by a machine. What Advanced Micro Devices calls reverse engineering is one of the best reasons why the mass market price of CPUs is now inexorably tending towards zero.

Free sources and open operating systems only have a chance because the computer industry always already undermines its own concept of property. Before Linux was ported to different hardware platforms, it was a highly specialized software that is said to have dismayed Andrew Tanenbaum (of Minix) by relying minimally on the Intel 386. Everything Linus Torvalds needed to that end was a publicly accessible programmer's manual, the software-abstraction of its hardware.

This may lead to a confident conclusion. "In the future," Bill Gates is supposed to have said in a perhaps not proprietary, but still internal, memo recently, "we will treat the end user as we treat computers: both are programmable." But as long as there are people who themselves are able to program instead of being programmed, this vision hopefully has no future.

Note

Originally delivered at "Wizards of OS," a conference organized by Volker Grassmuck and the Federal Office for Political Education [Bundeszentrale für Politische Bildung], House of World Cultures, Berlin, July 1999.

11

Cold War Networks or Kaiserstr. 2, Neubabelsberg

Friedrich Kittler
Translated by Peter Krapp

The task of introducing Cold War networks came unexpectedly. Usually such assignments, with the exception of those requested from the *Wolfsschanze*, are also possibilities. But when even contemporary historians prefer the baker's guilds in former zones of Soviet occupation to their own thing, only Beckettian imperatives remain. First: *"Il faut que le discours se fasse."* Second: *"Qu'importe qui parle?"* So I will start here, ideally even with the beginning.

Here in Potsdam, they say, is where the Cold War broke out and its networks began. Historians, whose mask I am going to wear for this lay engagement, seem to ignore the event, but other laypeople have reported it. Thomas Pynchon tells of Tyrone Slothrop, also known as Rocket Man, searching for the biggest chunk of pot of all times, who crosses the Avus and Lake Griebnitz, sneaks past Soviet guards through Neubabelsberg and finally excavates the hashish in the garden of a villa where a newly elected U.S. President has just arrived. Thus the first network, that of drugs, is already in place. Paul Virilio continues the story by recalling what that President talked about with his British counterpart, just before elections did away with Alan Turing's boss.

According to Virilio, Churchill told Truman that the European part of World War II had been decided not so much by blood, sweat, tears and similar things, but rather by eleven unassuming devices that were able to imitate other such unassuming devices perfectly. All operative and tactical command-lines of the *Wehrmacht*, translated by tens of thousands of encoders into apparently bug-proof, secret radio transmissions, were read in real time by those eleven British proto-computers. Thanks to the colossi of Bletchley Park, the British premier knew days ahead of time that paratroopers were going to jump over Crete, or Tiger tank divisions prepared to attack around Kursk. (Not to mention the police battalions whose radio communications were released only two months ago, for good reasons.) While the attack on Crete only concerned lieutenant general Freyberg and his New Zealanders, Operation Citadel concerned the entire Red Army in general and its commander in chief in particular. It was not for nothing that the Secretary General of the Soviet Communist Party had himself promoted to the rank of Marshall and Generalissimus. Rundstedt's attack plans, deciphered in time by Bletchley Park, had to be passed on to the ally, Stalin. But that could have meant telling Stalin how this information and the War fortunes of the British had come about. Despite all his historical materialism, it was hardly acceptable to inform the Generalissimus that since 1941, wars no longer needed men, whether as heroes or as spies, but were victories of machines over other machines. So the British Secret Service invented someone who was supposedly working for it as hero and spy in the highest ranks of the German military, passing copies of secret files such as those on Operation Citadel through Allen Dulles' Switzerland to London. From then on, all deciphered intercepts that landed on Stalin's desk were presented as heroic deeds of a certain General Werther, who could not be found even after the War ended, despite great efforts on the behalf of the Russians.[1] *Gravity's Rainbow* anticipated this as well: good old espionage is at its end. Under computing conditions, HumInt or human intelligence is only a cover for SigInt or signal intelligence. Slothrop's complaint that the spy's tradecraft depended too much on leg-work is met with Semyavin's wry comment: "It will get easier. One day machines will do it. Information machines. You are the bow wave of the future."[2]

Only Stalin could not fathom the future. For what Churchill and Truman agreed upon in Potsdam, behind the back of their ally in the East, amounted to a systematic and uninterrupted continuation of the cover-ups of all V-days. The eleven computers in Bletchley Park, the various listening posts all over Europe that fed them with secret transmissions, eventually also the U.S. American parallel efforts that were mostly directed against Japan—all that remained operational around the clock. What changed was merely the angle of their antennae: they no longer aimed at Rastenburg, Berlin, or Tokyo, but at Moscow, Murmansk, Vladivostok.

Thus the Cold War began, two years before it was officially declared, in the shape of a computer technology that remained as invisible as nature was to Heraclitus. For that reason, Stalin could declare all computer science a bourgeois aberration, in spite of the beautiful computing theories of Moscow academicians like Kolmogorov or Markov. In its effort to catch up with American nuclear technology and German rocket science, the Soviet Union failed to develop the controls necessary for the marriage of those two monstrous technologies, as if General Werther, that most simulated of all spies, had won twice. For years the payload, delivered by the Rosenbergs from Los Alamos, and the carrier rockets, procured from Peenemünde by Marshall Rokossowski, were lacking the proper computing power. Evidently not even Philby and McLean knew that it was available from Turing's Bletchley Park.

Just once, the secret is said to have trembled at its height. If it is true that only the machine cryptanalysis of Soviet radio transmission led to the Rosenbergs, if it is true that Turing still worked on the British interception schemes after the War, while he was supposedly already distracted by handsome young students and new research tasks, then it does not seem unlikely that Turing would have come across the names of some old acquaintances in those radio transmissions. Except those old Cambridge friends, homosexuals like him, were now serving the NKVD, the Soviet secret service.[3] So the security risk Turing may have seen no other way out than to become the first victim of Anglo-American computing relations—just like Snow White, he bit into an apple laced with cyanide.

This was 1954, twenty years before Turing machines even emerged from the cover of the Official Secret Act, and a year after the Cold War had bid adieu to its innocent predecessor. The winners of World War II, Turing and von Neumann, were finally allowed to die. For in 1953, the U.S. Air Force laid the groundwork for a network that, instead of merely linking listening posts to computer decryption, connected a system of distributed radar positions, computers fed with von Neumann's game theory, and strategic weapons. This network changed from a defensive stance as it was necessitated by Britain's need and chance in World War II to an offensive stance. SAGE, the Semiautomatic Ground Environment Air Defense System, was conceived as an answer to the Soviet atomic fleet, and it brought us everything today's computer users have come to love: from the monitor to networking to mass storage. "70 radar stations were in touch with 27 command centers, evenly distributed over the territory of the U.S."[4] The concept of the center itself lapsed into disuse, although it had only recently been cast in concrete in the designs for the Pentagon.

The great decentralization now celebrated as the civilian spin-off called information society began with the building of a network that connected sensors (radar), effectors (jet planes), and nodes (computers).

SAGE and its various successors gave rise to a series of problems. I mention only three aspects: hardware, software, networking. The hardware, to grant it its well-deserved primacy, was functioning, but its tube computers were too clunky and delicate to run the sensors and effectors themselves. Thus in contrast to the central computer, its terminals remained literally stupid. It was worse in the case of software: the data and routines expected to run on the semi-conductors to allow for a unified strategy only exacerbated the Babylonian difficulties of communication between the programmers on the one side and the military command structures on the other side. And then it proved fatal only too soon that strategic and operative data streams were still entrusted to telephone lines, even though they had served well in two World Wars. This fatality was paradoxically a side-effect of technological progress. By substituting transistors for tubes and integrated circuits for silicon transistors, the size of computers had shrunk by a factor of ten in the fifties and seventies, respectively, while their computing power had grown by a factor of more than ten. This miniaturization allowed the military, which had requested and financed these innovations, to use computers on land, in the air, and on the sea, before they landed on every desk today. But on the other hand, the Cold War consisted essentially of a simulation of hot war, which is to say of atomic explosions in the sky over a tropical atoll or the steppe of Kazakhstan. One of these test runs is said to have resulted in the vertiginous coincidence of simulation and reality over Hawaii, tens of thousands of miles away from the test center. Not that soldiers or anyone else had suffered—no, much worse: in addition to 300 vulgar street lights, top secret measuring devices and transistor computers were incapacitated as if by a ghost.[5] This was the experimental discovery of the electromagnetic impulse induced by every atmospheric nuclear explosion in semiconductors and copper cable, and it led to much nostalgia about the good old tubes that had been much too robust even to notice any such disruption, let alone fall prey to it. Both sides in the Cold War, even and especially the offensive party, ran the risk of being not only defeated by its own weapons, but robbed of all means of control.

The birth of the first strike doctrine was also the initiation of a new algebra which translated fundamental military terms into mathematical symbols, and finally received priority over all other weapons systems with Reagan's Presidential directive of October 1982.[6] In taking leave from its expensive armor, the military turned into probably the only subsystem of society that obeys Luhmann's systems theory literally, i.e., in empty self-reference. Communication, command, control, and intelligence were unified in the acronym C3I, until it recently shed the last vestiges of intelligence, in keeping with Churchill's fireside chat, and became C4: communication, command, control, computing.

C4 means nothing else than the permutation of these four elements, which have nothing in common with the four elements of ancient Greece. On the one hand, thanks to a Pentagon request for very high speed integrated circuits or VHSICs, contemporary microprocessors have been woken out of their leisurely megahertz pace and are beginning to run at frequencies that until recently had been reserved for radio communication.[7] On the other hand, where possible or whenever no enemy impact is anticipated, information technology leaves its domain of copper cable and ether waves in order to gird itself against the always threatening electromagnetic impulse. For the only connections that are atom-bomb proof are those that can, in contradistinction from metals, shield their inside by mirroring it against the outside, which is to say fiber-optic cable. And once computers communicating with each other over such connections are capable of appropriating all information to their own architecture and packet switching, the commands can pick their own route. Thus the kind of triumphal entries into cities like Moscow, Paris or Berlin that marked modern wars up to and including World War II are obsolete, since one fundamental rule of all strategy—the accumulation of one's own powers—is no longer relevant against distributed networks. While into the sixties the cables for early warning systems had led to buried bunkers, contemporary command centers can stand unprotected like holiday bungalows in greenery. In the spirit of magnanimity that the military inherited from its aristocratic predecessors, one only needs to open the networks to universities, programmers and finally entertainment concerns—and all dreams of technological democracy seem to come true. For that is how the Internet, favorite of the feature pages and philosophers of the day, was derived from the Pentagon's ARPAnet.

However, C4 has other worries. The result of all manner of permutations of communications, command, control and computers is a Babylonian tower of hardware architecture and programming languages, operating systems and net protocols. Evidently the silicon technology with its rate of evolution that according to Moore's law doubles every 18 months is far ahead of all soft- and wetware. That led in 1967 to the memorable declaration of a software crisis by NATO, presenting itself as scientific committee under the October sky of Garmisch-Partenkirchen. And that is why programmers today have to beg permission to use an old standby such as the GOTO command. Now there are whole programming languages that can only enter the market once a military authority has "validated" them, as they say. Cold War networks—this may be the best thing one can say about them—have given us a style that appears identical with Nietzsche's great style: dancing in shackles, as he put it, is now dancing in networks.

That is how the Cold War could end. The bourgeois aberration, as the clueless general secretary had been calling it, unaware that it was eavesdropping on him, proliferated no less than those weapons which were impossible without it. Yeltsin's team allegedly made it through the critical days (of the attempted coup that could have returned the new old Russia to the Soviet stone age) only because the insurgents had no idea of an operating system called UNIX. Having forgotten to interrupt a few Internet connections from Moscow to Helsinki was enough to break the rebels' news monopoly. And if Gorbachev's ghostwriter is not sponsored by Silicon Valley, the only reason the evil empire did not expire with a nuclear bang was that five-year plans were good for developing intercontinental missiles or space travel, but could not force Moore's fantastic evolutionary rate for microchips. Gordon Moore was co-founder of Intel, not of the VEB Robotron.[8] The Cold War that began in theory in the escalatory paradox of a certain Schiller who is only homonymous with the German dramatist, thus ended in proliferations of the kind of paradox that it had generated. The COCOM list (by which the NATO prevented the export of the silicon chips now operating on every desk to the East) was as ineffective as its heroic circumvention by Dr. Schalck-Golodkowski.[9] For as IBM realized, the pivot of the 30-year plan to computerize the Warsaw Pact countries was the cloning of every Intel gate and IBM operating system by the VEB Robotron, and this did not cause any consternation in Armonk—on the contrary: if Systemotekhnia as the leading technology of the Warsaw Pact worked on IBM standards, the domination of the world market of tomorrow was already guaranteed.

Even this was not the end of the proliferation, here called cloning; whatever the Warsaw Pact had to reverse-engineer as a last resort in the Cold War turned into the peace-time industry standard, and the successors to VEB Robotron are now companies like Cyrix and AMD who try to break Intel's patents—for the greater glory and market dominance of Intel. The industry ceased to think in the self-satisfied terms of customers as buyers, and finally learned from the Cold War about the concept of the enemy. In the same week the Potsdam conference dismantled the German military, the organization charts of the Prussian Chiefs of Staff were exported overseas as training materials for business schools, and now computer technology has brought reconnaissance and knowledge to coincide.[10] "Industry remains industry, regardless of its direction towards the destruction or creation of objects," as Friedrich Engels already knew.[11] Reverse engineering brings this identity to full coincidence, because henceforth generation requires destruction. To know what one does is to know first what the other does.

The huts of Bletchley Park, that analytic crypt of World War II, won a total victory. Turing's assumption that computers would be infinitely better suited for cryptanalysis than for physics has become reality. At this point, I could report the execution of my task, since the end has found its way back to the beginning. These days British historians are restoring the huts of Bletchley Park to their 1945 state, and the eleven colossi of Bletchley Park are able to run again. Their Russian, formerly Soviet colleagues have also signaled that the first computer of the Red Army will rise from the scrap heap in Novosibirsk, and it might even take a trip to Bletchley Park—a reunion, at last, of almost all the monsters of the Cold War.[12] To round off the mausoleum, Helmut Hoelzer, NASA veteran from Huntsville, Alabama and V2 veteran from Peenemünde/Usedom would have to place next to its digital colleagues the analog computer that replaced Vannevar Bush's languid mechanics with real time electronics.[13] Dwarves, as the pious Peter von Blois already knew, see farther on the shoulders of giants.

But the truth is—we know nothing. The museum-quality reconstruction of Bletchley Park only darkens the mirror in which we look for the heritage of the Cold War. The message of the digital computer, ending all media history in its universality, might support the Pax Americana for another while, but it cannot be verified, as Helmut Hoelzer may have suspected already. For it is quite possible that one day, wars will not be decided by C4 as they are today, but again by the physics for which Turing machines are somehow unfit. It is just feasible that someone somewhere invents a machine that is no longer based on reverse engineering and optical fiber proliferating from here to Baghdad, but turns a new page in the history of media. That would spell the end of the Pax Americana, for the entire military-industrial complex of silicon glow and fiber-optic light, of net topologies and end users, is only one side of the system, the side we can see. On the other side there is still poverty and darkness. I am here not referring to developing countries or migration, but to that unique resource that no distributed network can increase. As the head of the U.S. Chiefs of Staff, Admiral Moore, put it so unmistakably: victory in future Wars will go to the side which controls the electromagnetic spectrum. Carl Schmitt, who was talking of a very different sovereignty after the end of World War I, had to agree with the Admiral, without knowing it, after World War II—he revoked one of his best-known statements and decreed that sovereign is who commands the waves of space.

However, the electromagnetic spectrum is a principally limited resource, as the dominant military doctrine in the U.S. never ceases to emphasize. It extends to several frequency bands, from the almost unaffordable long waves used for submarine remote control to the quantum effects any rain can drown out. This finitude is brought home every time the net leaves its own high-tech infrastructure and takes steps beyond basically defensive war games. Silicon chips, even if they run in grenades or ICBMs instead of desktop computers or assembly lines, have the remarkable trait of self-destruction on either side of an acceptable temperature level. Evidently, fiber-optic cable is as easy to cut as the transatlantic telegraph was on the first summer day of World War I. For maintaining offensive capabilities in deserts of sand or water, the electromagnetic spectrum is the sole refuge. Like a dark mirror, its finitude reflects the bad infinity of the deserts that contemporary war implies, generates, and leaves behind.

In a book that appeared in the year 1832, one may read this: "Imagine a traveler who late in the day decides to cover two more stages before nightfall. Only four or five hours more, on a paved highway with relays of horses: it should be an easy trip. But at the next station he finds no fresh horses, or only poor ones; the country grows hilly, the road bad, night falls, and finally after many difficulties he is only too glad to reach a resting place with any kind of primitive accommodation. It is much the same in war. Countless minor incidents—the kind you can never really foresee – combine to lower the general level of performance, so that one always falls short of the intended goal."[14] Today, we have the electronic net instead of post horses and digital computers instead of paper, while the mechanically integrated theater of war prepares to replace the traveler as the metaphoric field commander. But in every world of limited resources Clausewitz's statement still stands: "Friction is the only conception that more or less corresponds to the factors that distinguish real war from war on paper."

Notes

Originally delivered at "Data Conflicts: Eastern Europe and the Geopolitics of Cyberspace," a conference organized by Thomas Keenan and Thomas Y. Levin at the Einstein Forum, Potsdam, December 1996.

1. See Jósef Garlinski, *The Enigma War* (New York: Scribner 1980), 138.
2. TN: See Thomas Pynchon, *Gravity's Rainbow* (New York: Penguin 1987), 258.
3. TN: the NKVD (Narodny Komissariat Vnutrennich Del, or People's Commissariat for Internal Affairs) was the Soviet secret service predecessor to the KGB, as it was renamed in 1953 after Stalin's death.
4. Claus Eurich, *Tödliche Signale. Die kriegerische Geschichte der Informationstechnik von der Antike bis zum Jahr 2000* (Frankfurt: Luchterhand 1991), 111.
5. Eurich, 142.
6. Eurich, 169.
7. Eurich, 128f.; see Marco de Arcangelis, *Electronic Warfare* (New York: Sterling 1985).
8. TN: The VEB Kombinat Robotron was a computer manufacturer founded in 1969 by the East German Government in Dresden; it was the main software developer and system integrator for Warsaw Pact countries. It was dismantled in 1990, and one of its privatized successor companies now maintains offices in Russia and Bellville, Texas (near Houston), focusing on Internet and Intranet communication, as Robotron-Projekt GmbH.
9. TN: Alexander Schalck-Golodkowski was a highly decorated East German politician who came under suspicion, right after the fall of the Berlin Wall, for his role in covert trade relations with other countries over the preceding decades. Accused of drug trafficking, embezzlement, espionage and treason, he was eventually tried and convicted for arms dealing only.
10. See Rolf Elble, ed., *Clausewitz in unserer Zeit* (Darmstadt: Wehr- und Wissen-Verlagsgesellschaft, 1971).
11. TN: cited after Karl Marx / Friedrich Engels, *Gesamtausgabe* (MEGA), vol. 20. Edited by the International Marx-Engels-Foundation (Berlin: Dietz 1991–1993), 155.
12. See Georg Trogemann, Alexander Y. Nitussov, and Wolfgang Ernst (eds.), *Computing in Russia: The History of Computer Devices and Information Technology revealed* (Braunschweig: Vieweg Verlag 2001).
13. See Helmut Hoelzer's contribution to Norbert Bolz and Friedrich Adolf Kittler (eds.), *Computer als Medium* (Munich: Fink 1994).
14. Carl von Clausewitz, *On War* (Princeton: Princeton University Press 1984, 119). See Carl von Clausewitz, *Vom Kriege* (Berlin: 1912), 60. TN: An older translation is available on the Web: "Suppose now a traveller, who, towards evening, expects to accomplish the two stages at the end of his day's journey, four or five leagues, with post horses, on the high road—it is nothing. He arrives now at the last station but one, finds no horses, or very bad ones; then a hilly country, bad roads; it is a dark night, and he is glad when, after a great deal of trouble, he reaches the next station, and finds there some miserable accommodation. So in war, through the influence of an infinity of petty circumstances, which cannot properly be described on paper, things disappoint us, and we fall short of the mark."—"Friction is the only conception which, in a general way, corresponds to that which distinguishes real war from war on paper." Carl von Clausewitz, *On War* (London: 1873), Book 1, Chapter 7; see http://www.clausewitz.com/CWZHOME/Waystatn.html.

12

Protocol vs. Institutionalization

Alexander R. Galloway

In this day and age, technical protocols and standards are established by a self-selected oligarchy of scientists consisting largely of electrical engineers and computer specialists. Composed of a patchwork of many professional bodies, working groups, committees and subcommittees, this technocratic elite toils away, mostly voluntarily, in an effort to hammer out solutions to advancements in technology. Many of them are university professors. Most all of them either work in industry, or have some connection to it.

Like the philosophy of protocol itself, membership in this technocratic ruling class is open. "Anyone with something to contribute could come to the party,"[1] wrote one early participant. But, to be sure, because of the technical sophistication needed to participate, this loose consortium of decision-makers tends to fall into a relatively homogenous social class: highly educated, altruistic, liberal-minded science professionals from modernized societies around the globe.

And sometimes not so far around the globe. Of the twenty-five or so original protocol pioneers, three of them—Vint Cerf, Jon Postel and Steve Crocker—all came from a single high school in Los Angeles's San Fernando Valley.[2] Furthermore during his long tenure as Request for Comments (RFC) Editor, Postel was the single gatekeeper through whom all protocol RFCs passed before they could be published.

Internet historians Katie Hafner and Matthew Lyon describe this group as "an ad-hocracy of intensely creative, sleep-deprived, idiosyncratic, well-meaning computer geniuses."[3]

There are few outsiders in this community. Here the specialists run the show. To put it another way, while the Internet is used daily by vast swaths of diverse communities, the standards-makers at the heart of this technology are a small entrenched group of techno-elite peers.

The reasons for this are largely practical. "Most users are not interested in the details of Internet protocols," Vint Cerf observes, "they just want the system to work."[4] Or as former IETF Chair Fred Baker reminds us: "The average user doesn't write code. [...] If their needs are met, they don't especially care how they were met."[5]

So who actually writes these technical protocols, where did they come from, and how are they used in the real world? They are found in the fertile amalgamation of computers and software that constitutes the majority of servers, routers and other internet-enabled machines. A signifigant portion of these computers were, and still are, Unix-based systems. A signifigant portion of the software was, and still is, largely written in the C or C++ languages. All of these elements have enjoyed unique histories as protocological technologies.

The Unix operating system was developed at Bell Telephone Laboratories by Ken Thompson, Dennis Ritchie and others beginning in 1969 and continuing development into the early '70s. After the operating system's release the lab's parent company, AT&T, began to license and sell Unix as a commercial software product. But, for various legal reasons, the company admitted they "had no intention of pursuing software as a business."[6] Unix was indeed sold by AT&T, but simply "as is" with no advertising, technical support or other fanfare. This contributed to its widespread adoption by universities who found in Unix a cheap but useful operating system that could be easily experimented with, modified and improved.

In January 1974, Unix was installed at the University of California at Berkeley. Bill Joy and others began developing a spin-off of the operating system which became known as BSD (Berkeley Software Distribution).

Unix was particularly successful because of its close connection to networking and the adoption of basic interchange standards. "Perhaps the most important contribution to the proliferation of Unix was the growth of networking,"[7] writes Unix historian Peter Salus. By the early '80s, the TCP/IP networking suite was included in BSD Unix.

Unix was designed with openness in mind. The source code—written in C, which was also developed during 1971–1973—is easily accessible, meaning a higher degree of technical transparency.

The standardization of the C programming language began in 1983 with the establishment of an American National Standards Institute (ANSI) committee called "X3J11." The ANSI report was finished in 1989 and subsequently accepted as a standard by the international consortium ISO in 1990.[8] Starting in 1979, Bjarne Stroustrup developed C++, which added the concept of classes to the original C language. (In fact, Stroustrup's first nickname for his new language was "C with Classes.") ANSI standardized the C++ language in 1990.

C++ has been tremendously successful as a language. "The spread was world-wide from the beginning," recalled Stroustrup. "[I]t fit into more environments with less trouble than just about anything else."[9] Just like a protocol.

It is not only computers that experience standardization and mass adoption. Over the years many technologies have followed this same trajectory. The process of standards creation is, in many ways, simply the recognition of technologies that *have* experienced success in the market place. One example is the VHS video format developed by JVC (with Matsushita), which beat out Sony's Betamax format in the consumer video market. Betamax was considered by some to be a superior technology (an urban myth, claim some engineers) because it stored video in a higher-quality format. But the trade off was that Betamax tapes tended to be shorter in length. In the late '70s when VHS launched, the VHS tape allowed for up to two hours of recording time, while Betamax only one hour. "By mid 1979 VHS was outselling Beta by more than 2 to 1 in the U.S."[10] When Betamax caught up in length (to three hours) it had already lost a foothold in the market. VHS would counter Betamax by increasing to four hours and later eight.

Some have suggested that it was the pornography industry, who favored VHS over Betamax, that provided it with legions of early adopters and proved the long term viability of the format.[11]

But perhaps the most convincing argument is the one that points out JVC's economic strategy which included aggressive licensing of the VHS format to competitors. JVC's behavior is pseudo-protocological. They licensed the technical specifications for VHS to other vendors. They also immediately established manufacturing and distribution supply chains for VHS tape manufacturing and retail sales. In the meantime Sony tried to fortify its market position by keeping Betamax to itself. As one analyst writes:

> Three contingent early differences in strategy were crucial. First, Sony decided to proceed without major co-sponsors for it Betamax system, while JVC shared VHS with several major competitors. Second, the VHS consortium quickly installed a large manufacturing capacity. Third, Sony opted for a more compact cassette, while JVC chose a longer playing time for VHS, which proved more important to most customers.[12]

JVC deliberately sacrificed larger profit margins by keeping prices low and licensing to competitors. This was in order to grow their market share. The rationale was that establishing a standard was the most important thing, and as they approached that goal, it would create a positive feedback loop that would further beat out the competition.

The VHS/Betamax story is a good example from the commercial sector for how one format can beat out another format and become an industry standard. This example is interesting because it shows that protocological behavior (giving out your technology broadly even if it means giving it to your competitors) often wins out over proprietary behavior. The Internet protocols function in a similar way, to the degree that they have become industry standards not through a result of propriety market forces, but due to broad open initiatives of free exchange and debate. This was not exactly the case with VHS, but the analogy is useful nevertheless.

This type of corporate squabbling over video formats has since been essentially erased from the world stage with the advent of DVD. This new format was reached through consensus from industry leaders and hence does not suffer from direct competition by any similar technology in the way that VHS and Betamax did. Such consensus characterizes the large majority of processes in place today around the world for determining technical standards.

Many of today's technical standards can be attributed to the Institute of Electrical and Electronics Engineers, or IEEE (pronounced "eye triple e"). In 1963 IEEE was created through the merging of two professional societies. They were the American Institute of Electrical Engineers (AIEE) founded in New York on May 13, 1884 (by a group which included Thomas Edison) and the Institute of Radio Engineers (IRE) founded in 1912.[13] Today the IEEE has over 330,000 members in 150 countries. It is the world's largest professional society in any field. The IEEE works in conjunction with industry to circulate knowledge of technical advances, to recognize individual merit through the awarding of prizes, and to set technical standards for new technologies. In this sense the IEEE is the world's largest and most important protocological society.

Composed of many chapters, sub-groups and committees, the IEEE's Communications Society is perhaps the most interesting area vis-à-vis computer networking. They establish standards in many common areas of digital communication including digital subscriber lines (DSLs) and wireless telephony.

IEEE standards often become international standards. Examples include the "802" series of standards which govern network communications protocols. These include standards for Ethernet[14] (the most common local area networking protocol in use today), Bluetooth, Wi-Fi, and others.

"The IEEE," Paul Baran observed, "has been a major factor in the development of communications technology."[15] Indeed Baran's own theories, which eventually would spawn the Internet, were published within the IEEE community even as they were published by his own employer, the RAND Corporation.

Active within the United States are the National Institute for Standardization and Technology (NIST) and American National Standards Institute (ANSI). The century old NIST, formerly known as the National Bureau of Standards, is a federal agency that develops and promotes technological standards. Because they are a federal agency and not a professional society, they have no membership per se. They are also non-regulatory, meaning that they do not enforce laws or establish mandatory standards which must be adopted. Much of their budget goes into supporting NIST research laboratories as well as various outreach programs.

ANSI, formerly called the American Standards Association, is responsible for aggregating and coordinating the standards creation process in the U.S. They are the private sector counterpart to NIST. While they do not create any standards themselves, they are a conduit for federally-accredited organizations in the field who are developing technical standards. The accredited standards developers must follow certain rules designed to keep the process open and equitable for all interested parties. ANSI then verifies that the rules have been followed by the developing organization before the proposed standard is adopted.

ANSI is also responsible for articulating a national standards strategy for the US. This strategy helps ANSI advocate in the international arena on behalf of United States interests. ANSI is the only organization that can approve standards as American national standards.

Many of ANSI's rules for maintaining integrity and quality in the standards development process revolve around principles of openness and transparency and hence conform with much of what I have said elsewhere about protocol. ANSI writes that:

- Decisions are reached through *consensus* among those affected.
- Participation is *open* to all affected interests. [...]
- The process is *transparent* — information on the process and progress is directly available. [...]
- The process is *flexible*, allowing the use of different methodologies to meet the needs of different technology and product sectors.[16]

Besides being consensus-driven, open, transparent and flexible, ANSI standards are also voluntary, which means that, like NIST, no one is bound by law to adopt them. Voluntary adoption in the marketplace is the ultimate test of a standard. Standards may disappear in the advent of a new superior technology or simply with the passage of time. Voluntary standards have many advantages. By not forcing industry to implement the standard the burden of success lies in the marketplace. And in fact, proven success in the marketplace generally preexists the creation of a standard. The behavior is emergent, not imposed.

On the international stage several other standards bodies become important. The International Telecommunication Union (ITU) focuses on radio and telecommunications, including voice telephony, communications satellites, data networks, television and in the old days, the telegraph. Established in 1865 they claim to be the world's oldest international organization.

The International Electrotechnical Commission (IEC) prepares and publishes international standards in the area of electrical technologies including magnetics, electronics and energy production. They cover everything from screw threads to quality management systems. IEC is comprised of national committees. (The national committee representing the U.S. is administered by ANSI.)

Another important international organization is ISO, also known as the International Organization for Standardization.[17] Like the IEC, ISO grows out of the electro-technical field and was formed after World War II to "facilitate the international coordination and unification of industrial standards."[18] Based in Geneva, but a federation of over 140 national standards bodies including the American ANSI and the British Standards Institution (BSI), their goal is to establish vendor-neutral technical standards. Like the other international bodies, standards adopted by the ISO are recognized worldwide.

Also like other standards bodies, ISO standards are developed through a process of consensus-building. Their standards are based on voluntary participation and thus the adoption of ISO standards is driven largely by market forces. (As opposed to mandatory standards which are implemented in response a governmental regulatory mandate.) Once established, ISO standards can have massive market penetration. For example the ISO standard for film speed (100, 200, 400, etc.) is used globally by millions of consumers.

Another ISO standard of far-reaching importance is the Open Systems Interconnection (OSI) Reference Model. Developed in 1978, the OSI Reference Model is a technique for classifying all networking activity into seven abstract layers. Each layer describes a different segment of the technology behind networked communication.

Layer 7 Application
Layer 6 Presentation
Layer 5 Session

Layer 4 Transport
Layer 3 Network
Layer 2 Data link
Layer 1 Physical

This classification helps organize the process of standardization into distinct areas of activity, and is relied on heavily by those creating data networking standards.

In 1987 the ISO and the IEC recognized that some of their efforts were beginning to overlap. They decided to establish an institutional framework to help coordinate their efforts and formed a joint committee to deal with information technology called the Joint Technical Committee 1 (JTC 1). ISO and IEC both participate in the JTC 1, as well as liaisons from Internet-oriented consortia such as the IETF. ITU members, IEEE members and others from other standards bodies also participate here. Individuals may sit on several committees in several different standards bodies, or simply attend as *ex officio* members, to increase inter-organizational communication and reduce redundant initiatives between the various standards bodies. JTC 1 committees focus on everything from office equipment to computer graphics. One of the newest committees is devoted to biometrics.

ISO, ANSI, IEEE, and all the other standards bodies are well established organizations with long histories and formidable bureaucracies. The Internet on the other hand has long been skeptical of such formalities and spawned a more ragtag, shoot from the hip attitude about standard creation.[19] I will focus the rest of this chapter on those communities and the protocol documents that they produce.

There are four groups that make up the organizational hierarchy in charge of Internet standardization. They are the Internet Society, the Internet Architecture Board, the Internet Engineering Steering Group, and the Internet Engineering Task Force.[20]

The Internet Society (ISOC), founded in January 1992, is a professional membership society. It is the umbrella organization for the other three groups. Its mission is "[t]o assure the open development, evolution and use of the Internet for the benefit of all people throughout the world."[21] It facilitates the development of Internet protocols and standards. ISOC also provides fiscal and legal independence for the standards-making process, separating this activity from its former U.S. government patronage.

The Internet Architecture Board (IAB), originally called the Internet Activities Board, is a core committee of thirteen nominated by and consisting of members of the IETF.[22] The IAB reviews IESG appointments, provides oversight of the architecture of network protocols, oversees the standards creation process, hears appeals, oversees the RFC Editor, and performs other chores. The IETF (as well as the Internet Research Task Force which focuses on longer term research topics) falls under the auspices of the IAB. The IAB is primarily an oversight board, since actually accepted protocols generally originate within the IETF (or in smaller design teams).

Underneath the IAB is the Internet Engineering Steering Group (IESG), a committee of the Internet Society that assists and manages the technical activities of the IETF. All of the directors of the various research areas in the IETF are part of this Steering Group.

The bedrock of this entire community is the Internet Engineering Task Force (IETF). The IETF is the core area where most protocol initiatives begin. Several thousand people are involved in the IETF, mostly through email lists, but also in face to face meetings. "The Internet Engineering Task Force is," in their own words, "a loosely self-organized group of people who make technical and other contributions to the engineering and evolution of the Internet and its technologies."[23] Or elsewhere: "the Internet Engineering Task Force (IETF) is an open global community of network designers, operators, vendors, and researchers producing technical specifications for the evolution of the Internet architecture and the smooth operation of the Internet."[24]

The IETF is best defined in the following RFCs:

- "The Tao of IETF: A Guide for New Attendees of the Internet Engineering Task Force" (RFC 1718, FYI 17)
- "Defining the IETF" (RFC 3233, BCP 58)
- "IETF Guidelines for Conduct"[25] (RFC 3184, BCP 54)
- "The Internet Standards Process — Revision 3" (RFC 2026, BCP 9)
- "IAB and IESG Selection, Confirmation, and Recall Process: Operation of the Nominating and Recall Committees" (RFC 2727, BCP 10)
- "The Organizations Involved in the IETF Standards Process" (RFC 2028, BCP 11)

These documents describe both how the IETF creates standards, but also how the entire community itself is set up and how it behaves.

The IETF is the least bureaucratic of all the organizations mentioned here. In fact it is not an organization at all, but rather an informal community. It does not have strict bylaws or formal officers. It is not a corporation (nonprofit or otherwise) and thus has no Board of Directors. It has no binding power as a standards creation body and is not ratified by any treaty or charter. It has no membership, and its meetings are open to anyone. "Membership" in the IETF is simply evaluated through an individual's participation. If you participate via email, or attend meetings, you are a member of the IETF. All participants operate as unaffiliated individuals, not as representatives of other organizations or vendors.

The IETF is divided up by topic into various Working Groups. Each Working Group[26] focuses on a particular issue or issues and drafts documents that are meant to capture the consensus of the group. Like the other standards bodies, IETF protocols are voluntary standards. There is no technical or legal requirement[27] that anyone actually adopt IETF protocols.

The process of establishing an Internet Standard is gradual, deliberate, and negotiated. Any protocol produced by the IETF goes through a series of stages, called the "standards track." The standards track exposes the document to extensive peer review, allowing it to mature into an RFC memo and eventually an Internet Standard. "The process of creating an Internet Standard is straightforward," they write. "A specification undergoes a period of development and several iterations of review by the Internet community and revision based upon experience, is adopted as a Standard by the appropriate body [...], and is published."[28]

Preliminary versions of specifications are solicited by the IETF as Internet-Draft documents. Anyone may submit an Internet-Draft. They are not standards in any way and should not be cited as such nor implemented by any vendors. They are works in progress and are subject to review and revision. If they are deemed uninteresting or unnecessary, they simply disappear after their expiration date of six months. They are not RFCs and receive no number.

If an Internet-Draft survives the necessary revisions and is deemed important, it is shown to the IESG and nominated for the standards track. If the IESG agrees (and the IAB approves), then the specification is handed off to the RFC Editor and put in the queue for future publication. The actual stages in the standards track are:

1. **Proposed Standard**—The formal entry point for all specifications is here as a Proposed Standard. This is the beginning of the RFC process. The IESG has authority via the RFC Editor to elevate an Internet-Draft to this level. While no prior real world implementation is required of a Proposed Standard, these specifications are generally expected to be fully-formulated and implementable.
2. **Draft Standard**—After specifications have been implemented in at least two "independent and interoperable" real world applications they can be elevated to the level of a Draft Standard. A specification at the Draft Standard level must be relatively stable and easy to understand. While subtle revisions are normal for Draft Standards, no substantive changes are expected after this level.

3. **Standard**—Robust specifications with wide implementation and a proven track record are elevated to the level of Standard. They are considered to be official Internet Standards and are given a new number in the "STD" sub-series of the RFCs (but also retain their RFC number). The total number of Standards is relatively small.

Not all RFCs are standards. Many RFCs are informational, experimental, historic, or even humorous[29] in nature. Furthermore not all RFCs are full-fledged Standards—they may not be that far along yet.

In addition to the STD subseries for Internet Standards, there are two other RFC subseries that warrant special attention: the Best Current Practice Documents (BCP) and informational documents known as FYI.

Each new protocol specification is drafted in accordance with RFC 1111, "Request for Comments on Request for Comments: Instructions to RFC Authors," which specifies guidelines, text formatting and otherwise, for drafting all RFCs. Likewise, FYI 1 (RFC 1150) titled "F.Y.I. on F.Y.I.: Introduction to the F.Y.I. Notes" outlines general formatting issues for the FYI series. Other such memos guide the composition of Internet-Drafts, as well as STDs and other documents. Useful information on drafting Internet standards is also found in RFCs 2223 and 2360.[30]

The standards track allows for a high level of due process. Openness, transparency and fairness are all virtues of the standards track. Extensive public discussion is par for the course.

Some of the RFCs are extremely important. RFCs 1122 and 1123 outline all the standards that must be followed by any computer that wishes to be connected to the Internet. Representing "the consensus of a large body of technical experience and wisdom,"[31] these two documents outline everything from email and transferring files to the basic protocols like IP that actually move data from one place to another.

Other RFCs go into greater technical detail on a single technology. Released in September 1981, RFC 791 and RFC 793 are the two crucial documents in the creation of the Internet protocol suite TCP/IP as we know it today. In the early '70s Robert Kahn of DARPA and Vinton Cerf of Stanford University teamed up to create a new protocol for the intercommunication of different computer networks. In September 1973 they presented their ideas at the University of Sussex in Brighton and soon afterwards completed writing the paper "A Protocol for Packet Network Intercommunication" which would be published in 1974 by the IEEE. The RFC Editor Jon Postel and others assisted in the final protocol design.[32] Eventually this new protocol was split in 1978 into a two-part system consisting of TCP and IP. (As mentioned elsewhere TCP is a reliable protocol which is in charge of establishing connections and making sure packets are delivered, while IP is a connectionless protocol that is only interested in moving packets from one place to another.)

One final technology worth mentioning in the context of protocol creation is the World Wide Web. The Web emerged largely from the efforts of one man, the British computer scientist Tim Berners-Lee. During the process of developing the Web, Berners-Lee wrote both the Hypertext Transfer Protocol (HTTP) and the Hypertext Markup Language (HTML), which form the core suite of protocols used broadly today by servers and browsers to transmit and display web pages. He also created the web address, called a Universal Resource Identifier (URI), of which today's "URL" is a variant, which is a simple, direct way for locating any resource on the Web.

Tim Berners-Lee:

> The art was to define the few basic, common rules of "protocol" that would allow one computer to talk to another, in such a way that when all computer everywhere did it, the system would thrive, not break down. For the Web, those elements were, in decreasing order of importance, universal resource identifiers (URIs), the Hypertext Transfer Protocol (HTTP), and the Hypertext Markup Language (HTML).

So, like other protocol designers, Berners-Lee's philosophy was to create a standard language for interoperation. By adopting his language, the computers would be able to exchange files. He continues:

> What was often difficult for people to understand about the design was that there was nothing else beyond URIs, HTTP, and HTML. There was no central computer "controlling" the Web, no single network on which these protocols worked, not even an organization anywhere that "ran" the Web. The Web was not a physical "thing" that existed in a certain "place." It was a "space" in which information could exist.[33]

This is also in line with other protocol scientists's intentions—that an info-scape exists on the net with no centralized administration or control. (But as I have pointed out elsewhere, it should not be inferred that a lack of centralized control means a lack of control as such.)

Berners-Lee eventually took his ideas to the IETF and published "Universal Resource Identifiers in WWW" (RFC 1630) in 1994. This memo describes the correct technique for creating and decoding URIs for use on the Web. But, Berners-Lee admitted, "the IETF route didn't seem to be working."[34]

Instead he established a separate standards group in October 1994 called the World Wide Web Consortium (W3C). "I wanted the consortium to run on an open process like the IETF's," Berners-Lee remembers, "but one that was quicker and more efficient. [...] Like the IETF, W3C would develop open technical specifications. Unlike the IETF, W3C would have a small full-time staff to help design and develop the code where necessary. Like industry consortia, W3C would represent the power and authority of millions of developers, researchers and users. And like its member research institutions, it would leverage the most recent advances in information technology."[35]

The W3C creates the specifications for Web technologies, and releases "recommendations" and other technical reports. The design philosophies driving the W3C are similar to those at the IETF and other standards bodies. They promote a distributed (their word is "decentralized") architecture, they promote interoperability in and among different protocols and different end systems, and so on.

In many ways the core protocols of the Internet had their development heyday in the '80s. But Web protocols are experiencing explosive growth today.

The growth is due to an evolution of the concept of the Web into what Berners-Lee calls the Semantic Web. In the Semantic Web, information is not simply interconnected on the Internet using links and graphical markup—what he calls "a space in which information could permanently exist and be referred to"[36]—but it is enriched using descriptive protocols that say what the information actually is.

For example, the word "Galloway" is meaningless to a machine. It is just a piece of information that says nothing about what it is or what it means. But wrapped inside a descriptive protocol it can be effectively parsed: "<surname>Galloway</surname>." Now the machine knows that Galloway is a surname. The word has been enriched with semantic value. If one makes the descriptive protocols more complex, then one is able to say more complex things about information, i.e., that Galloway is *my* surname, and my given name is Alexander, and so on. The Semantic Web is simply the process of adding extra meta-layers on top of information so that it can be parsed according to its semantic value.

Why is this significant? Before this, protocol had very little to do with meaningful information. Protocol does not interface with content, with semantic value. It is against interpretation. But with Berners-Lee comes a new strain of protocol: protocol that cares about meaning. This is what he means by a Semantic Web. It is, as he says, "machine-understandable information."

Does the Semantic Web, then, contradict the principle that protocol is against interpretation? I'm not so sure. Protocols can certainly *say* things about their contents. A checksum does this.

A file-size variable does this. But do they actually know the meaning of their contents? So it is a matter of debate as to whether descriptive protocols actually add intelligence to information, or if they are simply subjective descriptions (originally written by a human) that computers mimic but understand little about. Berners-Lee himself stresses that the Semantic Web is not an artificial intelligence machine.[37] He calls it "well-defined" data, not interpreted data—and in reality those are two very different things.

As this survey of protocological institutionalization shows, the primary source materials for any protocological analysis of Internet standards are the Request for Comments (RFC) memos. They began circulation in 1969 with Steve Crocker's RFC "Host Software" and have documented all developments in protocol since.[38] "It was a modest and entirely forgettable memo," Crocker remembers, "but it has significance because it was part of a broad initiative whose impact is still with us today."[39]

While generally opposed to the center-periphery model of communication—what some call the "downstream paradigm"[40]—Internet protocols describe all manner of computer-mediated communication over networks. There are RFCs for transporting messages from one place to another, and others for making sure it gets there in one piece. There are RFCs for email, for web pages, for news wires, and for graphic design.

Some advertise distributed architectures (like IP routing), some hierarchical (like the DNS). Yet they all create the conditions for technological innovation based on a goal of standardization and organization. It is a peculiar type of anti-federalism through universalism—strange as it sounds—whereby universal techniques are levied in such a way as ultimately to revert much decision-making back to the local level.

But during this process many local differences are elided in favor of universal consistencies. For example, protocols like HTML were specifically designed to allow for radical deviation in screen resolution, browser type and so on. And HTML (along with protocol as a whole) acts as a strict standardizing mechanism that homogenizes these deviations under the umbrella of a unilateral standard.

Ironically, then, the Internet protocols which help engender a distributed system of organization are themselves underpinned by adistributed, bureaucratic institutions—be they entities like ICANN or technologies like DNS.

Thus it is an oversight for theorists like Lawrence Lessig, despite his strengths, to suggest that the origin of Internet communication was one of total freedom and lack of control.[41] Instead, it is clear to me that the exact opposite of freedom, that is control, has been the outcome of the last forty years of developments in networked communications. The founding principle of the net is control, not freedom. *Control has existed from the beginning.*

Perhaps it is a different type of control then we are used to seeing. It is a type of control based in openness, inclusion, universalism, and flexibility. It is control borne from high degrees of technical organization (protocol), not this or that limitation on individual freedom or decision making (fascism).

Thus it is with complete sincerity that Web inventor Tim Berners-Lee writes:

> I had (and still have) a dream that the web could be less of a television channel and more of an interactive sea of shared knowledge. I imagine it immersing us as a warm, friendly environment made of the things we and our friends have seen, heard, believe or have figured out.[42]

The irony is, of course, that in order to achieve this social utopia computer scientists like Berners-Lee had to develop the most highly controlled and extensive mass media yet known. Protocol gives us the ability to build a "warm, friendly" technological space. But it becomes warm and friendly through technical standardization, agreement, organized implementation, broad (sometimes universal) adoption, and directed participation.

Protocol is based on a *contradiction* between two opposing machines, one machine that radically distributes control into autonomous locales, and another that focuses control into rigidly defined hierarchies. This essay illustrates this reality in full detail. The generative contradiction that lies at the very heart of protocol is that *in order to be politically progressive, protocol must be partially reactionary*.

To put it another way, in order for protocol to enable radically distributed communications between autonomous entities, it must employ a strategy of universalization, and of homogeneity. It must be anti-diversity. It must promote standardization in order to enable openness. It must organize peer groups into bureaucracies like the IETF in order to create free technologies.

To be sure, the two partners in this delicate two-step often exist in separate arenas. As protocol pioneer Bob Braden puts it, "There are several vital kinds of heterogeneity."[43] That is to say, one sector can be standardized while another is heterogeneous. The core Internet protocols can be highly controlled while the actual administration of the net can be highly uncontrolled. Or, DNS can be arranged in a strict hierarchy while users's actual experience of the net can be highly distributed.

In short, control in distributed networks is not monolithic. It proceeds in multiple, parallel, contradictory and often unpredictable ways. It is a complex of interrelated currents and countercurrents.

Perhaps I can term the institutional frameworks mentioned here a type of *tactical standardization*, in which certain short term goals are necessary in order to realize one's longer term goals. Standardization is the politically reactionary tactic that enables radical openness. Or to give an example of this analogy in technical terms: the Domain Name System, with its hierarchical architecture and bureaucratic governance, is the politically reactionary tactic that enables the truly distributed and open architecture of the Internet Protocol. It is, as Barthes put it, our "Operation Margarine." And this is the generative contradiction that fuels the net.

Notes

1. Jake Feinler, "30 Years of RFCs," RFC 2555, April 7, 1999. The RFCs cited in this article may be found online at http://www.faqs.org or by using a search engine.
2. See Vint Cerf's memorial to Jon Postel's life and work in "I Remember IANA," RFC 2468, October 1988.
3. Katie Hafner and Matthew Lyon, *Where Wizards Stay up Late: The Origins of the Internet* (New York: Touchstone, 1996), p. 145. For biographies of two dozen protocol pioneers see Gary Malkin's "Who's Who in the Internet: Biographies of IAB, IESG and IRSG Members," RFC 1336, FYI 9, May 1992.
4. Vinton Cerf, personal correspondence, September 23, 2002.
5. Fred Baker, personal correspondence, December 12, 2002.
6. AT&T's Otis Wilson who is cited in Peter Salus, *A Quarter Century of Unix* (New York: Addison-Wesley, 1994), p. 59.
7. Salus, *A Quarter Century of Unix*, p. 2.
8. See Dennis Ritchie, "The Development of the C Programming Language" in Thomas Bergin and Richard Gibson, eds., *History of Programming Languages II* (New York: ACM, 1996), p. 681.
9. Bjarne Stroustrup, "Transcript of Presentation" in Bergin & Gibson, p. 761.
10. S. J. Liebowitz and Stephen E. Margolis, "Path Dependence, Lock-In and History," *Journal of Law, Economics and Organization*, April 1995.
11. If not VHS then the VCR in general was aided greatly by the porn industry. David Morton writes that "many industry analysts credited the sales of erotic video tapes as one of the chief factors in the VCR's early success. They took the place of adult movie theaters, but also could be purchased in areas where they were legal and viewed at home." See Morton's *A History of Electronic Entertainment since 1945*, http://www.ieee.org/organizations/history_center/research_guides/entertainment, p. 56.
12. Douglas Puffert, "Path Dependence in Economic Theory," http://www.vwl.uni-muenchen.de/ls_komlos/pathe.pdf, p. 5.
13. *IEEE 2000 Annual Report* (IEEE, 2000), p. 2.
14. IEEE prefers to avoid associating their standards with trademarked, commercial, or otherwise proprietary technologies. Hence the IEEE definition eschews the word "Ethernet" which is associated with Xerox PARC where it was named. The 1985 IEEE standard for Ethernet is instead titled "IEEE 802.3 Carrier Sense Multiple Access with Collision Detection (CSMA/CD) Access Method and Physical Layer Specifications."
15. Paul Baran, Electrical Engineer, an oral history conducted in 1999 by David Hochfelder, IEEE History Center, Rutgers University, New Brunswick, NJ, USA.
16. ANSI, "National Standards Strategy for the United States," http://www.ansi.org, emphasis in original.

17. The name ISO is in fact not an acronym, but derives from a Greek word for "equal." This way it avoids the problem of translating the organization's name into different languages, which would produce different acronyms. The name ISO, then, is a type of semantic standard in itself.
18. See http://www.iso.ch for more history of the ISO.
19. The IETF takes pride in having such an ethos. Jeanette Hofmann writes: "The IETF has traditionally understood itself as an elite in the technical development of communication networks. Gestures of superiority and a dim view of other standardisation committees are matched by unmistakable impatience with incompetence in their own ranks." See "Government Technologies and Techniques of Government: Politics on the Net," http://duplox.wz-berlin.de/final/jeanette.htm.
20. Another important organization to mention is the Internet Corporation for Assigned Names and Numbers (ICANN). ICANN is a nonprofit organization which has control over the Internet's domain name system. Its Board of Directors has included Vinton Cerf, co-inventor of the Internet Protocol and founder of the Internet Society, and author Esther Dyson. "It is ICANN's objective to operate as an open, transparent, and consensus-based body that is broadly representative of the diverse stakeholder communities of the global Internet" (see "ICANN Fact Sheet," http://www.icann.org). Despite this rosy mission statement, ICANN has been the target of intense criticism in recent years. It is for many the central lightning rod for problems around issues of Internet governance. A close look at ICANN is unfortunately outside the scope of this article, but for an excellent examination of the organization see Milton Mueller's *Ruling the Root* (Cambridge: MIT, 2002).
21. http://www.isoc.org.
22. For a detailed description of the IAB see Brian Carpenter, "Charter of the Internet Architecture Board (IAB)," RFC 2850, BCP 39, May 2000.
23. Gary Malkin, "The Tao of IETF: A Guide for New Attendees of the Internet Engineering Task Force," RFC 1718, FYI 17, October 1993.
24. Paul Hoffman and Scott Bradner, "Defining the IETF," RFC 3233, BCP 58, February 2002.
25. This RFC is an interesting one because of the social relations it endorses within the IETF. Liberal, democratic values are the norm. "Intimidation or ad hominem attack" is to be avoided in IETF debates. Instead IETFers are encouraged to "think globally" and treat their fellow colleagues "with respect as persons." Somewhat ironically this document also specifies that "English is the de facto language of the IETF." See Susan Harris, "IETF Guidelines for Conduct," RFC 3184, BCP 54, October 2001.
26. For more information on IETF Working Groups see Scott Bradner, "IETF Working Group Guidelines and Procedures," RFC 2418, BCP 25, September 1998.
27. That said, there are protocols that are given the status level of "required" for certain contexts. For example the Internet Protocol is a required protocol for anyone wishing to connect to the Internet. Other protocols may be give status levels of "recommended" or "elective" depending on how necessary they are for implementing a specific technology. The "required" status level should not be confused however with mandatory standards. These have legal implications and are enforced by regulatory agencies.
28. Scott Bradner, "The Internet Standards Process—Revision 3," RFC 2026, BCP 9, October 1996.
29. Most RFCs published on April 1st are suspect. Take for example RFC 1149, "A Standard for the Transmission of IP Datagrams on Avian Carriers" (David Waitzman, April 1990), which describes how to send IP datagrams via carrier pigeon, lauding their "intrinsic collision avoidance system." Thanks to Jonah Brucker-Cohen for first bringing this RFC to my attention. Brucker-Cohen himself has devised a new protocol called "H2O/IP" for the transmission of IP datagrams using modulated streams of water. Consider also "The Infinite Monkey Protocol Suite (IMPS)" described in RFC 2795 (SteQven [sic] Christey, April 2000) that describes "a protocol suite which supports an infinite number of monkeys that sit at an infinite number of typewriters in order to determine when they have either produced the entire works of William Shakespeare or a good television show." Shakespeare would probably appreciate "SONET to Sonnet Translation" (April 1994, RFC 1605) which uses fourteen line decasyllabic verse to optimize data transmission over Synchronous Optical Network (SONET). There is also the self-explanatory "Hyper Text Coffee Pot Control Protocol (HTCPCP/1.0)" (Larry Masinter, RFC 2324, April 1998), clearly required reading for any under-slept webmaster. Other examples of ridiculous technical standards include Eryk Salvaggio's "Slowest Modem" which uses the U.S. Postal Service to send data via diskette at a data transfer rate of only 0.002438095238095238095238 kb/s. He specifies that "[a]ll html links on the diskette must be set up as a href='mailing address' (where 'mailing address' is, in fact, a mailing address)" ("Free Art Games #5, 6 and 7," *Rhizome*, September 26, 2000), and Cory Arcangel's "Total Asshole" file compression system that, in fact, enlarges a file exponentially in size when it is compressed.
30. See Jon Postel and Joyce Reynolds, "Instructions to RFC Authors," RFC 2223, October 1997, and Gregor Scott, "Guide for Internet Standards Writers," RFC 2360, BCP 22, June 1998.
31. Robert Braden, "Requirements for Internet Hosts — Communication Layers," RFC 1122, STD 3, October 1989.
32. Milton Mueller, *Ruling the Root* (Cambridge: MIT, 2002), p. 76.
33. Tim Berners-Lee, *Weaving the Web* (New York: HarperCollins, 1999), p. 36.
34. Ibid., p. 71.
35. Ibid., pp. 92, 94.
36. Ibid., p. 18.
37. Tim Berners-Lee, "What the Semantic Web can represent," http://www.w3.org/DesignIssues/RDFnot.html.
38. One should not tie Crocker's memo to the beginning of protocol per se. That honor should probably go to Paul Baran's 1964 RAND publication "On Distributed Communications." In many ways it serves as the origin text for the RFCs that would follow. Although it came before the RFCs and was not connected to it in any way, Baran's memo essentially fulfilled the same function, that is, to outline for Baran's peers a broad technological standard for digital communication over networks.

Other RFC-like documents have also been important in the technical development of networking. The Internet Experiment Notes (IENs), published from 1977 to 1982 and edited by RFC editor Jon Postel, addressed issues connected to the then-fledgling Internet before merging with the RFC series. Vint Cerf also cites the ARPA Satellite System Notes and the PRNET Notes on packet radio (see RFC 2555). There exists also the MIL-STD series maintained by the Department of Defense. Some of the MIL-STDs overlap with Internet Standards covered in the RFC series.

39. Steve Crocker, "30 Years of RFCs," RFC 2555, April 7, 1999.
40. See Nelson Minar and Marc Hedlund, "A Network of Peers: Peer-to-Peer Models Through the History of the Internet," in Andy Oram, Ed., *Peer-to-Peer: Harnessing the Power of Disruptive Technologies* (Sebastopol, CA: O'Reilly, 2001), p. 10.
41. In his first book, *Code and other Laws of Cyberspace* (New York: Basic Books, 1999), Lessig sets up a before/after scenario for cyberspace. The "before" refers to what he calls the "promise of freedom" (6). The "after" is more ominous. Although as yet unfixed, this future is threatened by "an architecture that perfects control" (6). He continues this before/after narrative in *The Future of Ideas: The Fate of the Commons in a Connected World* (New York: Random House, 2001) where he assumes that the network, in its nascent form, was what he calls free, that is, characterized by "an inability to control" (147). Yet "[t]his architecture is now changing" (239), Lessig claims. We are about to "embrace an architecture of control" (268) put in place by new commercial and legal concerns.

Lessig's discourse is always about a process of becoming, not of always having been. It is certainly correct for him to note that new capitalistic and juridical mandates are sculpting network communications in ugly new ways. But what is lacking from Lessig's work, then, is the recognition that control is endemic to all distributed networks that are governed by protocol. Control was there from day one. It was not imported later by the corporations and courts. In fact distributed networks *must* establish a system of control, which I call protocol, in order to function properly. In this sense, computer networks are and always have been the exact opposite of Lessig's "inability to control."

While Lessig and I clearly come to very different conclusions, I attribute this largely to the fact that we have different objects of study. His are largely issues of governance and commerce while mine are technical and formal issues. My criticism of Lessig is less to deride his contribution, which is inspiring, than to point out our different approaches.

42. Cited in Jeremie Miller, "Jabber," in Oram, Ed., *Peer-to-Peer*, p. 81.
43. Bob Braden, personal correspondence, December 25, 2002.

13

Reload
Liveness, Mobility, and the Web

Tara McPherson

Part I: Convergence on the Digital Coast

During the heyday of the dot.com years, otherwise known as the late 1990s, I spent some time attending many of the "Digital Coast" events which the Los Angeles new media industry frenetically sponsored, events most often framed around a rhetoric of convergence that insisted on the inevitability of a collision between the internet and television, a vision of the future of the screen that in many ways wed the two media tightly together. While the earliest of these events framed television as the bad object to be overcome by all things web-like, a more symbiotic relationship between the two media quickly emerged.[1] Drawing on the tropes of television and the channel surfer, some "convergence" executives began promoting what they called a "lean back interactivity," which, in their words, provided "little snippets of interactivity to enhance the broadcast experience."[2] Further described as a "minimal interactivity," this mode was promoted as an "enhancement" to conventional broadcast which offered consumers a wider array of click-and-buy shopping. A "give the buyer what she wants" logic buttressed the move, as Wink Communications CTO and Chairman, Brian Dougherty, maintained that "if the interactivity is so complex... consumers aren't going to want it." Corporate CEOs proclaimed that "the really cool digital application turns out to be about TV," while the Pseudo Web site suggested that the Web just may end up "anointing talk shows as the killer app for next generation, two-way broadband networks."

Such talk framed the Web as a "better" version of TV, stressing particular aspects of the medium which illustrate its superiority to television while simultaneously linking the two media in a seemingly natural convergence. Here, the rhetoric revolved around notions of personalization and empowerment, focusing, in the words of Rob Tercek, the former VP of Digital Media at Columbia's Tristar Television Group, on the Web as "software that gets familiar with you." He also insisted that "controlling an audience" is an old idea more suited to broadcast than the niche markets of netcasting, which privilege "a consumer-centric point of view." Pseudo.com, a now-defunct New York interactive TV company that until recently produced over sixty Web-based television shows a week, promoted their programming as "the next logical step in the development of entertainment media," describing this "major deconstruction of television into niche programming" as opening up the possibility for "deeper, focused, interactive content tailored to individual interests, style and taste" (buzzwords courtesy of the old Pseudo Web site.) DEN, the Digital Entertainment Network, another crashed and burned internet TV venture that was LA's answer to Pseudo, included on its Web site a promotional video which presented DEN as a "media revolution" intent on providing "more interactive," "participatory entertainment." The clip went on to castigate television's essentially passive format while celebrating "the DEN" as providing "what you want to watch when you want to watch it. It's completely in your control." Chairman and CEO Jim Ritts championed both the customization the Web would allow as well as DEN's capacity to meld a "click and buy" element to more traditional modes of TV viewing.

Now, it's fairly easy to simply mock this rhetoric and certainly, after listening to DEN's president David Neuman talk about how "empowering" and interactive it would be to click and buy Jennifer Anniston's sweater while watching an episode of *Friends*, I couldn't help doing so. I've written elsewhere about the degree to which this industrial rhetoric of convergence can actually work as self-fulfilling prophecy, obscuring larger questions about whether or not the internet is really (or really should be) tied to corporate traditions of U.S. television while framing the internet as essentially a commercial medium, intent on servicing consumers rather than citizens. In his work on early radio, Tom Streeter reminds us that a similar logic of functionality and inevitability worked to close down alternative, grass-root forms of radio, bringing broadcast firmly under corporate control in less than thirty years and lessening its potential as a democratizing technology. With this history in mind, it is certainly important to question corporate rhetoric, querying the seemingly natural links being forged between television and the internet by companies ranging from the now-defunct DEN or Pseudo to the increasingly prevalent corporate mergers manifested in sites like MSNBC or CNN.[3]

Yet, as I surfed DEN or Pseudo and eavesdropped at Hollywood cocktail parties, I did notice a certain connect between the corporate rhetoric and my own experiences of the Web, suggesting that there's a level of accuracy within the corporate business plans, a glimmer of possibility and promise buried deep within their hype. "Choice," "presence," "movement," "possibility" are all terms which could describe the experiential modalities of Web surfing. In fact, as I'll argue, a phenomenology of the Web might focus on its capacity to structure three closely-related sensations, sensations I call *volitional mobility, the scan-and-search,* and *transformation*. It's crucial to think of these modes as both specific to the medium of the Web itself, as related to its materiality and, in some ways, independent from content, and also as ideologies packaged and promoted within certain Web sites, i.e., as corporate strategies of narrative and structural address. What a medium like the Web is or will be, in its very form, is not separate from the discourse which surrounds it and which structures particular conditions of possibility. Yet, if these discourses shape what the Web might become, they are also shaped by the medium and its particular material forms (even as it's sometimes difficult to think of the virtual realms of the digital as material).

For now, I want to turn away from considerations of corporate hype and rhetoric and instead look at the Web itself, trying to describe and understand the experiences it structures. In an article entitled "Print is Flat, Code is Deep," N. Katherine Hayles argues for the importance of media specific analyses, noting that "it is time to turn again to a careful consideration of what difference the medium makes."[4] Her concern is to investigate the insights a specific look at hypertext might reveal for literary theory, a field Hayles describes as "shot through with unrecognized assumptions specific to print."[5] I am interested in how a look at the specificity of the Web as a broader cultural form might illuminate certain aspects of both that medium and of television theory, perhaps suggesting the limits of these theories for analyses of new media while also limning their usefulness to an analysis of the Web.

This is not to imply, despite the conjectures of the new media executives at DEN and Pseudo to the contrary, that one medium is structurally and inherently superior to the other, that the Web is indeed "better." Rather, TV and the Web do reference each other, and, as Hayles maintains, "media-specific analysis attends both to the specificity of the form... and to citations and imitations of one medium in another. [These analyses are] attuned not so much to similarity and difference as to simulation and instantiation."[6] What follows is an investigation of the Web as an interface between users and digital data, the ones and zeros of the infosphere. In some regards, I take this notion of interface quite literally, and, thus, my methodology builds on Hayles in one key respect. Rather, than simply cataloging a typology of digital data focused primarily on its formal elements, I am also interested in exploring the specificity of *the experience of using the Web, of the Web as mediator between human and machine, of the Web as a technology of experience.* Put differently, I am interested in how the Web constitutes itself in the unfolding of experience. This necessarily entails an appreciation of the electronic form of the Web: after all, a Web browser is a interpreter of digital data, a translator of code, and this relation to digital data profoundly shapes how the user experiences the Web and what it promises. A media specific analysis can move beyond a certain formalism to explore what's before us in the moment that we are in. This exploration will finally return us to the realm of the corporate and the economic, for any understanding of the forms of and the experiences provided by the Web must necessarily account for the role new media technologies play in the changing economic landscape characteristic of neo-Fordism and transnationalism. The Web's ability to structure certain experiential modalities for the user also helps to situate that user within particular modes of subjectivity and within the networks of capital. While the political possibilities of these emergent modes of being cannot be specified in advance or in the abstract, their relation to corporate capital must be taken seriously.

Part II: Tara's Phenomenology of Web Surfing

When I explore the Web, I follow the cursor, a tangible sign of presence implying movement. This motion structures a sense of liveness, of immediacy, of the now. I open up my "personalized" site at MSNBC: via "instant" traffic maps (which, the copy tells me, "agree within a minute or two" to real time), synopses of "current" weather conditions, and individualized news bits, the Web site repeatedly foregrounds its currency, its timeliness, its relevance to me. A frequently changing tickertape scroll bar updates both headlines and stock quotes, and a flashing target floats on my desktop, signaling "breaking news" whenever my PC's on, whether or not a Web browser is open. The numerous polls or surveys that dot MSNBC's electronic landscape (they're called "live votes") promise that I can impact the news in an instant; I get the results right away, no need to wait for the 10 p.m. broadcast. Just click. Immediate gratification. Even the waiting of download time locks us in the present as a perpetually unfolding now.

This sense of being in the moment is further enhanced by the chat rooms included in many TV-centered Web sites, forums intended to fuse the sites more clearly to the television schedule, allowing the computer user to join the television audience by posing questions to talk show guests as live shows unfold on dual platforms. From E-Bay to E-Trade to ESPN, the Web references the unyielding speed of the present, linking presence and temporality in a frenetic, scrolling now. We hit refresh. We feel time move. We wait for downloads. We still feel time move, if barely. Processors hum, marking motion.

Of course, we know liveness from television studies. In prescient theoretical investigations of television in the 1980s, work intent on distinguishing television from film, Jane Feuer observed that "the differences between TV and ... cinema are too great not to see television as a qualitatively different medium, but granted this," she pursued what *was* specific to TV, both as a form and as an ideological and industrial practice.[7] Liveness (or, more crucially, its illusion) was her answer, and she skillfully illustrated how liveness was continually represented as a core ontological form of television when it might more accurately be seen as an ideology used in the promotion of television and its corporate manifestations. Liveness remains a key dimension of our experiences of the internet, a medium which also promotes itself as essentially up-to-the-minute (one need only hit "reload" or follow the scrolling updates), ideology once again masquerading as ontology.

Of course, as with television, this much touted liveness is actually the *illusion* of liveness: though the weather conditions may indeed be up to date, most of the "breaking news" I access via my personalized MSNBC front page is no more instant than the news I would watch at 6:00 p.m. on KTLA. Indeed, many Web sites display a marked inability to keep up with the present, recycling older stories in order to take advantage of the vast databases which underwrite the Web, old content repackaged as newness. But, as with television, what is crucial is not so much the *fact* of liveness as the *feel* of it. Many TV-centric Web sites capitalize on television's historic ties to liveness and thus present liveness as a given, as an essential element of the medium.

We might say, to paraphrase Feuer, that the Web "positions the [surfer] into its imaginary of presence and immediacy."[8] Yet, as I've noted in my earlier work on convergence, this is not just the same old liveness of television: this is liveness with a difference. This liveness foregrounds volition and mobility, creating a liveness on demand. Thus, unlike television which parades its presence before us, the Web structures a *sense of causality* in relation to liveness, a liveness which we navigate and move through, often structuring a feeling that our own desire drives the movement. The Web is about presence but an unstable presence: it's in process, in motion. Interestingly, as we imagine ourselves *navigating* sites, following the cursor, the Web *feels* more mobile than television, even though it relies more often on text and still images than on the moving video of TV. Furthermore, this is a sense of a connected presence in time. The Web's forms and metadiscourses thus generate a circuit of meaning not only from a sense of immediacy but through yoking this presentness to a feeling of choice, structuring a mobilized liveness which we come to feel we invoke and impact, in the instant, in the click, reload. I call this sensation *volitional mobility*.

If television, in the words of Bob Stam, obliges the telespectator "to follow a predetermined sequence" exhibiting "a certain syntagmatic orthodoxy,"[9] the Web appears to break down the preordained sequencing of TV, allowing the user to fashion her own syntagmas, moving from link to link with a certain illusion of volition. Our choices, perhaps our need to know, our epistemophilia, seem to move us through the space and time of the Web, and this volitional mobility implies our transformation, shimmering with the possibility of change, difference, the new and the now. From the dress-up mannequins of the Gap to the instant quizzes and horoscopes on sites like BabyCenter or Pseudo, the click propels us elsewhere and along. Volitional mobility is more about momentum than about the moment. The extensive database capacities of the internet structure the field upon which this sense of volition and movement unfolds, permitting the Web surfer to move back and forth through history and geography, allowing for the possibility (both real and imagined) of accident and juxtaposition to an even greater degree than television.

While this sense of volitional mobility seems to reside on the relatively analog surface of our monitor screens, a function of Web site design, the very form of digital data also helps underwrite this sensation. As Hayles notes, due to its very form, digital data is "intrinsically more involved in issues of mapping and navigation" than are most other media.[10] Web browsers translate code on the fly, structuring a kind of mobility which does indeed respond to the click. Computers are processors, in a sense, mobile machines. There's a fluidity to digital data: processing involves data in motion. These processes of navigation or motion relate to the depth of electronic forms. At a simple level, there's code "behind" a Web page, underwriting a kind of perceived depth between code and the programs visible upon our screens, coding underwriting image and movement. A relatively simple program may be hundreds of functions deep, yet the computer remembers and navigates these functions. As we roam the Web, the computer remembers where we've been, even if we don't.

At the level of the interface, this sense of movement through space is most obvious in the various Quicktime VR applications which dot our computer screens. A concrete example of the Web's capacity to structure a sense of volitional spatial and temporal mobility is found on MSNBC's "Kennedy Remembered" page, part of a multimedia repackaging of MSNBC's TV program, *Time and Again*. At this site, a real-time plug-in called SurroundVideo allows me to move around Dallas' grassy knoll in a 3-D representational space via a fairly seamless patching together of digital photographs, navigating actual footage of the area. Once the image loads (waiting is also one of the Web's temporalities), I am able to explore this Texan geographic terrain moving back and forth between the road, the book depository, the grassy arena. I am able to choose my own path with a click and drag of the mouse, zooming in and out for different perspectives and "edits." The sense of spatiality and mobility is fairly intense and certainly feels driven by my own desire. An even odder experience is created by clicking a button which maps black and white images of the 1963 assassination of the president *over* the color images, a slightly surreal collapse of space and time, still navigable. An archive of video and audio clips, various articles about JFK, transcripts of debates and speeches, and Web visitors' own stories structure a roam-able space of JFK, evoking mediated memories of Camelot and a poignant affect of national loss and nostalgia. I am able to be both here (in LA) and there (in Dallas), both then (1963) and now (2002), but I am always present, moving, live, in command. For those not moved by mobile history, other SurroundVideo sites at MSNBC allow users to surf the solar system and tour the Whitehouse, each positioning the national via specific moments of geography and movement. Other Web sites tackle less hallowed ground: DEN offered up a virtual frat house in Quicktime format, designed to accompany the live (or replayed) webcast of its episodic series focused on campus life. While a given episode played out in a small video window, the Web surfer could cruise the empty corridors of the frat house in a separate section of the browser, checking out sloppy rooms and communal showers, or post to an online chat which also shared the screen space. Here, the click-and-move mobility familiar from video games collided with the narrative world of television's teen dramas, all mapped out for maximum user navigability and choice. If early television promised to bring us the world, on the Web, our own volition in relation to this travel gets foregrounded. Microsoft asks, "Where do you want to go today?"

This sense of directed movement through space need not be so literal. The Web is also a fly-through infoscape, a navigable terrain of spatialized data. The windows, folders and bookmarks which populate our desktops create individualized architectures for the infospheres of the Web, building structures which allow us to inhabit realms of information, managing (or at least feeling as if we do) the vast database structures underwriting our Web browsers. Search engines move through realms of data, more or less responding to our command. Chat structures information as a collaborative performance. Programs like Flash allow our cursors to activate lively sequences of motion via a simple rollover, charting movement in a colorful, pixilated dance and visualizing our mobility before we even click. Again, the cursor seems to embody our trajectory, an expression of our movement and our will. We are increasingly aware of ourselves as databases, as part and parcel of the flow of information.

Pseudo and Den archived their episodic series, allowing a movement back through their "broadcast" histories. The Pseudo Web site insisted that "you can search for and play any episode you want, any time you want." This movement felt temporal, an aspect of the "on-demand" nature of the Web, as well as of its more material forms: its lack of fixity, its mutability, its variability. There's a sense of process to the Web that does not simply equate to liveness but also to promises of change. E*Trade, e-mail and eBay all manage time, producing and transforming temporalities; we feel connected to others within these temporal zones. There's a sense of presence with strangers. Community on the Web, via chat but also auctions, is as much about meeting times as meeting places, as the empty chat rooms of Pseudo's archived shows suggested. This temporality can be multidirectional and also simultaneous, both forward and backward at once, taking timeshifting to a different level. Recycling on demand. Michael Nash has said that "temporality connects our bodies to the computer" and joins us in digital space via the "dynamic of a connected presence."[11] We might see volitional mobility as the experience of choice (or its illusion) within the constraints of Web space and Web time.

This aspect of choice, of volition, is closely tied to what I categorize as a second modality of Web experience, the *scan-and-search*. Writing in the 1970s, British scholar Raymond Williams proposed the concept of "planned flow" as "the defining characteristic of broadcasting, simultaneously as a technology and as a cultural form."[12] Flow unites the disparate bits of information, advertising, and narrative comprising an evening's television into a seamless whole, establishing a planned sequence which is more important than the individual segments which might seem to categorize TV programming. Thus, as viewers, we are as likely to say that we watched television as to say what we watched, indicating the power of televisual flow. As a conceptual framework, flow has been amply explored and debated within television theory, with Jane Feuer arguing that television might more accurately be seen as a dialectic of segment and flow[13] and with John Caldwell similarly challenging the notion that TV watching might be characterized by flow's boundlessness.[14]

While Web surfing might seem to operate in a manner similar to flow, bringing the vast array of data that categorizes the Web into an experiential sequence, segmentation on the Web—what we might more accurately call "chunking"—is not identical to the segmentation of television. The Web's chunking is spatial as much as temporal; our experience of moving through these chunks may seem akin to our experience of TV's flow, but this is also a boundlessness we feel we help create or impact. It structures a different economy of attention than that underwritten by flow. We move from the glance-or-gaze that theorists have named as our primary engagements with television (or film) toward the scan-and-search.[15] The scan-and-search is about a fear of missing the next experience or the next piece of data. Whereas this fear of missing something in the realm of television may cause the user to stay tuned to one channel, not to miss a narrative turn, this fear of missing in the Web propels us elsewhere, on to the next chunk, less bound to linear time and contiguous space, into the archive and into what feels like navigable space that responds to our desire. We create architectures as we move through the Web via bookmarks and location bars, structuring unique paths through databases and archives. This is not just channel-surfing: it feels like we're wedding space and time, linking research and entertainment into similar patterns of mobility. The scan-and-search feels more active than the glance-or-gaze.

The Web, less strictly a time-based and time-moving medium than broadcast, combines sequencing with discrete bits more robustly than TV, encouraging the scan and the search as modes of engagement, structuring a spatialized and mobile subjectivity which feels less orchestrated than the subject hailed by televisual flow (a subject moving forward in time—or back with the VCR—but less likely to feel a movement elsewhere, spatially, a kind of sideways or lateral movement). With the Web, we feel we create the sequences rather than being programmed into them. Feuer sees television's use of flow as imposing unity over fragmentation (including spatial fragmentation), but the Web is less invested in such fictions of unity. While the Web certainly is about structuring movement, particularly in sites like MSNBC or DEN, with careful attention paid to information architectures which strive to orchestrate the visitor's path through a site with precision, it does so while also structuring a heightened sense of choice and mobility through navigable spaces. The solicitation of our interaction overcomes a sense of disparate, chunked information, creating a feeling of mobility across data. DEN's site included the tagline "All Available On Demand," and its crowded browser windows demanded a different kind of attention than that of the glance while rarely sustaining a fixed gaze. We move through such sites searching and scanning, looking for the next thing.

The Web's activation of our desire for what's next hints at a third modality of Web experience, the promise of *transformation*. Janet Murray notes that transformation is a "characteristic pleasure of digital environments."[16] She goes on to frame this particular pleasure via its relation to narrative structures (and narrative structures of a very particular kind), but we might instead think of transformation as endemic to the Web in a broad sense, motivating an extensive variety of narrative and non-narrative forms. Of course, popular culture has long traded on the lure of transformation, from the glimmer of hope embodied in each sexy tube of MAC lipstick to the mighty morphing power rangers to the promise of the makeover in *Glamour*, *Oprah*, or *This Old House*. But computer culture introduces a new level of personalization and sense of choice in relation to transformation in forms as diverse as architectural cd-roms, the ill-fated Microsoft BOB, endless pink Barbie products, and the flash-enabled dress-up spaces of e-commerce. Even my MSNBC homepage or My Yahoo turn on transformation, as the faceless dataspaces of the Web are made-over via my demands. Personalization holds out the tantalizing lure of transformation, remaking information into a better reflection of the self.

From the VR frat house of DEN to the countless "live" chats which populate the internet, the movement of the Web harbors hopes of transformation. Regardless of content, there's a haptic potential to these spaces, both the literal 3-D spaces of Quicktime VR and the seemingly flat spaces of chat, of scrolling text. When one enters the space of chat, the dialogue that unfolds can equal a loss of self, structuring a transformative space. To borrow a phrase from Amelie Hastie (writing about doll houses), these environments are consumed by both the mind and the eye, an imagined space of possibility and change.[17]

Again, this sensation is tied to the actual form of digital data, to programmability, to the fragmentation and recombination which Hayles notes as intrinsic to the medium.[18] Digital code is malleable and subject to manipulation, at levels both accessible and inaccessible to the average user. In a language as simple as HTML (which my UNIX-coding husband refuses to even call a programming language), changing the descriptor "FFFFFF" to "FF6600" on a lengthy block of code seemingly transforms the page from a predictable white to the bold orange so hot circa 2000. Likewise, a lone missing comma can override thousands of lines of code, producing only error messages and frustration. As Web browsers render pages on the fly, transformation's literalized; code is broken down and reassembled; new forms appear possible; recombination rules.

But, before I get too carried away in the heady realms of possibility, it is well to sound a cautionary note. Both Marsha Kinder and Susan Willis have alerted us to the often illusory status of promises of transformation. As Willis notes in relation to transforming toys, there is always the risk "that everything transforms but nothing changes." She describes toys that "weld transformation to consumption" and ascribes the fascination with transforming toys to a "utopian yearning for change which the toys themselves then manage and control."[19] Thus, while the Web may indeed foster the related sensations of volitional mobility, scan-and-search, and transformation, our understanding of these modalities needs another working through in order to discern how they underwrite particular spatialities and temporalities, enabling specific selves and particular publics.

Part III: On Sensation and the Corporation in the Age of Neo-Fordism

While volitional mobility, the scan-and-search, and transformation are at least partially structured by the very forms of digital data, our experience of these modes is also shaped by the more analog representations on our screens. For example, the MSNBC Web site is highly controlled, severely curtailing the user's movement in subtle yet limiting ways, yet the promise and feeling of choice, movement, and liveness powerfully overdetermine its spaces. MSNBC.com self-consciously constructs itself as a projected fulfillment of what seems missing in the status quo (both on TV and in real life), becoming a solution to the oft-voiced dilemma of having 100 channels and still nothing to watch. We could say it promises everything and changes nothing.

The illusion of a mobilized liveness in a Web site like MSNBC actually masks the degree to which the site already stages a linear, largely uni-directional model of the internet, a model predicated on television's broadcast modes of information delivery and encouraged in Web design manuals which illustrate modes of information architecture orchestrated to move a user through a site in very predetermined fashion. Many entertainment executives have taken to pitching a model of Internet access based on TV's network structure. This model would limit internet access to three or four providers who would function much like TV networks, offering their own programming, directing users to approved parts of the Web, and limiting the capacity to post a home page or Web program to specialized producers. While this may sound far-fetched, small steps in this direction are already underway. For instance, if you want to cruise around the grassy knoll in MSNBC's recreated Dallas, you had better be using a Microsoft Internet Explorer browser. Netscape can't take you there.

The interfaces deployed by MSNBC (and most other commercial Web sites) suggest a sense of liveness and movement even while the very programming which underwrites them works to guide and impede the user's trajectory. The increasing popularity of "portal" sites leads to a Web architecture which works to constrict the surfer's movement, effectively detouring users along particular paths or containing them within particular sites. For instance, both MSNBC and AOL work as portal sites which make it hard to leave their confines, functioning as the kind of locked-in channel television executives have long dreamed about. The increasing vertical monopolies characterizing the mediascape as well as the death of smaller (if well funded) players like DEN and Pseudo take the meaning of convergence to new level, naturalizing the relationships between TV and the Web. Rather than simply accepting the link between these two media, theorists need to investigate the ideological implications of actual interfaces and other programming choices; we need to foreground the political effects of burying the author function within the code. The standardization of temporality and style via channels, regular programming, and published schedules are a central part of the history of television and radio's commercialization. Television's much-heralded "flow" worked to move viewers through segments of televisual time, orchestrating viewership, and Web programming could allow for an even more carefully orchestrated movement, all dressed up in a feeling of choice.

Another example of the illusory nature of the Web's modalities could be drawn from search engines, powerful programs which promote the illusion that one is actively surfing the Web. Of course, when you use a search engine, you're not really moving through the Web but through fairly limited databases. You might say that these databases structure volitional mobility to mask their own algorithmic structure, giving users the sense of control and movement through cyberspace when really you don't even touch the Web when you initiate a search. Rather, you remain within a contained database, usually cataloguing less than thirty to forty percent of the Web as a whole, processes which increasingly privilege commercial sites, enacting a very particular politics of information and design.[20] All of which introduces questions of representation, underscoring that the analog representations on our screen are powerfully connected to life offscreen: certain constructions of space enable certain spheres of Domination. Digital metaphors and representations are powerful processors.

In corporate structure, technological form and modes of experience, the Web and TV increasingly interact in mutually supportive modes reinforcing what Margaret Morse has called the institutions of mobile privatization.[21] If, as she maintains, freeways, malls, and TVs exist in a "kind of sociocultural distribution and feedback system,"[22] the Web operates within this circuit of exchange, albeit with slightly adjusted modalities. Choice, personalization, and transformation are heightened as experiential lures, accelerated by feelings of mobility and searching, engaging the user's desire along different registers which nonetheless still underwrite neo-Fordist feedback loops. Eric Alliez and Michel Feher characterize the neo-Fordist economy as a shift away from the massive scale of factory production in the Fordist era toward a regime marked by a more supple capitalism. There is a move toward flexible specialization, niche marketing, service industries, and an increasing valorization of information, which is now awarded a status "identical to the one assigned labor by classic capitalism: both a source of value and a form of merchandise."[23] The separation of the spaces and times of production from those of reproduction (or leisure) which was central to an earlier mode of capitalism is replaced by a new spatio-temporal configuration in which the differences between work and leisure blur. This leads to a "vast network for the productive circulation of information," structuring people and machines as interchangeable, equivalent "relays in the capitalist social machine." Rather than being *subjected* to capital, the worker is now *incorporated* into capital, made to feel responsible for the corporation's success.

While Alliez and Feher first described this mode in 1987, locating its emergence in the late 1960s, their description of neo-Fordism brilliantly predicts the logic of the dot.com era. The fanatic and frenetic work habits of the denizens of Silicon Valley and the Digital Coast modeled the incorporation of the worker within capital, while the proliferation of networked existence via the internet, pagers, and cellular phones helped fuel the dissolution between the spatio-temporal borders of work and leisure. In the new networked economy, "regular" readers help drive the databases of Amazon.com by freely posting their book or movie reviews and avid video game players help fuel corporate capital by posting homegrown game add-ons to corporate sites without compensation, succinctly illustrating their incorporation into capital and its flows. Likewise, we might see our Web-enhanced experiences of volitional mobility, scan-and-search, and transformation as training us for a new neo-Fordist existence. Old (narrative) strategies of identification and point of view give way to information management and spatial navigation, underwriting the blur (or convergence) between research and entertainment that so characterizes much of life under the conditions of virtuality.

Thus, it's important to recognize that these emergent modes of experience are neither innocent nor neutral, simple expressions of the material forms of the digital. They model particular modes of subjectivity which can work all too neatly in the service of the shifting patterns of global capital. Yet, even if the mobility offered by a search engine or a corporate Web site is both technically limited and central to our incorporation into capital, this does not mean that search engines (or MSNBC for that matter) aren't experienced by their users as offering up choice and possibility; rather, it highlights the degree to which these experiences are doubly constructed, an element of both the very forms of the digital data and the ideology of mobility and change created by the sites themselves.

In conclusion, we might ask why, in a culture increasingly subject to simulation, volition (or its illusion) emerges as such a powerful modality of experience, such a visceral desire. If Walter Benjamin reminds us that early film served to drill the viewer in the modes of perception structured by the mechanical era, how do Web spaces function as instructions for our bodily adaptation to virtuality? Mark Hansen has characterized the two main forms of experiential alienation of the digital age as the "ubiquitous encounter with estranged, rootless images . . . and the loss of agency ensuing from the increasing distribution of perceptual and cognitive tasks into systems centrally involving non-human components."[24] In the face of these forces, then, the volitional mobility, the scan-and-search, and the transformation promised by the Web might offer a glimmer of hope, a hope not entirely foreclosed by corporate rhetoric, the will of interactive companies like DEN and Pseudo, and the hegemony of Microsoft. While the "click and buy" logic of DEN certainly overrides the ontology of volitional mobility with an illusory ideology of volition, that these modalities are also part of the *forms* of the Web suggests a redemptive possibility, if only in the ways they activate our very desire for movement and change, a desire that might be mobilized elsewhere.

Notes

1. I trace the emergence and stakes of this rhetoric of convergence in my "TV Predicts Its Future: On Convergence and Cybertelevision" in *Virtual Publics: Policy and Community in an Electronic Age*, Ed. Beth Kolko (New York: Columbia UP, 2003) My theorization about the web's modalities took initial shape in that essay, and "Reload" draws from and expands this earlier work.
2. Charles Pillar. "Improved Technology May Finally Make Interactive TV a Go" *Los Angeles Times* (16 November 1998): C1.
3. During the dot-com boom, Pseudo was hailed by many as an excellent example of the Web's capacity to leverage and improve upon conventional television. While the site, begun in 1994, did offer some innovative and interesting programming in a format which in actuality drew as much from radio as from television, it still was a victim of the new

technology slowdown and of uncontrolled spending, having burned through in excess of thirty-two million in funding without becoming profitable (Jason Blair. "Remains of Pseudo.com Bought for Fraction of What It Spent." *New York Times* (January 25, 2001). The site was officially shut down in September 2000, a closure described by the *New York Times* as "one of the most visible casualties of the dot-com sector's downturn" (Peter Meyers. "In a Crisis, a Wireless Patch," *New York Times* (October 4, 2001).

The company's assets were purchased in January 2001 by INTV (huttp://www.intv.tv), another "interactive television" company which maintains a circumscribed site at http://www.pseudo.com. This site promised an imminent (and unrealized) re-launch of Pseudo throughout Fall, 2001. DEN, an acronym for the "Digital Entertainment Network," launched to incredible hype in Los Angeles and more clearly drew from television's series format. At its height, it had over 300 employees as well as investments by Microsoft, Dell Computer, NBC and other large technology and entertainment companies. After speeding through more than sixty million in capital in about two years, DEN closed shop in June, 2000 amid heavily-reported scandals, both financial and sexual. [Menn, Joseph. "DEN Investigated on Alleged Fraud in Liquidation" *Los Angeles Times* (August 2, 2000).] Streeter, Thomas. *Selling the Air: A Critique of the Policy Of Commercial Broadcasting in the United States* (Chicago: University of Chicago Press, 1996): 49-64).

4. N. Katherine Hayles. "Print is Flat, Code is Deep." Paper presented at the Interactive Frictions Conference, USC. Los Angeles, June 4, 1999: 2. (Forthcoming in *Interactive Frictions*, Eds. Marsha Kinder and Tara McPherson (Berkeley CA: University of California Press, 2006).
5. Hayles, "Print is Flat," 1.
6. Hayles, "Print is Flat," 1.
7. Feuer, Jane. "The Concept of Live Television: Ontology as Ideology." In *Regarding Television: Critical Approaches, An Anthology*, edited by E. Ann Kaplan (Los Angeles: AFI, 1983): 12.
8. Feuer, "Liveness," 14.
9. Stam, Robert. "Television News and Its Spectator." In *Regarding Television: Critical Approaches, An Anthology*, edited by E. Ann Kaplan (Los Angeles: AFI, 1983): 32.
10. Hayles, "Print is Flat," 15.
11. Michael Nash. Remarks at the Interactive Frictions Conference, USC. Los Angeles, June 6, 1999.
12. Raymond Williams. *Television: Technology and Cultural Form* New York: Schocken, 1975): 86.
13. Feuer, "Liveness," 15.
14. John T Caldwell. *Televisuality: Style, Crisis and Authority in American Television*. (New Brunswick: Rutgers UP, 1995): 163.
15. In *Visible Fictions*, John Ellis argued that television viewing was more accurately characterized by the glance than by the sustained gaze film theorists had posited as key to cinema's power and allure. This reading was at least partially based on television's domestic setting, a context likely to encourage distracted viewing. As John Caldwell notes, "glance theory" now functions as a truism of television theory, although Caldwell argues that newer modes of televisuality do demand a more concentrated gaze and a less distracted viewer. I'd suggest that TV might be seen as playing to both the glance and the gaze, depending on particular modes of programming, and that neither of these modes adequately explains how we engage with our web browsers (John Ellis. *Visible Fictions: Cinema, Television, Video.* London: Routledge and Kegan Paul, 1982).
16. Murray, Janet. *Hamlet on the Holodeck: The Future of Narrative in Cyberspace* (New York: The Free Press, 1997): 154.
17. Hastie, Amelie. "History in Miniature: Colleen Moore's Dollhouse and Historical Recollection." Paper presented at the 1999 Society for Cinema Studies Conference, Palm Beach, Florida, April 15–18, 1999.
18. Hayles, "Print is Flat," 9.
19. Cited in Marsha Kinder. *Playing with power in movies, television, and video games : from Muppet Babies to Teenage Mutant Ninja Turtles* (Berkeley: University of California Press, 1991): 136.
20. The particularity of this politics of information and design is further driven home when one realizes that over fifty percent of the registered domain names for the Internet are in the United States and that eleven percent are in California. English is by far the predominant language on the Web.
21. Margaret Morse. *Virtualities: Television, Media Art, and Cyberculture* Bloomington: Indiana University Press, 1998): 118.
22. Morse, *Virtualities*, 119.
23. Eric Alliez and Michel Feher. "The Luster of Capital," trans. Alyson Waters, *Zone 1 and 2* (1987): 316.
24. Mark Hansen. "Performing Affect: Interactive Video Art and the Cybernetic Body." Paper presented at the Interactive Frictions Conference, USC. Los Angeles, June 4, 1999: 9. (Forthcoming in the *Interactive Frictions* anthology.)

14

Generation Flash

Lev Manovich

This essay, which comprises a number of self-contained segments, looks at the phenomenon of Flash graphics on the Web that attracted a lot of creative energy in the last few years. More than just a result of a particular software/hardware situation (low bandwidth leading to the use of vector graphics), Flash aesthetics exemplifies the cultural sensibility of a new generation.[1] This generation does not care if their work is called art or design. This generation is no longer interested in "media critique," which preoccupied media artists of the last two decades; instead it is engaged in software critique. This generation writes its own software code to create its own cultural systems, instead of using samples of commercial media.[2] The result is the new modernism of data visualizations, vector nets, pixel-thin grids and arrows: Bauhaus design in the service of information design. Instead of the Baroque assault of commercial media, the Flash generation serves us the modernist aesthetics and rationality of software. Information design is used as a tool to make sense of reality while programming becomes a tool of empowerment.[3]

Turntable and Flash Remixing

[Turntable is a Web-based software that allows the user to mix in real-time up to 6 different Flash animations, in addition manipulating color palette, size of individual animations and other parameters. For www.whitneybiennial.com, the participating artists were asked to submit short Flash animations that were exhibited on the site both separately and as part of Turntable remixes. Some remixes consisted of animations from the same artist while others used animations by different artists.]
URL: http://www.whitneybiennial.com

It has become a cliché to announce that "we live in remix culture." Yes, we do. But is it possible to go beyond this simple statement of fact? For instance, can we distinguish between different kinds of remix aesthetics? What is the relationship between remixes made with electronic and computer tools and earlier forms such as collage and montage? What are the similarities and differences between audio remixes and visual remixes?

Think loop. The basic building block of an electronic sound track, the loop also conquered a surprisingly strong position in contemporary visual culture. Left to their own devices, Flash animations, QuickTime movies, and the characters in computer games loop endlessly—until the human user intervenes by clicking. As I have shown elsewhere, all nineteenth century pre-cinematic visual devices also relied on loops. Throughout the nineteenth century, these loops kept getting longer and longer—eventually turning into a feature narrative... Today, we witness the opposite movement—artists sample short segments of feature films or TV shows, arrange them as loops, and exhibit these loops as "video installations." The loop thus becomes the new default method to "critique" media culture, replacing the still photograph of 1980s post-modern critique. At the same time, it also replaces the still photograph as the new index of the real: since everybody knows that still photography can be digitally manipulated, a short moving sequence arranged in a loop becomes a better way to represent reality—for the time being.

Think Internet. What was referred to in post-modern times as quoting, appropriation, and pastiche no longer needs any special name. Now this is simply the basic logic of cultural production: download images, code, shapes, scripts, etc., modify them, and then paste the new works online—send them into circulation. (Note: with the Internet, the always-existing loop of cultural production runs much faster: a new trend or style may spread overnight like a plague.) When I ask my students to create their own images by making photographs or by shooting video, they have a revelation: images do not have to come from Internet! Shall I also reveal to them that images do not have to come from a technological device that records reality—that instead they can be drawn or painted?

Think image. Compare it to sound. It seems possible to layer many many many sounds and tracks together while maintaining legibility. The results just keep getting more complex, more interesting. Vision seems to work differently. Of course commercial images we see everyday on TV and in cinema are often made from layers as well, sometimes as many as thousands—but these layers work together to create a single illusionistic (or super-illusionistic) space. In other words, they are not heard as separate sounds. When we start mixing arbitrary images together, we quickly destroy any meaning. (If you need proof, just go and play with the classic The Digital Landfill.[4]) How many separate image tracks can be mixed together before the composite becomes nothing but noise? Six seems to be a good number—which is exactly the number of image tracks one can load onto Turntable.

Think sample versus the whole work. If we are indeed living in a remix culture does it still make sense to create whole works—if these works will be taken apart and turned into samples by others anyway? Indeed, why painstakingly adjust separate tracks of a Director movie or After Effects composition to get it just right if the "public" will "open source" them into individual tracks for their own use using some free software? Of course, the answer is yes: we still need art. We still want to say something about the world and our lives in it; we still need our own "mirror standing in the middle of a dirt road," as Stendhal called art in the nineteenth century. Yet we also need to accept that for others our work will be just a set of samples, or maybe just one sample. Turntable is the visual software that makes this new aesthetic condition painfully obvious. It invites us to play with the dialectic of the sample and the composite, of our own works and the works of others. Welcome to visual remixing Flash style.

Think Turntable.

Art, Media Art, and Software Art

Recently "software art" has emerged as the new dynamic area of new media arts. Flash's ActionScript, Director's Lingo, Perl, MAX, JavaScript, Java, C++, and other programming and scripting languages are the medium of choice of a steadily increasing number of young artists. Thematically, software art often deals with data visualization; other areas of creative activity include the tools for online collaborative performance/composition (Keystroke), DJ/VJ software, and alternatives to/critiques of commercial software (Auto-illustrator), especially the browsers (early classics like Netomat, Web Stalker, and many others since then). Often, artists create not singular works but software environments open for others to use (such as Alex Galloway's Carnivore.) Stylistically, many works implicitly reference visual modernism (John Simon seems to be the only one so far to weave modernist references in his works explicitly).

Suddenly, programming is cool. Suddenly, the techniques and imagery that for two decades were associated with SIGGRAPH geekness and were considered in bad taste—visual output of mathematical functions, particle systems, RGB color palette—are welcomed on the plasma screens of the gallery walls. It is no longer *October* and *Wallpaper* but Flash and Director manuals that are the required read for any serious young artist.

Of course, since the early days of 1960s, computer artists have always written their own software. In fact, until the mid-1980s, writing one's own software or at least using special very high-end programming languages designed by others (such as Zgrass) was the only way to do computer art.[5] So what is new about the recent phenomenon of software art? Is it necessary?

Let's distinguish between three figures: an artist; a media artist; and a software artist.

A romantic/modernist artist (the nineteenth century and the first half of the twentieth century) is a genius who creates from scratch, imposing the phantoms of his imagination on the world.

Next, we have the new figure of a media artist (the 1960s–the 1980s), which corresponds to the period of post-modernism. Of course modernist artists also used media recording technologies such as photography and film but they treated these technologies similar to other artistic tools: as means to create an original and subjective view of the world. In contrast, post-modern media artists accept the impossibility of an original, unmediated vision of reality; their subject matter is not reality itself, but representation of reality by media, and the world of media itself. Therefore these media artists not only use media technologies as tools, but they also use the content of commercial media. A typical strategy of a media artist is to re-photograph a newspaper photograph, or to re-edit a segment of TV show, or to isolate a scene from a Hollywood film/TV show and turn it into a loop (from Nam June Paik and Dara Birnbaum to Douglas Gordon, Paul Pffefer, Jennifer and Kevin McCoy). Of course, a media artist does not have to use commercial media technologies (photography, film, video, new media)—s/he can also use other media, from oil paint to printing to sculpture.

The media artist is a parasite who lives at the expense of the commercial media—the result of collective craftsmanship of highly skilled people. In addition, an artist who samples from/subverts/pokes at commercial media can ultimately never compete with it. Instead of a feature film, we get a single scene; instead of a complex computer game with playability, narrative, AI, etc., we get just a critique of its iconography.

Thirty years of media art and post-modernism have inevitably led to a reaction. We are tired of always taking existing media as a starting point. We are tired of being always secondary, always reacting to what already exists.

Enter a software artist—the new romantic. Instead of working exclusively with commercial media—and instead of using commercial software—the software artist makes his/her mark on the world by writing the original code. This act of code writing itself is very important, regardless of what this code actually does at the end.

A software artist re-uses the language of modernist abstraction and design—lines and geometric shapes, mathematically generated curves and outlined color fields—to get away from figuration in general, and the cinematographic language of commercial media in particular. Instead of photographs and clips of films and TV, we get lines and abstract compositions. In short, instead of QuickTime, we use Flash. Instead of the computer as a media machine—a vision being heavily promoted by the computer industry (and most clearly articulated by Apple which promotes a MAC as a "digital hub" for other media recording/playing devices), we go back to computer as a programming machine.

Programming liberates art from being secondary to commercial media. A similar reason may be behind the recent popularity of "sound art." While commercial media now use every possible visual style, commercial sound environments still have not appropriated all sound space. While rock and roll, hip-hop, and techno have already become standard elevator music (at least in more hip elevators such as the Hudson Hotel in NYC), it seems that the rhythm-less regions of sound space are still untouched—at least for now.

UTOPIA in Shockwave

[UTOPIA is a Shockwave project by Futurefarmers for Tirana Biennale 01 Internet section.]
[Futurefarmers: Amy Franceschini and Sascha Merg]
URL: http://nutrishnia.org/level/

UTOPIA is playful and deceitful—because it pretends to be more innocent, more simple, and more light than it actually is. At first glance it can be taken for something made for children—or for adults whose references are not Karl Marx, Sigmund Freud, Rem Koolhaas, and Philip Stark, but text messaging, gnutella, retro Atari graphics, and nettime. This is the new generation that emerged in the 1990s. In contrast to visual and media artists of the 1960s–1980s, whose main target was media—ads, cinema, television—this new generation does not waste its energy on media critique. Instead of bashing commercial media environment, it creates its own: Web sites, mixes, software tools, furniture, cloves, digital video, Flash/Shockwave animations and interactives.

The new sensibility, which Utopia exemplifies so well, is soft, elegant, restrained, and smart. This is the new software intelligentsia. Look at the thin low-contrast lines of UTOPIA, praystation.com, and so many Flash projects included in Tirana Biennale 01. If the images of the previous generations of media artists, from Nam June Paik to Barbara Kruger, were screaming, trying to compete with the intensity of the commercial media, the work of the new data artists such as Franceschini/Merg whisper in our ears. In contrast to media's arrogance, they offer us intelligence. In contrast to the media stream of endless repeated icons and sound bytes, they offer us small and economical systems: stylized nature, ecology, or the game/music generator/Lego-like parade in UTOPIA.

Futurefarmers are among the few Flash/Shockwave masters who use their skills for a social, rather than simply a formal, end. Their project theyrule.net is a great example of how smart programming and smart graphics can be used politically. Instead of presenting a packaged political message, it gives us the data and the tools to analyze it. It knows that we are intelligent enough to draw the right conclusion. This is the new rhetoric of interactivity: we get convinced not by listening/watching a prepared message but by actively working with the data: reorganizing it, uncovering the connections, becoming aware of correlations.

UTOPIA does not have explicitly political content; instead it presents its message through a visual allegory. Like *SimCity* and similar sims, the program presents us with a whole miniature world, which runs according to its own system of rules. (The animation in UTOPIA is result of code execution—nothing is hand animated.) The cosmogony of this world reflects our new understanding of our own planet—post-Cold War, Internet, ecology, Gaia, and globalization. Notice the thin barely visible lines that connect the actors and the blocks. (This is the same device used in theyrule.net.) In the universe of UTOPIA, everything is interconnected, and each action of an individual actor affects the system as a whole. Intellectually, we know that this is how our Earth functions ecologically and economically—but UTOPIA represents this on a scale we can grasp perceptually.

The lines also serve another purpose. Despite CNN, Greenpeace, the glass roof of Berlin's Reichstag and other institutions and devices that work to make the functioning of modern societies transparent to their citizens, most of it is not visible. This is not only because we don't know the motives behind this or that Government policy or because advertising and PR constantly work to make things appear differently from what they really are—society's functioning is not visible in a literal sense. For instance, we don't know where the cells are which make our cell phones work; we don't know the layout of private financial networks that circle the Earth; we don't know what companies are located in a building we pass everyday on the way to work; and so on. But in UTOPIA, we do know—because the links are made visible. UTOPIA is Utopia because it is a society in which cause and effect connection are rendered visible and comprehensible. The program re-writes Marxism as vector graphics; it substitutes the figure of "connections" for the old figure of "unveiling."

UTOPIA is serious business behind its playful façade—but it is not all business. Drawing on our current fascination with computer games and interactive image-sound software, UTOPIA is a visual and intellectual delight; UTOPIA draws on the current fascination with computer games and interactive image-sound software. It is Tetris that meets Marx that meets data mining that meets the dance club floor. It is a game for the new generation that knows that the world is a network, that the media is not worth taking very seriously, and that programming can be used as a political tool.

The Unbearable Lightness of Flash

[Tirana Biennale 01 Internet section was organized by Miltos Manetas/Electronic Orphanage. The exhibition consisted of a few dozen projects by Web designers and artists, many of whom work in Flash or Shockwave. Manetas commissioned me, Peter Lunenfeld, and Norman Klein to write the analysis of the show. This text is my contribution; many ideas in it developed out of the conversations the three of us had about the works in the show. The names in brackets below refer to the artists in the show; go to the show site to see their projects.]
URL: http://www.electronicorphanage.com/biennale

Biology

Flash artists are big on biological references. Abstract plants, minimalist creatures, or simply clouds of pixels dance in patterns which to a human eye signal "life"' (Geoff Stearns: deconcept.com, Vitaly Leokumovich: unclickable.com, Danny Hobart: dannyhobart.com; uncontrol.com). Often we see self-regenerating systems. But this is not life as it naturally developed on Earth; rather, it looks like something we are likely to witness in some biotech laboratory where biology is put in the service of industrial production. We see hyper-accelerated regeneration and evolution. We see complex systems emerging before our eyes: millions of years of evolution are compressed into a few seconds.

There is another feature that distinguishes life à la Flash from real life: the non-existence of death. Biological organisms and systems are born, they develop, and eventually they die. In short, they have a teleology. But in Flash projects, life works differently: since these projects are loops, there is no death. Life just keeps running forever—more precisely, for as long as your computer maintains a Net connection.

Amplification: Flash aesthetics and Computer Games

Abstract ecosystems in Flash projects have another characteristic that makes playing so pleasurable (Joel Fox). They brilliantly use the power of the computer to amplify the user's actions. This power puts a computer in line with other magical devices; not accidentally, the most obvious place to see it is in games, although it is also at work in all of our interactions with a computer. For instance, when you tell Mario to step to the left by moving a joystick, it initiates a small delightful narrative: Mario comes across a hill; he starts climbing the hill; the hill turns out to be too steep; Mario slides back onto the ground; Mario gets up, all shaking. None of these actions required anything from us; all we had to do is just to move the joystick once. The computer program amplifies our single action, expanding it into a narrative sequence.

Historically, computer games were always a step ahead of the general human computer interface. In the 1960s and 1970s users communicated with a computer using non-graphical interfaces: entering the program onto a stack of punch cards, typing on a command line, and so on. In contrast, since their beginnings in the late 1950s, computer games adopted interactive graphical interfaces—something that only came to personal computers in the 1980s.

Similarly, today's games already use what many computer scientists think will be the next paradigm in HCI: active amplification of the user's actions. In the future, we are told, agent programs would watch our interactions with a computer, notice the patterns, and then automate many tasks we do regularly, from backing up the data at regular intervals to filtering and answering our email. The computer would also monitor our behavior and attention level, adjusting its behavior accordingly: speeding up, slowing down, and so on. In some ways this new paradigm is already at work in some applications: for instance, an Internet browser offers us the list of sites relevant to the topic we are searching on; Microsoft Office Assistant tries to guess when we need help. However, there is a crucial problem with expanding such active amplification to the whole of HCI. The more power we delegate to a computer, the more we lose control over what it is doing. How do we know that the agent program identified a correct pattern in our daily use of email? How do we know that a commerce agent we send on the Web to negotiate with other agents the lowest price for a product was not corrupted by them? In short, how do we know that a computer amplified our actions correctly?

Computer games are games, and the worst that may happen is that we lose. Therefore active amplification is present in practically every game: Mario embarking on mini-narratives of its own with a single move of a joystick; troops conducting complex military maneuvers while you directly control only their leader in Rainbow Six; Lara Croft executing whole acrobatic sequences with a press of a keyboard key. (Note that in "normal" games this amplification does not exist: when you move a single figure on a chessboard, this is all that happens; your move does not initiate a sequence of steps.)

Flash projects heavily use active amplification. It gives many projects the magical feeling. Often we are confronted with an empty screen, but a single click brings to life a whole universe: abstract particle systems, plant-like outlines, or a population of minimalist creatures. The user as a God controlling the universe is something we often encounter in computer games; but Flash projects also give us the pleasure of creating the universe from scratch.

Active amplification is not the only feature Flash projects share with games. More generally, as Peter Lunenfeld has suggested, computer games are for the Flash generation what movies were for Warhol. Cinema and TV colonized the unconscious of the previous generations of media artists, who continue to use the gallery as their therapy coach, spilling bits and pieces of their childhood media archives in public (for instance, Douglas Gordon). Flash artists are less obsessed with commercial time-based media. Instead, their iconography, temporal rhythms, and interaction aesthetics come from games (Mike Clavert: mikeclavert.com). Sometimes user participation is needed for the Flash game to work; sometimes the game just plays itself (UTOPIA by futurefarmers.com; dextro.org).

Flash Versus Net Art

Tirana Biennale 01 Internet exhibition: this title is deeply ironic. The exhibition did not include any projects from Albania, or any other post-communist East European country for that matter. This was quite different from many early Net art exhibitions of the mid-1990s whose stars came from the East: Vuc Cosic, Alexei Shulgin, Olga Lialina. 1990s Net art was the first international art movement since the 1960s that included Eastern Europe in a big way. Prague, Ljubljana, Riga, and

Moscow counted as much as Amsterdam, Berlin, and New York. Equally including artists from the West and the East, Net art perfectly corresponded to the economic and social utopia of a new post-Cold War world of the 1990s.

Now this utopia is over. The power structure of the global Empire has become clear, and the demographics of Tirana Biennale 01 Internet section reflected this perfectly. Many of the artists included in Tirana Biennale 01 Internet exhibition work in key IT regions of the world: San Francisco (Silicon Valley), New York (Silicon Alley) and Northern Europe.

What happened? In the mid 1990s, Net art relied on simple HTML that ran well on both fast and slow connections—and this is enabled the active participation of the artists from the East. But the subsequent colonization of the Web by multimedia formats—Flash, Shockwave, QuickTime, and so on—restored the traditional West/East power structure. Now Web art requires fast Internet connections for both the artist and the audiences. With its slow connections, the East is out of the game. The Utopia is over; welcome to the Empire.

(Tirana Biennale 01 did include one artist from China who contributed a beautiful animation of martial arts fighters. But we never found out who he was. All we knew about him was his email address: zhu_zhq@sohu.com. Maybe he did not even live in China.)

Generation Flash: FAQ

After I posted the preceding segments on popular mailing lists dealing with new media art and cyberculture (rhizome.org and nettime.org), I received lots of responses. Here are my answers to the two most commons questions.

Question:

Is not the "soft modernism" you describe simply a result of particular technological limitations of multimedia on the Net? You seem to mistake the particular features of Flash designed to deliver animation over the narrow bandwidth for a larger zeitgeist.

Answer:

Now that the new release of Flash (Flash MX) allows for the import and streaming of video, it is possible that soon "Flash generation"/"soft modernism" aesthetics will leave Flash sites. This is fine. My concern in this essay is *not* with Flash software and its limitations/capabilities per se, but with the new sensibility that during the last couple of years manifested itself in many Flash projects. In other words, I am interested in a "generation Flash" that is quite different from the Flash software/format.

Therefore the number of people who after reading my text accused me of confusing a technical standard with aesthetics missed my argument. The vector-oriented look of "soft modernism" is not simply a result of narrow bandwidth or a nostalgia for 1960s design—it *always* happens when people begin to generate graphics through programming and discover that they can use simple equations, etc. This is also why the "soft modernism" of Flash projects and other software artists replays, sometimes in amazing detail, the aesthetics of early computer art (1950s–1970s) when people were only able to create images and animations through programming.

Question:

There is no reason software art cannot use representational images or any other form. Why do you associate software art with non-representational, abstract vector-based graphics?

Answer:

Of course software artists can use representational images or any other "conventional" form or media. It was not accidental that soon after his arrival at Xerox PARC in the 1970s, Alan Kay and his associates created a paint program and an animation program, along with overlapping windows, icons, Smalltalk and other principles of modern interactive graphical computing. The ability to manipulate and generate media is not an after-thoughts to a modern computer—it is central to its identity as a "personal dynamic medium" (Alan Kay.) To put this differently: the computer is a simulation machine, and as such it can and should be used to simulate other media.

So I have nothing against software artists using/creating media, but I hope that the "Flash generation" will extend its programming work to representational media! In other words, if in the early 1970s the paint program and the animation program were revolutionary in changing people's idea of a computer away from computation and towards a (creative) medium, after almost two decades of menu-based media manipulation programs and the use of computers as media distribution machines (greatly accelerated by World Wide Web), a little programming can be quite revolutionary! In short, we have now become so used to thinking of a computer as a "personal dynamic medium," that we need to remind ourselves and others that it is also a programmable machine.

Now, think about how programming has been used so far to create/use still images, animation and film/video. There are three trajectories that can be traced historically. One trajectory extends from the earliest works of computer art—the films by the Whitney Brothers made with an analog computer in the mid 1950s (who were the students of Oscar Fischinger and thus represent a direct link with early twentieth century modernism)—to the "soft modernism" of today's Flash projects and data visualization artwork. In other words, this is the use of programming to generate and control abstract images.

The second trajectory begins in the 1980s when Hollywood and TV designers started to use computer-generated imagery (CGI). Now, programming was put in the service of traditional cinematic realism. Particle systems, formal grammars, AI and other software techniques became the means to generate flying bats, hilly landscapes, ocean waves, explosions, alien creatures, and other figurative elements integrated in the photorealistic universe of a narrative film.

What about using algorithms not simply to generate the figurative elements of a narrative but to control the whole fictional universe? This is the third trajectory: programming in computer games (1960–). Here algorithms may control the narrative events, the behavior of characters, camera movement, and other characteristics of the game world—all in real time. Unfortunately, as we all know, aesthetically revolutionary computer- and player-driven game worlds feature formula-driven content that makes even a bad Hollywood film appear original and inspiring by comparison. (*Grand Theft Auto 3* is no exception here—despite its breakthroughs in simulating a more compelling open universe.)

I think this brief survey shows that there is still an untouched space completely open for experimentation and creative research—using programming to generate and/or control figurative/fictional media. For instance, in the case of a movie, programming can be used to generate characters on the fly, to composite in real-time characters shot against a blue screen with backgrounds, to control the sequence of scenes, to apply filters to any scene in real-time, to combine a pre-recorded scene with the imagery generated on the fly, to have characters interact with the viewer, etc., etc. In short, programming can be used to control *any* aspect of a fictional media work.

Of course, once in a while one encounters projects moving in this direction at places like SIGGRAPH or ISEA, but they are typically research demos created in universities that do not reach the culture at large. Of course, you can object that having an algorithmically-controlled complex fictional universe requires the kind of programming investment only possible in a commercial game company or in a university. After all, this is not the same as writing a script that draws a few lines that keep moving in response to user input…yes, but why do our fictional/figurative works have to follow the formulas of commercial media? If one accepts that characters do not have to be "photorealistic," that the fictional world does not have to be exclusively three-dimensional, that chance and randomness can co-exist with narrative logic, or that stick figures can co-exist with 3-D characters and video footage, etc., programming figuration/fiction becomes less formidable. In short, while I welcome programming Flash, I think it is much more challenging to program QuickTime.

Postscript: On The Lightness of Flash

When I first visited the most famous Flash site—praystation.net—I was struck by the lightness of its graphics. More quiet than a whisper, more elegant than Dior or Chanel, more minimal than the 1960s minimalist sculptures of Judd, more subdued than the winter landscape in heavy fog, the site pushed the contrast scale to the limits of legibility. A similar lightness and restraint can be found in many projects included in the Biennale 01 show. Again, the contrast with the screaming graphics of commercial media and the media art of the previous generations is obvious.

The lightness of Flash can be thought of as the visual equivalent of electronic ambient music. Every line and every pixel counts. Flash appeals to our visual intelligence—and cognitive intelligence. After the century of RGB color which begun with Matisse and ended with aggressive spreads of Wired, we are asked to start over, to begin from scratch. Flash generation invites us to undergo a visual cleansing—this is why we see a monochrome palette, white and light gray. It uses neo-minimalism as a pill to cure us of post-modernism. In Flash, the rationality of modernism is combined with the rationality of programming and the affect of computer games to create the new aesthetics of lightness, curiosity and intelligence. Make sure your browser has the right plug-in: welcome to generation Flash.

I am not advocating a revival of modernism. Of course we don't want simply to replay Mondrian and Klee on computer screens. The task of the new generation is to integrate the two key aesthetic paradigms of the twentieth century: (1) a belief in science and rationality, an emphasis on efficiency and basic forms, the idealism and heroic spirit of modernism; (2) skepticism, an interest in "marginality" and "complexity," deconstructive strategies, the baroque opaqueness and excess of post-modernism (1960s–). At this point all the features of the second paradigm have become tired clichés. Therefore a partial return to modernism is not a bad first step, as long as it is just a first step towards developing the new aesthetics for the new age.

Of course this aesthetics should also fully engage with the difficult questions of globalization. The *remix culture* we are living in now is not only engaged in remixing all previous cultural forms and texts of but also in remixing various features which come from what used to be called national cultures as well as from already existing remixes between immigrant populations and their "host" cultures. The solution offered by multinational conglomerates—a composite which takes certain signifiers from a few national cultures—for instance, a French idea of elegance, Japanese *manga* iconography, "cool Britannia" references, and so on, and integrates it all into a rather bland and monolithic text which is then being sent back to all the places around the world—is obviously not a satisfactory solution. (It reminds me of a Soviet-style centralized economy in which all the output of collective farms was sent to the center where it was decided how it was to be distributed nationally.) Luckily, numerous remixes which follow different logics are being explored around the world by musicians, theatre groups, dancers, designers, architects, and so on. Nobody knows what will emerge from this global cultural laboratory—and this is what makes our times so interesting.

Although most of my arguments are about visual culture and visual aesthetics, it is relevant at this point to evoke a different practice. Music historically has been the artistic field that was always been ahead of other fields in using computers to enable new aesthetic paradigms. The whole practice of popular electronic music in the last three decades is a testament to how empowering new technologies are in welding new complex and rich remixes between different cultures, styles, and sensibilities. Without electronic and computing technologies—from a turntable and a tape recorder to peer-to-peer file sharing networks and music synthesis software running on a regular laptop—most of this culture would never come to be. The field of electronic sound (which pretty much means most sounds today) with its multitude voices and a real bottom-up, "emergent" logic, is a powerful alternative to the "top-down" cultural composites sold by global media conglomerates around the world. Let us hope that other artists and designers in other fields will follow music's lead in using a computer to enable similarly rich remix cultures.

Notes

1. This article, written in 2002, is about the "Flash Generation" and not about the Web sites made with Flash software. Many of the sites which inspired me to think of "Flash aesthetics" are not necessarily made with Flash; they use Shockwave, DHTML, Quicktime and other Web multimedia formats. Thus the qualities I describe below as specific to "Flash aesthetics" are not unique to Flash sites.
2. For instance, the work of Lisa Jevbratt, John Simon, and Golan Levin.
3. "Generation Flash" incorporates revised versions of the texts commissioned for www.whitneybiennial.com and www.electronicorphanage.com/biennale. Both exhibitions were organized by Miltos Manetas/Electronic Orphanage. "On UTOPIA" was commissioned by Futurefarmers.
4. See http://www.potatoland.org/landfill/
5. After GUI-based applications such as Hypercard, Director, Photoshop and others became commonplace, many computer artists continued to do their own programming: writing custom code to control an interactive installation, programming in LINGO an interactive multimedia work, etc. This was not referred to as software art; it was taken for granted that even in the age of GUI-based applications a really serious artistic engagement with computers requires getting one's hands dirty in code.

15

Viruses Are Good for You

Julian Dibbell

What scares you most about getting that virus?

Is it the prospect of witnessing your system's gradual decay, one nagging symptom following another until one day the whole thing comes to a halt? Is it the self-recrimination, all the useless dwelling on how much easier things would have been if only you'd protected yourself, if only you'd been more careful about whom you associated with?

Or is it not, in fact, something deeper? Could it be that what scares you most about the virus is not any particular effect it might have, but simply its assertive, alien presence, its intrusive otherness? Inserting itself into a complicated choreography of subsystems all designed to serve your needs and carry out your will, the virus hews to its own agenda of survival and reproduction. Its oblivious self-interest violates the unity of purpose that defines your system as yours. The virus just isn't, well, you. Doesn't that scare you?

And does it really matter whether the virus in question is a biological or an electronic one? It should, of course. The analogy that gives computer viruses their name is apt enough to make comparing bioviruses and their digital analogs an interesting proposition, but it falls short in one key respect. Simply put, the only way to fully understand the phenomenon of autonomously reproducing computer programs is to take into account their one essential difference from organic life forms: they are products not of nature but of culture, brought forth not by the blind workings of a universe indifferent to our aims, but by the conscious effort of human beings like ourselves.

Why then, after a decade of coexistence with computer viruses, does our default response to them remain a mix of bafflement and dread? Can it be that we somehow refuse to recognize in them the traces of our fellow earthlings' shaping hands and minds? And if we could shake those hands and get acquainted with those minds, would their creations scare us any less?

These are not idle questions. Overcoming our fear of computer viruses may be the most important step we can take toward the future of information processing. Someday the Net will be the summation of the world's total computing resources. All computers will link up into a chaotic digital soup in which everything is connected—indirectly or directly—to everything else. This coming Net of distributed resources will be tremendously powerful, and tremendously hard to harness because of its decentralized nature. It will be an ecology of computing machines, and managing it will require an ecological approach.

Many of the most promising visions of how to coordinate the far-flung communication and computing cycles of this emerging platform converge on a controversial solution: the use of self-replicators that roam the Net. Free-ranging, self-replicating programs, autonomous Net agents, digital organisms—whatever they are called, there's an old fashioned word for them: computer viruses.

Today three very different groups of heretics are creating computer viruses. They have almost nothing to do with each other. There are scientists interested in the abstract behaviors of self-replicating codes, there are developers interested in harnessing the power of self-replicating programs, and there are unnamed renegades of the virus-writing underground.

Although they share no common experience, all these heretics respect a computer virus for its irrepressible mobility, for the self-centered autonomy it wrests from a computer environment, and for the surprising agility with which it explores opportunities and possibilities. In short, virus enthusiasts relate to the virus as a fascinating and powerful life form, whether for the fertile creation of yet more powerful digital devices, as an entity for study in itself, or, in the case of one renegade coder, for reckless individual expression.

Getting a Buzz from the Vx

One computer virus writer in his early 20s lives on unemployment checks in a white, working-class exurb of New York City. He tends to spend a fair amount of his leisure time at the local videogame arcade playing Mortal Kombat II, and would prefer that you didn't know his real name. But don't let the slacker resume fool you: the only credential this expert needs is the pseudonym he goes by in the computer underground: Hellraiser.

Hellraiser is the founding member of the world-renowned virus-writers' group Phalcon/Skism. He is also creator of 40Hex, an electronic zine whose lucid programming tips, hair-raising samples of ready-to-run viral code, and trash-talking scene reports have done more to inspire the creation of viruses in this country than just about anything since Robert Morris Jr.'s spectacularly malfunctional worm nearly brought down the Internet.

And as if all this weren't enough, Hellraiser also comes equipped with the one accessory no self-respecting expert in this cantankerous field can do without—his very own pet definition of computer viruses. Unlike most such definitions, Hellraiser's is neither very technical nor very polemical, and he doesn't go out of his way to make it known. "Sure," he'll say, with a casual shrug, as if tossing you the most obvious fact in the world: "Viruses are the electronic form of graffiti."

Which would probably seem obvious to you too, if you had Hellraiser's personal history. For once upon his teenage prime, Hellraiser was also a hands-on expert in the more traditional forms of graffiti perfected by New York City youth in the 1980s. Going by the handle of Skism, he roamed the city streets and train yards with a can of spray paint at the ready and a Bronx-bred crew of fellow "writers" at his side, searching out the sweet spots in the transit system that would give his tag maximum exposure—the subway cars that carried his identity over the rails, the truck trailers that hauled it up and down the avenues, and the overpasses that announced it to the flow of travelers circulating underneath.

In other words, by the time Hellraiser went off to college and developed a serious interest in computers, he was already quite cozy with the notion of infiltrating other people's technology to spread a little of himself as far and wide as possible. So when he discovered one day that his PC had come down with a nasty little digital infection, his first thought was not, as is often customary, to curse the "deviant hackers," "sociopaths," and "assholes" who had written the program, but to marvel at the possibilities this new infiltration technique had opened up. Street graffiti's ability to scatter tokens of one's identity across the landscape of an entire metropolis looked provincial in comparison. "With viruses," Hellraiser remembers thinking, "you could get your name around the world."

He was right. The program that had infected his own computer in late 1990, the so-called Jerusalem virus, had spread from Italy to Israel to North America before finally making its way into the pirated copy of Norton Utilities that brought it to Hellraiser's hard drive. And though Jerusalem's author remained uncredited, other programmers from nearly every corner of the globe were pulling off feats of long-distance self-aggrandizement that dwarfed anything within the reach of America's spray-paint commandos. A kid who called himself Den Zuk had launched a virus that was flashing his handle on computer screens all over Europe, the U.S., and South America. Early speculation placed its origin in Venezuela, but the virus was eventually tracked to its true source in Bandung, Indonesia, when a researcher in Iceland guessed that some enigmatic characters in the source code were in fact a ham-radio call sign; contact was made with the call sign's registered operator, who immediately copped to his authorship of the program.

Equally far-ranging was the journey of the Joshi virus, which spread from India to parts of Africa and on to the rest of the world, popping up every January 5th to command computer users to type "Happy Birthday Joshi" if they wanted control of their systems back.

What impressed Hellraiser as much as the vast geographic distances covered by viruses, however, was their long range over time. After all, a painted graffiti tag would only last as long as it took to fade away or be painted over, but viruses, it seemed, might replicate forever in the wild. Indeed, the Jerusalem virus had been doing so for three years before Hellraiser encountered it, and four years later it remains one of the world's most commonly reported viruses. Likewise, Den Zuk is still reproducing on computers worldwide six years after it first left the island of Java; Joshi continues for the fifth year in a row to extort international birthday wishes. Dozens of other viruses from the U.S., Canada, Eastern Europe, Taiwan, Australia, Turkey, Malta, and other far-flung locales thrive globally. (This despite the fact that the antivirus industry spends tens of millions of dollars a year to eradicate them.) Bearing encoded bits of their authors' souls—clever jokes, crude graphics, friendly greetings, and, of course, occasionally, malicious intentions (though in fact the majority of viruses found in the wild are designed to do no damage)—viruses roam the earth in apparent perpetuity.

For Hellraiser, steeped as he was in graffiti culture's imperative to "get the name across," there was only one possible response to this new technology of self-projection: he had to get in on the action. But how? Virus writing wasn't exactly a standard subject in computer-science courses, and even the computer underground—with its loose-knit network of bulletin boards and e-zines proffering instruction in the illicit arts of hacking and phone phreaking—wasn't the most dependable source of virus lore. Occasionally, a hack and phreak board might offer a small collection of cryptic viral source code for brave souls to experiment with, but as far as Hellraiser knew, the only system exclusively devoted to viruses at the time was a place called the Virus Exchange, operating out of what was then the world's epicenter of virus production: post-Communist Bulgaria, where the Cold War's endgame had left a lot of overtrained programmers with time on their hands and anarchy on their minds.

Lacking the money or the phreaking skills to dial in to the Virus Exchange, Hellraiser made do with what he did have: a live specimen of the Jerusalem virus, replicating furiously inside his desktop system and poised to trash every program file he tried to run on any upcoming Friday the 13th. Carefully, Hellraiser extracted all copies of the virus from the computer and holed up in his dorm room to examine its workings. He studied it for weeks, and then finally, tentatively, he produced a virus of his own. It was a shameless hack really, essentially just the Jerusalem code with the tag line "SKISM-1" inserted in place of a few of the original characters. But after infecting as many computers as he could and subsequently finding his creation enshrined in antivirus literature as the "Skism-1" virus, Hellraiser swelled with a pride he would later recall with some amusement: "Shit, I thought I was the man back then."

Hooked on that buzz, he dove deeper into his studies, aiming for proficiency in DOS assembly language, the formidably austere low-level programming dialect in which Jerusalem was written (like the vast majority of computer viruses then and now). He quickly acquired the ability to produce viruses he could truly say were his, and along with this ability, he picked up the beginnings of a rep among New York-area denizens of the underground. Gradually, through the hack/phreak (h/p) bulletin-board scene, he made contact with other isolated virus writers—subculture orphans compared with the h/p crowd and its Legions of Doom, MODs, Chaos Clubs, and other constantly forming and re-forming groups and factions.

Hellraiser started wondering why he shouldn't put together a group of his own. Soon enough, the retired graffiti bomber was again running with a crew, formally known as Smart Kids Into Sick Methods (Skism for short) and dedicated to sharpening the virus-writing skills of both its members and the virophilic public at large.

And it was to serve more or less those lofty ends that Skism's electronic house journal 40Hex was born. Named for the assembly-language function by which viruses copy themselves, the publication hit the boards of the Vx underground with an infectiousness all its own. (Vx, short for virus exchange, denotes all boards devoted, like their Bulgarian namesake, to virus discussion and traffic in viral source code.) Its unapologetic bad attitude was a brash wake-up call to the still-embryonic virus-writers' community. "This is a down and dirty zine [which] gives examples on writing viruses and...contains code that can be compiled to viruses," wrote Hellraiser in the introductory file of 40Hex's March 1991 première. "If you are an antivirus pussy, who is just scared that your hard disk will get erased so you have a psychological problem with viruses, erase these files. This aint for you."

The warning scared off no one, of course, least of all the alleged pussies of the antivirus industry, who took to scouring every new issue for a peek inside the mind of the enemy, getting up close and personal at last with the phantoms they'd been battling for years. Not that the life of the virus hunter was a lonely one. In fact, the antivirus community was already in many ways a more advanced subculture than that of the virus writers, complete with local color and a mystique all its own: the industry pioneer and media darling John MacAffee was famed for his giddy morning-after overestimation by a factor of 10 of the Internet worm's damage; then there were those Bulgarians, the notorious and proud Dark Avenger—who signed, and even dedicated, his viruses—and his driven nemesis, Vesselin Bontchev. Endlessly revising and debating the burgeoning taxonomy of virus species, nervously policing the boundary between the great unwashed and those trustworthy enough to handle "live" specimens, the world of antivirus research offered its initiates a thrill somewhere between the delightful romance of butterfly collecting and the grim camaraderie of working for the National Security Agency.

In comparison, virus writing—while obviously not without its kicks—lacked community. But in the months and years following 40Hex's début, that began to change. The previously inchoate and virtually invisible virus-writing underground at last coalesced and shifted into high gear. Various groups proliferated and crossbred: Skism merged with another New York posse called Phalcon to form the Phalcon/Skism supergroup, while the pan-European TridenT team and the Canadian-Australian-Swiss-Taiwanese-multinational NuKE crew quickly rose to challenge Phalcon/Skism's prestige and programming skills. Zines multiplied, too: NuKE's Info Journal and West Coast virus writer Urnst Kouch's Crypt Newsletter challenged 40Hex's hegemony, as did the number of so-called Vx bulletin boards that rocketed from a handful worldwide to rough estimates of as many as 200 at present.

Amid all the rapid growth it helped set in motion, 40Hex has kept pace. After the first four raucous issues, Hellraiser handed over the editorial reigns to Phalcon's designated archivist, Garbage Heap, who has steadily increased the circulation of the zine while slowly steering it toward something suspiciously like respectability. Available now in a crisp, desktop-published paper edition as well as good old-fashioned e-text, today's 40Hex still brims with the gnarliest of viral code and remains a feisty defender of the right to create and publish viruses. But it frowns on anyone who looses viruses into the wild and is more likely to solicit guest editorials from antivirus types than to hurl obscenities at them.

The young hellion who founded the zine would probably not approve—that is, if the same young hellion were still around to say anything about it. But he isn't. Not really. Hellraiser has undergone some changes of his own lately. Once quite cavalier about releasing viruses that intentionally deleted files or otherwise "fucked people's shit up" (after all, what better way to make your tag linger on in their memory?), he eventually decided that creating destructive programs just gave virus writing a bad name and resolved thenceforth to produce viruses with more or less benign payloads only. And then one day, not too long ago and without much fanfare, he simply called it quits. Partly, he was starting to chafe at the limited range of programming challenges involved in virus creation, he says, but more to the point, his evolving young world view had somehow gotten infected by a creeping respect for the right of others to control what goes into their own digital back yards. Destructive payload or no destructive payload, Hellraiser reached the conclusion that it was just plain "wrong to 'pollute' other people's systems with viral garbage."

Which isn't to say he's gone over to the ranks of his old antivirus nemeses. Hardly. He's still too tight with all his Phalcon/Skism homeboys for that. Even if he weren't, he's been a virus writer for too long to feel comfortable with the easy demonizations that are the stock in trade of antivirus rhetoric. For the rest of us, of course, it's easy enough to accept the standard caricature of the underground virus writer as a low-grade sociopath. After all, what else but antisocial perversity could lead someone to produce a mechanism we encounter principally as contamination in the digital environment, as noise on the line?

Yet Hellraiser's career path—from graffiti writing to virus writing and beyond—demands a more complicated understanding of the virus phenomenon. It asks us to recognize that viruses, like graffiti, are just as much signal as noise—that they are in fact an irreducible confusion of the two. As Hellraiser came to recognize, the noisiness of viruses is built in—they are by definition information that subverts control. But as the subculture Hellraiser helped build will always remember, every virus turned out into the computer wilds—like every tag sprayed onto the hard urban landscape—is also a carrier for the purest and strongest signal a human being can send. "Remember my name," the virus says, which—after all—is another way of saying: "I'm alive."

This is about as far as most discussions of virus writing get: ignorant kids thrashing about in codes, creating horribly simple but efficient digital bombs. And even if you take a very generous view that the underground virus writers are inadvertently creating new forms of life, the discussion of beneficial viruses would have to stop here if it weren't for folks like Dr. Mark A. Ludwig.

The Mutator in the Desert

Mark Ludwig lives in a desert, and compared to Hellraiser's background, seems to hail from an entirely different planet. But Ludwig, too, is chasing the elusive nature of computer viruses.

A married man with three young children, Ludwig lives in Tucson, Arizona, where barrens of sand and sun and saguaro cactus shimmer not too far beyond the sump-cooled confines of his home. But the desert where he wanders is someplace else entirely: it's the lonely intellectual wilderness reserved for those who practice science on the fringe, outside the cozy realms of institutional affiliation, professional consensus, or methodological decorum.

He doesn't have to be there. With his PhD in physics from the University of Arizona (and his prior course work at Cal Tech and MIT), Ludwig could easily return to the fold of respectable researchers if he chose. All he'd have to do is let go of his somewhat obsessive scholarly pursuit of the wild computer virus, and pick a slightly more conventional object of study. Or maybe just pursue his present subject with a little more sober attention to devising antivirus countermeasures and a lot less gleeful fascination with viruses in and of themselves. Or maybe just tone down the florid libertarian rhetoric and sweeping philosophical claims in which he tends to couch his otherwise gruelingly meticulous analyses of viral performance and technique.

Really, it wouldn't take much.

But Ludwig isn't likely to do any of these things, because he actually seems to prefer the hardships of the fringe to the rewards of a life on the techno-scientific inside.

He didn't always. "Once I was a scientist of scientists," writes Ludwig in the introduction to his latest self-published treatise, *Computer Viruses, Artificial Life, and Evolution*. "Born in the age of Sputnik, and raised in the home of a chemist, I was enthralled with science as a child. If I wasn't dissolving pennies in acid, I was winding an electromagnet, or playing with a power transistor, or doing a cryogenics experiment—like freezing ants—with liquid propane." Eager to work his way into the company of "the great men of science" and join their noble quest for objective Truth (he'd read about it in textbooks), Ludwig rushed through his undergraduate work at MIT in two years, then plunged into his graduate course of studies with equal enthusiasm. By the time he got his doctorate, however, he'd seen enough of the political infighting and blind prejudice that structure the real work of contemporary scientific investigation to sour the romance permanently. Disillusioned, he dropped out of the hard-sci grind and into a job working with computers, a field that at least provided some of the wide-open pioneering spirit that the textbook histories of science had promised, even if it moved him further from pure science's intimacy with the mysteries of nature.

But not long after that, around 1988, he started picking up reports of contagious programs running loose among the machines he now made his living from, and the course of his life changed yet again. For Ludwig, viruses came bearing the same mind-expanding message-in-a-bottle they would not much later be bringing to Hellraiser. Except that Ludwig decoded the message a little differently. Where Hellraiser heard the signal "I'm alive" coming from the virus's creator, Ludwig understood the message as coming directly from the virus itself. Viruses behaved like living things: self-reproducing and autonomous. Might we not understand life a little better, he wondered, if we could create something similar, and study it, and try to understand it? The mysteries of nature, in other words, now loomed closer than ever—right there on the wide-open technological frontier to which he'd fled from the wreckage of his scientific aspirations—and Ludwig couldn't resist the temptation to go questing after them once more.

His initial attempts to acquire specimens to observe were frustrating. Today's teeming ecology of one-stop Vx trading posts didn't exist. When Ludwig approached the antivirus community for access to its shared research collections, he found himself shut out: then as now, the A-V crowd refused to release captured virus code to anyone outside a trusted inner circle. So, true to his style, Ludwig decided to go it alone. He set up a BBS, announced a bounty of US$25 for every virus uploaded, and sat back while the code rolled in. After building up a representative cross section of the wild virus population, he set about examining his haul, and within a few months his research bore its first fruit: *The Little Black Book of Viruses*, a technical primer on the essentials of virus writing, complete with scrupulously annotated source code for four virus programs of his own creation.

The Little Black Book made something of a name for Ludwig, but it wasn't an especially pretty one. Though the tutorial viruses were pointedly nondestructive and came surrounded by warnings against their misuse and instructions on how to keep them from getting loose, the book was roundly condemned as an incitement to digital vandalism. In the three years of steady sales since *The Little Black Book*'s original publication in 1991, various mainstream computer magazines have summarily dropped Ludwig's advertisements for the book as inappropriate subject matter for their audiences. And when the book was recently released in France (as *Naissance d'un Virus*, or *Birth of a Virus*), its publishers there were immediately slapped with a legal injunction against distributing it with the infectious source code intact.

But Ludwig has remained undaunted in the face of the world's virophobia. If anything, its vehemence has only sharpened his determination to share the wealth of his knowledge. "People think of viruses as an invasion from Mars," he says, "and that hurts research into these things. My aim is to change people's attitudes, to cut down some of the fear."

To that end he has established an annual international virus-writing competition, flying cheerfully in the face of the "swarming hordes of antivirus developers." (One year's contest rewarded the smallest functional DOS virus submitted.) Ludwig also publishes a newsletter now, *Computer Virus Developments Quarterly*, in which he mingles detailed technical discussion of viral code with rants against the tyrannical tendencies of American government, the moral bankruptcy of contemporary Western culture, and (last but not least) the evils of repressing detailed technical discussion of viral code. Occasionally he even gets a sign that the general public is starting to come around to his pro-knowledge agenda: after five months of wrangling its way through the French courts, for instance, the suit against *Naissance d'un Virus* was finally thrown out by a tribunal arguing, as Ludwig proudly reports, that "trying this case was like putting Galileo on trial again."

Yet amid all of Ludwig's busy agitation in defense of viruses, what ever became of the intellectual mysteries that first drew his attention to them? His pleasure at being compared to Galileo, the archetype of the politically incorrect scientist, certainly suggests that he never lost his sense of scientific mission. But the proof of Ludwig's abiding interest in viruses as tools of natural philosophy lies in his sequel to *The Little Black Book*: the aforementioned *Computer Viruses, Artificial Life, and Evolution*. Published late in 1993, the book is a dense and daunting 373 pages' worth of charts, differential equations, and tightly reasoned arguments in support of Ludwig's intuition that self-reproducing computer code bears deep lessons about the workings of life.

As the title's nod to the fashionable new scientific discipline of artificial life makes plain, however, Ludwig is clearly aware that other researchers, backed by the imprimatur of Official Science, have been building on the very same intuition for some time now. The first two volumes of the Santa Fe Institute's *Proceedings on Artificial Life,* published in 1989 and 1992, devote several papers to the idea of computer viruses as synthetic life. But taking the idea further, Ludwig argues that computer viruses, unlike such other forms of artificial life as cellular automata, mobots, or genetic programming, are the only form of artificial life not biased by the hope of their creators. Because computer viruses must exist in an environment (DOS in particular) that was designed without any thought of the digital organisms that might come to inhabit it, they are free from any accusation that the environment's "physics" were written to support the emergence of their lifelike behavior. Or to put it more bluntly, feral viral ecologies (versus the controlled experiments in university labs) represent the only known simulation of life that does not implicitly (and quite unscientifically) build God into the system.

Having carefully constructed this ambitious claim, Ludwig proceeds to test drive it straight into the heart of biology's most vexing questions: How did life get here in the first place? How did the staggering diversity of life forms that exists today come to be? He sics viruses on the theory of evolution itself, in other words, sending them in to illuminate with their logical simplicity the still murky depths of Darwin's grand hypothesis. It's a bold move, but a puzzling one at first glance. Although the viruses found in the wild may exhibit a wide range of lifelike features, they've never been known, after all, to evolve.

Or have they? Not too long after the first virus was written, the first antivirus program was written as a countermeasure. Once anti-virus software was introduced into the cybernetic ecology, viruses and the programs that stalk them have been driving each other to increasing levels of sophistication. This is nothing less than the common coevolutionary arms race that arises between predators and prey in organic ecosystems.

Step one in this quasi-Darwinian dance took place when security-minded programmers developed what has since become the standard defense against viruses for most PC owners—scanning software that looks for telltale code fragments of known viruses (often some scrap of graffiti-esque text) and alerts the user when it finds any. In time, virus hackers responded by wrapping their programs in a blanket of encryption impenetrable to scanners. But since the built-in subroutines that decrypt the programs for execution cannot themselves be enciphered, antivirus programmers simply retooled their scanners to look for the decryption code. Later, in step two, the legendary Bulgarian writer Dark Avenger came up with a clever innovation known as a mutating, or polymorphic, virus. A mutating virus randomly reorganizes its decryption algorithm every time it replicates to outsmart the policing of the scanner. In step three, antivirus engineers devised "heuristic" scanners, built to sniff out all but an insignificant percentage of a virus' mutants through educated pattern recognition.

Surveying the fossil record of this game, Ludwig found himself pondering a logical next move: what if someone were now to develop a strain of polymorphs with a genetic memory, so that rather than completely reshuffling their structure with every generation, the few mutants that escape discovery by heuristics could pass their undetectable code on to their offspring?

The prospect of virus populations able to autonomously build up immunity to any scanning techniques thrown at them thoroughly depressed antivirus programmers. To Ludwig, however, the possibility proved too intriguing to wait around for some random underground hacker to realize it, and he resolved to do the job himself. The result: Ludwig's "Darwinian Genetic Mutation Engine," a programming utility that turns any normal DOS virus into a souped-up, genetically evolving polymorph, complete with an option for sexual gene-swapping between individuals that come into contact in the wild. Curious hackers can find the Darwinian Genetic Mutation Engine's complete source code in the pages of *Computer Viruses, Artificial Life, and Evolution*, along with detailed experimental results demonstrating the ability of Darwinian Genetic Mutation Engine-enhanced viruses to run rings around existing scanners. But the program's deeper significance, of course, lies in its potential to transform viruses' heretofore hacker-driven pseudo-evolution into something very like the real thing: a finely tuned interaction of variety and natural selection that allows the environment itself to shape the internal code of the organisms dwelling in it.

The Darwinian Genetic Mutation Engine is all Ludwig needs, in other words, to prove viruses capable of meaningful evolution, and incidentally, test Darwin's theory. And it's no surprise perhaps, given Ludwig's hard-earned distrust of anything smacking of intellectual orthodoxy, that he has found that Darwin's venerable theory fails the test. Running his beloved viruses through assorted experimental hoops and mazes, Ludwig followed them to the conclusion that Darwinian evolutionary mechanisms alone are just not mathematically fertile enough to have created and shaped life as we know it. This is a well-worn scientific heresy, of course, but it's not without its small but respectable following within the ivory walls Ludwig so proudly dismisses.

To be fair, though, Ludwig is not asking to be ranked among his boyhood heroes—those scientific greats whose unique insights clear broad new vistas of understanding in a single bound. All he wants from the rest of the world is a modicum of respect for the wild computer virus as a legitimate subject of scientific investigation. Or at least acknowledgment that this enduringly lifelike wonder could be useful if we but understood it, rather than the casting of it as the ultimate technological taboo.

Ludwig managed a remarkable intellectual shift. He elevated the computer virus from the digital equivalent of a can of spray paint to an object capable of perhaps infinite variations and almost lifelike behavior. He transformed a tool of vandals into a field of scientific study by emphasizing a computer virus' biological affinity. But by the time Ludwig began publishing, the computer virus was already well on its way from the fringes of science to the seat of honor at research symposiums.

Booting Up the Cambrian Explosion

"I'll be out at my place in the jungle over the weekend," said the message, posted in May 1994 from an obscure Internet site in Central America, "so I'll be out of e-mail contact till Monday."

And just like that, University of Delaware ecologist Tom Ray (now visiting scholar at the Advanced Telecommunications Research Institute International in Kyoto, Japan) disappeared once more into the rain forests of Costa Rica, leaving behind the clean conveniences of the digital world for an organic riot of plant and animal life. As promised, though, he would be back. Ray's passion for the unkempt splendor of the jungle has remained unabated after nearly two decades of intermittent research there, but in the last few years, it's the digital world that has claimed his closest attentions. Since late 1989, Ray has done his most important fieldwork seated in front of a computer, observing the busy fruits of an activity that has come to define his career: he breeds viruses.

Or to put it more precisely, he breeds worms, since that's the stickler's term for software that is both self-reproducing and able to execute its code independent of any host program. Ray, convinced that his programs are as good as alive, calls them simply "organisms," or "creatures." Whatever they are, though, he's been breeding quite a lot of them. He's been breeding them with the full support of his university employers, with the financial backing of major corporations, and with the steadily growing curiosity and respect of fellow researchers in the fields of both biology and computer science. And if all goes according to plan, he will keep on breeding them until he has achieved a goal far more adventurous than anything yet attempted by other virus programmers—infusing the vast unused spaces of the global computer networks with a roiling digital ecology as complex, as fascinating, and ultimately as beneficial to humankind as the rain forests that he has long sought to protect and understand.

In short, by infecting the Net with self-replicating code, Ray aims to turn it into a jungle.

He didn't start out so ambitious. In the beginning there was just a lone drive of a Toshiba laptop to populate, one tiny digital germ to do it with, and a hunch Ray had been kicking around for a decade or so to spur him on. The hunch was that experiments with self-replicating programs (Ray had first heard about them as a Harvard undergrad in the late '70s) might add some theoretical rigor to eco-science's essentially anecdotal attempts at explaining the abstract processes that gave rise to the complex interspecies relationships he had observed in the field. "I was frustrated," he would later tell a group of colleagues, "because I didn't want to study the products of evolution—vines and ants and butterflies. I wanted to study evolution itself."

In this, Ray's attraction to self-reproducing programs differed little from that of Mark Ludwig (who in fact was not unfamiliar with Ray's work by the time he set out to write his magnum opus on computer viruses and evolution). Unlike Ludwig, however, Ray felt neither philosophically obliged nor ethically disposed to work with viruses able to thrive in already existing computer environments. Not that he never considered the option. In fact, his initial plan was to set mutating machine-language organisms loose in a single computer and watch their evolution as they competed against one another for direct access to the computer's core memory, a strategy that might have evolved viruses superbly adapted to any system based on the same instruction set as the original petri chip. But Ray soon scrapped this idea—the risk of accidentally releasing his specimens into the wild seemed too great. Instead, he decided, he would evolve his organisms inside a virtual computer, modeled inside a real one in much the same way some operating systems today can model working emulations of other OSes, allowing DOS programs (for instance) to run in Macintosh environments. The difference, in Ray's scheme, was that his simulated system would be the only environment of its kind; thus, any program that escaped into other computers would find itself a fish out of water, unable to function anywhere but in its birthplace.

While the security benefits of this approach were obvious, its contribution to the scientific effectiveness of the experiment was even more significant: now that Ray was working with an imaginary computer, he was free to shape the system's design to create an environment more hospitable to life. And there was one key change to be made in that regard, for as Ray had come to recognize (and Ludwig would later set down in hard math), today's digital environments simply weren't built with mutant programs in mind. Typical operating systems might let a program randomly move some of its algorithms around with impunity (as the polymorphic viruses do), but at the fine-grained level of individual bit-flipping most closely analogous to genetic variation, even a single chance alteration almost always results in a system-crashing bug. Nature's tolerance of random code revisions is much greater, and if Ray wanted a more "natural" computer, then one way to get there would be to give it an instruction set in which nearly any sequence of bits would make some kind of sense to the system's virtual CPU.

So he gave it that instruction. He also equipped his phantom computer with a death function, a "Reaper," which would terminate any individual program sooner or later—but would always get to the oldest or most error-prone programs first. Thus primed to carry out the requisite natural selections, Ray's digital ecosphere was nearly complete. He called it *Tierra* (Spanish for "earth") and started preparing the final touch: an inhabitant. Later dubbed "the Ancestor," it was the first worm Tom Ray ever created—an 80-byte-long self-replicating machine written in Tierra's quirky assembly language—and as it happens, it was also the last. Once loosed into the Tierra environment installed on Ray's laptop, the creature's offspring quickly spread to the new world's every corner, within minutes displaying the evolutionary transformations that would "write" Ray's organisms from then on.

A 79-byte variation appeared, rapidly displacing its slightly clunkier predecessors, then smaller descendants followed—a 45-byter, a 51, eventually even a 22—entering a taxonomy that would grow to accommodate hundreds of subspecies as Ray played with Tierra in the months and years to follow. The swift and drastic size reductions of those first runs startled Ray, but even more remarkable were the survival strategies these variants encoded. The 45- and 51-byte creatures, it turned out, were not worms but bona fide parasitic viruses, achieving their leanness by borrowing reproductive code from larger programs when they needed to copy themselves. In turn, host programs acquired an immunity from parasites by failing to register their location in the virtual computer's memory, thus foiling the parasites' attempts to find them.

To the casual student of computer viruses, it's interesting to observe that despite the wide-open and neutral terrain into which the first Tierrans were placed, they swiftly and spontaneously adopted the same techniques built into wild viruses to ensure survival in an environment thick with hostile users and their software: parasitism and stealth. But to the serious scholars of biology who soon began to take note of Ray's work, such developments were more than just interesting. Out of the barest simulation of environmental forces, some of life's more sophisticated interrelationships were emerging entirely unbidden, and while the Mark Ludwigs of the world might object that Ray's initial fine-tuning of Tierran "physics" tainted the experiment, Ray was more than satisfied with its scientific implications. Here, in the unexpectedly colorful diversity bred from a single simple program, was a compelling model of evolution's creative power.

"In my wildest dreams, that was what I wanted," Ray later told author Steven Levy. "I didn't write the Ancestor with the idea that it was going to produce all this."

As much as this bustling ecology-in-a-box thrilled and surprised Ray, however, it soon began to dawn on him that the Ancestor had produced something even more unexpected: high-quality software. Almost all of the Ancestor's progeny displayed some improvement in the efficiency of their code, but in a few cases, evolution seemed to have attained a level of tight-wound optimization difficult for even the most wizardly of human software engineers to achieve, and Ray couldn't help wondering if there was a way to yoke this inhuman skill to the development of practical applications.

It wasn't an unheard-of notion. As long ago as the early '60s, for instance, cutting-edge programmers had begun experimenting with what they called "genetic algorithms"—pools of software subroutines repeatedly multiplied, mutated, and weeded according to how well they performed a given task.

Two decades later, in the same ground-breaking work that established the ability of digital viruses to penetrate nearly any system defenses, computer scientist Fred Cohen also proved that viruses are potentially useful as all-purpose computing devices. As Cohen later put it, "anything a Turing machine can compute, a virus can evolve." Since then, Cohen has tested the proposition that viruses can create useful code in a number of applications. One notable experiment of his is a network-maintenance ecosystem in which survival of the most needed cleanup tasks ensures maximum efficiency—in which, for instance, self-replicating programs designed to delete unwanted files randomly mutate their file-chasing strategies, with those strategies least wasteful of system resources being spared the Reaper's blade.

But the benefits realized in these experiments were limited, as Ray saw it, by their dependence on artificial rather than natural selection—that is, the software was allowed to evolve only in the direction of a particular function chosen by the programmer. In Tierra, on the other hand, organisms evolved according to criteria that they themselves created collectively, constrained only by the "natural" imperative to reward the thriftiest use of existing resources. Tierra gave evolution a free hand, in other words, and Ray felt certain that the creativity thus unleashed had the potential to tackle software-writing challenges far beyond the reach of human programmers. In particular, the difficulties involved in writing the most productive code for the parallel-processing machines that will take us into of the next century of computing seem to cry out for an evolutionary approach. "We will probably never be able to write such software, as it is way too complex," Ray observes. "Yet we know that evolution can handle that kind of problem."

The reason we know that, of course, is that we—and all other multicellular organisms—are wetware embodiments of frightfully complex parallel processes. But that fact posed a new challenge for Ray. Despite the great variety of digital forms Tierra had generated, it remained an ecology of one-celled organisms, none much larger or much more complicated than the 80-byte Ancestor. In fairness it should be pointed out that the terrestrial biosphere spent its first 3 billion years or so in a similar state before finally exploding into multicellular diversity at the dawn of the Cambrian era (a mere 600 million years ago). Yet if Tierra was ever to prove its full value as a software-writing machine—or indeed as a scientific model of evolution—sooner or later it would have to cough up a Cambrian explosion of its own. And since the key to this burst of complexity seemed to Ray to lie in challenging his evolving creatures with more intricate problems than the simple bit-copying tasks they'd grappled with thus far, he decided that the explosion wouldn't happen nearly soon enough if Tierra remained stuck inside conventional computers, and he began looking into the possibility of installing Tierra on a parallel-processing system.

But then one day in early 1994, Ray had a minor epiphany: "I realized that the global network is just a loosely connected parallel computer, and much larger and more powerful than anything that will ever exist as a single machine."

And thus was born Ray's plan to colonize the Net. He wrote it up soon thereafter in a document plain-spokenly entitled "A Proposal To Create a Network-Wide Biodiversity Reserve for Digital Organisms" (see *Wired* 2.08, page 33), the text of which outlines a vast collective enterprise devoted to hastening the arrival of the digital Cambrian. Ray envisions a Tierran subnetwork spread across thousands of volunteer Net nodes, each of them running the environment as a low-priority background process sustained only by unused (and otherwise wasted) CPU cycles. He is confident that once his "one-celled" simple self-replicating organisms encounter the immensity, the topological intricacy, and the fluid instability of the Net, they will quickly rise to the occasion and evolve into tightly coordinated multicellular conglomerates, thus setting off the dreamed-of Big Bang of complex digi-biotic diversity.

Ray foresees digital naturalists like "modern day tropical biologists exploring our organic jungles. However, occasionally these digital biologists will spot an interesting information process for which they see an application. At this point, some individuals will be captured and brought into laboratories for closer study, and farms for breeding." Harvested, domesticated and then neutered of their self-replicating properties, these prize specimens of code could then be translated from Tierran language into standard programming languages and set to work at any number of tasks. Ray suspects some form of intelligent network agents would be the likeliest first applications to be culled, but he prefers to emphasize that the most useful products of the digital jungle would be as difficult to predict as rice, pigs, penicillin, and silkworms might have been for an observer of the pre-Cambrian ooze of early carbon-based life.

There's a whiff of science fiction rising from all this, of course, but Ray is hardly indulging in idle speculation. Already a team of computer scientists has gathered under his supervision to work full-time on hammering out the technical details of the plan. He's accustomed by now to dealing with his listeners' occasional anxieties about the prospect of Tierran viral-like pests infiltrating the workaday network environment. "I explain why the things can't escape," he says, "and that quiets the nervous people, but some of them continue to look nervous."

But when the time comes to put their systems where their mouths are, how many site administrators will do so? Not enough, fears Danny Hillis, founder and chief scientist of Thinking Machines Corporation, the former manufacturer of massively parallel computers that had been supporting Ray's work. For all the tricky engineering involved in running Tierra on a Netwide scale, Hillis believes, the greatest challenge facing Ray "turns out to be more of a political issue than a technical issue. People are not necessarily going to want to give up their processing cycles for this"—even if those cycles will otherwise rot on the vine—simply because of a deep-seated reluctance to cede so much as a fragment of administrative control over system resources to a program whose internal processes serve no immediate ends but their own.

But even if computer users ultimately reject the deliberate presence of a global wilderness reserve for computer viruses woven neatly into the fabric of the Net, they may yet fail to keep the computer landscape from turning to jungle. After all, the same personal and subcultural imperatives that drove Hellraiser's career will continue to inspire underground virus writers. And the digital terrain continues to get more interesting. If the Darwinian innovations introduced by Mark Ludwig are any indication of coming trends in viral technique, then it's not inconceivable that a vital ecology might someday flourish in the midst of our daily routines, unplanned, uncontained, ill-comprehended, and irrepressible. It's an unnerving prospect. Yet it wouldn't have to be—not if we prepared for it by actively cultivating a digital biodiversity of the sort Tom Ray proposes. This is a niche that will be filled, whether we fill it deliberately or not.

"We're just going to have to live with them," artificial life researcher Chris Langton says of computer viruses. Our global web of digital systems, he predicts, is fast unfolding towards a degree of complexity rich enough to support a staggering diversity of autonomously evolving programs.

Viruses in a Suit and Tie

But the future of beneficial viruses is not only in the hands of eccentrics such as Hellraiser, Ludwig, or Ray. The good folks at General Magic corporation are eager to put viral code on a firmer and decidedly more lucrative footing. Not that they like to hear it said that they have anything to do with viruses, mind you.

General Magic manufactures a hand-held communication device that relies on a nifty new network-streamlining program language called Telescript. Announced earlier this year with the very visible backing of such info-dollar heavyweights as AT&T, Apple, Sony, and Matsushita, Telescript proposes to do good things. Its intelligent agents, General Magic co-founder Bill Atkinson promises, will soon be flitting about cyberspace on your behalf, visiting remote commercial sites to buy, sell, and trade information for you, and generally behaving themselves with all the decorum you'd expect from a personal digital valet.

Still, despite rather severe restrictions on the agents' ability to replicate, it's hard to deny certain broad similarities between intelligent agents and the offerings of your typical Vx board. Both wild viruses and Telescript agents routinely copy themselves from one computer to another. Both viruses and Telescript agents can run themselves on the computers they travel to, and, for those same reasons, raise differing degrees of concern about their security. "A virus never does anything good for you, it only does things to you," says hacker legend Bill Atkinson, nervously reaching for a fine semantic distinction between computer wildlife and Telescript's semi-autonomous "intelligent agent" programs.

More intriguing, though, are Telescript's close similarities with Tom Ray's digital diversity reserve and the experiments of Fred Cohen. Cohen, now happily self-exiled from academia and in business for himself as a computer-security guru, is experimenting with a distributed database in which self-reproducing query agents scurry throughout a network, much like the Telescript scheme. And like the sprawling biosphere of global Tierra, Telescript's bustling marketplace depends on a broad base of local interpreter programs installed wherever its agents go to do their business. This has two significant implications. For one thing, the fact that the mobile organisms of both Telescript and Tierra interact only with their interpreters, incapable of functioning in their absence or of bypassing them to directly affect the host environment, obviates many of the security concerns surrounding their autonomy. (Telescript, additionally, makes use of a battery of cryptographically secured restrictions to ensure that its agents don't subvert control of the host machine, either by accident or by malicious design).

And for another thing, the fact that all the interpreters speak the same programming language regardless of the underlying operating system and hardware means that, as the base of interpreters approaches omnipresence on the world's computer networks, the Net approaches the condition of a single, vast, and unmappable supercomputer, with each wandering digital organism a process in one worldwide parallel computation.

Taken together, these two features represent something of a watershed in the history of computing. It has long been observed, rather wistfully, that in principle the world's computers sum up to one gigantic parallel processor, and that the crushing bulk of that metacomputer's CPU cycles goes to waste, unused. Only now, however, with the advent of protocols like Telescript and Tierra, do we have the means to deploy such processes that treat the Net as one machine, safely and sensibly. This, then, is the real significance of these endeavors.

The Dark Side of Benefits

Trying to imagine the marvels that pour forth once you've successfully tapped a computer as elaborate as the Net is as futile as trying to map the future of a society, or of a life—or of life itself.

Of course, trying to foresee the risks that could emerge from that same computer is an equally hopeless task. But as it happens, we are bound to face those risks whether or not we seek to harness the full power of the Net, since the teeming and inevitable population of uncaged digital organisms will in any case plow forward with its own relentless exploration of the Net's capabilities. All we would miss by failing to orchestrate a more manageable viral exploration of our own, therefore, would be the potential benefits—including quite possibly some antidotes to the worst depredations visited on us by the viruses of the wild. And including also, perhaps, something even more precious. For if there is any purpose legible at all in the millennia of human history, it is in the unflagging persistence with which we add to the complexity of the universe. So, if we were to shrink from the chance to actively participate in transforming the Net into the single most complex information entity since the emergence of the human brain, would we not then be shirking a duty of almost cosmic proportions?

It could happen. It's hard to say which is really the more characteristically human trait—our drive toward complexity or our sometimes irrational fear of it. In the matter of computer viruses, fear could well gain the upper hand. It has already shown itself, after all, in our human tendency to overly reduce the multifaceted motivations of the virus writer to a caricature of hooliganism. Likewise it seems to lurk behind the urge to deny that viruses can be anything but lethally dangerous.

But we'd better think long and hard before we let it stand between us and the epic opportunities that globally distributed viral programming presents us with. Because in the end, the meaning of our long-term coexistence with computer viruses may prove difficult to distinguish from the meaning of our own existence.

Bibliography

Cohen, Fred, *It's Alive! The New Breed of Living Computer Programs* (Wiley & Sons, 1994).
Ludwig, Mark, *The Little Black Book of Viruses* (Tucson: American Eagle Publications, 1991).
———. *Computer Viruses, Artificial Life, and Evolution* (Tucson: American Eagle Publications, 1993).
Tom Ray: The Tierra home page can be found at http://www.his.atr.jp/~ray/tierra/.

16

The Imaginary of the Artificial

Automata, Models, Machinics—On Promiscuous Modeling as Precondition for Poststructuralist Ontology

Anders Michelsen

1. "The Image of Man:" Approaching the Imaginary of the Artificial

In the book *L'image de l'homme*, Philippe Breton proposes that an unacknowledged creative imagination circumscribes the historical invention of the computer.[1] Even when "progressional" histories of the computer acknowledge the imagination, the real importance of the creative imagination—of creativity *eo ipso*—is most often glossed over by the need to grasp the momentous "determinable" impetus of the inaugural technology: logic, computation, programming, signal processing, components, engineering, etc. Also, constructivist accounts of science and technology leave untouched the issue of invention in the sense of constitutive creativity, of an ontological "ordre de contenu."[2]

Breton re-situates the question of creation primarily through an inquiry into the "foundational narrative" that "grounds" this artifact in an ontological parallel to man. The computer is *created* in "the image of man," he argues, indirectly in the early "parallelism"of first order cybernetics, and explicitly in the agenda of AI-programs' "android epistemology."[3] More importantly, however, Breton somewhat unintentionally introduces the much more radical problem of constitutive creativity. He does so in relation to a particular *misconception* that complicates our ontological understanding of this machine, and, more radically, in the notion of the machinic *eo ipso* and *in extenso* in the postwar era. The "imagining" of the machine is constructed on a *paradox*:

> There is not anywhere in the world a form of intelligence which can not be considered human and no contemporary computer program can pretend to be assimilated to the human brain's functionality [functionnement]. This leads to a paradoxical situation: for each time *artificial intelligence obtains results it ceases to be of concern to this field, to the extent that it achieves a significance in another sense…* [italics mine][4]

I believe that this paradox holds a significant if not crucial position in the history of computing, during the early years and today, to the extent that it problematizes many inherited notions, from AI-programs to basics such as machine, interface, peripherals, etc. To put it differently: from the 1930s to the 1950s (from Alan Turing's universal machine to the milieu of the Macy-conferences co-defining the development of computer science, technology and applications), the early group of mathematicians, engineers, scientist, psychologist, etc., throwing themselves into the new issues of the computer did not necessarily celebrate a manifest form of the machinic becoming gradually more transparent, they struggled with something *more*, or to be precise something *different*. The "image of man" would lead—and still leads—elsewhere.

For one thing, the genealogy of the computer in the postwar era is without precedent because it establishes the artificial in a new manner, as related to the piecemeal and still more comprehensive issuing of a "cosmology of information,"[5] ranging from cybernetics to contemporary complexity science: a new and different view of the world in ontological as well as epistemological terms. Secondly, the computer becomes "genealogically" distributed in a variety of creative forms not necessarily restricted to—or focused on—computer science, technology and applications.[6] The issues of machine, information, code, communication, etc., surge forth within other "ontological regions," Marshall McLuhan's Canadian medium theory still being one of the most influential, and in their changed capacities, they underwrite an important impetus in postwar history still vaguely understood. The early struggle was a *critical struggle* over how to conceive of an artificial form of Being, over how to conceive the image of an object "X" in the double sense indicated by Breton: as a process of creative positing according to a certain image apprehended from a tradition (e.g., "man" in a flat and unmitigated sense), yet also as something different *ex nihilo*, as a "force without precedent,"[7] which immediately returns to question the very creation that posited it.

From the vantage point of man—the android "definition"—the issue of "imagery" or "imagining" may lead to a creative *imaginary* that is substantially distanced from inherited biases of creation.[8] From the vantage point of the mechanical, it may lead to an approach to the artificial that is *de facto* displaced from inherited definitions of the artifact as an "object" or a "system of objects," more or less determinable and distinguishable. The net result is a different relationship between the imaginary and the artificial, which I call the *imaginary of the artificial*, an inexplicit and poorly understood impetus for the creative articulation of the artificial:

> (a) While the android definition has a long history preceding the computer, within "that intermediate zone, that shadow realm"[9] which unfolds between the dream of humanizing the inhuman and of making man into a transparent entity through the production of a man-machine, it gathers a particular force at the time of the computer as a *qualitative* repercussion of a *gradual "sedimented"* creation of made artifacts with a "well defined" regime (e.g., the automobile transport system).
>
> In this sense of a "cumulative" history, the *imaginary of the artificial* is not the making transparent of a "shadow realm," but rather the ongoing creation of artificial forms, which take on a specific *raison d'être* as an "artificial environment"[10] towards the end of the 20th century.
>
> (b) Against this background, the notion of the computer relates what we, for lack of better expression may term a *quantitative* consequence of a *qualitative* turn in the specificity and variety of the created—a *critical radicality* of the complexification of made artifacts, of the artificial *eo ipso*. Or, in more cautious terms: the crisis of the artificial will not be solved within the inherited sense of a more or less simple "invention" of "mechanism." The critical radicality of the qualitative turn questions the very notion of invention and creation, and lends a new misunderstanding—a new heteronomy—to the term genesis.
>
> In this territory, this "other-spatial" dispersion—the *imaginary of the artificial*—indicates an explicitly novel "shadow realm." To put it differently: we may change the terms of early

> debates on the computer by taking them as our point of departure for a specific study of the artificial as something not *primarily* divided by the inherited dichotomies of man-machine, mechanism-organism, etc., but rather *imagined*, thus created, from the standpoint of an organizational *novum* with an *ontological contingency beyond inherited determinations and constraints*.

The imaginary of the artificial may thus be understood precariously as a new form of heteronomy. It is not at all clear what the "determinability" of this "being" is and the determinability appointed to it may be mistaken, one consequence of which being the confusion of the artificial and the imaginary, what I discuss below as neo-cybernetic "auto-imagination."

What I am talking about is this: the early debates on the computer exhibit a new attitude in relation to the *possible* and the *feasible* that increasingly comes to settle within unclear but unquestionably explicit schemes of the artificial, such as "cyberspace," "the network society," "the cyborg" in current everyday language and culture, or somewhat differently in still more comprehensive projects for e.g., nanotechnology, "weather modification" and biotechnology within science and technology. Increasingly, the artificial—the machine—the "machinic"—in a broad, yet specific sense, the lines of force instrumented by artifacts and by artificial instrumentation—becomes a reference for creative conceptions, for creativity *eo ipso*, that is, for ontological constitution. This is true whether one views the tragic schemes for the commercial marketing of *artificially* cloned children, or the tragicomic ideas to move the Earth by *artificial* means in order to solve the Greenhouse-effect.

This leads to the thesis of the present paper: the imaginary of the artificial attains its radicality because it is predicated on a schism between formal machinic organization (in a post-objective artificial sense) and creative constitution (in a new imaginary sense). In what follows, I focus on three aspects of the imaginary of the artificial: (1) Machinic self-production: promiscuous modeling. I first discuss W. Ross Ashby's focus on spontaneous self-organization and the ambiguity within John von Neumann's work on models for self-reproduction of automata, pointing to extended issues of the artificial. (2) Modeling and artifact: simulative reality? By the 1990s, the early ideas of modeling had fed into the impressive momentum of complexity—self-organization, connectionism, networking. However, the imaginary of the artificial also draws from the early ambiguity of "the model." (3) The third order and neo-cybernetics: auto-imagination? From the '70s and '80s onwards this is manifested in a "neo-cybernetic" impetus inherent in poststructuralist ontology, stipulating a new form of machinically biased "auto-imagination," e.g., in poststructuralist manifestos for the assumed "network-age" such as Félix Guattari's *Chaosmosis* (1992).

2. Machinic Self-Production: Promiscuous Modeling

The complexity of the computer is obvious to anybody who can (or cannot) use an ordinary word processing program. The functionality of such a program seems an endless maze of machinic options, which in strange ways bring something "alive" (often because the support of the GUI-design is highly questionable). However, such trivial problems are suggestive of different, more generic and complex issues.

In a note from 1946 (published in 1947), W. Ross Ashby, who would later become one of the first proponents of "second order cybernetics" and "complexity," discusses the possibility of a "self-organizing dynamic system."[11] He argues that the widespread denial "that a machine" can be "self-organizing," that a machine can be determinate and yet able to undergo spontaneous change of internal organization, can be critically countered by looking at the human nervous system.[12] This system is both a strictly determinate physico-chemical system and able to undergo self-induced "internal reorganizations resulting in changes in behavior."[13] Thus, a machine can "be at the same time... strictly determinate in its action, and (b) yet demonstrate a self-induced change in organization."[14] Moreover, this proposed machine may be defined with the particularity of "some real, material dynamic system which we can examine objectively" and yet also be specified mathematically.[15] Ashby goes on to demonstrate this by describing the behavior of what he terms an "absolute system" where "substitutions converting one configuration to the next must form a finite continuous group. . . . "[16] Such a system, such a machine, can undergo a "spontaneous change or organization" when one of its variables, "by its physical nature perhaps," Ashby suggests, is "restricted to taking one of two values,"[17] thus resulting in different "fields" of organization corresponding to a spontaneous change of "certain configurations."[18] In other words, a system with absolute characteristics may yet change spontaneously: a machine may "produce" spontaneity under well-described conditions.

Ashby's early proposition is of interest because it is one of the first to depict the artificial as capable of self-organizational forms. Moreover, this spontaneity has nothing—or at least little—to do with the shadow realm of the human. It is defined as a set of *relata* between abstract description and material device: it has the defining characteristic of a model, as models are understood from the invention of the computer onwards. It introduces the idea of a *model for artificial self-production*. Thus Ashby is not only questioning, the inherited division of the machinic and the organic (that is, producing what Norbert Wiener called "badly posed questions"[19] opposing vitalism and mechanism). He is also *de facto* conjecturing a certain division in the view of how the world is set by indicating a new machinic order beyond the "badly" posed questions. Spontaneity is to be understood radically as the *being of a Being* that is not—not yet, perhaps not at all—determinable, or in any case not easily determinable, e.g., within the inherited nature-culture dichotomy.

The automaton seems prone to the kind of spontaneous "creativity" suggested by John von Neumann from 1945 onwards as a wholly new type of generic artificial form with a logic and semi-material capacity of self-production.[20] Even if von Neumann starts from the parallelism of early cybernetics, and even if he never completes his theory and takes great care not to overstep what he sees as the line between scientist and demiurge, he appears quite aware of the prospects of such a singularly new "body of experience" (as a form of mathematics) in his last attempt at "automata theory," the posthumously published "Silliman Lectures," *The Computer and the Brain*.[21]

Von Neumann first reveals his ideas in his well-known study of the electronic calculator, the ENIAC. The famous "First Draft of a Report on the EDVAC" outlines the architecture of a serial and stored-program computer—the "von Neumann architecture"—through an analogy to human neurobiology.[22] Inspired by Walter Pitts and Warren McCulloch's idea of the possibility of a "logical calculus" of ideas "immanent in nervous activity," Von Neumann outlines the architecture of the EDVAC via organs in human neurobiology, such as "memory" (much to the chagrin of the machine's engineering fathers J. Presper Eckert and John Mauchly).[23] However, "The Draft" also reveals that from the very beginning, von Neumann's conjecture is peculiarly ambiguous. On the one hand, he does not hesitate to apply aspects of Pitts and McCulloch's neurobiology to stricter terms of logical definition; on the other hand, he engages something that achieves "significance in another sense" (Breton).

While he notes that the significant parallel between neurobiology and computing is at a specific, yet quite abstract level (primarily in the "all-or-none character"—the digital "nature"—of both neuron and digital computing components), he nevertheless writes, "Following W. Pitts and W.S. MacCulloch ('A logical calculus of the ideas immanent in nervous activity') we ignore the more complicated aspects of neuron functioning: Thresholds, temporal summation, relative inhibition, changes of the threshold by after effects of stimulation beyond the synaptic delay, etc."[24] Von Neumann is thus only interested in the neurobiological analogy to the extent that it devises aspects of the artificial, which linger ambiguously *in between* something more or less well defined (e.g., in relation to the "all-to-none character") and something more precarious. After all, the parallel between neuron and vacuum tube in the ENIAC is not—and cannot be—conceived clearly "outside" the realm of the "digitalism" of von Neumann's model proposed in "The Draft" (as indeed the later history of component-miniaturization from transistors to microprocessors will show). More importantly, how this model will appear as a "real" embodied apparatus with an assumed parallelism to a neurobiological "object" (e.g., as instrumented with "organs," "neurons," etc.) outside this digitalism, is not—and cannot—be clear.

But von Neumann does not hesitate. He expands on this problematic through the idea of a "general and logical theory of automata," first presented at the 1948 Hixon Symposium.[25] He first compares computers and biological information processing systems[26] in more general terms, then suggests a "Broadening of the Program to Deal with Automata That Produce Automata" in the concluding sections.[27] He asks, "Can one build an aggregate out of such elements in such a manner that if it is put into a reservoir, in which there float all these elements in large numbers, it will then begin to construct other aggregates, each of which will then at the end turn out to be another automaton exactly like the original one?"[28] In the 1949 draft of a "theory of and organization of complicated automata,"[29] he pursues these questions by aligning the automaton with complication and complexity and with debates over automata reproduction. More specifically, these debates examine automata reproduction as the emergence of a threshold, "inconceivable" from the initial state, that issues a higher degree of complexity, in abstract as well as material form:

> There is thus this completely decisive property of complexity, that there exists a critical size below which the process of synthesis is degenerative, but above which the phenomenon of synthesis, if properly arranged, can become explosive, in other words, where syntheses of automata can proceed in such a manner that each automaton will produce other automata which are more complex and of higher potentialities than itself.[30]

Arthur Burks, who continues aspects of von Neumann's work, especially on cellular automata, reports that von Neumann in the summer of 1948, before the Hixon Symposium, considered a range of components or parts in self-reproducing automata. Indeed, he contemplated "eight kind of parts:"[31] a "stimulus organ," a "coincidence organ," an "inhibitory organ," a "stimuli producer," a "rigid member," a "fusing organ, " a "cutting organ," and a "muscle." If von Neumann's sketch is to be taken seriously, the actual status of these parts is unclear. For instance, it is not clear what status a "muscle" of the kind devised will have in itself (will it be organic, will it be mechanical, will it be . . . ?), and vis-à-vis an "inhibitory organ" (will it be organic, will it be mechanical, will it be . . . ?), even if Burks and von Neumann define parts such as the inhibitory organ as logical and the muscle as material.

Arthur Burks sees this 3-dimensional kinematic model as one of von Neumann's most detailed (together with the 2-dimensional cellular model which he goes on to develop), but it actually points towards a new territory of abstract/concrete modeling.[32] Burks specifies and generalizes von Neumann's conjectures in five models of "self-reproducing automata": the "robot model," the "cellular model," the "excitation-threshold-fatigue model," the "continuous model," and the "probabilistic model of self-reproduction and evolution."[33] Even if these models are carefully defined through known entities and definitions, it is quite clear that these models offer a step-by-step entry into a *new creative territory*—a new "body of experience" as von Neumann puts it. Moreover, they point towards a new range of embodied forms, towards the modeling of a new class of complex phenomena: that is, as a possible *new "species" or better, type, or, token of Being.*

Overall, these manuscripts present a comprehensive and *promiscuous* (in a non-moralistic sense) vision of how the organic-machinic (analogical) parallelism of early cybernetics can be developed in radical ways. By mixing ideas of new artifacts with notions of digital computation, the "digital procedure"[34] outlines a potential—virtual we would say today—for the self-production of artifacts within a "coherent body of concepts and principles concerning the structure and organization of both natural and artificial systems, the role of language and information in such systems, and the programming and control of them."[35] Burks summarizes this in his introduction to von Neumann's papers on computing and computer theory:

> Von Neumann thought that science and technology would shift from a past emphasis on the subjects of motion, force, energy, and power, to a future emphasis on the subjects of communication, organization, programming and control. He began a theory of automata that would contain the general principles common to artificial automata (computers, robots, complex automated systems) and natural goal-directed systems (cells, organisms, evolution).[36]

It is clear that this conjecture of self-reproduction is also one of self-production, that it leads to the "actualization" of, or "interaction" with, something properly "virtual." Although von Neumann in his drafts often situates this as primarily a problem of a "general mathematical theory"[37] that may "alter the way in which we look on mathematics and logic proper"[38] and explicitly attaches less importance to the material, he cannot help but emphasize the artificial as ambiguous and ontologically contingent.

As is clear from Ashby's note and von Neumanns' work, as well as from conjectures made by Norbert Wiener and others within the founding milieu of computer science and technology, the idea of formal machinic self-organization is not simply, and perhaps not even primarily, an abstract form (i.e., a mathematical artifact). It is also an outline of a state of self-production, circumscribing notions of structure, organization, organism, machine, program, control, communication, redundancy, etc. "Automaton" thus should not be taken "literally" in any one sense of "artificiality," since it may lead anywhere, or in fact not at all, since it has little to do with the inherited idea of automaton. However, it is not really a well-defined ideal or mathematical issue either. To be more precise, the relation between mathematics and materiality is predicated on a whole range of notions and applications. This relation stimulates and articulates the conjecture of an *ontological contingency beyond inherited determinations and constraints*, leading to a schism between formal machinic organization (in a post-objective artificial sense) and creative constitution (in a new imaginary sense). Thus, it is never clear exactly what/who will result from production of this artificiality and what/who is to be seen as constitutive.

3. Modeling and Artifact: Simulative Reality?

In the contemporary science of complexity, the emergence of spontaneous behavior within a system is also the production of something new, un-expected, or "surprising" (John L. Casti). This emerging behavior may be modeled in computer-created "would-be worlds,"[39] systems

> that are completely inexplicable by any conventional analysis of the systems' constituent parts. These phenomena, commonly referred to as emergent behavior, seem to occur in many complex systems involving living organisms, such as a stock market or the human brain... Complex systems are not new, but for the first time in history tools are available to study such systems in a controlled, repeatable, scientific fashion... with today's computers, complete silicon surrogates of these systems can be built, and these "would-be worlds" can be manipulated in ways that would be unthinkable for their real-world counterparts.[40]

While modelers like Casti observe a Kantian distinction between "an sich" and "für sich" (Casti discusses a principal circular diagrammatic of "encoding" and "decoding" of complexity),[41] the modeling of spontaneity does not *in principle* leave out the option of something being "in excess" of the more or less strict formal modeling dynamics of the "science of surprise," e.g., emergence, catastrophe, chaos, connectionism, etc.

We may thus argue that complexity-modeling *continues* the options and the implications of von Neumann's drafts in the 1940s and 1950s. In his book on the origins of cognitive science, Jean-Pierre Dupuy considers the importance of modeling to early cognitive science and artificial intelligence, but also to the question of what sort of artifact the computer could be seen to be.[42] The importance of "modeling" lies in the sense of "*verum et factum convertuntur,*" meaning that humans can "have rational knowledge only about that of which we are the cause, about what we have ourselves produced."[43] The production of a model is at the same time a product of and a transcendence of human finitude, because it produces something:

> A model is an abstract form . . . that is embodied or instantiated by phenomena. Very different domains of phenomenal reality . . . can be represented by identical models, which establish an equivalence relation among them. A model is the corresponding equivalence class. It therefore enjoys a transcendent position, not unlike that of a platonic Idea of which reality is only a pale imitation. But the scientific model is man-made. It is at this juncture that the hierarchical relation between the imitator and the imitated comes to be inverted. Although the scientific model is a human imitation of nature, the scientist is inclined to regard it as a "model," in the ordinary sense, of nature. Thus nature is taken to imitate the very model by which man tries to imitate it.[44]

The model abstracts from phenomenal reality "the system of functional relations" putting aside everything else.[45] Models obtain a life of their own, "an autonomous dynamic independent of phenomenal reality."[46]

With the invention of the computer from the 1930s onwards, the principle of *verum et factum* gains a particular emphasis. Alan Turing and Alonso Church's alignment of computation and mechanics issue a new significance to the machinic conjectured as "effective computability:"

> It seems plain to us now that the notion of effective computability that was being sought [in the 1930s], involving only blind, "automatic" execution procedures, was most clearly illustrated by the functioning of a *machine*. It is due to Turing that this mechanical metaphor was taken seriously. In his remarkable study of the development of the theory of automata, Jean Mosconi makes an interesting conjecture about the nature of the resistance that this idea met with in the 1930s: "Considering the calculating machines that existed at the time—the perspectives opened up by Babbage having been forgotten in the meantime—any reference to the computational possibilities of a machine was apt to be regarded as arbitrarily narrowing the idea of computability . . . If for us the natural meaning of "mechanical computability" is "computability by a machine," it seems likely that until Turing came along "mechanical" was used in a rather metaphorical sense and meant nothing more than "servile" (indeed the term "mechanical" is still used in this sense today to describe the execution of an algorithm).[47]

Thus Turing and Church *not only* expand on the notion of computability, they also expand *on the notion of the machine*. A machine will henceforth have all the options of computation at its disposal in more than one sense: one, the demonstration of logic, modeling in stricter mathematical terms, another, the actual mechanics of the computing machine, i.e. the computer as working artifact. The ambiguity of the model, of modeling *in extenso*, thus derives directly from the issues of constructing a real computer, hardware and software-wise, if one likes, but it also derives from the issue of application, that is *how* and to *what ends* such a working artifact may be set: the machine may, as a concrete instance of the artificial, turn back on the real as "an autonomous dynamic independent of phenomenal reality." It may create something, not in its capacity for incorporated mathematics or in its capacity of calculating mechanics, but in its capacity as a *mathematical mechanism*, i.e. in the sense of effective computation, thus *foregrounding* "*effect*." The machine might just do more than allowed for within the inherited servitude. The machine could, as in Ashby's note, be seen to establish a spontaneity, which means that this application had the capacity for spontaneous organization, self-organization, as applied abstract/concrete modeling.

The notion of the model thus gains a much wider application: the computer could in ontological terms be seen to be an artifact with a disposition towards creativity, as indeed the artificial intelligence programs would be portrayed in the 1960s and later—"Machines will be capable, within twenty years, of any work that a man can do," Herbert Simon states in 1965.[48] Most contentiously, the computer was a model brought alive outside the laboratory, not as some sorcerers apprentice gone wild, but as the laboratory set free, the equipment come alive, wandering about in the world. Dupuy points to two important implications of this problematic:(a) the principle of *verum et factum* comes to embody a constructive condition for simulative reality. Since experiences with computers support the notion that we can only know what we can construct, *factum as* manifested model becomes the condition of truthful knowing. The principle is thus turned upside down and the fact that science is "making up" something is qualified as a new form of the real. (b) perhaps more importantly, this leads to a broader philosophical and historical evaluation of the idea of complexity (and cybernetics as a founding impetus). Dupuy makes clear that notions of machine, simulation, modeling, etc., can be affirmed on a new level, one that considers the proposition that mechanics as complexity make up a higher principle of "mechanization" of the real, that is the disclosure of a third "type of order"—"non reductionist without having to accept holism."[49] Following Friedrich von Hayek, Dupuy argues that the idea of complexity points to an order that, albeit humanly created, is too complex to be humanly governed, "human beings bring society into existence through their actions," but the ensuing order, is "beyond their control, because it is (infinitely) more complex than they are:"[50]

> spontaneous social order constitutes a third type of order, along with natural order and artificial order. It signifies an emergence, an effect of composition, a system-effect. The "system" is obviously not a subject, endowed with consciousness and will. The knowledge that the system exploits is irreducibly distributed over the set of its constituent elements: it cannot be synthesized in one place, for the system has no "absolute knowledge" about itself that is localized somewhere within it. This collective knowledge resides in the social order of the system insofar as it is the "result of human action but not of human design."[51]

4. The Third Order and Neo-Cybernetics: Auto-Imagination?

For Dupuy, there is no question that the cybernetic "heritage" pervades and informs many contemporary conjectures that bring new meaning to the machinic. Even if this heritage has been unacknowledged, Dupuy discusses a series of cases and arguments from French structuralism, notably from Levi-Strauss, Lacan, and Derrida, which he sees as mistaken elaborations of cybernetics—as mistakenly focused on a "symbolic" level," "structured like a language."[52] For example, Lacan stipulates that "the symbolic world is the world of the machine," when in fact the issues of the machine placed within language by the structuralists pertain to a different, may we say, "machinic," third type of order.

Klaus Bartels presents a similar critique, which furthers Dupuy's insights by broadening the perspective of how the machinic and the informational are put to work in postwar French thought.[53] According to Bartels, cybernetics has the status of "basso continuo" for structuralism together with McLuhan's media theory (itself drawing heavily on information science and cybernetics) up to the mid-1970s and the early 1980s, constituting what Bartels calls a "philosophy of information" culminating in the "crypto theory" of Jean Baudrillard.[54] Baudrillard's writing, especially towards the late 1970s (e.g., "The Precession of Simulacra"), is certainly under the influence of cybernetics.[55] The peculiar order or "precession" of "simulacra" culminates in "the generation by models of a real without origin or reality: a hyperreal."[56] This idea thus not only reflects on the prospects of modeling, but also substitutes the real with modeled fact: the "image" procured by models will become a sort of self organized constructive problematic which "bears no relation to any reality whatever: it is its own pure simulacrum."[57] The factum of computer-generated effects lends credibility to a new metaphysic of the "hyperlevel," conceived by Baudrillard as a "magnetic field of events"[58] with recourse only to simulation: "Simulation is characterized by *a precession of the model*, of all models around the merest fact—the models comes first, and their orbital... circulation constitutes the genuine magnetic field of events."[59] However, Bartels views Baudrillard as only one instance, testifying to the ongoing influence of cybernetics as an organizing "metaphor" to poststructuralism. Bartels suggests a "history of the cybernetic-metaphor,"[60] which runs from a *first* phase characterized by the significance of Claude Levi Strauss and Jacques Lacan's interpretations of cybernetics in the 1950s,[61] to a *second* phase in the mid-60s characterized by Jacques Derrida's ideas of writing—Gramma—as an instantiation of a "universal writing machine" referring to Norbert Wiener's cybernetic machine,[62] to a *third* phase from the late 1960s to the mid 1970s characterized by the idea of a "poetic machine," in the work of Julia Kristeva and culminating in the poststructuralist idea par excellence of a "desiring-machine" put forward in Gilles Deleuze and Félix Guattari's 1972 *Anti-Oedipus*,[63] which emphatically proclaims the machinic to be "at work everywhere."[64]

Bartels argues that one must see "cybernetics as metaphor" as complementary to the advancement of media in the postwar era—the information society, which starts to be debated in France in the '70s; e.g., Simon Nora and Alain Minc's *The Computerization of Society* (1978).[65] Cybernetics thus becomes implicated not only in a certain history, but more importantly, it takes on a particular significance as a stimulating broad yet specific agenda of the machinic. Edgar Morin's account of the machine—"les être-machines"—in *La méthode. I. La Nature de la Nature* (1977)[66] where he embraces the machine as a "poietic"[67] instance, offers a sense of the intensity of this agenda (Morin makes explicit his inspiration from "second order cybernetics," e.g., Heinz von Foerster, Humberto Maturana, Francisco Varela, etc.):

> We have been captives of the idea of mechanical repetition, of the idea of standardized fabrication. The word machine must also "be given" the meaning we find in pre- or extra-industrial significations, where it designates the set or complex agencies [agencements] wherein the market is both regulated and regulating:... the political machine, the administrative... It is necessary above all to give it meaning in its poietic dimension as a term which in the machine connects creation and production, praxis and poetry.... In the machine is found not only the machinic [le machinal] (the repetitive) but also the fabricating [le machinant] (the inventive).[68]

Moreover, Morin talks about "the family of machines," the "arche machine: the sun," "proto machines and wild engines", the "living poly machines" with an autopoietic capacity, the "social mega machine," the "artificial machines,"[69] and not least, Wiener's cybernetic machine. The cybernetic automat transforms the mechanical machine's "externality" into an organized internality by means of its program whereby it stands forward as "comparable with the living... by means of its organization of behavior."[70]

In his well-known 1979 introduction to French philosophy,[71] Vincent Descombes dedicates a section to the issue of "communication" followed by a section on "structures"[72] where he attempts to demonstrate how code, message, sender, receiver are closely related to the philosophical prospects of semiology, and most of all, to the concept "structure," especially with respect to the critique of consciousness and subject formation—that is, to anti-humanism:

> The paradox of structuralism is as follows. It announces its project (to combat "the philosophy of consciousness") by showing that the signifier is not at the service of the subject, nor entrusted by the latter with his "significative intentions" (as phenomenologists say). It

> wishes to show man's subjection to signifying systems (which precede each of us individually). But this demonstration has recourse to concepts from information theory, i.e., from the thinking of engineers whose goal (so the word "cybernetics," as they have called their science, suggest) is to invest human beings with total control by means of better communications techniques.[73]

One may debate Descombes' somewhat naïve assumption of information science as "well-meaning" and allegedly humanist engineering, but one should not overlook (1) the substantial influence of cybernetics in relation to semiology and structuralism from the '50s to the '70s and (2) more importantly the notion of paradox, that is, the liaison between a general influence of cybernetics, and a remaining problem of a probable "agency" beyond cybernetics. The role of the engineers becomes crucial: whereas structuralism wants to apply communication as a model of the machines' effect on man, on the extra-machinic, the engineers come to stand for something extra-machinic with a constitutive value (making "communications techniques" function "better").

In the context of this paper, the correspondence of the "cybernetic metaphor" with the imagination (of a philosophical, engineering, etc., nature) related to artifacts (e.g., communications techniques) emphasizes Dupuy's argument that an ongoing cybernetic impulse leads to neo-cybernetics, as I will return to in a moment. But it also introduces another factor in the problematic: while underlining Dupuy's argument, the embrace of the machinic is under the influence of exactly this embracement as something imaginary. Thus the assumed impact of cybernetics challenges Dupuy in the following manner: How is the relationship between the artificial order and the third type of order to be understood, once the artificial is seen as something potentially creative? And moreover, if this potential is viewed as related to the imaginary, how will this then affect the idea of a third order? Or to put it differently: How is the third type of order, the social, to be seen when questioned from the hypothesis of the imaginary of the artificial? That is, when complexity is not only the property of the factual *as* model, but an expression of a reality permeated with these very same principles. Or: Is the abstraction of the third order, discernible in its effects, but not in its capacity of "effectuator" able to overcome the dichotomy between natural order and artificial order, set up by Dupuy? The factuality of the second—artificial—order points to the complexity of the third order, but only by recourse to the second order: the factual is in this sense producing the reality, or better, is implicating the reality of the third order in the workings of the second.

Insofar as the third type of order is the social and historical complement of complexity proper, the argument may soon be settled and the artificial viewed as merely an application, or an appendix, of this spontaneous order, as in contemporary arguments for the network society, socio-cybernetics, etc. Thus in *Chaos, Complexity and Sociology* (1997),[74] Frederick Turner argues that the use of dynamic non-linear models of complexity not only circumscribes "the experience of creating universes [which] however limited in scope, put at the disposal of even fairly ordinary thinkers the sort of imaginative extrapolation formerly available only to genius,"[75] and in this sense "diminish" the role of any extra-machinic imagination. But, moveover, this modeling imputes the prospect of a "complexification" of imaginary forms, of what he terms "an instrument of moral judgment of history" by way of the operation of strange attractors—derivations of cybernetics if one likes—on e.g., religious morality.[76]

However, if we acknowledge that the conception of the complex is inherently relayed by the "factum" *of* modeling, that is as an "arte-factum," the solution may become much less transparent. Why these artifacts, why this perspective, why at this time, we may ask: how, and why, is the "ordre de contenu" of complexity actually instantiated, not in the capacity of modeled organizational efficiency (e.g., vis-à-vis "data," e.g., "religious morality"), but in its capacity of constitution, of deliberate human design (at some level at least)? This, it appears to me, is exactly what Descombes hints at—perhaps too modestly—in his introduction of "real" engineers into the crystalline systems of structuralism. The real engineers stand for something else, which cannot be accounted for in the systemic inclusiveness of early structuralism, but more importantly, they point towards an issue of constitution in the sense of creative articulation.

Dupuy acknowledges this by touching on the problem of methodological individualism within von Hayek's approach; but in the light of the artificial, we might argue somewhat differently. The objectionable structuralists become an interesting case, because they point to another dimension. As Bartels' brief history of cybernetics as metaphor indicates, the genealogy leading to poststructuralist ontologies of the artificial in the 1990s may be neo-cybernetic, not just in a metaphorical sense, but in a painstaking strict sense, since neo-cybernetic means modeling in the sense of effective computability, of effective "incorporated" action. Perhaps the most far reaching conjecture is to be found in *A Thousand Plateaus* (1980) by Gilles Deleuze & Félix Guattari.[77] It is well known that Deleuze and Guattari's work is heavily influenced by ideas from mathematics and physical science, including complexity (in their ideas of multiple systemicity, one powerful "vehicle" of which is the "abstract machine"). This peculiar instance of the machinic continues and revises the idea of a "desiring machine" from the early 1970s, but more importantly, it is also a high point in the proliferation of cybernetics: what we somewhat cautiously could call the invigoration of a neo-cybernetic problematic. We may highlight at least two prospects: (1) First the abstract machine implies the Deleuzean-Guttarian idea of an ontology based on strata with immanent ratios and dynamics, clearly distanced from any human agency, and shares in this with the third order discussed earlier. Thus issues of imagination must be stipulated as effects of the abstract machine, of the machinic, that is the *neo-cybernetic auto-imagination* (interestingly, Deleuze and Guattari dedicate substantial space in the introduction to *A Thousand Plateaus* to dissociating themselves from the idea of a book as an *image*). (2) Second the abstract machine comes to play an important role in the later Guattari's *Chaosmosis* (1992)[78] which becomes one of the "manifestos" for the explicitly neo-cybernetic prospects associated with mass-computing and the Internet in the 1990s, as "cyberspace," "the network society," "the cyborg," and so forth. The notion of "machinic heterogenesis" from *Chaosmosis*[79] particularly draws on a number of the issues discussed here (including explicit references to Wiener). When one reads this book with Guattari's earlier work *Molecular Revolution* (1972–1984),[80] especially the section "Towards a New Vocabulary"[81] with themes such as "Machine and Structure," "The Plane of Consistency," "Subjectless Action," "Machinic Propositions," and "Concrete Machines," an emphatic flow of argument in favor of a new "machinics" with an ontological contingency beyond inherited constraints appear.

It is important to understand that neo-cybernetics does not abolish man in a "crude" way. It disperses an imaginary that forcefully stipulates different—new—ways of conceiving of the world. Machines may wander about among men, they disperse themselves among men, and in doing so they in turn change the entire setting of man, as the "humanist controversy" related to structuralism makes clear: they render man unnecessary, or better they "erase" man (Foucault)[82]—like a symbol at the mercy of a Turing-machines's tape head—"like a face drawn in sand at the edge of the sea" (Foucault).[83] Thus the machinic is not really concrete (a central point in Guattari's thought is to abstract values, concepts, sensations from the cybernetic heritage). They make up something post-objective, an all-encompassing artifice bound to certain artifacts, but, nevertheless bound on a new ontology.

We nevertheless need to be extremely careful: is this the appearance of the machinic *cum* third order conditioning auto-imagination, or is it something created by an imaginary positing of artificiality with an ontological contingency beyond inherited constraints. To revert to this paper's thesis: Does the ambiguity of modeling persist in the third order, *because its ultimate "recourse" lies somewhere else?* To put it differently: may the ambiguity within cybernetic modeling—"cleared" as the affirmation of *factum* as a new form of simulative reality leading to a third order by Dupuy—translate into an idea of auto-imagination and in turn disclose that this translation needs constituting—thus questioning the creativity of the third order explicitly, by way of the artificial, *not*—to paraphrase the Deleuze and Guattari of *Anti-Oedipus*—because the machinic is "at work everywhere," *but* because the imaginary of the artificial seems to be at work everywhere.

Now this imagery clearly departs from the mathematical modeling of systems, but it is no less a step into the uncertain. It aspires not "to certain points where cybernetics impinges on religion" (such as *God & Golem, Inc.*—the title of Wiener's book from 1964),[84] but to a creative act of translative imagination, an imaginary of the full-fledged convergence of technology and non-technology, in the neo-cybernetic sense, but without necessarily leading to neo-cybernetic conclusions. Within such schemata this is what *may* and *can* be thought and done: the machinic is seen as "auto-imagination," i.e., as an auto-creation set at large, or as a convergence between artificial imagination and the imaginary of the artificial. The lesson to be learned—and questioned—seems to be:

(a) Self-production. From the early postwar years onwards the issue of artificial self-production is not infeasible or preposterous. The artificial may be created without any—or at least very many—preconditions whatsoever; it may initiate a field of equally post-human and post-natural "originary" creations so that it comes to linger with ever-increasing intensity from the 1980s onwards, as a machinic "at work everywhere." For instance, the artificial lingers in the focus on "cyborgs, "monsters," etc.: a radical "reconfiguring" of and by technology, in which "...posthuman creatures equal to but different than humans..."[85] populate a world seen to be equal, but different: but in what sense different and with what specificities?

(b) Cross-breeding. This issue is developed by a continuous and extended "cross-breeding" of reflections on organic and mechanical phenomena inspired by the pioneers of the computer, establishing a new type of simulative relation between the abstract and the concrete. One example: in 1992, M.V. Gandhi and B.S. Thompson report that "smart materials and structures"[86] (resonating with von Neumann) will be largely defined through a "mimesis," "biomimetics": "exhibiting nervous systems, brains, and muscular capabilities" including organic features such as "Self-repair, self-diagnosis, self-multiplication, and self-degradation"[87] But how are these 'creatures' to be acknowledged except by the fact that they proliferate as an organizational *novum* with an *ontological contingency beyond inherited determinations and constraints*.

(c) Ambiguity. Perhaps, the state of nature—humans and non-humans alike—one of two 'essential' starting points for the early cyberneticians, may be the most interesting indicator of the ambiguity following five decades of promiscuous modeling. Gernot Böhme writes in the early 1990s that nature's self evidence [*Selbstverständlichkit*] is disappearing due to artificial reproduction in an expanded yet unclear sense. We still use "classical dichotomies [*Entgegensetzungen*]"[88] such as "nature and the established [*Setzung*]," "nature and technique," "the natural versus the artificial and contaminated," "the original versus the civilized," and the "outer and inner," Böhme argues, and yet it has become unclear what nature is, what we will designate as such "whether what we consider as nature, is nature at all, what nature we desire":[89]

"In the dimensions of a Terrestial history, a colonization [*Besiedelung*] of the World's space is possible without further notice, i.e. the idea of a separation of the human species in artificially adapted life conditions for subspecies or even a dissolution of man as species. It is possible to conceive of living beings that only reproduce themselves in a continuous symbiosis with machines. Within such perspectives the expression "artificial nature" in fact comes to designate an intermediate phenomenon, a boundary [*Grenze*], and perhaps also the point of evolutionary decision."[90]

I believe that these points circumscribe what I have proposed as a schism between formal machinic organization and creative constitution: It is not clear, or better, it is not necessarily a given premise that the self-production resulting from the heritage of the automaton in cybernetics can overcome this schism. But, neither is it very clear that the notion of the human can easily be 'transcended' as creative instance. In fact, the schism seems to appear in a number of modalities: (a) as the self-production of something with unclear effects; (b) in the complication or complexification by cross-breeding in various ways; (c) in the impossibility to decide what comes from the artificial and what comes from the natural, in turn leading to a need for rephrasing the relation between human and "arti-factual" creativity.

To conclude: I hope to have shown that across five decades of promiscuous modeling, the computer may be seen as an inexplicit and poorly understood impetus for the creative articulation of the artificial—from von Neumann's automaton to the neo-cybernetics of poststructuralist ontologies, with a number of important implications for our understanding of the computer, and for our "handling" of the artificial. Our brief excursion into details of a possible "origin" (alas) of poststructuralist assumptions, seems to indicate a highly difficult schism between a proliferating number of machinic organizations and their creative constitution(s), which in their ultimate definition become attached, or better, coupled with the ontological form of the "human strata of the real" (Cornelius Castoriadis) in order for this creativity to be possible: to *give or project meaning related to us*. If the *inverse projection* of the imaginary of the artificial in artificial imagination makes sense in such a context, it is only as a certain expression of an ontological contingency beyond inherited determinations and constraints, elaborating on a peculiar autonomy pertaining to this strata.

Notes

1. Philippe Breton, *A l'image de l'homme. Du Golem aux creatures virtuelles* (Paris: Éditions du Seuil, 1995).
2. Paul Dumouchel & Jean-Pierre Dupuy (dir.) *Colloque de Cerisy: L'auto-organisation. De la physique au politique* (Paris: Éditions du Seuil, 1983), 17.
3. Cf. Kenneth M. Ford, Clark Glymour & Patrick Hayes (eds.): *Android Epistemologi,* (Menlo Park/Cambridge/London: AAAI Press/MIT Press, 1995).
4. Breton, *op. cit.*, p.102.
5. Cf. Daniel Parrochia, *Cosmologie de l'information. Pour une nouvelle modélisation de l'univers informationnel* (Paris: Edition Hermés, 1994).
6. "Cybernetics as metaphor" as Klaus Bartels demonstrates in a study of the connections between cybernetics and structuralism and poststructuralism. Cf. Klaus Bartels, "Kybernetik als Metapher. Der Beitrag des französischen Strukturalismus zu einer Philosophie der Information und der Massemedien," in Helmut Brackert und Fritz Wefelmayer, *Kultur Bestimmungen im 20. Jahrhundert* (Frankfurt a.M.: Suhrkamp Verlag 1990), 441ff.
7. Breton, *op. cit.*, p.101.
8. My argument in this paper leans substantially on the philosophical background given by Cornelius Castoriadis's work with the philosophy of "the imaginary institution of society." Castoriadis sees creativity as the ontological distinction of the human, of the "human strata of the real." In a summary from the early '90s he distinguishes between "two connotations" to be applied to the word "imagination," (1) imagination as the formation of images—forms—in the most general sense, including this formation's connection with the idea of invention and creation, and (2) a "radical imagination" which is "*before* the distinction between 'real' and 'fictitious'": "(...) it is because radical imagination exists that 'reality' exist *for us*—exists *tout cour*—and exists *as* it exists." Cf. Cornelius Castoriadis, "Radical Imagination and the Social Instituting Imaginary," in Cornelius Castoriadis, *The Castoriadis Reader*, edited by D. Curtis (Oxford: Blackwell Publishers Ltd., 1997), 319ff, 321. See also my article "Afirmación Technológia Postmoderno/Postmodern Technology-Affirmation," in *En Reconocimiento de la Historia/ In Recognition of History. Atlantica Revista de Arte y Pensamiento, Numéro 30 Otõno 2001* (Las Palmas: Gran Canaria): Centro Atlantico de Arte Moderno – CAAM). Web version on http://www.caam.net/en/atlantica.htm—search on author or title.
9. Carsten Thau: "Menneske-automaten. Levende statuer mellem barok og romantik" [The man-machine automaton. Living statues between the baroque and the romantic], in *Kritik* 105 (Copenhagen: Gyldendal, 1993), 46 ff., 47.
10. Ezio Manzini: *Artefacts. Vers une nouvelle écologie de l'environnement artificiel.* Les Essais (Paris: Centre Georges Pompidou, 1991).
11. W. Ross Ashby, "Principles of the self-organizing dynamic system," in *The Journal of General Psychology*, Vol. 37 (1947), 125ff.
12. Ibid., 125.
13. Ibid.
14. Ibid.
15. Ibid.
16. Ibid., 126.
17. Ibid., 127.
18. Ibid., 128.
19. Cf. Norbert Wiener, *Cybernetics: or Control and Communication in the Animal and the Machine* (Cambridge, MA: The MIT Press, 1991 (1948)), 44.

20. Cf. Arthur Burks, "Introduction", in *Papers of John von Neumann on Computing and Computer Theory*, edited by William Aspray and Arthur Burks (Cambridge, MA: The MIT Press 1987), 367. The collection consists of original papers and notes from von Neumann, as well as text edited and supplemented by Arthur Burks. The chronology of the papers is clarified in "Biographical Notes," xiiiff.
21. John von Neumann, *The Computer and the Brain* (New Haven and London: Yale University Press 1969), 2.
22. John von Neumann, "First Draft of a Report on the EDVAC," in Burks et al., 17ff.
23. Ibid., 24.
24. Ibid.
25. Cf. John von Neumann, "The General and Logical Theory of Automata," in Aspray et al., *op. cit.*, 391ff.
26. Ibid., 399ff.
27. Ibid., 418ff.
28. Ibid., 418.
29. Cf. John von Neumann, "Theory and Organization of Complicated Automata," in Aspray et al., 432ff.
30. Ibid., 483.
31. Ibid., 484ff, 484.
32. Ibid., 485ff.
33. Cf. Burks, *op. cit*, 374ff.
34. Cf. John von Neumann, *The Computer and the Brain*, 6ff.
35. Burks, *op. cit.* 364.
36. Ibid., 365.
37. Ibid., 363.
38. Von Neumann, *The Computer and the Brain*, 2.
39. Cf. John L. Casti, *Complexification: Explaining a Paradoxical World through the Science of Surprise* (New York: HarperPerennial, 1994); John L. Casti, *Would-Be Worlds* (New York: John Wiley & Sons, Inc., 1997); John F. Casti, *Reality Rules: I. Picturing the World in Mathematics—The Fundamentals* (New York: John Wiley & Sons, Inc., 1992); John F. Casti, *Reality Rules: II. Picturing the World in Mathematics—The Frontier* (New York: John Wiley & Sons, Inc., 1992).
40. John L. Casti, "Complexity". *Encyclopedia Britannica* [database online] http://search.eb.com/eb/article?eu=108252 [Accessed May 14, 2002]. © 2002 Encyclopedia Britannica Inc.
41. C.f. Casti, *Reality Rules: I. Picturing the World in Mathematics—The Fundamentals*, 28ff, 29.
42. Jean-Pierre Dupuy, *The Mechanization of the Mind. On the Origins of Cognitive Science* (Princeton: Princeton University Press, 2000).
43. Ibid., 27ff, 28.
44. Ibid., 29–30.
45. Ibid.
46. Ibid., 31.
47. Ibid., 35.
48. Ibid., 39.
49. Ibid., 157.
50. Ibid.
51. Ibid.
52. Ibid., 158
53. Bartels, *op. cit.*, 441, 441ff.
54. Ibid.
55. Jean Baudrillard, "The Precession of Simulacra" in), in *Art After Modernism. Rethinking Representation*, edited by Brian Wallis (New York: The New Museum of Contemporary Art 1984), 253ff.
56. Ibid., 253.
57. Ibid., 256.
58. Ibid., 264.
59. Ibid.
60. Bartels, *op. cit.*, 458ff.
61. Ibid., 458–461.
62. Ibid., 461–465.
63. Ibid., 465–467.
64. Gilles Deleuze and Félix Guattari, *L'anti-oedipe. Capitalisme et Schizoprhénie* (Paris: Les Édition de Minuit, 1972), 7ff. Here quoted from English excerpt in Lawrence E. Cahoone (ed.), *From Modernism to Postmodernism. An Anthology* (Cambridge MA: Blackwell Publishers, 1996), 401.
65. Bartels, *op. cit.*, 454f.
66. Edgar Morin, *La méthode. I. La Nature de la Nature.* (Paris: Éditions du Seuil, 1977).
67. Ibid., 160–161.
68. Ibid.
69. Ibid., 161ff.
70. Ibid., 169.
71. Vincent Descombes, *Modern French Philosophy* (Cambridge: Cambridge University Press 1980).
72. Ibid., 92ff.
73. Ibid., 103.
74. Raymond A. Eve, Sara Horsfall and Mary E. Lee (eds.), *Chaos, Complexity and Sociology. Myths, Models, and Theories* (Thousand Oaks, CA: SAGE Publications, 1997).

75. Frederick Turner, "Foreword. Chaos and Social Science," in Eve et al. (eds.), *op. cit*, xxvi.
76. Ibid., xxv.
77. Gilles Deleuze and Félix Guattari, *A Thousand Plateaus. Capitalism and Schizophrenia* (Minneapolis: The University of Minneapolis Press, 1987).
78. Félix Guattari, *Chaosmosis. An Ethco-aesthetic Paradigm* (Bloomington and Indianapolis: Indiana University Press, 1995).
79. Ibid., 33ff.
80. Félix Guattari, *Molecular Revolution. Psychiatry and Politics* (Harmondsworth: Penguin, 1984).
81. Ibid., 109ff
82. Michel Foucault, *The Order of Things. An Archeology of the Human Sciences* (London: Routledge, 1992), 387 (... that man would be erased ...).
83. Ibid.
84. Cf. Norbert Wiener, *God & Golem, Inc. A Comment on Certain Points where Cybernetics Impinges on Religion* (London: Chapman & Hall, 1964).
85. Chris Hables Gray et al.: "Cyborgology", in *The Cyborg Handbook*, edited by Chris Hables Gray, Heidi J. Figueroa-Sarriera and Steven Mentor (New York/London: Routledge 1995), 3.
86. M.V. Ghandi and B.S. Thompson, *Smart Materials and Structures* (London: Chapman & Hall, 1992).
87. Ibid., ix, 58ff..
88. Gernot Böhme, *Natürlich Natur. Über Natur im Zeitalter ihrer technischen Reproduzierbarkeit* (Frankfurt a.M.: Suhrkamp Verlag, 1992), 11ff.
89. Ibid., 15.
90. Ibid., 196.

Part IV

Network Events

17

Information, Crisis, Catastrophe

Mary Ann Doane

The major category of television is time. Time is television's basis, its principle of structuration, as well as its persistent reference. The insistence of the temporal attribute may indeed be a characteristic of all systems of imaging enabled by mechanical or electronic reproduction. For Roland Barthes, the *noeme* of photography is the tense it inevitably signifies—the "*That-has-been*" which ensures both the reality and the "pastness" of the object "photographed."[1] The principal gesture of photography would be that of embalming (hence Barthes' reference to Andre Bazin). In fixing or immobilizing its object, transforming the subject of its portraiture into dead matter, photography is always haunted by death and historicity. The temporal dimension of television, on the other hand, would seem to be that of an insistent "present-ness"—a "*This-is-going-on*" rather than a "*That-has-been*," a celebration of the instantaneous. In its own way, however, television maintains an intimate relation with the ideas of death and referentiality Barthes finds so inescapable in his analysis of the photograph. Yet, television deals not with the weight of the dead past but with the potential trauma and explosiveness of the present. And the ultimate drama of the instantaneous—catastrophe—constitutes the very limit of its discourse.

According to Ernst Bloch, "Time *is* only because something happens, and where something happens, there time is."[2] Television fills time by ensuring that something happens—it organizes itself around the event. There is often a certain slippage between the notion that television covers important events in order to validate itself as a medium and the idea that because an event is covered by television—because it is, in effect, deemed televisual—it is important. This is the significance of the media event, where the referent becomes indissociable from the medium. The penetration of everyday life by the media is a widely recognized phenomenon. But it is perhaps less widely understood that television's conceptualization of the event is heavily dependent upon a particular organization (or penetration) of temporality which produces three different modes of apprehending the event—information, crisis, and catastrophe. Information would specify the steady stream of daily "newsworthy" events characterized by their regularity if not predictability. Although news programs would constitute its most common source, it is also dispersed among a number of other types of programs. Its occasion may be politics, science, or "human interest." Information is noteworthy but is not shocking or gripping—its events are only mildly eventful, although they may be dramatized. The content of information is ever-changing but information, as genre, is always *there*, a constant and steady presence, keeping you *in touch*. It is, above all, that which fills time on television—using it up. Here time is flow: steady and continuous. The crisis, on the other hand, involves a condensation of temporality. It names an event of some duration which is startling and momentous precisely because it demands resolution within a limited period of time. Etymologically, crisis stems from the Greek *krisis*, or decision, and hence always seems to suggest the necessity of human agency. For that reason, crises are most frequently political—a hijacking, an assassination, the take-over of an embassy, a political coup, or the taking of hostages. There is a sense in which information and catastrophe are both subject-less, simply there, they *happen*—while crisis can be attributed to a subject, however generalized (a terrorist group, a class, a political party, etc.). The crisis compresses time and makes its limitations acutely felt. Finally, the catastrophe would from this perspective be the most critical of crises for its timing is that of the instantaneous, the moment, the punctual. It has no extended duration (except, perhaps, that of its televisual coverage) but, instead, happens "all at once."[3]

Ultimately, the categories of information, crisis, and catastrophe are only tenuously separable in practice. There are certainly phenomena which seem to annihilate the distinctions between them—a flood, for instance, which has elements of both the crisis (duration) and the catastrophe (it takes many lives), or an assassination which, although it may be experienced as a catastrophe, is a political action which must be attributed to a subject. But what is more striking in relation to this inevitable taxonomic failure is that television tends to blur the differences between what seem to be absolutely incompatible temporal modes, between the flow and continuity of information and the punctual discontinuity of catastrophe. Urgency, enslavement to the instant and hence forgettability, would then be attributes of both information and catastrophe. Indeed, the obscuring of these temporal distinctions may constitute the specificity of television's operation. The purpose of this essay is to investigate the implications and effects of this ambivalent structuration of time, particularly in relation to the categories of information and catastrophe.

Television overall seems to resist analysis. This resistance is linked to its sheer extensiveness (the problem of determining the limits or boundaries of the television text has been a pressing one), its continual barrage of information, sensation, event together with its uncanny ability to assimilate, appropriate, or recuperate all criticisms of the media. A story on the March 7, 1988 CBS Evening News detailed how the presidential candidates of both parties produced increasingly provocative or scandalous commercials in order to generate additional television coverage. The commercials would be shown several times in the regular manner and then, depending upon the level of their shock quotient, would be repeated once or more on local or national news, giving the candidates, in effect, free publicity. CBS News, in airing the metastory of this tendency, demonstrates how television news reports on, and hence contains through representation, its own exploitation. Its recuperative power is immense, and television often seems to reduce and deflate, through its pervasiveness and overpresence, all shock value.

Televisual information would seem to be particularly resistant to analysis given its protean nature. Not only does television news provide a seemingly endless stream of information, each bit (as it were) self-destructing in order to make room for the next, but information is dispersed on television among a number of genres and forms, including talk shows, educational/documentary type programs such as *Nova*, *National Geographic Specials*, and *Wide, Wide World of Animals*, "how-to" programs such as *The Frugal Gourmet*, *This Old House*, and *Victory Garden*, news "magazines" such as *60 Minutes* and *Chronicle*, children's shows (*Sesame Street*), sports, etc. Furthermore, even the two generic forms which are most consistently associated with the concept of information—news and the educational/documentary program—exhibit diametrically opposed formal characteristics. Documentary programs such as *Nova* tend to activate the disembodied male voice-over whose authority has long ago lapsed in the realm of the cinema (it is a voice which, as Pascal Bonitzer points out, has irrevocably "aged"[4]). News programs, on the other hand, involve the persistent, direct, embodied, and personalized address of the newscaster. Information, unlike narrative, is not chained to a particular organization of the signifier or a specific style of address.

Antithetical modes reside side by side. Hence, information would seem to have no formal restrictions—indeed, it is characterized by its very ubiquity. If information is everywhere, then the true scandal of *disinformation* in the age of television is its quite precise attempt to *place* or to *channel* information—to use its effects. Even if it is activated through television, it uses broadcasting in a narrowly conceived way. Disinformation loses credibility, then, not only through its status as a lie but through its very directedness, its limitation, its lack of universal availability. The scandal is that its effects are targeted. Disinformation abuses the system of broadcasting by invoking and exploiting the automatic truth value associated with this mode of dissemination—a truth value not unconnected to the sheer difficulty of verification and the very entropy of information.

Yet, in using this concept of information, I am accepting television's own terms. For the concept carries with it quite specific epistemological and sociological implications associated with the rise of information theory. As Katherine Hayles points out, the decisive move of information theory was to make information quantifiable by removing it from the context which endowed it with meaning, and, instead, defining it through its own internal relations. According to Hayles, this results in what is, in effect, a massive decontextualization: "Never before in human history had the cultural context itself been constituted through a technology that makes it possible to fragment, manipulate, and reconstitute informational texts at will. For postmodern culture, the manipulation of text and its consequently arbitrary relation to context *is our* context."[5]

From this point of view, television could be seen as *the* textual technology of information theory. Insofar as a commercial precedes news coverage of a disaster which in its own turn is interrupted by a preview of tonight's made-for-TV movie, television is the preeminent machine of decontextualization. The only context for television is itself—its own rigorous scheduling. Its strictest limitation that of time, information becomes measurable, quantifiable, through its relation to temporality. While the realism of film is defined largely in terms of space, that of television is conceptualized in terms of time (owing to its characteristics of "liveness," presence, and immediacy). As Margaret Morse notes, television news is distinguished by the very absence of the rationalized Renaissance space we have come to associate with film—a perspectival technique which purports to *represent* the truth of objects in space.[6] Instead, the simultaneous activation of different, incongruous spaces (the studio, graphics, footage from the scene, interviews on monitor) is suggestive of a writing surface and the consequent annihilation of depth. Television does not so much *represent* as it *informs*. Theories of representation painstakingly elaborated in relation to film are clearly inadequate.

Conceptualizing information in terms of flow and ubiquity, however, would seem to imply that it lacks any dependence whatsoever upon punctuation or differentiation. Yet even television must have a way of compensating for its own tendency toward the leveling of signification, toward banalization and nondifferentiation—a way of saying, in effect, "Look, this is important," of indexically signaling that its information is worthy of attention. It does so through processes that dramatize information—the high seriousness of music which introduces the news, the rhetoric of the newscaster, the activation of special effects and spectacle in the documentary format. Most effective, perhaps, is the crisis of temporality which signifies *urgency* and which is attached to the information itself as its single most compelling attribute. Information becomes most visibly information, becomes a televisual commodity, on the brink of its extinction or loss. A recent segment of *Nova*, "The Hidden Power of Plants," chronicles the attempt to document the expertise of old medicine men who, when they die, take their knowledge with them (it is "worse than when a library burns down," the anonymous voice-over tells us). Similarly, the numerous geographic specials demonstrate that the life of a particular animal or plant becomes most *televisual* when the species is threatened with extinction. The rhetoric of impending environmental doom is today applicable to almost any species of plant or animal life given the constant expansion and encroachment of civilization on territory designated as still "natural." In this way, television incessantly takes as its subject matter the documentation and revalidation of its own discursive problematic. For information is shown to be punctual; it inhabits a moment of time and is then lost to memory. Television thrives on its own forgettability. While the concept of information itself implies the possibilities of storage and retrieval (as in computer technology), the notion of such storage is, for television, largely an alien idea. Some television news stories are accompanied by images labeled "file footage," but the appellation itself reduces the credibility of the story. Reused images, unless carefully orchestrated in the construction of nostalgia, undermine the appeal to the "live" and the instantaneous which buttresses the news.

The short-lived but spectacular aspect of information is revealed in the use of special effects sequences where the drama of information is most closely allied with visual pleasure. In a *National Geographic* special entitled "The Mind," an artist's conception of the brain curiously resembles the *mise-en-scene* of information theory. The brain is depicted as an extensive network of neurons, synapses, and neuro-transmitters regulating the flow of information. In one cubic inch of the brain there are 100 million nerve cells connected by 10,000 miles of fibers (laid end to end, the voice-over tells us, they would reach to the moon and back). The amount of information is so enormous that cells must make instantaneous decisions about what is to be transmitted. The sequence is organized so that music announces the significance of these data and an almost constantly moving camera suggests the depths of the representation. The camera treats what is clearly a highly artificial, technologically produced space as the experienced real while the voice-over provides verbal analogues to real space (the fibers which reach to the moon and back, the pinch of salt in a swimming pool which helps one to grasp what it would be like to look for a neuro-transmitter in the brain). Yet, there is no pretense that an optical representation of the brain is adequate—it is simply *necessary* to the televisual discourse. The voice-over announces, "If it could be seen, brain cell action might look like random flickering of countless stars in an endless universe. Seemingly an infinite amount of information and variety of behaviors in an unlikely looking package," while the visuals mimic such a sight with multi-colored flickering lights. Television knowledge strains to make visible the invisible. While it acknowledges the limits of empiricism, the limitations of the eye in relation to knowledge, information is nevertheless conveyable only in terms of a *simulated visibility*—"If it could be seen, this is what it might look like." Television deals in potentially visible entities. The epistemological endeavor is to bring to the surface, to expose, but only at a second remove—depicting what is not available to sight. Televisibility is a construct, even when it makes use of the credibility attached to location shooting—embedding that image within a larger, overriding discourse.

The urgency associated with information together with the refusal to fully align the visible with the dictates of an indexical realism suggests that the alleged value of information, like that of television, is ineluctably linked with time rather than space. And, indeed, both information and television have consistently been defined in relation to the temporal dimension. According to Walter Benjamin, the new form of communication called information brought about a crisis in the novel and in storytelling: "The value of information does not survive the moment in which it was new. It lives only at that moment; it has to surrender to it completely and explain itself to it without losing any time. A story is different. It does not expend itself. It preserves and concentrates its strength and is capable of releasing it even after a long time."[7] Information must be immediately understandable, graspable—it is "shot through with explanation." Meaning in storytelling has time to linger, to be subject to unraveling. It has "an amplitude that information lacks."[8] This tendency to polarize types of discourses with respect to their relation to temporality is evident also in Jonathan Culler's activation of Michael Thompson's categories of transience and durability: "we are accustomed to think—and tradition urges us to think—of two sorts of verbal, visual compositions: those which transmit information in a world of practical affairs—utilitarian and transient—and those which, not tied to the time or use value of information, are part of the world of leisure, our cultural patrimony, and belong in principle to the system of durables."[9] Benjamin might say that the loss of aura associated with electronic reproduction is a function of its inability to *endure*. In other words, there are things which last and things which don't. Information does not. It is expended, exhausted, in the moment of its utterance. If it were of a material order, it would be necessary to throw it away. As it is, one can simply forget it.

Television, too, has been conceptualized as the annihilation of memory, and consequently of history, in its continual stress upon the "nowness" of its own discourse. As Stephen Heath and Gillian Skirrow point out, "where film sides towards instantaneous memory ('everything is absent, everything is *recorded*—as a memory trace which is so at once, without having been something else before'), television operates much more as an absence of memory, the recorded material it uses—including the material recorded on film—instituted as actual in the production of the television image."[10] This transformation of record into actuality or immediacy is a function of a generalized fantasy of "live broadcasting." Jane Feuer pursues this question by demonstrating that a certain ontology of television, defined in terms of a technological base which allows for instantaneous recording, transmission, and reception, becomes the ground for a pervasive ideology of "liveness."[11] Although, as she is careful to point out, television rarely exploits this technical capability, minimalizing not only "live" transmission but preservation of "real time" as well, the ideology of "liveness" works to overcome the excessive fragmentation within television's flow. If television is indeed thought to be inherently "live," the impression of a unity of "real time" is preserved, covering over the extreme discontinuity which is in fact typical of television in the U.S. at this historical moment.

From these descriptions it would appear that information is peculiarly compatible with the television apparatus. Both are fully aligned with the notion of urgency; both thrive on the exhaustion, moment by moment, of their own material; both are hence linked with transience and the undermining of memory. But surely there are moments which can be isolated from the fragmented flow of information, moments with an impact which disrupts the ordinary routine—moments when information bristles, when its greatest value is its shock value (in a medium which might be described as a modulated, and hence restrained, series of shocks). These are moments when one stops simply *watching* television in order to *stare*, transfixed—moments of catastrophe. But what constitutes catastrophe on television? And what is the basis of the widespread intuition that television exploits, or perhaps even produces, catastrophe? To what extent and in what ways is the social imagination of catastrophe linked to television?

Etymologically, the word "catastrophe" is traceable to the Greek *kata* (over) plus *strephein* (turn)—to overturn. The first definition given by *Webster's* is "the final event of the dramatic action esp. of a tragedy" (in this respect it is interesting to note that the etymology of the term "trope" also links it to "turn.") Hence, although the second and third definitions ("2. a momentous tragic event ranging from extreme misfortune to utter overthrow or ruin 3. a violent and sudden change in a feature of the earth") attempt to bind catastrophe to the real, the initial definition contaminates it with fictionality. Catastrophe is on the cusp of the dramatic and the referential and this is, indeed, part of its fascination. The etymological specification of catastrophe as the overturning of a given situation anticipates its more formal delineation by catastrophe theory. Here, catastrophe is defined as unexpected discontinuity in an otherwise continuous system. The theory is most appropriate, then, for the study of sudden and unexpected effects in a gradually changing situation. The emphasis upon suddenness suggests that catastrophe is of a temporal order.

The formal definition offered by catastrophe theory, however, points to a striking paradox associated with the attempt to conceptualize televisual catastrophe. For while catastrophe is designated as discontinuity within an otherwise continuous system, television is most frequently theorized as a system of discontinuities, emphasizing heterogeneity. Furthermore, the tendency of television to banalize all events through a kind of leveling process would seem to preclude the possibility of specifying *any* event as catastrophic. As Benjamin pointed out in a statement which seems to capture something of the effect of television, "The concept of progress is to be grounded in the idea of catastrophe. That things 'just go on' *is* the catastrophe."[12] The news, in particular, is vulnerable to the charge that it dwells on the catastrophic, obsessed with the aberrant, the deviant. According to Margaret Morse, "The news in the West is about the *a*normal. It is almost always the 'bad' news. It is about challenges to the symbolic system and its legitimacy."[13] Furthermore, in its structural emphasis upon discontinuity and rupture, it often seems that television itself is formed on the model of catastrophe.

Given these difficulties, is it possible to produce a coherent account of events which television designates as catastrophe? What do these moments and events have in common? One distinctive feature of the catastrophe is that of the *scale* of the disaster in question—a scale often measured through a body count. By this criterion, Bhopal, the Detroit Northwest Airlines crash of August 1987, and the Mexican earthquake could all be labeled catastrophes. However, other events which are clearly presented as catastrophic—Chernobyl, the explosion of the *Challenger*—do not involve a high number of deaths while wartime body counts (Vietnam, the Iran-Iraq war), often numerically impressive, do not qualify as catastrophic (undoubtedly because war makes death habitual, continual). Evidently, the scale which is crucial to catastrophe is not that of the quantification of death (or at least not that alone).

Catastrophe does, however, always seem to have something to do with technology and its potential collapse. And it is also always tainted by a fascination with death—so that catastrophe might finally be defined as the conjuncture of the failure of technology and the resulting confrontation with death. The fragility of technology's control over the forces it strives to contain is manifested most visibly in the accident—the plane crash today being the most prominent example. Dan Rather introduced the CBS story about the August 1987 Detroit Northwest Airlines crash with the rhetoric of catastrophe—the phrase "aftershocks of a nightmare" accompanying aerial images of wreckage strewn over a large area. The inability of television to capture the precise moment of the crash activates a compensatory discourse of eyewitness accounts and animated re-enactments of the disaster—a simulated vision. Eyewitnesses who comment upon the incredible aspects of the sight or who claim that there were "bodies strewn everywhere" borrow their authority from the sheer fact of being there at the disastrous moment, their reported presence balancing the absence of the camera. What becomes crucial for the act of reportage, the announcement of the catastrophe, is the simple gesture of being on the scene, *where* it happened, so that presence in space compensates for the inevitable temporal lag. Hence, while the voice-over of the anchor ultimately organizes the event for us, the status of the image as indexical truth is not inconsequential—through it the "story" touches the ground of the real. Nevertheless, the catastrophe must be immediately subjected to analysis, speculation, and explanation. In the case of the airplane crash, speculation about causes is almost inevitably a speculation about the limits and breaking points of technology (with respect to Northwest flight 255, the history of the performance of the engine was immediately a subject of interrogation).

As modes of transportation dependent upon advanced and intricate technologies become familiar, everyday, routine, the potential for catastrophe increases. The breakdown of these technologies radically defamiliarizes them by signaling their distance from a secure and comforting nature. As Wolfgang Schivelbusch points out, this was the case for the railroad in the 19th century, its gradual acceptance and normalization subjected to the intermittent shock of the accident.

> One might also say that the more civilized the schedule and the more efficient the technology, the more catastrophic its destruction when it collapses. There is an exact ratio between the level of the technology with which nature is controlled, and the degree of severity of its accidents. The preindustrial era does not know any technological accidents in that sense. In Diderot's *Encyclopedie*, "Accident" is dealt with as a grammatical and philosophical concept, more or less synonymous with coincidence. The preindustrial catastrophes are natural events, natural accidents. They attack the objects they destroy from the outside, as storms, floods, thunderbolts, hailstones, etc. After the industrial revolution, destruction by technological accident comes from the inside. The technical apparatuses destroy themselves by means of

> their own power. The energies tamed by the steam engine and delivered by it as regulated mechanical performance will destroy that engine itself in the case of an accident.[14]

In the late twentieth century, the potential for technological collapse is more pervasive, characterizing catastrophes as diverse as Bhopal, Chernobyl, the *Challenger* explosion, earthquakes which science and technology fail to predict, as well as railway and plane crashes. But this massive expansion is perhaps not the decisive difference. After the Detroit crash, airport authorities spray-painted the burned-out grass green in order to conceal all traces of the accident and enable other travelers to avoid the traumatic evidence. Yet, this action was then reported on radio news, indicating that what is now at stake in the catastrophe, for us, is *coverage*. While the vision of catastrophe is blocked at one level, it is multiplied and intensified at another. The media urge us now to obsessively confront catastrophe, over and over again. And while the railway accident of the 19th century was certainly the focus of journalistic inquiry, its effects were primarily local. Television's ubiquity, its extensiveness, allows for a global experience of catastrophe which is always reminiscent of the potential of nuclear disaster, of mass rather than individual annihilation.

Catastrophe is thus, through its association with industrialization and the advance of technology, ineluctably linked with the idea of Progress. The time of technological progress is always felt as linear and fundamentally irreversible—technological change is almost by definition an "advance," and it is extremely difficult to conceive of any movement backward, any regression. Hence, technological evolution is perceived as unflinching progress toward a total state of control over nature. If some notion of pure Progress is the utopian element in this theory of technological development, catastrophe is its dystopia, the always unexpected interruption of this forward movement. Catastrophic time stands still. Catastrophe signals the failure of the escalating technological desire to conquer nature. From the point of view of Progress, nature can no longer be seen as anything but an affront or challenge to technology. And so, just as the media penetrates events (in the media event), technology penetrates nature. This is why the purview of catastrophe keeps expanding to encompass even phenomena which had previously been situated wholly on the side of nature—earthquakes, floods, hurricanes, tornadoes. Such catastrophes no longer signify only the sudden eruption of natural forces but the inadequacy or failure of technology and its predictive powers as well.

On the ABC Evening News of September 15, 1988, Peter Jennings stood in front of a map tracking the movements of Hurricane Gilbert for the first fifteen minutes of the broadcast. A supporting report detailed the findings of a highly equipped plane flying into the eye of the hurricane. The fascination here was not only that of the literal penetration of the catastrophic storm by high technology but also that of the sophisticated instruments and tracking equipment visible inside the plane—a fetishism of controls. Our understanding of natural catastrophe is now a fully technological apprehension. Such incidents demonstrate that the distinction made by Schivelbusch between preindustrial accidents (natural accidents where the destructive energy comes from without) and post-industrial accidents (in which the destructive energy comes from within the technological apparatus) is beginning to blur. This is particularly the case with respect to nuclear technology which aspires to harness the most basic energy of nature itself—that of the atom. And in doing so it also confronts us with the potential transformation of that energy into that which is most lethal to human life.

While nuclear disaster signals the limits of the failure of technology, the trauma attached to the explosion of the *Challenger* is associated with the sheer height of the technological aspirations represented by space exploration. The *Challenger* coverage also demonstrates just how nationalistic the apprehension of catastrophe is—our own catastrophes are always more important, more eligible for extended reporting than those of other nations. But perhaps even more crucial here was the fact that television itself was on the scene—witness to the catastrophe. And the played and replayed image of the *Challenger* exploding, of diverging lines of billowing white smoke against a deep blue Florida sky—constant evidence of television's compulsion to repeat—acts as a reminder not only of the catastrophic nature of the event but also of the capacity of television to record instantaneously, a reminder of the fact that television was *there*. The temporality of catastrophe is that of the instant—it is momentary, punctual, while its televisual coverage is characterized by its very duration, seemingly compensating for the suddenness, the unexpected nature of the event.

A segment of Tom Brokaw's virtually non-stop coverage on NBC contained a video replay of the explosion itself, a live broadcast of the president's message to the nation, Brokaw's reference to an earlier interview with a child psychiatrist who dealt with the potential trauma of the event for children, Chris Wallace's report of Don Regan's announcement and the press's reception of the news during a press briefing, a mention of Mrs. Reagan's reaction to the explosion as she watched it live on television, Brokaw's speculation about potential attacks on Reagan's support of SDI ("Star Wars"), and Brokaw's 1981 interview with one of the astronauts, Judy Resnick. The glue in this collection of disparate forms is Brokaw's performance, his ability to *cover* the event with words, with a commentary which exhausts its every aspect and through the orchestration of secondary reports and old footage.[15] Brokaw is the pivot, he mediates our relation to the catastrophe. Furthermore, as with television news, it is a direct address/appeal to the viewer, but with an even greater emphasis upon the presence and immediacy of the act of communication, with constant recourse to shifters which draw attention to the shared space and time of reporter and viewer: terms such as "today," "here," "you," "we," "I." Immediately after a rerun of the images documenting the *Challenger* explosion, Brokaw says, clearly improvising, "As I say, we have shown that to you repeatedly again and again today. It is not that we have a ghoulish curiosity. We just think that it's important that all members of the audience who are coming to their sets at different times of the day have an opportunity to see it. And of course everyone is led to their own speculation based on what happened here today as well." The "liveness," the "real time" of the catastrophe is that of the television anchor's discourse—its nonstop quality a part of a fascination which is linked to the spectator's knowledge that Brokaw faces him/her without a complete script, underlining the alleged authenticity of his discourse. For the possibility is always open that Brokaw might stumble, that his discourse might lapse—and this would be tantamount to touching the real, simply displacing the lure of referentiality attached to the catastrophe to another level (that of the "personal" relationship between anchor and viewer).

There is a very striking sense in which televisual catastrophe conforms to the definition offered by catastrophe theory whereby catastrophe represents discontinuity in an otherwise continuous system. From this point of view, the measure of catastrophe would be the extent to which it interrupts television's regular daily programming, disrupting normal expectations about what can be seen and heard at a particular time. If Nick Browne is correct in suggesting that, through its alignment of its own schedule with the work day and the work week, television "helps produce and render 'natural' the logic and rhythm of the social order,"[16] then catastrophe would represent that which cannot be contained within such an ordering of temporality. It would signal the return of the repressed. The traumatic nature of such a disruption is underlined by the absence of commercials in the reporting of catastrophe—commercials usually constituting not only the normal punctuation of television's flow, but, for some, the very text of television.

That which, above all, cannot be contained within the daily social rhythms of everyday life is death. Catastrophe is at some level always about the body, about the encounter with death. For all its ideology of "liveness," it may be death which forms the point of televisual intrigue. Contemporary society works to conceal death to such an extent that its experience is generally a vicarious one through representation. The removal of death from direct perception, a process which, as Benjamin points out, was initiated in the nineteenth century, continues today:

> In the course of the nineteenth century bourgeois society has, by means of hygienic and social, private and public institutions, realized a secondary effect which may have been its

> subconscious main purpose: to make it possible for people to avoid the sight of dying. Dying was once a public process in the life of the individual and a most exemplary one.... There used to be no house, hardly a room, in which someone had not once died... Today people live in rooms that have never been touched by death...[17]

Furthermore, the mechanization of warfare—the use of technologically advanced weapons which kill at a greater and greater distance—further reduces the direct confrontation with death. Consistent with its wartime goal of allaying the effects of death and increasing the efficiency with which it is produced, technology also strives to hold death at bay, to contain it. Hence, death emerges as the absolute limit of technology's power, that which marks its vulnerability. Catastrophe, conjoining death with the failure of technology, presents us with a scenario of limits—the limits of technology, the limits of signification. In the novel, according to Benjamin, death makes the character's life *meaningful* to the reader, allows him/her the "hope of warming his shivering life with a death he reads about."[18] What is at stake in televisual catastrophe is not meaning but reference. The viewer's consuming desire, unlike that of the novel reader, is no longer a desire for meaning but for a referentiality which seems to have been all but lost in the enormous expanse of a television which always promises a contact forever deferred. Death is no longer the culminating experience of a life rich in continuity and meaning but, instead, pure discontinuity, disruption—pure chance or accident, the result of being in the wrong place at the wrong time.

And it is not by coincidence that catastrophe theory, on an entirely different level, seeks to provide a means of mapping the discontinuous instance, the chance occurrence, without reducing its arbitrariness or indeterminacy. Catastrophe theory is based on a theorem in topology discovered by the French mathematician René Thom in 1968. Its aim is to provide a formal language for the description of sudden discontinuities within a gradually changing system. The points of occurrence of these discontinuities are mapped on a three-dimensional graph. In 1972, E. D. Zeeman developed an educational toy called the "catastrophe machine" to facilitate the understanding of Thom's theory. (The appeal of this toy is that you can make it yourself with only two rubber bands, a cardboard disk, two drawing pins and a wooden board). The point of the catastrophe machine is the construction of an apparatus which is guaranteed to *not* work, to predictably produce unpredictable irregularities. For catastrophe theory is, as one of its proponents explains, "a theory about singularities. When applied to scientific problems, therefore, it deals with the properties of discontinuities directly, without reference to any specific underlying mechanism."[19] It is, therefore, no longer a question of explanation. Catastrophe theory confronts the indeterminable without attempting to reduce it to a set of determinations. Thom refers to "islands of determinism separated by zones of instability or indeterminacy."[20] Catastrophe theory is one aspect of a new type of scientific endeavor which Lyotard labels "postmodern"—a science which "by concerning itself with such things as undecidables, the limits of precise control, conflicts characterized by incomplete information, 'fracta,' catastrophes, and pragmatic paradoxes—is theorizing its own evolution as discontinuous, catastrophic, nonrectifiable, and paradoxical. It is changing the meaning of the word *knowledge*, while expressing how such a change can take place."[21]

Television is not, however, the technology of catastrophe theory or, if it is, it is so only in a highly limited sense. The televisual construction of catastrophe seeks both to preserve and to annihilate indeterminacy, discontinuity. On the one hand, by surrounding catastrophe with commentary, with an explanatory apparatus, television works to contain its more disturbing and uncontainable aspects. On the other hand, catastrophe's discontinuity is embraced as the mirror of television's own functioning and that discontinuity and indeterminacy ensure the activation of the lure of referentiality. In this sense, television is a kind of catastrophe machine, continually corroborating its own signifying problematic—a problematic of discontinuity and indeterminacy which strives to mimic the experience of the real, a real which in its turn is guaranteed by the contact with death. Catastrophe thrives on the momentary, the instantaneous, that which seems destined to be forgotten, and hence seems to confirm Heath's and Skirrow's notion that television operates as the "absence of memory." But because catastrophe is necessary to television, as the corroboration of its own signifying problematic, there is also a clear advantage in the somewhat laborious construction and maintenance of a memory of catastrophe. The spectator must be led to remember, with even a bit of nostalgia, those moments which are preeminently televisual—the explosion of the *Challenger*, the assassination of John F. Kennedy (the footage of which was replayed again and again during the time of the recent twenty-fifth anniversary of the event). What is remembered in these nostalgic returns is not only the catastrophe or crisis itself but the fact that television was there, allowing us access to moments which always seem more real than all the others.

Catastrophe coverage clearly generates and plays on the generation of anxiety. The indeterminacy and unexpectedness of catastrophe seem to aptly describe the potential trauma of the world we occupy. But such coverage also allows for a persistent disavowal—in viewing the bodies on the screen, one can always breathe a sigh of relief in the realization that "that's not me." Indeed, the celebrity status of the anchorperson and of those who usually appear on television can seem to justify the belief that the character on the screen is always—dead or alive—is always definitively *other*, that the screen is not a mirror. Such persistent anxiety is manageable, although it may require that one periodically check the screen to make sure. But this is perhaps not the only, or even the most important, affect associated with catastrophe coverage.

Something of another type of affective value of catastrophe can be glimpsed in Slavoj Žižek's analysis of the sinking of the *Titanic* and its cultural and psychical significance. At the end of the nineteenth century, "civilized" Europe perceived itself as on the brink of extinction, its values threatened by revolutionary workers' movements, the rise of nationalism and anti-Semitism, and diverse signs indicating the decay of morals. The grand luxury transatlantic voyage incarnated a generalized nostalgia for a disappearing Europe insofar as it signified technological progress, victory over nature, and also a condensed image of a social world based on class divisions elsewhere threatened with dissolution. The shipwreck of the *Titanic* hence represented for the social imagination the collapse of European civilization, the destruction of an entire social edifice—"Europe at the beginning of the century found itself confronted with its own death."[22] The contradictory readings by the right and the left of the behavior of first-class "gentlemen" with respect to third-class women and children corroborate this reading of a social imagination seized by the shipwreck and treating it as an index to the maintenance or collapse of former class differences.

But Žižek goes on to claim that there must be something in excess of this symbolic reading. For it is difficult to explain satisfactorily the contemporary fascination with images of the wreck at the bottom of the sea: "The mute presence of wrecks—are they not like the congealed residue of an impossible *jouissance*?... One understands why, notwithstanding technical problems, we hesitate to raise the wreckage of the Titanic to the surface: its sublime beauty, once exposed to daylight, would turn to waste, to the depressing banality of a rusted mass of iron."[23] It would be problematic to bring the *Titanic* too close—it is there to be watched in its "proper" grave, to be regarded as a monument to catastrophe in general, a catastrophe which, in its distance, makes you feel real. According to Žižek, the two aspects of the *Titanic*—the "metaphorical one of its symbolic overdetermination and the real one of the inertia of the thing, incarnation of a mute *jouissance*"—represent the two sides of the Freudian symptom.[24] For although the symptom can be interpreted as a knot of significations, it is also always more than that. There is a remainder, an excess not reducible to the symbolic network (in the words of Jacques-Alain Miller, one "loves one's symptom like oneself.") This is why, according to Žižek, "one remains hooked on the real of one's symptom even after the interpretation has accomplished its work."[25]

It is this remainder, this residue, which televisual catastrophe exploits. The social fascination of catastrophe rests on the desire to confront the remainder, or to be confronted with that which is in excess of signification. Catastrophe seems to testify to the inertia of the real and television's privileged relation to it. In the production and reproduction of the metonymic chain—the body-catastrophe-death-referentiality—television legitimates its own discourse. This is why it is often difficult to isolate and define catastrophe, to establish the boundary which marks it off from ordinary television. Information and catastrophe coexist in a curious balance. According to Susan Sontag, "we live under continual threat of two equally fearful, but seemingly opposed, destinies: unremitting banality and inconceivable terror."[26] Television produces both as the two poles structuring the contemporary imagination.

This relation to catastrophe is by no means an inherent or essential characteristic of television technology. Rather, it is a feature which distinguishes television and its operations in the late capitalist society of the United States where crisis is produced and assimilated as a part of the ongoing spectacle—a spectacle financed by commercials and hence linked directly to the circulation of commodities. What underlies/haunts catastrophe but is constantly overshadowed by it is the potential of another type of catastrophe altogether—that of the economic crisis. According to Schivelbush, "If the nineteenth century perceives the cause of technological accidents to be the sudden disturbance of the uncertain equilibrium of a machine (i.e., the relationship between curbed energy and the means of curbing it), Marx defines the economic crisis as the disruption of the uncertain balance between buying and selling in the circulation of goods. As long as buying and selling work as a balanced and unified process, the cycle goes on functioning, but as soon as the two become separated and autonomous, we arrive at a crisis."[27] Of course, economic crisis does not appear to meet any of the criteria of the true catastrophe. It is not punctual but of some duration, it does not kill (at least not immediately), and it can assuredly be linked to a notion of agency or system (that of commodity capitalism) if not to a subject. Yet, for a television dependent upon the healthy circulation of commodities, the economic crisis can be more catastrophic than any natural or technological catastrophe. Ironically, for this very reason, and to deflect any potentially harmful consequences, it must be disguised as catastrophe and hence naturalized, contained, desystematized. The economic crisis as catastrophe is sudden, discontinuous, and unpredictable—an accident which cannot reflect back upon any system.

In comparison with the lure of referentiality associated with catastrophe "proper," the economic crisis confronts us as an abstraction. Yet, the abstraction of catastrophe is difficult since catastrophe seems to lend itself more readily to an account of bodies. Hence, the reporting of the Wall Street crash of October 1987 strives to restore the elements of catastrophe which are lacking—the iconography of panic becomes the high angle shot down at the milling crowd of the stock exchange, bodies in disarray. An interviewee claims, "It's fascinating, like a bloodbath." Furthermore, a catastrophe which seems furthest removed from the concept of a failure of technology is rebound to that concept through the oft-repeated claim that a major cause of the crash was computer trading gone awry. Economic crisis is also tamed by naturalizing it as a cyclical occurrence, like the change of seasons. This is a containment of a catastrophe which, unlike the others, potentially threatens television's own economic base, its own mechanism for the production of commodity-linked spectacle. And perhaps this is why catastrophe has become such a familiar, almost everyday, televisual occurrence. According to Ernst Bloch, "the crises of the accident (of the uncontrolled things) will remain with us longer to the degree that they remain deeper than the crises of economy (of the uncontrolled commodities)."[28] The depth which television accords to the catastrophes of things is linked to the lure of referentiality which they hold out to us. Catastrophe makes concrete and immediate, and therefore deflects attention from, the more abstract horror of potential economic crisis. For the catastrophe, insofar as it is perceived as the *accidental* failure of technology (and one which can be rectified with a little tinkering—O-rings can be fixed, engines redesigned) is singular, asystematic—it does not touch the system of commodity capitalism.

The concept of crisis is linked to temporal process, to a duration of a (one can hope) limited period. This is why the time of crisis can coincide with that of politics, of political strategy. Crisis, *krisis*, is a decisive period insofar as it is a time when decisions have to be made, decisions with very real effects. The televisual representation of catastrophe, on the other hand, hopes to hold onto the apolitical and attach it to the momentary, the punctual. Here time is free in its indeterminacy, reducible to no system—precisely the opposite of televisual time which is programmed and scheduled as precisely as possible, down to the last second. Television's time is a time which is, in effect, wholly determined. And this systematization of time is ultimately based on its commodification (time in television is, above all, not "free"). As both Stephen Heath and Eileen Meehan point out, what networks sell to advertisers is the viewing time of their audiences.[29] Here the commodification of time is most apparent (and perhaps this is why, in the reporting of catastrophes, there are no commercials).

The catastrophe is crucial to television precisely because it functions as a denial of this process and corroborates television's access to the momentary, the discontinuous, the real. Catastrophe produces the illusion that the spectator is in direct contact with the anchorperson, who interrupts regular programming to demonstrate that it can indeed be done when the referent is at stake. Television's greatest technological prowess is its ability to be there—both on the scene and in your living room (hence the most catastrophic of technological catastrophes is the loss of the signal). The death associated with catastrophe ensures that television is felt as an immediate collision with the real in all its intractability—bodies in crisis, technology gone awry. Televisual catastrophe is thus characterized by everything which it is said not to be—it is expected, predictable, its presence crucial to television's operation. In fact, catastrophe could be said to be at one level a condensation of all the attributes and aspirations of "normal" television (immediacy, urgency, presence, discontinuity, the instantaneous, and hence forgettable). If information becomes a commodity on the brink of its extinction or loss, televisual catastrophe magnifies that death many times over. Hence, catastrophe functions as both the exception and the norm of a television practice which continually holds out to its spectator the lure of a referentiality perpetually deferred.

Postscript (2003)

Although this essay was written fifteen years ago, I believe that the tendencies it describes have only intensified and deepened. The televisual coverage of September 11, 2001 seemed to obscenely corroborate the idea that "the death associated with catastrophe ensures that television is felt as an immediate collision with the real in all its intractability—bodies in crisis, technology gone awry." It further blurred the already fragile opposition between catastrophe and crisis outlined here, transforming a political act into something with the proportions of a monumental natural disaster (or a grandiose battle between an abstractly defined good and evil), at the expense of any more nuanced attempt at historical explication. The concept of catastrophe has been systematically broadened since 1988 and this steady encroachment is enabled by the fact that catastrophe is defined not so much by any stable or finite content but by the ideological inflection of a technological potential—that of the "liveness" of representation. Hence, events as varied as the O.J. Simpson car chase and later trial, the death of Lady Diana, Princess of Wales, the 1991 Persian Gulf War, the Washington sniper killings, the loss of the space shuttle *Columbia* as well as, most prominently, the collapse of the twin towers on September 11, 2001 and the ensuing Iraq War have all been covered as catastrophic, invoking all the technical skills that have been honed and refined continually since (at least) the 1980s. Those skills above all allow television to be there, on the scene, "live," to stress the urgency, indeed inescapability, of our attending to the event. Catastrophe takes on the logic of innovation associated with the commodity system and especially with fashion—its constant demand for newness, difference, uniqueness and its consequent "forgetting" of yesterday's styles and yesterday's catastrophes (except in the form of the quotable, of "retro" —one might note the existence of "retro catastrophes," replayed somewhat nostalgically on anniversaries and other appropriate moments.)

The role of televisual "liveness" has been made more critical by the emergence and rapid dissemination of digital media that lay claim to an even more desirable temporality—"real time." The use of "real time" in the digital register both appropriates the meanings of this term for film (continuity, lack of disruption) and for television (the instantaneity and immediacy of liveness) and adds the additional connotations of interactivity and 24-hour availability. Computer "real time" not only allows you to remain connected, to be *in touch* with what is happening now, but to allegedly interact with the source of information as well, expanding your choices within a commodity driven economy while leaving intact the restricted, corporate-produced definition of choice. In television's continual effort to assimilate everything—including all other forms of representation, news anchors frequently exhort their viewers to keep up with the news in "real time" by visiting the station's or network's Web site online. A Web site called "freerealtime.com," whose motto is "Turning knowledge into wealth," provides its visitors with direct and instantaneous access to action on the stock market. The site promises its visitors the opportunity to "see what's happening right now, not 15 minutes later." The Web site of Lynuxworks aptly proclaims that "there is no tolerance for delays in a telecom environment."

No tolerance for delays. What is at stake in this continual technological celebration of instantaneity, in this insistence upon identifying the real with the "now"? In a sense, it signifies the social abjection of representation itself in a highly mediated society. Bazin, in writing about what must strike most people now as a fairly antiquated medium—film—links cinematic specificity to a scandal, that of the repeatability of the unique ("I cannot repeat a single moment of my life, but cinema can repeat any one of these moments indefinitely before my eyes").[30] While all moments are unique, according to Bazin, the paradox of their cinematic repetition is muted by our acceptance of this status as a form of memory. But there are two moments that are so intensely unique that their repetition in film must be *obscene*—death and the sexual act: "Each is in its own way the absolute negation of objective time, the qualitative instant in its purest form. Like death, love must be experienced and cannot be represented (it is not called the little death for nothing) without violating its nature. This violation is called obscenity. The representation of a real death is also an obscenity, no longer a moral one, as in love, but metaphysical. We do not die twice."[31] Death and sex mark the limits of representation, the point at which no difference, no splitting of the instant, is acceptable. For Bazin, this scandal was specific to the cinema and did not exist for photography because only cinema could represent the passage from one state to another, from life to death. (The supreme perversion, in his view, would be the projection of an execution in reverse.) Bazin wrote, of course, before television (or at least its widespread dissemination), video, and digital media, other visual time-based media which could share in the scandal of death's representation.

Bazin's formulation of representational obscenity is presented in the course of an analysis of a film about a bullfight. The death at issue is the ever-present potentiality of the death of either the bull or the toreador. Spectatorial cognizance of this risk is heightened by a refusal of editing, by allowing the bull and the man to occupy the same frame. Hence, the obscenity is very much a function of the continuity of cinematic "real time," of the protection of cinema's indexicality from the violence of editing. Instead, as Serge Daney has pointed out, violence, difference, or heterogeneity are internalized as subject matter in order to preserve the spatio-temporal unity of the representation. It is representation which is truly at risk ("*To intern difference means saving representation*").[32] This violence is eroticized, given Bazin's view of the intimacy of sex and death as apogees of the unique.

For Bazin, obscenity is the repetition of the absolutely unique, the fact that death could be made to happen over and over again, made possible by time-based mechanical representation and by filmic "real time," which acts as a kind of proof of the process and its integrity. "Real time" today, in its televisual and digital forms, is less about continuity (the refusal of editing) and more about instantaneity (the adherence of the time of representation to the time of the event). It makes possible a repetition that threatens to annihilate the temporal gap between the event and its representation—in the live telecast, the event is virtually its own repeatability. The scandal would be the disappearance of the very idea of the unique, the loss of death as a measure of singularity. Death happens over and over again on the television screen and there is a general hemorrhaging of the notion of the catastrophic. The media thrive on this disruption, this discontinuity, because it promises to stave off the boredom of the banal, of television's unrelenting flow. Incorporated within that flow, disruption becomes "reality television."

Yet, each catastrophe is somehow new despite its repetitiveness, for the catastrophic is also the unscripted, the tinge of referentiality which seems to cling to mechanical and electronic reproduction. The dialectic of repeatability and the unique associated with catastrophe works both to affirm representation (to "save" it in Bazin's terms) and to allow for a hope in its effacement, the assurance of an access to the real. This is, perhaps, the supreme paradox of a media-saturated society.

Notes

1. Roland Barthes, *Camera Lucida: Reflections on Photography*. trans. Richard Howard (New York: Farrar, Strauss and Giroux, Inc., 1981): 77.
2. Ernst Bloch. *A Philosophy of the Future*, trans. John Cumming (New York: Herder and Herder, 1970):124.
3. The time proper to catastrophe might be thought of as compatible with that of the digital watch where time is cut off from any sense of analogical continuity, and the connection between moments is severed. One is faced only with the time of the instant—isolated and alone.
4. Pascal Bonitzer, "The Silences of the Voice," trans. Philip Rosen and Marcia Butzel. In *Narrative, Apparatus, Ideology*, edited by Philip Rosen (New York: Columbia University Press, 1986): 328.
5. N. Katherine Hayles, "Text Out of Context: Situating Postmodernism Within an Information Society," *Discourse*, no. 9 (Spring/Summer 1987): 26.
6. Margaret Morse, "The Television News Personality and Credibility: Reflections on the News in Transition," in *Studies in Entertainment*, edited by Tania Modleski (Bloomington: Indiana University Press, 1986): 70.
7. Walter Benjamin, "The Storyteller." In *Illuminations*, trans. Harry Zohn, edited by Hannah Arendt (New York: Schocken Books, 1969): 90.
8. Benjamin, "Storyteller," 89.
9. Jonathan Culler, "Junk and Rubbish," *Diacritics*, vol. 15, no. 3 (Fall 1985): 10.
10. Stephen Heath and Gillian Skirrow, "Television, a World in Action," *Screen*, vol. 18, no. 2 (Summer 1977): 55–56.
11. Jane Feuer, "The Concept of Live Television: Ontology as Ideology." In *Regarding Television: Critical Approaches—An Anthology*, edited by E. Ann Kaplan (Frederick, MD: University Publications of America, Inc. and the American Film Institute): 13–14.
12. Benjamin, "Central Park," trans. Lloyd Spencer, *New German Critique*, no. 34 (Winter 1985): 50.
13. Morse, "The Television New Personality," 74.
14. Wolfgang Schivelbusch, *The Railway Journey: Trains and Travel in the 19th Century*, trans. Anselm Hollo (New York: Urizen Books, 1979): 133.
15. This performance could also be seen as a masculinist discourse which attempts to reestablish control over a failed masculinized technology. In this sense, catastrophe is feminized insofar as it designates the re-emergence of the nature technology attempts to repress and control. Brokaw's performance is thus a discursive management of catastrophe. Such a reading is problematic insofar as it equates nature and the feminine, technology and the masculine, but gains a certain amount of historical force from an influential mythology. It was, after all, a woman (Pandora) who unleashed catastrophe upon the world.
16. Nick Browne, "The Political Economy of the Television (Super)Text." In *Television: The Critical View*, Fourth Edition, edited by Horace Newcomb (New York: Oxford University Press, 1987): 588.
17. Benjamin, "The Storyteller," 94–95.
18. Benjamin, "The Storyteller," 101.
19. P. T. Saunders, *An Introduction to Catastrophe Theory* (Cambridge: Cambridge University Press, 1980): 1.
20. René Thom, "Topological models in biology, "Topology, vol. 8 (1969), cited in Michael Thompson, *Rubbish Theory: The Creation and Destruction of Value* (Oxford: Oxford University Press, 1979): 142.
21. Jean-Francois Lyotard, *The Postmodern Condition: A Report on Knowledge*, trans. Geoff Bennington and Brian Massumi (Minneapolis: University of Minnesota Press, 1984): 60.
22. Slavoj Žižek, "Titanic-le-symptome," *L'Ane*, numero 30 (Avril-Juin 1987): 45.
23. Ibid.
24. Ibid.
25. Ibid.
26. Susan Sontag, "The Imagination of Disaster," *Against Interpretation* (New York: Dell Publishing Co., Inc., 1969): 227.
27. Schivelbusch, *The Railway Journey*, 134.
28. Cited in Schivelbusch, *The Railway Journey*, 131.
29. Stephen Heath, "Representing Television." In *Logics of Television*, ed. Patricia Mellencamp (Bloomington and Indianapolis: Indiana University Press, 1990): 267–302; Eileen R. Meehan, "Why We Don't Count: The Commodity Audience." *In Logics of Television*, 117–137.
30. André Bazin, "Death Every Afternoon," trans. Mark A. Cohen, *Rites of Realism: Essays on Corporeal Cinema*, ed. Ivone Margulies (Durham and London: Duke University Press, 2003): 30.
31. Bazin, 30.
32. Serge Daney, "The Screen of Fantasy (Bazin and Animals)," trans. Mark A. Cohen, in Margulies, 34.

18

The Weird Global Media Event and the Tactical Intellectual [version 3.0]

McKenzie Wark

1. Media Times

"The almost insoluble task is to let neither the power of others, nor our own powerlessness, stupefy us."[1] Theodor Adorno was writing of the intellectual's challenge in comprehending Hitler, but perhaps the same injunction might apply to events of more recent times. As with Hitler, so with Osama Bin Laden: both might be, to a psychologist, pathological cases, but "people thinking in the form of free, detached, disinterested appraisal" are "unable to accommodate within those forms the experience of violence which in reality annuls such thinking."

It is a characteristic of traditional scholarship that it assumes a certain kind of time within which the scholarly enterprise can unfold. Scholarship is knowledge occupying an abstract, homogenous, formal time. Indeed scholarship might be defined as the production of precisely this kind of time. A scholar's primary duty is the patient working through of the consequences attendant on what one's predecessors and colleagues deposit for us in the archive.

As a consequence, scholarship has difficulties with those images which, as Walter Benjamin said, "flash up in a moment of danger."[2] Such images interrupt the time of scholarship, breaking the thread of its apparent continuity. There are always parallel times—the news media ticks over at a faster rate than scholarship. The time of everyday life takes its distance and insists on its own rhythms. These times may occasionally synchronize, but mostly they follow their own beat.

Every now and then there is an event which interrupts all such discrete and parallel times, cutting across them and marking them all with the image of a moment of danger. We know that September 11 interrupted the time of news media. The evidence is there in videotapes of CNN and other live news feeds. The news story suddenly confronted its opposite, which I would call the event. A routine news story has a narrative structure, which pre-exists any given circumstances. Facts, when they emerge, can be fitted into a story. An event as an irruption of raw facticity into the news, for which a story is not ready to hand.

The event, when it occurs in news media, opens up a certain abyss. One stares at the evidence of an event for which the story is lacking, or rather, lagging. News media respond with a range of coping strategies, with which to paper over the evident fact that events have violated the narrative control and management of the news media, at least for a moment.

One coping strategy is repetition. News feeds reiterate a cluster of images and sounds over and over, as if only through repetition could the facticity of the event be acknowledged. Exploratory attempts will be made using file footage to construct a beginning to the event. Events always irrupt into news as if in the middle. News responds by speculating on the beginning point for the story. As the narrative arc of the event is unknown or unstable, wise old white haired gentlemen are recruited to provide a speculative trajectory, a template, which might serve to reduce the event to some familiar variant on the common stock of stories.

The event now has the capacity to synchronize many very diverse local times, spilling over into the living rooms, bars, bazaars and places of worship of many different kinds of people. Local and communal rhythms suddenly appear as connected to global forces and relations. Yet for all that, it proves remarkably difficult to think back from one's experience to the causes of the event itself. *The New Yorker* put some of the most distinguished writers in town on the job of recording their experiences of September 11. The results were remarkably banal. Star writers from Jonathan Franzen to Adam Gopnik could all provide richly detailed versions of their whereabouts on the day, connected to nothing but trivial remarks about the more abstract forces at work.

As Fredric Jameson notes, this is an era in which the forces that determine one's life chances are abstract and global, yet the means by which one would usually communicate about one's life chances with others, one's immediate experience, appears as merely an effect of unseen forces. "There comes into being, then, a situation in which we can say that if individual experience is authentic, then it cannot be true; and that if a scientific or cognitive model of the same content is true, then it escapes individual experience."[3] This is a problem, as Jameson notes, for art; it is also a problem, as he doesn't note, for critical theory.

While I agree with Jameson on the disconnect between appearances and relations, which in art is the disconnect between naturalism and realism, I think there is a solution. One needs to displace the terms a little. The disconnect can be expressed as a difference between kinds of time. The time of everyday life not only differs from the time of news media and the time of scholarship, it differs from the time of capital flows and global power. The latter appear in everyday life as images that flash up, not just in moments of danger, but as moments of danger. The moment when they flash up is the moment of the event. The event opens a critical window onto the disjuncture between different kinds of time precisely because it is the moment when times suddenly connect, even if, in connecting, the usual means of making sense of time within the horizon of a specific temporal narrative is obliterated.

So if one is not to be stupefied by the power of others, or one's own powerlessness, one needs to know something of the time in which power operates. But this is a temporality to which one usually does not have access, either in everyday life, or in scholarship, or in art—it is even doubtful if the news media is all that proximate to the most effective times of power and powers of time. But there are moments, interruptions in the polyrhythmic flow, in which a kind of knowledge is possible.

These moments are events. Or to give them the full specification I have given them elsewhere, "weird global media events."[4] They are events because they interrupt routine time. They are media events because they happen within a space and time saturated in media. They are global media events because they traverse borders and call a world into being. They are weird global media events because each is singular and none conform to any predetermined narrative. They introduce a new quality of time.

The event not only breaches the separation among what we might call after Marx superstructural times, but between them and what we might call infrastructural times of political and economic power. As Jameson notes, Marx borrowed this terminology from the railways. Superstructure and infrastructure are the rolling stock and the rails. In these terms, the event might be the juncture at which both the track and the train change paths.

2. Media Spaces

Where do events come from? Do they fall from the sky? Yes they do. From the Comsat angels in orbit overhead, or thrown from a truck onto the ground in front of your local news stand. Robert McChesney points out that these vectors from whence we get the information to form an ongoing map of the world and its histories become increasingly concentrated in fewer and fewer corporate hands. These corporate owners are increasingly integrating diverse media holdings to more profitably co-ordinate print and audio-visual flows.[5] No matter how many channels we can get, our main news feed comes from few hands indeed.

Herbert Schiller once argued that the growth of transnational corporations, who seek rich offshore markets and cheap offshore labor forces, necessitates an internationalization of media vectors. The deregulation of economic flows during the Reagan years went hand-in-hand with a deregulation of information flows and attacks on public control and access to information.[6] The media that feed us are not only more and more concentrated, but increasingly global in both ownership and extent. Since business consumes a vast amount of media information, and business is increasingly global, so too are the information providers. Three developments come together: the globalization of business communication, the communication of global business and the business of global communication.

The global media vector does not connect us with just anywhere. It connects us most frequently, rapidly and economically with those parts of the world which are well integrated into the major hubs of the vector. It comes as no surprise that New York is a major media hub, as it is a major business hub, but so too is the Middle East. Hamid Mowlana points out that the Middle East has a long history of integration into the international media vector. At the turn of century, Lord Curzon described British interests in the Persian Gulf as "commercial, political, strategical and telegraphic."[7] Some of the world's first international telegraph lines passed through there. British communications with India flowed along this route. With the recognition of the strategic value of oil for propelling the mechanized vectors of war from 1914 on, the region became important in its own right.

An event that connects an expatriate Saudi to New York so spectacularly is then not surprisingly an event that punctures the time of everyday life with a major impact. One should, however, add Tariq Ali's caveat: "To accept that the appalling deaths of over 3,000 people in the USA are more morally abhorrent than the 20,000 lives destroyed by Putin when he razed Grosny or the daily casualties in Palestine and Iraq is obscene."[8] In proposing that September 11 is a weird global media event, I am not assuming that the violence of that moment somehow trumps these other instances of violence. The point is rather that the globalization of media flows is subject to very uneven development. One of the characteristics of the event is precisely to reveal the uneven topography of the vectoral landscape along which media messages speed.

One of the striking things about September 11 is that the event happened in a major node in the media network, and hence was rapidly and thoroughly reported, thus provoking remarkably different responses around the world. Ali records some of the range of responses: "In the Nicaraguan capital, Managua, people hugged each other in silence.... There were celebrations in the streets of Bolivia... In Greece the government suppressed the publication of opinion polls that showed a large majority actually in favor of the hits... In Beijing the news came too late in the night for anything more than a few celebratory fireworks."[9] The centralization and concentration of media has some effect on what events may spark across the vector field of time and space, but does not necessarily determine how they may be interpreted, which still depends on the tempos of everyday life and of local media envelopes. The people make meaning, but not with the media of their own choosing.

The "global village" is a fractious and contentious place, particularly when the lightning strike of an event gives way to the thunder of a thousand pundits explaining it away. Local interpretive strategies and authorities invariably script the event in terms which make it appear as if it were meant to make sense within the dominant local framework. John Hartley suggests that "news includes stories on a daily basis which enable everyone to recognize a larger unity or community than their own immediate contacts, and to identify with the news outlet as 'our' storyteller."[10] The protocols of everyday life appear here as the imagined categories of a far more vast and unevenly global terrain of what I call telesthesia, of perception at a distance. This world of telesthesia is organized temporally in terms of "visible, distant visions of order," but where these are highlighted negatively by "the fundamental test of newsworthiness," namely, "disorder—deviation from any supposed steady state."[11] Telesthesia is organized spatially by what Hartley calls Theydom. "Individuals in Theydom are treated as being all the same; their identity consists in being 'unlike us', so they are 'like each other.'"

Slavoj Žižek and Edward Said offer a general and a specific theory respectively that may help us reconstruct, after the event, our own narrative about how the narrative of Theydom works. To start with the specific theory: Said proposes the category of Orientalism to account for the doubling of an Wedom with a Theydom, in which the defining characteristics of Wedom come into focus against the background of a Theydom. The opening up of the Middle East to European trade, conquest and most importantly communication opens up a vector field in which information may flow across boundaries for the purpose of commerce or colonization, but where that flow produces an anxious desire for a sense of border or boundary. That boundary is defined by Orientalism, a discourse by, for and secretly about Wedom, sustained by the image of a Theydom, in which it is axiomatic that the "attributes of being Oriental overrode any countervailing instance."[12]

For Žižek, the Orientalist image of Theydom might count as a local and specific variant on a general structure: "We always impute to the 'other' an excessive enjoyment; s/he wants to steal our enjoyment (by ruining our way of life) and/or has access to some secret, perverse enjoyment."[13] As if to illustrate such a theory, one of the more popular images to circulate via e-mail shortly after September 11 was a Photoshop collage of Osama Bin Laden sodomizing President George W. Bush. For Žižek, the other is dangerous because Theydom either pursues enjoyment too much, or too little. In the construction of a Theydom in the wake of September 11, the focus is usually on terrorist as denier of pleasure, as a fanatic, a militant. But curiously, this image keeps flipping over into its other. The terrorist is also the one panting after the 70 virgins promised in paradise, or putting liquor and lap dances on the al-Qaida credit card.

So far we have two things defining the space of September 11. One is the presence of a vector from where the World Trade Center is to wherever you are. The other is a set of everyday conventions operating to make the fate of its victims, who belong to Wedom, the subject of sympathy or mourning, and an evil Theydom. There is a connection and a convention, in time and space, making those fatal flights fall from the sky into our lives.

Whatever the virtues of the work of Said and Žižek, neither really offers a narrative of the dialectic of Wedom and Theydom that takes full account of the role of the time of the event in creating and recreating the boundaries, nor do they highlight the role of telesthesia in the formation of Wedom and Theydom on a global scale. The weird global media event is more than an anomaly in the "normal" functioning of culture; it is the moment which disrupts its normal functioning, and in the wake of which a new norm will be created.

How then can such a weird global media event be conceptualized? The event as I define it is something that unfolds within the movement of telesthesia along media vectors. These media vectors connect the site at which a crisis appears with the sites of image management and interpretation. Vectors then disseminate the flows of images processed at those managerial sites to the terminal sites of the process, so they fall from the sky into our lives. In this instance the vector connects a bewildering array of places: New York, Managua, Beijing. Into the vision mix went images hauled off the global satellite feed, showing us file footage of Osama Bin Laden one second and live footage of Mayor Giuliani the next, as if the Mayor were responding to that absent figure. The vector creates the space of telesthesia where one can appear quite "naturally" to respond to the other, in the blink of an edit. We witnessed the montaging of familiar and surprising sites into the seamless space and staccato time of the media vector. The terminal site of the vector is the radio, television or internet terminal within reach—directly or indirectly—of almost everyone almost everywhere.

3. Vectors and Antipodes

A word on this word vector. I've adapted it from the writings of Paul Virilio. It describes the aspect of the development of technology that interests him most and the style of writing he employs to capture that aspect. It is a term from geometry meaning a line of fixed length and direction but having no fixed position. Virilio employs it to mean any trajectory along which bodies, information or warheads can potentially pass. The satellite technology used to beam images from Afghanistan to America can be thought of as a vector. This technology could link almost any two such sites, and relay video and audio information of a certain quality along those points at a given speed and at a certain cost. It could just have easily linked Copenhagen to Chiapas, or quite a few other combinations of points. Yet in each case the speed of transmission and its quality could be essentially the same. (That it often is not points to the politics and economics that shape the infrastructure of the vector field, but which it in turn also shapes).

This is the sense in which any particular media technology can be thought of as a vector. Media vectors have fixed properties, like the length of a line in the geometric concept of vector. Yet that vector has no necessary position: it can link almost any points together. This is the paradox of the media vector. The technical properties are hard and fast and fixed, but it can connect enormously vast and vaguely defined spaces together and move images, and sounds, words and furies between them.

In every weird global media event, new dimensions to the vector field are "discovered," and new technical properties of the vector implemented. After September 11, the Western world discovered—as if for the first time—the significance of al-Jazeera satellite television.[14] During the Gulf War, most of the Middle East was more or less effectively contained within state controlled national media envelopes, at least as far as television was concerned. Al-Jazeera changed all that. Or to take a more poignant instance: it seems that while people all over the world knew that one of the WTC towers had collapsed, the firefighters in the other tower did not know it, as the vectors along which information might pass to them was disrupted by the collapse of the tower itself. Telesthesia failed at the point where it was most pressingly required.

In the analysis of the weird global media event, a theoretical approach that highlights the technical, such as the concept of the vector, is crucial, but must be handled as a critical tool. Everyone marvels at what the latest media technologies make possible in the moment of the event. It is one of the most immediate ways of constructing a narrative for it. But then the material means by which the space in which the event happens is constructed tends to be pushed to the background. The knowledge of the vector that the event highlights passes imperceptibly into an unacknowledged part of the information landscape we take for granted. Victor Shklovsky one said that the real reveals itself in culture in much the same way as gravity reveals itself to the inhabitants of a structure when its ceiling caves in on them.[15] That might stand as a good emblem for the event.

It is not only media technologies that have this vectoral aspect. The high-jacked 767s were also a vector. So too are the bombs and missiles rained down on Afghanistan in what Ali calls the "lightly disguised war of revenge."[16] All if these vectors had certain fixed technical properties: payload, range and accuracy. Yet they could be launched at any point within a given radius. On the other hand, one could think of the entire U.S. invasion force that mobilized for what President Bush initially called Operation Infinite Justice as a vector too. The fixed properties here have to do with the length of time it takes to deploy a force of a given size. Yet that force could be deployed almost anywhere. Indeed, in an age of proliferating media vectors, perhaps the public spectacle of a threat to the interests of imperial powers will provoke the deployment of this other kind of vector. The alternative, something we also saw on TV during the war in Afghanistan, is the vector of diplomacy: diplomats can shuttle between any series of points negotiating an apparently limitless range of demands with seemingly limited results. The time pressures introduced by the military and media vectors pose a serious problem for the tactful tempo of diplomacy.

The beauty of Virilio's concept of vector is that it grasps the dynamic, historical tendency of weird global media events, but it is not a concept limited to media technologies alone. It also provides a way of thinking about the other aspects of such events. Virilio homes in on the apparent tendencies that seem to result from the relentless, competitive development of vectors. For instance, the tendency towards a homogenization of the space of the globe. Its tendency to become an abstract, geometric space across which powerful vectors can play freely, producing new differentials of Wedom and Theydom.

Virilio grasps the novel kinds of crisis this seems to engender: "An imperceptible movement on a computer keyboard, or one made by a 'skyjacker' brandishing a cookie box covered with masking tape, can lead to catastrophic chains of events that until recently were inconceivable. We are too willing to ignore the threat of proliferation resulting from the acquisition of nuclear explosives by irresponsible parties. We are even more willing to ignore the proliferating threat resulting from the vectors that cause those who own or borrow them to become just as irresponsible."[17]

There is a limit to the way Virilio conceptualizes the vector, in that he doesn't distinguish the vectors of telesthesia, which move information, from those that move bodies and things, labor and commodities, subjects and objects. Thus he loses focus on the way telesthesia creates a space for the logistical tracking of objects and subjects in movement, and for ordering that movement. The second nature of labor and commodities, of work and leisure, of private and public worlds, is traversed by an emergent space composed of vectors capable of moving information more quickly than people or things can move. Just as second nature is built out of the historical transformation of the raw materials of nature, so too a third nature arise, built out of the historical transformation of second nature by the vectors of telesthesia.

Perhaps it is worth hitting the video pause-button at this point in the replay, just as the image of the 767 hitting the WTC comes into view. Here we have a vector of second nature, the ubiquitous passage of the 767 through the skies, which is only made possible by the existence of a third nature, of radio and radar and global positioning technology. And here we have the rerouting of the aircraft, using that same technology of telesthesia, to new coordinates, bringing about an event in the most built up part of second nature, New York City, which in turn disrupts the third nature of the news media.

What bears critical attention is the way telesthesia is part and parcel of what killed people in both New York and subsequently in Afghanistan. The event takes place at the level of the physical vector and the media vector conjointly. In terms of vectoral power in general, the media are part of the problem of power, not merely a separate space of reportage or critique of emergent forms of power that exist elsewhere. Needless to say, this essay too is a part of that problematic, and does not exist outside, in a neutral space. It is in the worst of all possible worlds: within the regime of power created by the media vector, but relatively powerless there, within. What is indeed stupefying is that the ability to think critically about the event depends on the same vectoral power that produces its violence.

Reading the critical coverage of September 11 and the subsequent war in Afghanistan in journals such as *The Nation*, I am struck firstly by the double bind its correspondents found themselves in, and secondly by the curious way that the critical response to imperial power nevertheless participated in the same way of seeing the world. As Michel Feher notes, the leftist response to such events is caught between two desires. One desire is to oppose American imperial power, in which case it can appear to lend support to dictatorial anti-western regimes. The other desire is to overturn tyranny in dictatorial anti-western regimes, in which case it ends up lending support to American imperial power.[18] Either way, the rhetorical structure of Wedom versus Theydom is reproduced, without really addressing the vectoral power that underlies the production of their relation in the first place.

The massive presence in the media flow of American stories, images, faces, voices, is sometimes all that stabilizes the flow of meaning in third nature. Take away America's imaginary domination and the domination of the imaginary of America, and meaning would drift and eddy, caught in impossible turbulence and glide.[19] Not only the instant media coverage, but also the critical coverage relies on this stabilization of the referents, either positively or negatively. The frightening paradox of September 11 is how this attack on actual human lives in New York and Afghanistan is at the same time merely an attack on abstract signifiers of Wedom and Theydom. The trick, if this is not to stupefy us, is to look for a way of displacing the terms within which the event is understood.

4. Nightly Chimeras

By starting with the appearance of the vector in everyday life, we can trace it back to a general problematic of the velocity of power. The "departure lounge" for this is not some abstract concept of everyday life in general, not the life of others, under the microscope, but this life, these events. A vectoral writing strategy considers the production of events within the media as the primary process that nevertheless gives the appearance of merely reflecting "naturally occurring" moments outside all such apparatus.

This may sound a little counter-intuitive, since we all tend to take it for granted that regardless of how much the media constructs a particular view of an event the media still reports something outside of itself. While not disputing the fact that violent and momentous conjunctures arise whether the media report them or not, once the media takes up such conjunctures they assume a quite different character. A vectoral approach looks at movements of information transgressing the boundaries between what were once historically distinct sites. It looks at the effect of this movement on the outcomes of conjunctures. It looks at the event as a peculiar and historically emergent form of communication—or rather of non-communication.

In writing about September 11 as an event happening in a network of global vectors, which made it that much more instant, that much more deadly, writing struggles to recall that we are not just spectators. The whole thing about the media vector is that its tendency is towards implicating the entire globe. Its historic tendency is towards making any and every point a possible connection—everyone and everything is a potential object and/or subject of a mediated relation, realized instantly. In September 11, to see it was to be implicated in it. There is no safe haven from which to observe, unaffected. Nor is there a synoptic vantage point, above and beyond the whole process for looking on in a detached and studious manner. We are all, always, already—there, in third nature.

As the possibility of an extensive war of revenge increased, the media's role changed, ever so imperceptibly. No longer did it exist in a relation to an audience assumed to be a mass of consumers or a public to be educated.[20] The event turns the media into part of a feedback loop connecting the spectator to the action via the vagaries of 'opinion' and the pressures of the popular on political elites. The media user becomes a vague and quixotic, unpredictable yet manipulatory "delay" in the circuit of power.[21]

This is the curious thing about telesthesia. It can make events that connect the most disparate sites of public action appear simultaneously as a private drama filled with familiar characters and moving stories. The vector blurs the thin line between political crisis and media sensation; it eclipses the geographical barriers separating distinct cultural and political entities; and it transgresses the borders between public and private spheres both on the home front and the front line. There is no longer a clear distinction between public and private spaces, now that the vector transgresses the boundaries of the private sphere.

As Donna Haraway suggests, "we are all chimeras, theorized and fabricated hybrid of machine and organism."[22] Our chimerical confusion may result from the dissolution of the spaces which kept aspects of the social order separate. Indeed, one of the defining characteristics of the event is that it exposes the ironic ability of the vector to disrupt all seemingly stable distributions of space and the more or less water-tight vessels that used to contain meaning in space and time. As September 11 unfolded, the hallowed ground bled into the profane domain—of media. One keeps the sense of what it means to be in public life as opposed to private life by keeping them spatially separate. The horror of bodies jumping from the towers—a rare image, quickly edited out—has a layer to it which draws on the horror of the separate and excluded part reappearing in the everyday sphere of "normality."

The reasons why these interpretations should spring to mind stems from another sense of separation, the separation of such things off from Wedom and their projection into an other. Yet here they are, returned to haunt us, in an uncontrollable way. Here they are in everyday life, intersected by the rays of the screen. To adapt a line from William Burroughs, in an incongruous yet strikingly apt context: "These things were revealed to me in the Interzone, where East meets West coming around the other way."[23] The interzone is this space where chimerical and monstrous images become a part of everyday life. The interzone is the experience, in everyday life, of the ironizing impact of the event.

The media weave a Wedom and a vast map of Theydoms together as the light and dark strands of a narrative distinction within the event as it of threads its way across these other kinds of border. In breaking down solid old boundaries, the vector creates new distinctions. Flexible distinctions airily flow through the story-time realm of information. They selectively replace the heavy walls and barriers that compartmentalized information in days when vectors were less rapid and less effective. This cruder narrative structure can be applied to more sudden and diverse events to produce the same effect of apparent narrative seamlessness. The application by the media of simple temporal structures, in a flexible fashion, produces more rigid and uniform stories about events.

There are many analyses of these war-time bed-time stories that expose the interests of capital and empire that lie behind them.[24] What matters is telling convincing stories, which show others ways to account for the facts—and for the way facts are produced. Or persuasive stories, which help as many people as possible to credit this version of the event over other ones. The democratic forces that want to rewrite this event as a chapter in the story of, say, American imperialism or Orientalist racism, must learn the tools and the tricks of the story trade—and prevail.

But as the technology of persuasion grows more complex, the art of telling stories in the wake of events grows both more complex and more instantaneous. If this essay is less concerned with telling these alternative stories it is not because such things are not important. It is because it is also important to understand the nature of weird global media events and the power field of the vector. This is the field of becoming within which a certain kind of power is immanent. A field in which democratic forces need to speak, and attempt at least to make good sense, for and with, the many against the few. But the tools for doing so may have less to do with the hypocritical earnestness of Wedom and more to do with pushing the ironic spatial and temporal displacements of vectors to the limit.

5. Tactical Media and Tactical Knowledge

As Montaigne remarked, there are certain viewpoints that expose us to our own fundamental state of ignorance. Confronting an event in the media is such a viewpoint. This is not to celebrate stupidity, merely to recognize that there are no authorities one can evoke when genuine, full-blown, out-of-control events occur. (And this is precisely why outlets like CNN wheel out the white haired authorities at the first whiff of a weird global media event.) There is, however, always a store of useful information and sets of conceptual tools that might help. Access to these is a form of power that can be very unevenly distributed. The vector is a form of power. Rapid and effective access to useful information is a vector. Not all vectors are extensive ones, seeking to cover the span of the globe. Some are intensive. They seek microscopic paths through the labyrinthine mazes of data stored in the cores of the information-rich archives of the West.

Some of the really useful information is "classified." It will be released very slowly and to few people. On the other hand, conceptual tools for extracting the most out of the information that is freely available about any actual or potential event are available to a much wider pool of people. I believe this "tactical response" to the media vector to be a worthwhile skill to learn, to teach, to practice and communicate. But there is a caveat. When responding in a timely fashion to events that stupefy, it is important not to respond stupidly, reactively, with a reflexive negation that merely reproduces the dialectical terms of Wedom and Theydom. Rather, one has to deploy tactics that display a certain ironic knowledge about how the vector works, and which attempts to reach that everyday interzone where, in the wake of the event, boundaries seem to dissolve, and irony finds its intemperate time.

Geert Lovink and David Garcia speak of a tactical media that might free itself from the dialectic of being an alternative or an opposition, which merely reproduces the sterile sense of a Wedom versus a Theydom in the media sphere.[25] They claim that the "identity politics, media critiques and theories of representation" that were the foundation of oppositional media practices "are themselves in crisis." They propose instead an "existential aesthetic" based on the temporary "creation of spaces, channels and platforms". Lovink and Garcia's seminal text on tactical media doesn't entirely succeed in extracting itself from the oppositional language of Wedom versus Theydom, but it points towards an alterative strategy to the negation that paradoxically unites Osama Bin Laden, George W. Bush and the writers of *The Nation* as purveyors, not of the same world view, but of world views constructed the same way. It is a question of combining tactical media with a tactical knowledge, of using the extensive vector of the media in combination with the intensive vector of the scholarly archive.

In a nominally democratic country, one acts as part of a public sphere in the sense Alexander Kluge give to the term.[26] A public sphere—a matrix of accessible vectors—acts as a point of exchange between private experience and public life; between intimate, incommunicable experience and collective perception. Public networks are arenas where the struggle to communicate takes place. Two aspects of this concept are relevant here. For Kluge, writing in post war Germany, the problem revolves around the historic failure in 1933 of the public sphere to prevent the rise of fascism. "Since 1933 we have been waging a war that has not stopped. It is always the same theme—the noncorrelation of intimacy and public life—and the same question: how can I communicate strong emotions to build a common life?"[27] For Kluge, the public sphere is a fundamentally problematic domain, caught between the complexities of the social and the increasing separation of private life.

One has to ask: for whom does Kluge imagine he speaks? Perhaps there are other experiences of the relation between the time of intimate experience and the time of the public sphere, buried out there in popular culture. Perhaps it is only intellectuals who feel so estranged from the time of information in the era of telesthesia. After all, the mode of address adopted by most popular media doesn't speak to a highly cultured intellectual like Kluge—or even a provincial one like me. We were trained in slower ways of handling information, and have a repertoire of quite different stories with which to filter present events. How could we claim to know what goes on out there in the other interzones, in quite other spaces where different flows from different vectors meet quite other memories and experiences of everyday life? After all, we intellectuals keep finding more than enough differences amongst ourselves.

A tactical knowledge of media may have among its merits the fact that it takes these other interzones seriously. It tries to theorize the frictions between Kluge's intimate experience and the network of vectors, or it actually tries to collect and interpret accounts of such experiences.[28] It is necessary to at least attempt to maintain a self-critical relation to the codes and practices of the interzone specific to intellectual media experiences. After all, "our" training, "our" prejudices in relation to the vector might be part of the problem. Nothing exempts 'our' institutions and interests from the war of the vector, the struggle to control the trajectories of information.

With the spread of the vector into the private realm, a window opens that might be used to create a line along which the communication of intimate experience and collective feeling might take place, in those eventful moments when their separation collapses. The protocols of tactical media are not given in advance. As Gilles Deleuze says: "Experiment, never interpret."[29] What is at stake is not the recreation of the public grounds for a universal reason, but finding the tactical resources for a far more differentiated and diverse struggle to communicate, that simple thing so hard to achieve.[30]

The maintenance of democracy requires a practice within the public networks for responding to events that it was never quite designed to handle. Virilio asks whether democracy is still possible in this era of what he calls "chronopolitics." Perhaps democracy succumbs to "dromocracy"—the power of the people ploughed under by the power to technological speed.[31] Well perhaps, but the only way to forestall such pessimism is to experiment with tactics for knowing and acting in the face of events. One has to experiment with relatively freely available conceptual tools and practices and base a democratic knowledge on them. This may involve moving beyond the techniques and procedures of the academy. In Antonio Gramsci's terms, the academic intellectual risks becoming merely a traditional intellectual, one of many layers of cultural sediment, deposited and passed over by the engine capital and the trajectory of the vector, caught up in a temporality that is not even dialectically resistant, but is merely residual. One has to make organic connections with the leading media and cultural practices of the day.[32]

Nevertheless, the historic memory and living tissue of scholarship stores resources that are useful and vital. In studying an event like September 11, a tactical knowledge can build on the best of two existing critical approaches. To the schools that concentrate on the structural power of transnational capital flows and military coercion it adds a close attention to the power of transgressive media vectors and the specific features of the events they generate. To the schools that study the space of the media text in the context of periodic struggles for influence with the national-popular discourse it adds an international dimension and a closer attention to the changing technical means that produce information flows. The event is a phenomena a little too slippery for either of these approaches. Hence the need to examine it in a new light, as the chance encounter of the local conjuncture with the global vector—on the operating table.

The chance encounter of Osama Bin Laden with CNN, like the meeting of the umbrella with the sewing machine, has a surreal, "surgical" logic specific to it. It is not entirely reducible to the long term temporalities of capital or military power and lies in the spaces between national-popular discourses. Writing the vector is not really something that can be practices with the tools of the Herbert Schiller school of political economy or the Stuart Hall school of cultural studies, alone, although a tactical knowledge might owes something to both.[33] A tactical intellectual practice that uses the moment of the event to cross the divide between infrastructural and superstructural time.

The event is not reducible to the methods of the "areas specialists." When studying events from the point of view of the site at which the originate, they always remain the province of specialists who deal with that particular turf. Events often generate valuable responses from area specialists, but these usually focus on the economic, political or cultural factors at work in the area the specialists know first hand. They do not often analyze the vectoral trajectories via which the rest of the world views the event. A tactical knowledge borrows from area studies without being caught within its territorial prerogatives.

In an age when transnational media flows are running across all those academic specialties, perhaps it is time to construct a discourse that follows the flow of information (and power) across both the geographic and conceptual borders of discourse. Perhaps it is time to start experimenting, as Kluge has done, with modes of disseminating critical information in the vector field. Perhaps it is time to examine intellectual practices of storing, retrieving and circulating knowledge. Without wishing to return to the practice of the "general intellectual," it may be worth considering whether the development of the vector calls for new ways for playing the role of the tactical intellectual.[34] The tactical intellectual would combine the practices of tactical media and tactical scholarship, while being careful not to fall into the temporality of either journalism or the academy, but rather remain alert to the moments in which such distinct times are brought into crisis by the time of the event.

6. Afghan eXplorer

The Afghan eXplorer is described on its Web site as "a tele-operated, robotic war reporting system, able to provide images, sound, and interviews in real time."[35] It bears an uncanny resemblance to the Mars Explorer. As the Web site notes, "One central advantage of Afghanistan over Mars is that Afghanistan features tens of thousands of miles of functioning roadways." Its makers note "the system may be retrofitted, with only minor software modifications, to work in other potential hotspots, such as Palestine, Israel, Iraq, Syria, Sudan, Lebanon, Indonesia, Pakistan and Qatar." These might all qualify, in the eXplorer's subtle and ironic displacement, as alien landscapes to Western journalism and its audience.

Chris Csikszentmihályi, who led the team that designed it at MIT's Media Lab, reports that when journalists started to hear about the eXplorer, interest rapidly snowballed.[36] Journalists love to write about themselves, and journalists tend to write about what other journalists are writing about. So Csikszentmihályi found himself fielding calls from journalists in a wide range of media, all interested in the eXplorer. The eXplorer touches on the interzone of journalistic experience.

Csikszentmihályi says he studied Noam Chomsky's approach to responding to interviews, and learned from Chomsky the practice of ignoring the journalist's questions and hammering away at one's own agenda. The agenda as far as he was concerned was to emphasize the closure of the field of conflict to fair and unbiased reporting by the military, and the use of what he calls "robotic killing machines" in Operation Infinite Justice. The eXplorer calls attention to the effect of the vector in a double sense: the robotic war vector appears in a displaced form as the robotic journalism vector, which in turn refers to the absence of journalists from infrastructural deployment of military vectoral power.

While Csikszentmihályi would not necessarily embrace the term, I want to use the Afghan eXplorer as a striking instance of tactical intellectual work. Csikszentmihályi was able to exploit mainstream media's fascination with its own practices of reporting, and also a fascination with technological solutions to political problems to his advantage, inserting a point of view into the media feed that is not oppositional, but which cuts across Wedom and Theydom at an ironic tangent, displacing the terms within which one may think about the event. The eXplorer manages to reconnect the naturalism of the experience with its quirky form and function, with the realism of the abstract relations of vectoral power for which it is so ironic, and iconic, an interzone.

Csikszentmihályi was able to insert at least some mention of this other perspective into interviews with journalists not only in the United States, but also in Pakistan, and at the BBC World Service. He notes that live radio and television interviews were particularly good tactical opportunities. Print media journalists usually plug the facts of the Afghan eXplorer story into pre-existing scripts. The eXplorer provides the tactical leverage for a fact gathering mission into what for many artists or scholars is the alien world of news media time.

One way of disentangling this practice of the tactical intellectual from opposititional or alternative media strategies is to see it as being a kind of micro-event in itself. The media tactician presents an image that endangers the conventions of journalistic narrative time, yet which is capable of inserting itself into it. This kind of tactical media ironically displaces the boundaries drawn by the machine of the news story. The moment when such a tactic is most likely to be successful is when news media time has itself already been disrupted by an event of a much larger scale—a weird global media event, for instance. In that moment of instability, the ironic displacement of a tactical media micro-event may find its purchase on media time.

Notes

1. Theodor Adorno, *Minima Moralia* (London: New Left Books, 1974): 57.
2. Walter Benjamin, "Theses on the Philosophy of History."
3. Fredric Jameson, "Cognitive Mapping," in *Marxism and the Interpetation of Culture*, edited by Cary Nelson and Lawrence Grossberg (Urbana: University of Illinois Press, 1988): 349.
4. McKenzie Wark, *Virtual Geography: Living With Global Media Events* (Bloomington: Indiana University Press, 1994): 21–24.
5. Robert McChesney, *Rich Media, Poor Democracy* (New York: New Press, 20000.
6. Herbert Schiller, *Culture Inc: The Corporate Takeover of Public Expression* (New York: Oxford University Press, 1989).
7. Quoted in Hamid Mowlana, "Roots of War: the Long Road to Intervention," in *Triumph of the Image: The Media's War in the Persian Gulf—A Global Persepctive*, edited by Mowlana et al (Boulder, CO: Westview, 1992): 30–50, 36.
8. Tariq Ali, *The Clash of Fundamentalisms* (London: Verso, 2002): 290.
9. Ali, ibid, 2.
10. John Hartley, *The Politics of Pictures: The Creation of the Public in the Age of Popular Media* (London: Routledge, 1992): 207.
11. Hartley, op cit, 140.
12. Edward Said, *Orientalism* (Harmondsworth: Penguin Books, 1978): 231.
13. Slavoj Žižek, "Eastern Europe's Republics of Gilead," in *New Left Review*, no. 183 (September 1990): 53–54.
14. Mohammed El-Nawawy et al, *Al Jazeera: The the Free Arab News Network Scooped the World and Changed the Middle East* (Boulder: Westview Press, 2002).
15. Victor Shklovsky, *Mayakovsky and His Circle* (London: Pluto Press, 1978).
16. Tariq Ali, op cit, 3.
17. Paul Virilio, *Speed and Politics*, Semiotext(e) (1986): 143–144. See also McKenzie Wark, "On Technological Time," *Arena*, No. 83 (1988).
18. Michel Feher, *Powerless By Design: The Age of the International Community* (Durham, NC: Duke University Press, 2000).
19. See McKenzie Wark, "From Fordism to Sonyism: Perverse Readings of the New World Order," in *New Formations*, no. 15 (1991).
20. Ien Ang, *Desparately Seeking the Audience* (London: Routledge, 1991).
21. Gilles Deleuze, *Bergsonism* (New York: Zone Books, 1988). Deleuze's thinking takes the form of a reading of the first chapter of Henri Bergson, *Matter and Memory* (New York: Zone Books, 1991).
22. Donna J. Haraway, 'A Cyborg Manifesto', in her book *Simians, Cyborgs and Women: The Reinvention of Nature* (New York: Routledge, 1991): 150.
23. William Burroughs, 'Word', in *Interzone* (London: Picador, 1989): 137.
24. See for instance Noam Chomsky, *9-11* (New York: Seven Stories Press, 2001); Michael Parenti, *The Terrorism Trap* (San Francisco: City Lights Books, 2002).
25. Geert Lovink and David Garcia, "The ABC of Tactical Media," retrieved from: http://www.ljudmila.org/nettime/zkp4/74.htm.
26. On the public sphere, see Alexander Kluge and Oskar Negt, "The Public Sphere and Experience: Selections," and Stuart Liebman, "On New German Cinema, Art, Enlightenment, and the Public Sphere: An Interview with Alexander Kluge," *October*, no. 46 (Fall 1988).
27. Stuart Leibman, "Interview with Alexander Kluge," *October*, no. 46 (Fall 1988): 45.
28. See, for example, John Fiske,*Television Culture* (London: Methuen, 1989) and Ien Ang, *Watching Dallas* (London: Methuen, 1985).
29. Gilles Deleuze and Claire Parnet, *Dialogues*, 48.
30. Kluge and Oskar Negt's writings on the public sphere, such as "The Public Sphere and Experience: Selections," in *October*, no. 46 (Fall 1988) are a critical response to Jürgen Habermas, in this case *The Transformation of the Public Sphere* (Cambridge: Polity Press, 1989).
31. Paul Virilio, *Pure War*, Semiotext(e) Foreign Agents Series (New York, 1983): 58.
32. Antonio Gramsci, "The Intellectuals," *Selections From the Prison Notebooks* (New York: International Publishers, 1980).
33. Herbert Schiller, *Culture Inc*, op cit; Stuart Hall, *The Hard Road to Renewal* (London: Verso, 1988).
34. Russell Jacoby laments the decline of the public intellectual in *The Last Intellectuals* (New York:Farrar, Straus and Giroux, 1989). He argues that while radical academics have inserted new content into academic discourses, they have not changed the form. While I agree with this assessment, I am less inclined towards a nostalgia for the public intellectuals of the past.
35. http://www.afghanexplorer.net.
36. Presentation by Chris Csikszentmihályi at the Blur conference, Parsons School of Design, 12th April, 2002. http://www.nsu.newschool.edu/blur.

19

Imperceptible Perceptions in Our Technological Modernity

Arvind Rajagopal

When a new technological medium enters the world, we tend to think the world of it.[1] We identify it with the world, and imagine it brings the different parts of the world together like never before. You might say that a new medium provokes a certain boundary confusion. And since boundaries remain important as a way of making sense of things, in everyday life as in politics, conscious defenses against impurity are constantly erected, even while borders are being dissolved via markets and technology.[2] A now little-read volume provides engaging reflections on the character of this misdirected defense.

The Absolute at Large, a novel by Czech writer Karel Capek published in 1922, describes the invention of a machine, the "Karburator," that can convert matter completely into energy. The inventor discovers new aspects to the pantheistic doctrine that God is everything. When matter is destroyed, a small quantity of divine principle ("the absolute") is released. Those exposed to it become religious and begin to proselytize. But each person affected espouses a different religion, and is prepared to do battle to advance their views. Heedless of the consequences, an unprincipled entrepreneur mass-produces the Karburators and creates a religious war of all against all. Written at the close of World War I, the novel is a satire that mocks the pretensions of a technological age. Capek hints that technology might ally with fanaticism rather than with enlightenment.[3] Neither alien nor instrumental, it transforms perception imperceptibly, manifesting not as something new, but as something human and familiar. The failure of recognition here arises from our mode of submission to the machine. Capek does not suggest a straightforward identification with the machine and its power. Rather, he recounts a story of infection and contagion where the effects appear to multiply by themselves while the cause is ignored.

Capek presents a penetrating account of how technology's effects manifest. The question remains as to how we might unlearn such habits of thought. If technology appears as one or other form of traditional power, multiplied by the zeal of its individual adherents, how would we identify it? Are we any closer to doing so, nearly a century after Capek? If anything the clarity of his insight is harder to grasp because we are taken up with a host of related but distinct issues, e.g., about surveillance, disinformation, centralized control, uneven development, and as well, religious fanaticism, and gender and racial bigotry. In consequence, questions of technology and its fetishism are layered over by growing concerns of domination, as if the latter could be understood independently.

We can notice the instant formation of a field of influence around a successful new technology, such as Capek describes. Tracing the interference patterns that emerge at the intersection of different force fields perhaps offers an access route to thinking about ignored forms of materiality. If every machine is willy-nilly a medium of communication, those designed explicitly as media communicate multiply: denotative and connotative levels each have material and symbolic dimensions. Interference patterns therefore tend to proliferate around an old medium when seen from the perspective of new media, and render it more noticeable. What appears as noise can therefore be reconsidered to provide insights into the work of technology.

Media do not only affect us, they also affect each other. For example, when print becomes the vehicle for middle classes, oral media such as gossip and rumor become under-the-radar vehicles for mobilization, potentially undermining the power of print to consolidate opinion. To take another instance, once television and newer media become prominent, audio-cassettes and radio become propaganda tools that experience little surveillance. Those who rely on a specific medium tend to be unselfconscious about their cognitive limits, making them vulnerable to insurrections from older media and from life-worlds and language groups considered beneath the pale. By the same token of course, a generation whose consciousness is shaped by the Internet may be able to thwart existing policing capacity, as we saw in worldwide protests against the WTO, in Seattle and elsewhere.

Technics and technology are not the same (technology means knowledge of technics), but the two words have collapsed into each other, implying an object adequate in itself, both inviting and resisting proper understanding. The increasingly technologized character of the world thus presents a paradox. Technology is obtrusively present, in new and constantly changing ways. At the same time, it is everywhere and invisible, and it provides the representational apparatus through which to understand itself. There is no "off" switch for technology, no place unaffected by it.[4] Even the absence of new media in a given place is now marked by their presence elsewhere. Perceptions are transformed as well, so that no place is what it used to be. There is no there anywhere, if we follow this logic. That is, our cognitive dependence on technology has rendered it into a second nature.

The resulting paradox can be seen, for instance, in the events of September 11, 2001 and thereafter. Islamic zealots5 with box-cutters succeeded in converting a civilian medium, the passenger plane, into a bomb, and struck at global centers of financial and military power. No one had imagined such a thing would happen, and there were few barriers once the attackers' plan was set in motion. An advertisement for Microsoft explains: "[I]t came to light that various agencies had clues to the intentions of the 9/11 hijackers, but no one connected the dots, *in part because of incompatible information systems*" (emphasis added).[6] We might say, in fact, that for all its pervasiveness and power, technology becomes a cultural envelope promising safety from those without. But if those within cannot conceive of intimacy with outsiders, the converse is not true. The detailed understanding displayed by the attackers, of the relevant civilian systems and their operations, and of the media and publicity, underlined that they were not, or not simply, mad mullahs from afar, but savvy insiders as well.

Part of the myth about technology is that it has helped refined the application of power, so that the brutality of earlier regimes, such as in the era of sovereign states, are superseded by sophisticated and unobtrusive methods of surveillance and control. But "soft power" functions together with the will to exercise overwhelming force. As such the idea of a new era of refined power is, at a minimum, incomplete and misleading.

Against the conspicuously authoritarian initiatives witnessed in so-called counter-terrorism, there are more positive effects attributed to technology as markers of progress and human endeavor. Previously, political economy foresaw the worlding of the world through commerce, materialized in the forces and relations of production. Today technology accomplishes this performance more effectively without requiring initiation into a specialized language. For example, pictures can stand in for any theory of global interconnection. Scenes of a Masai herdsman on a cell phone or of a laptop-wielding Buddhist monk may be juxtaposed with similar combinations of distant and familiar images to imply a world already globalized, projecting a future that will be mysterious but safe.

Images like these present unequal global exchange as technological facticity, where consumption frames the limits to understanding. If political economy, grounded in utilitarian assumptions, falls short in its critique of sensuous forms of capital, technology itself has appeared an obvious referential basis for scholars attempting to map these new times. It thus often appears as the driver of change, linked obviously to economy and politics, but able nevertheless to explain social change on its own. Whether blatant and intrusive or discreet and ubiquitous, technology has acquired doxic status, making a critical understanding of the forms of its influence difficult.

Thus when President Bush depicted the perpetrators of the Sept 11, 2001 attack as cave dwellers, he understood them as primitive despite their evident ability to maneuver and exploit modern sensibilities. Technological backwardness became a means of expressing racial and cultural difference, to say nothing of moral backwardness.

As we relinquish master narratives, "globalization" acts as a placeholder for the questions that we don't answer, a seemingly neutral name by which different societies are ordered. Technology fulfills a similar function; its use appears matter of fact, but it is value-laden. As Bush's allusion suggests, the West is thought to symbolize a technological modernity that others do not share. Such a view misstates the character of the enmity as well as the form of technological culture. There is indeed no outside any longer to technology, no cultural insulation adequate to stop its effect; the idea of civilizational conflict is therefore nostalgic. The persistent invocation of difference in these terms, however (not only by President Bush, but also, e.g., by Samuel Huntington and others), indicates that the west's relationship to technology has become narcissistic. We should recollect that Narcissus's problem was not simply self-love; rather, he failed to recognize that the image he loved was his reflection. Not surprisingly, other cultures find ways of identifying with technology as well.[7]

In this paper, I will consider the materiality of mediation as it surfaces in the interaction of old and new media, in three ways: forms of exchange and of property; differential effects on sense perceptions; and linguistic variation. First, as instrumentalities of human interaction, communication media shape the character of social and economic exchange by virtue of what they enable and forbid, and this in turn shapes the form of property, i.e., the kind of power that takes shape in their wake, as I will explain. Second, each medium impacts on the sense ratios in different ways, with different implications for historically located sensory formations. If the effects of a medium are available to perception, perception itself has a history that can complicate the relation between medium and its message, in ways that require to be understood. Finally, although a monolingual imagination dominates most thinking about the media, each medium reproduces or bridges linguistic boundaries. Hence it is salient to inquire what effect media have on existing cultural imaginaries located in language, and on the rules regulating their boundaries. Specifically, I will inquire into the working of a linguistically fragmented print public alongside an electronic public that can bridge such divisions.

Exploring these issues allows us to specify the social and cultural aspects around which particular identities tend to crystallize, and to demystify forms of exclusion that characterize even recent critical theories. The extent to which an unembarrassed Eurocentric imaginary gets reproduced might be explained by the fetishism of new technologies, and the dismissive attitude sanctioned towards zones harboring putatively older technology. An older Marxian political economy, for all of its Enlightenment conceits and productivist fallacies, recognized and sought to account for spaces of difference. Theorists of new technology, by contrast, seldom acknowledge such alterity, much less confront it. I will examine one instance here.

Power in Control Societies, and the Power of Control Societies

Deleuze is not a theorist of "new" technology, but his insights have been widely influential amongst those writing on the subject. As is well known, he writes in his essay "Postscript on Control Societies" about the successor form to disciplinary power as theorized by Foucault. In societies characterized by disciplinary power, the production of institutional sites of confinement such as the asylum, the factory and the prison is based on molding specific subject-positions for individuals. By contrast, "control societies," in Deleuze's adaptation of William Burroughs's term, no longer focus on individuals, but produce multiple subjectivities in changing patterns. Imposing debt rather than confinement, they represent a shift in the modes of exercising power. The shift accompanies the move from analog to digital technology; power is thus more flexible, short-term and multi-form; it is continuous and recognizes no borders. Whereas disciplinary power was long-term, confined to specific spaces, and thus discontinuous. Power in control societies is therefore harder to name, and to resist.

As a heuristic, Deleuze's account is valuable in alerting us to the repressive potential of developments in "information society" that are often welcomed uncritically. He relates it to the growth of "meta-production," where capitalism is "no longer directed toward production but toward products, that is toward sales or markets." And what happens to production itself? "[It] is often transferred to remote parts of the Third World, even in the case of complex operations like textile plants, steelworks and oil refineries."[8]

Although we were told control societies recognize no borders, the sense of place continues to divide the world into intimate and remote spaces. We might assume that older forms of power obtain in those remote parts, corresponding to the obsolete forms of capitalist production carried on there that "no longer" characterize capitalism as such. But in this connection, Deleuze notes:

> "One thing, it's true, hasn't changed—capitalism still keeps three quarters of humanity in extreme poverty, too poor to have debts and too numerous to be confined: control will have to deal not only with vanishing frontiers, but with mushrooming shantytowns and ghettos."[9]

Deleuze does not clarify how control works or would work in these distant places; his argument is addressed to those regions that have succeeded in shifting production to remote places in the Third World. His inclusion of the Third World at times appears like a postscript to his Postscript ("One thing, it's true, hasn't changed ..."), but it is nevertheless integral to the claim he makes. Deleuze is theorizing the form of power specific to "metaproduction" but curiously, regards this as separable from "production."

Deleuze's argument reflects, even if it disavows, the biopolitics of control society in relation to the world's "three-fourths;" in this respect we find here a consolidated rather than a shifting identity for the west. The logic of power operative within control society does not appear to be the same as that without.

Now, the global spread of capitalist production is clearly layered and uneven; sovereign power is well and alive in many parts of the globe, even if it co-exists with newer forms of power. What kind of power is exercised and for whom requires to be considered if the invocation of the term "power" is not to become a theodicy. As Foucault noted, new forms of power do not always or entirely replace older ones;[10] rather they subsume them, in ways that can hardly be told in advance.

What was naïve optimism in Marx regarding the future of non-western countries, after a century and a half appears in Deleuze as naïve pessimism. I'm not interested in dwelling on Deleuze's dismissive and indeed incoherent account of the world's three-fourths. The question is rather, how can we open up this racialized account of technological modernity, which even a savant reproduces so casually? If we agree with Hardt and Negri that communication has become the central element in the relations of production, "guiding capitalist development and also transforming productive forces," communication then requires to be theorized together with the work of capital so as to "demonstrate all the contradictions at the heart of it."[11]

Communication as Property

The rendering of communication into a thing makes it possible to submit it to a rigorous analysis in terms of forms of exchange. Although it occurs within market relations, mass mediated communication can be described in terms of what anthropologists have described as gift exchange. Mauss acknowledges, in his essay on the subject, that there is in fact no historical practice corresponding to the present meaning of the term, i.e., a voluntary transference of property without consideration. What anthropologists term gift exchange is part of a network of mutual transfers, where the giving of a gift entails the compulsion to reciprocate. With mass communication, we have the enactment however of what is *experienced* as a free gift, albeit on terms enabled by capitalist exchange. Within the private space of print communication, it is possible to contemplate society as an object of criticism without the threat of surveillance and reprisal, as would be customary with face-to-face communication, for example.

As I noted at the outset, technology reshapes perception at the same time as its productive capacities are harnessed, that is, technology is always also a communicative force. The conversion of a good into a commodity, of use value into exchange value, necessitates a communicational component. As technologies of communication are themselves commodified, the productive power of communication begins to be built into the process of valuation. The forms of valuation, however, are premised on economic theories that hold scarcity as central, and social interaction as modeled on a zero sum game. The media, in circulating images and information that can theoretically be shared by everyone, indicate the possibility of economies premised on abundance rather than scarcity, and in this respect invoke what have been theorized as characteristic of so-called primitive economies. As Bataille has observed, primitive economies are premised on abundance rather than scarcity, and indicate different ways of thinking about the social mediation of needs.[12]

Anthropological accounts of the forms of exchange specific to these economies center, of course, on the gift and gift exchange, rather than on the commodity. Gift exchange was believed to be the model of primitive economies, where goods were circulated not according to individual interest but dictated by rituals of social obligation and the compulsion to reciprocate. The conceit of economists and indeed of most social scientists was that modern capitalist exchange was normatively centered on rational interest, and thus distinct from gift exchange. In economic terms, we can understand the effects created by modern communications through what Balibar has called "universal property," sustaining modes of participation distinct from the competitive, zero-sum activity of markets.[13] Thus the expansion of markets and media leads to the circulation not only of commodities but of commodity images, that is, of non-material forms of property that are inexhaustible and undiminished by use, and yet are treated in other respects like conventional property, that is, they can be owned and transferred. These new non-rivalrous forms of property, and the kinds of solidarity they enable, are already being mobilized, and require to be more accurately understood.[14] I suggest that they cast light on the explosive salience of the imagination, and the resulting importance of identity politics in recent times.[15]

Property, for the philosopher John Locke, was an attribute of the human ability to labor freely and independently, and as such was an individual quality.[16] Property socialized individuals into civil society, and as such, expressed extant forms of social regulation.[17] Prevailing understandings of property are constantly being re-politicized, however, and new forms eventually mutate and develop to express conflicts irresoluble through older property forms. A given state regime gives rise to historically determinate property forms, and in turn provides the crucible for the elaboration and refinement of different types of property, which then demand new modes of regulation and generate new kinds of political contestation.

Tracking the changing forms of property is thus a way of tracing the changing forms of power, in the departure from state absolutism and the shift through disciplinary to control society. In Deleuze's account, the continuous and modulated character of power in control society renders it distinct from its more bounded form in disciplinary society. We can clarify his account in terms of the expansion of communications and the resulting proliferation of non-material forms of property, which have come to be designated as intellectual property. These forms of property, whether as code, commodity images, or other information, increasingly help shape the individual's mode of participation in social life. Access to different social realms is regulated through the particular kinds of intellectual property one has gained access to, and is thus both a means of empowerment and of control. Unlike a form of power that works through direct perception, surveillance here is indirect and mediated, and hence can be continuous across different bounded spaces. It is able to insinuate a more intimate cooperation with its subjects, since the mechanisms it utilizes are non-rivalrous. That is to say, there is no inherent necessity that the prevailing codes of private property, i.e., of scarcity and the zero-sum game, are mapped onto the forms of intellectual property being utilized.

Machinic Mystery, Sensory History

A new communicating medium is first of all a cultural object to which habits and perceptions must be accommodated. For most people, it usually arrives already sanctified, as something capable of superseding existing technology. Its power has a magical rather than a purely objective quality at this point; it does not simply execute a function but has a special charge. It replaces or transcends a familiar entity and therefore acquires a distinct energy in accomplishing this task.[18]

For example, when the computer came to India, it did not of course arrive as a foreign object. It was long preceded by news of itself, as a fearsome calculating machine capable of doing the work of hundreds and indeed of replacing them. The IBM 1620, when it came to my college in Madras, was housed in a special air-conditioned room, and in its own building. To enter the wing it was contained in, it was necessary to take one's shoes off at the entrance, whether as a gesture of sanctity or a sanitary gesture it was uncertain. A doormat might have sufficed to rub the dust off one's feet, but a more powerful ritual was required to contain and maintain the aura. Today of course, computers have become more commonplace, and this is no longer considered necessary.

Television too was no ordinary object as it made its entry in India. Why do they call such a beautiful thing "TB," an old woman asked me, pronouncing "v" like "b," as many native speakers do, and so rendering television into tuberculosis. It was not just a beautiful object, but a special one. Protective devices such as an add-on screen would be used to guard the front of the machine, and also insulate viewers from what was believed to be its radiation.

This was not simply an appreciation for the materiality of the object or the effulgence of its appearance. There was a larger and more complex context in which linguistic and political forces were at work, into which the reception of television played. Briefly outlining this context and its interaction with the medium is helpful in understanding the kinds of misrecognition that may occur with the arrival of a new technology. My example is from the recent history of Indian television.

A Linguistically Split Public and the Politics of Sensory Perception

A few years after nationwide television programming was inaugurated in India, Hindu epics began to be aired on state-run television, in 1987, violating the existing ban on devotional programming. The serials were a critical failure, but not only attracted immense audiences, they brought everything to a halt while they were on air.

The English language press was for the most part embarrassed by the "outing" of what they saw as an idolatrous and superstitious national culture, and by the fact that the epics were not produced as classics, but rather as low-budget melodrama. Here television was drawing on indigenous traditions of mythological realism in the cinema, as well as of vision itself as a participatory and tactile collective ritual, of *darsan*.[19] The Indian language press, by contrast, gave the epics, which were broadcast in Hindi, a relatively rapturous welcome, and promoted the serials as returning to the values of Indian, or Hindu culture. Hindi and English are together, the national languages in India. It appeared like a sudden coming together of an immense and usually discordant nation. That the occasion was the broadcast of an ancient epic allowed myopia about the technology of collective assembly. Television was like a hinge, swiveling between time zones. Audiences could travel back to an age when the culture was harmonious; sponsors for their part heralded the onset of market liberalization.

In the ambiguity of whether audiences were going back to a golden age or forward to a globalized future was contained the kernel of controversy. Since India became independent, a secular government had gone to some lengths to limit the possibility of minority oppression by the Hindu majority. Over time, secularism's opponents succeeded in identifying it as an English language view and a neocolonial prejudice. Secularists themselves usually understood that theirs was a minority position, and relied on state power to keep opponents such as Hindu nationalists at bay. But it was a secular ruling party, the Congress, which sanctioned the broadcast of Hindu epics on state television, violating a decades-old taboo on religious programming to court the Hindu vote. Ultimately, it was Hindu nationalists, long awaiting such an opportunity, who were able to capitalize on the event. A violent campaign that left thousands dead, mainly Muslims, arose in the wake of the serial, and ultimately brought the Hindu nationalists to power in India in 1998.

The movement advanced through visual and print media working in tandem. Television spanned regional and language divides, and provided a spark to Hindu nationalist consciousness, but it was through the Hindi language press that Hindu nationalists were able to obtain the most effective support. The English language press anchored its authority in the state, and saw the Hindu nationalist agitation as a threat to law and order. For the Hindi press, it was first and foremost a popular issue, and they reported therefore as a cultural as well as a political matter. The English language media was handicapped in exploring the agitation's motives, by contrast, since its news values ensured a social distance from the movement. The criticism of the English press had the opposite of the intended effect, since it confirmed to Hindu nationalism's Indian language audiences that the neocolonial elite was apprehensive of the movement's success. More importantly, it prevented secularists from making effective interventions into the campaign, and enshrined their position as hostile outsiders to the culture itself. Thus the Indian and English language news media themselves worked together as well, the former expanding the support for the agitation, and the latter providing the friction necessary for its forward movement.

Friedrich Kittler has argued that the sense of loss that haunts writing is erased by new media, that render the past into an accessible presence.[20] If new media make information "want" to be free, they seem also to create pasts that "want" to be restored.[21] But this dynamic was not compartmentalized by medium. Nor was there any naivety in the political act of invoking of a bygone era. An opportunistic political program brought together stratified and re-organized ways of perceiving, and institutional differences in language news media, under a single visual regime and an icon of high tech modernity. If secularists had believed that a technology for unifying the nation was at hand, they were confounded by the material force of a communication medium that allowed a regressive politics to be mobilized. But if Hindu nationalists for their part assumed that with the past on their side, the future was theirs, matters would prove more complicated. The embodied history of a caste-divided society could also be "the Real" summoned by new media, setting a different dynamic in motion, slow but sure in its effects, and spotlighting the fabricated character of Hindu unity.

The specialization of the senses via technical media, and their separation and fragmentation from embodied sensations, lead to disadvantaging communities that do not have control over the means of representing themselves in mass media.[22] Prevailing and more deeply rooted sensory histories become interleaved with mass-mediated perceptual formations. To be attentive to this layered formation contributes to understanding how media help shape our environment.

Virtual Touch and the Quality of Social Space

Marshall McLuhan observes that the contact between cultures based on different media, e.g., oral and print cultures, is an explosive event, accompanied by the release of tremendous hybrid energies.[23] The force of his point derives from emphasizing the aesthetic over the historical dimension of these encounters, and directs our attention to the material media through which the perceptions of such encounters are framed. The point is applicable within as well as across societies. But we should note that the perception of explosive energy may correspond to patterns of activity whose rules are not discernible to the observer. The advantage here may belong to those who can play upon the illegibility of their actions. We observed this advantage in the ability of Hindi language print media to advance a movement subversive of the existing political dispensation partly because of the false sense of power generated by television's introduction. McLuhan could not have predicted the outcome that occurred here, however, because he tends to homogenize the space in which media effects occur, and to treat these effects as literal rather than as themselves mediated, e.g., by linguistic and sensory histories.

We should then consider the implications of McLuhan's dictum that old media appear as the content of new media. To be sure, the telegraph subsumes print in this way; similarly, television incorporates the cinema, radio and print, and the computer envelopes all the rest. But is the content of older media so unstructured as to dissolve into the forms of newer media? What are the relationships of the contents of the older media within and amongst themselves? Here we can think broadly of oral and print media, and the differences and tensions that they may carry, between various communities of oral and print culture. These differences could hardly be nullified within electric media while the communities themselves remained in similar circumstances. How do we map the complex transactions that occur here?

Oral media present the immediacy and unpredictability of the face-to-face encounter, where spontaneity can complicate the routines of market rationality. In being folded into a market economy, the promise of such subversion tends to be important as part of the appeal of oral media, although of course this could redound to the benefit or the misfortune of interlocutors. It is important to recall here that oral media tend to reproduce the power of the communities within which they occur. Expression in oral media is anchored and circumscribed through the senses of belonging and obligation, reciprocity and surveillance. Oral media thus mediate the social dynamics of dominance and subordination, with market rationality as but one of the attendant factors of interaction.

Print has been described as figuring/ushering in a homogenization of space and time. It may be more accurate to say that this account reflects a specific age of capitalist rationality, since capitalism came to self-consciousness in the age of print.[24] Printed goods were the first mass-marketed commodities, after all. With print media, a different imagination of the social compact is possible, because communication can be entertained in isolation from the supervision of others, in the private space afforded by commodity consumption. Social membership is thus made possible without the burden of community obligation and scrutiny that attend direct interaction. As such, print communication provides its users a means of re-imagining existing community relations.

Electronic media exponentially increase the capacity to project the experience of autonomy for its users, again, within the space made available by consumption. We should note that the speed of its circulation alters the dynamics of communication, since large numbers can experience the same events via media. If print redefined the boundaries of a community of experience in abstract linguistic terms, divorced from the sensory immediacy of a given context, electronic media redefine the boundaries again. They can transcend a given linguistic field with sounds and images that recreate the sense of presence with oral communication. McLuhan referred to the effect of such media as "retribalization," in a radical break from the abstract, linear rationality of print and a return to the direct and unmediated character of oral culture.

But each new medium changes the sense ratios: print emphasizes the visual to the exclusion of other senses; electric media emphasize sound and image. Perception and its effect are not necessarily on the same plane; the senses refer to each other, and work through the body.

Touch is the guiding sensation in infancy, along with, to a lesser extent, sound and smell; it is touch that orients the sense of sight. Together with the social inculcation of spatial cues, touch enables the otherwise incomprehensible quilt of light and dark colors comprising the visible world to acquire meaning and distinction.[25] Sight is not a pure datum, but is structured through rules regulating objects in social space, and one learns these rules in order to be able to acquire vision. To see is also to confirm one's bodily relation to social space. McLuhan's point about the effect of "electric media" being primarily tactile rather than cognitive or visual becomes clearer in this context. The media form an exteriorized nervous system and thus a collective prosthetic, through which mediated sense perceptions then refer to each other. The abolition of distance through sound and images induces a sense of intimacy, reflected and reinforced through the shrinking social distance governing norms of publicity. One "tunes in" to get "in touch." How is this mediated sense of touch distinct?

Aniket Jaaware, in his important work on the phenomenology of touch, has pointed out that the sense of touch is extremely difficult to deceive; contact is both the form and the content of touch, and gives reality to it.[26] In the case of actual contact, figural meanings accrete around literal touch. With electric media, it is the opposite: a figural, imaginative sense of touch occurs, which may then generate the feelings associated with literal touch. Elias Canetti begins his famous work on the crowd invoking the universal fear of being touched. Electric media present the opportunity to be touched without fear; one seeks being touched and it becomes a regular and sought-after form of intimacy. Jaaware has observed that touch is a sensation that cannot be stopped; hence the vulnerability of the body to torture for example. But the unstoppability of touch takes on a different meaning with electric media; to describe it in terms of shock and anaestheticization retains a literalist frame of reference that is superseded here. Touch becomes virtual.

In societies dominated by strict rules of social distance, of racial, class and caste divisions, electric media insinuate the touchability of one and all. A culturally insular imaginary is thus liable to be challenged as the rules of visceral touch are subverted by those regulating virtual touch.

The world of print media has often received nostalgic treatment, as an era of rationality that monopoly capital and consumerism have torn us away from. Electronic media might be ranked here as instruments of seduction; at any rate they place a question mark against older assumptions that technological mediation could only promote enlightenment.[27] Print extruded the authority that previously was secret and located in privileged bodies, and rendered it abstract and diffusible. But the printed book has a doubled character, a publicly circulated good with a declared exchange value, and an interior life disclosing unforeseen imaginary worlds. The fetish character of the book could be concealed through its normalization, and its cultural sanction as a "rational" power; this sanction was of course retrospective, occurring after the heyday of print, not during it. If books require to be bought and read, and so limit their direct constituency, electronic media create environments from which no one is excluded; indeed exclusion becomes the new privilege. In this sense, there are no longer any have-nots in the earlier meaning of the word. Electronic media democratize the imagination of intimacy, but technological fetishism obscures our perception of this process; an essentialist understanding of media influence is the counterpart to such thinking. McLuhan's proposition of a global village resulting from electronic media is an example of this error, just as his dictum, "the medium is the message," sanctions fetishistic thinking. Thinking about the interface between older and newer forms of mediation can help identify the forms of misrecognition and displacement that, as Karel Capek has described, allow us to treat technology as world-transforming, but nevertheless transparent to human will. A racialized identification of the west with technological modernity is currently rampant. But the irony of course is that others are capable of the same errors as ourselves in this respect. Hence, we err at our mutual risk.

Notes

1. Portions of this paper have been presented at the University of Colorado in Boulder, in March 2004, and at the University of Hyderabad, India in September 2004.
2. For an extended discussion of this topic see Allen Feldman's essay in "America and Its Others." Special Issue of *Interventions: International Journal of Postcolonial Studies* 7, no. 3, edited by Arvind Rajagopal.
3. Karel Capek, *The Absolute at Large*. New York: Hyperion Press, 1989. I am indebted to Thomas Ort for drawing my attention to this book, and for making me aware of its significance.
4. Untitled presentation by Avital Ronell at a conference titled America and Its Others, New York University, Steinhardt School of Education, Dec 13, 2002.
5. *The Concise Oxford Dictionary* notes that zealot referred originally to a Jewish sect resisting the Romans in the first century C.E. As such, the nomenclature confers minority status on the members of this category, we may note; zealotry was located in the refusal to accede to ruling power.
6. "Open Society and its Allies," Advertisement, *New York Times*, July 9, 2003, A21. Posted also at http://www.microsoft.com/issues/essays/2003/07-08security.asp
7. For instance we have witnessed invocations of a Hindu bomb (in India) and of an Islamic bomb (in Pakistan).
8. Gilles Deleuze, *Negotiations. 1972–1990*. Trans. Martin Joughin. New York: Columbia University Press, 1995, 181.
9. Ibid.
10. Michel Foucault, "Governmentality," in Colin Gordon et al., eds., *The Foucault Effect*. Chicago: University of Chicago Press, 87–104.
11. Michael Hardt and Antonio Negri, *Empire*. Cambridge: Harvard University Press, 2000, 348.
12. Marshall Sahlins, *Stone Age Economics*. Chicago: Aldine Atherton, 1972; Georges Bataille, *The Accursed Share* (2 vols.). Trans. H. Robert Hurley. New York: Zone Books, 1988.
13. Etienne Balibar, "'Rights of Man' and 'Rights of the Citizen': The Modern Dialectic of Equality and Freedom" in *Masses, Classes, Ideas*. Trans. James Swenson. New York: Routledge, 1994, 52–53.
14. See Lawrence Lessig, *The Future of Ideas: The Fate of the Commons in a Connected World*. New York: Vintage, 2001, e.g., 115–116 on non-rivalrous resources. See Helen Nissenbaum on the qualification of private property rights when applied to intellectual property in her essay "Should I Copy My Neighbor's Software?" in Deborah G. Johnson and Helen F. Nissenbaum eds., *Computers, Ethics and Social Values*, Englewood Cliffs, NJ: Prentice-Hall, Inc., 1995, 200–212. On the need to "thin" extant notions of copyright, see Siva Vaidhyanathan, *An Anarchist in the Library*. New York: Basic Books, 2003.
15. On this point, see Arjun Appadurai, *Modernity at Large*, Minneapolis: University of Minnesota Press. For an elaboration of the argument I present here, see my *Politics After Television: Hindu Nationalism and the Reshaping of the Public in India*. Cambridge, UK: Cambridge University Press, 2001, Introduction.
16. John Locke, *Second Treatise of Government*. New York: Hackett Publishing Company, 1980.
17. Etienne Balibar, "'Rights of Man' and 'Rights of the Citizen': The Modern Dialectic of Equality and Freedom" in *Masses, Classes, Ideas*. Trans. James Swenson. New York: Routledge, 1994, 50–51.
18. See the discussion of contesting regimes of perception in Michael Taussig's *Mimesis and Alterity*. New York and London: Routledge, 1995. Christopher Pinney provides a valuable account of North Indian negotiations between the optic and haptic registers in his essay, "The Indian Work of Art in the Age of Mechanical Reproduction: Or, What Happens When Peasants Get Hold of Images," in *Media Worlds*, 355–369
19. I have discussed these ideas at length in *Politics After Television: Hindu Nationalism and the Reshaping of the Public in India*. Cambridge, UK: Cambridge University Press, 2001.
20. F. Kittler, *Gramophone, Film, Typewriter*. Trans. Geoffrey Winthrop-Young and Michael Wutz. Stanford, CA: Stanford University Press, 1999.
21. Rosalind Morris, "A Room With a Voice: Mediumship in Thailand," in Faye Ginsburg, Lila Abu-Lughod and Brian Larkin, eds., *Media Worlds: Anthropology on New Terrain*. Berkeley: University of California Press, 2002: 383–397.
22. Allen Feldman, "From Desert Storm to Rodney King: On Cultural Anesthesia," in C. Nadia Seremetakis, ed., *The Senses Still: Perception and Memory as Material Culture in Modernity*, University of Chicago Press, 1998.
23. "Culture contact" is not usually an event, of course; more often it is a mediating structure of perception, a zone of negotiation and an indefinite period of interaction. See, e.g., Mary Louise Pratt, *Imperial Eyes: Travel Writing and Transculturation*. New York: Routledge, 1992.
24. Elizabeth Eisenstein's work has complicated this Enlightenment self-image of print culture, showing how the shift from writing to print led to the separation of scientific and religious discourses, which grew apart with the development of print. It is scientific thinking that has been identified with print culture, but if religious literature, which mystified rather than rationalized authority, grew alongside it, then we cannot accept the received view of print culture. See, e.g., her volume *The Printing Revolution in Early Modern Europe*, Cambridge, 1983.
25. See Otto Lowenstein, *The Senses*. Penguin Books. Cited in Marshall McLuhan and Quentin Fiore, *War and Peace in the Global Village*. New York: Simon and Schuster, 1968, 10.
26. Aniket Jaaware, unpublished manuscript.
27. For a provocative discussion, see Todd Gitlin, *Media Unlimited: How the Torrent of Images and Sounds Overwhelms Our Lives*. New York: Owl Books, 2003.

20

Deep Europe
A History of the Syndicate Network

Geert Lovink

The inner life of a mailing list reveals more than discursive threads and communication patterns. There are sophisticated forms of silence, repressed messages and unanswered remarks. Because of the intimacy of e-mail and the immediacy of open, unmoderated channels, lists foreshadow events. As antennas of culture they do more then merely discuss current affairs: online communities do not just reflect events but have the potential to create their own auto-poetic systems and provoke events. For mainstream media and its professional critics, discussion lists are almost invisible cultural phenomena, yet they play a key part in the life of their participants. Many incidents happen on lists, which become visible and emerge later in different forms. Founded in early 1996 as a "post-89" East-West exchange network between new media artists, Syndicate grew into a network of 500 members Europe-wide and beyond. Built as an informal new media arts network, the Syndicate network was suddenly polarized by political debate, which it did not survive. Its open architecture was vulnerable to the challenges of hackers, "trolls" and quasi-automatic bots. These challenges eventually brought down Syndicate in August 2001. The story of Syndicate is a didactic one because the hatred that appeared in a medium, which originally was meant to be democratic, can tell us something about upcoming extreme cultures that operate beyond rational consensus.

Syndicate, founded in January 1996 during the second "tactical media" Next Five Minutes conference, was the brainchild of Andreas Broeckmann, a German new media critic and curator who worked out of the Rotterdam-based V2 new media arts organization. In the autumn of 1995, Andreas Broeckmann started a new initiative called V2_East, which sought to create a network of people and institutions involved with, or interested in, media art in Eastern Europe: "V2_East wants to create an infrastructure that will facilitate cooperation between partners in East and West, and it will initiate collaborative media art projects." Syndicate was going to be the vehicle for V2_East.[1] During the early to mid-nineties most of the exciting media (arts) initiatives didn't come from the recession-plagued West but from the "wild" East which had only recently opened up. Before 1989, creating a network with new media artists and organizations throughout the fifteen countries in the East would have been next to impossible. This was the time to do it. But how would an equal East-West network function, especially if it was run from Western Europe? Conspiracy theories thrived, especially in an environment flocked with money from Wall St. speculator/philanthropist George Soros. Was there a hidden neo-colonialist agenda, with new media arts as its forerunners?[2]

As well, there was unspoken skepticism about exchanges planned from above—and good intentions in general. "Community" was a contaminated concept that came dangerously close to "communism."[3] On the other hand, this was not the time to be dogmatic and reject opportunities. For decades, many had sincerely desired a "normalization" of East-West relations.

Deep Europe

The term "Deep Europe," with which Syndicate became associated, stems from Syndicate's participation in the 1997 Hybrid Workspace project, a temporary media lab that was part of the Documenta X art exhibition in Kassel/Germany. Syndicate was one of twelve groups that organized its own ten-day workshop, partly open to the public. A group of twenty artists, mainly from the former East, held debates, screenings and performances. The highlight was the "visa department" performance, in which all Syndicalists participated: the DX exhibition visitors had to stand in a long queue and be interrogated before obtaining a Deep Europe visa. The announcement stated: "The new lines that run through Europe are historical, political, cultural, artistic, technological, military. The role of the EU and its institutions, the notion of Mittel (central) Europa, old and new ideologies, messianic NGOs and late-capitalist profiteers contribute to a cultural environment in which we have to define new strategies and new tools, whether as artists, activists, writers or organizers."[4] The text warns us not to ascribe too much meaning to the Deep Europe concept—but that's exactly what happened.

The exact origin of the Deep Europe term, before early 1997, remains unclear. It may have had multiple sources. I can only provide the reader with my interpretation. Deep Europe was a precise, timely and productive label precisely because of its ambiguity, being neither geographic (East-West) nor temporal (old-new). Deep Europe was proposed as the opposite of fixed identities. The overlapping realities were there to be explored.[5] Caught in-between regions, disciplines, media and institutions, the V2_East/Syndicate network was open to those interested in "Becoming Europe," working with "Becoming Media." Of course, Deep Europe ironically underscored essential values in opposition to superficial simulations. There was nothing "deep" about the twentieth century tragedy called Europe. Deep Europe would grow out of the tension produced by the crisis of the ethnic nation state and the promising poverty of globalism. I would reconstruct it as a blend of Continental Europe (a notion used by English islanders) and the astronomic/science fiction term "deep space." It was an unknown, yet to be discovered part of Europe, way beyond the bureaucratic borders drawn by the EU, the Schengen agreement, NATO and Russia. Europe in this context had to be understood as an open and inclusive, lively translocal network. Deep Europe was an alternative, imaginative mental landscape, a post-1989 promise that life could be different. Rejecting both superficial Western mediocrity and backward Eastern despotism, Deep Europe could be read as a desire to knit webs and tell stories about an unrealized, real *and* virtual world.

For moderator Inke Arns, Deep Europe expressed "a new understanding of Europe, an understanding which leads away from a horizontal/homogeneous/binary concept of territory (e.g., East/West) and—by means of a vertical cut through territorial entities—moves towards a new understanding of the different heterogeneous, deep-level, cultural layers and identities which exist next to each other in Europe."[6] Lisa Haskel, describing the possibilities of Deep Europe, writes:

> Not a political position, a utopia or a manifesto, but rather a digging, excavating, tunneling process toward greater understanding and connection, but which fully recognizes different starting points and possible directions: a collaborative process with a shared desire for making connection. There may be hold-ups and some frustrations, quite a bit of hard work is required, but some machinery can perhaps aid us. The result is a channel for exchange for use by both ourselves and others with common aims and interests.[7]

Concepts such as tunnels, channels and rhizomes were used here to indicate how informal, decentralized networks with their "subterranean connections" (Deleuze/Guattari) could cut through existing borders.

Syndicate as a Network

Meetings were essential to building a post East-West network. To produce real outcomes, Syndicate required considerable trust among its participants. Trust was never going to be achieved just via e-mail. Syndicate in 1996–99 was a traveling social network, moving from event to workshop to conference, from office to café to club, and further to the next airport, train station and bus terminal. Syndicate existed as an accumulation of meetings, collaborations and "peer to peer" exchanges, with the list as a secondary tool for exchange. Three readers edited by Inke Arns, in which the most important texts from the mailing list were collected, were published on the occasion of some of the meetings.[8]

Unlike the usual Internet lists, the first Syndicate years generated hardly any debate or response. The one or two postings a day were mainly festival and project announcements. As long as the offline community was organizing meetings and collaborations, there was nothing wrong with a list focused on the exchange of practical information. By 1998, Syndicate had reached 300 subscribers: the list had reached its critical mass and started to become more lively. Traffic went up. Typical Syndicate topics would be access, connectivity, collaboration and most of all the exchange of information about upcoming festivals, possible grants and new projects. The "no border" campaign, which focused on migration issues, turned out to be an important topic. The intensity of the list traffic exploded during the 1999 Kosovo crisis. The debates over the NATO bombing of Yugoslavia would be a turning point for the larger new media arts community.

Net Activism in Wartime

On March 22, 1999 the Serbian nationalist net artist Andrej Tisma, the source of earlier controversies on Syndicate, posted: "Message from Serbia, in expectation of NATO bombing. Could be my last sending. But I don't worry. If I die, my web site will remain."[9] Peace talks in Rambouillet between NATO, Yugoslav authorities and the Kosovo-Albanians had failed to produce an agreement. With mass killings and armed resistance spiraling out of control, Kosovo was well on the way to becoming the next Bosnia. In Bosnia it had taken Western powers three and a half years to intervene in a serious manner, after years of half-hearted diplomacy, broken cease-fires and limited UN mandates. In Kosovo, with the spring season close and parties on both sides gearing up for the next big killing spree, NATO took action in a decisive manner, causing a spiral of effects. On March 24, "the most serious war in Europe since 1945" (Michael Ignatieff) started. The NATO bombings on Yugoslavia were going to last for 78 days, until the Yugoslav army withdrew from Kosovo in early June 1999.

By the first day, the independent radio station B92 was closed and its director, Veran Matic, arrested by the Serbian police.[10] Local radio transmission no longer worked, but B92 continued its radio casts via Web. Not long afterward, the radio signal was retransmitted via satellite. News bulletins in both Serbian and English could be read on the B92 Web site. The Internet strategy to "rout around" the Milosevic regime worked for a good nine days, with 15 million visitors hitting B92. On April 2, B92 was permanently silenced. In the early hours, police officers arrived to seal the station's offices, and ordered all staff to cease work and leave the premises immediately. The final closure of B92 was a serious to blow to the tactical media strategy with which so many Syndicate members identified.

During the NATO bombings, Syndicate turned into a unique unfiltered citizens' channel, crossing geographic and political borders turned enemy lines. It had taken three years to build up the community; its direction had been unclear at times. This time proved to be Syndicate's finest hour. One day into the event, political postings started to appear on the list. Nikos Vittis, writing from Greece, pointed at the possible oil in Balkans as the reason for the US-intervention.[11] Andreas Broeckmann, in Berlin, summed up the Western position: "The only person responsible for the attacks is Milosevic—this is not a war against the Yugoslav people—the military objective is to stop the killing and humanitarian catastrophe in Kosovo and to force the Serb leadership to sign the Rambouillet agreement—this agreement cannot be negotiated any further—the attacks will be stopped as soon as the Serb leadership commits itself to signing the Rambouillet agreement."[12] From Skopje, Macedonia, Melentie Pandilovski reported anti-American demonstrations ("Let's hope things stay calm").[13] Nina Czegledy wrote about similar demonstrations in her hometown, Toronto.

Internet-based support initiatives sprung up in Budapest, Spain, the Californian Bay Area, Portugal, London, and even Tokyo and Taipei.[14] Groups translated texts, put up Web links, produced radio programs on joining Help B92. In these first days, people sought to stay informed and to keep the communication channels open. The emphasis on Syndicate was freedom of speech, thus tactically avoiding taking sides in the political conflict over the moral and strategic usefulness of the NATO bombardments. Both NATO commanders and Serb nationalists already had their war propaganda channels—and used them accordingly. The call for media freedom positioned itself as a "third way," a long-term contribution to resolve ethnic hatred. The position could be roughly described as such: We are not pro- or anti-NATO, pro- or anti-Serbian, we live in cyberspace. We come from the future and offer you hope, to drag you out of the nightmare called history. Global communication is not just a tool for reconciliation—it is part of the solution. In this view, new media did not just diffuse tensions in order to impose a manufactured consensus—the digital devices would lead the (online) participants into a new world. The independent-media-as-part-of-the-solution argument would be developed over the next three months in a variety of actions, worldwide.

ASCII-Art & the Serbian Revolution

By August 1999, the traffic on Syndicate was back to normal. Syndicate postings had jumped from 87 in February 1999 to 417 in March, 400 in April, down to 237 in May, 250 in June, and were back to previous levels of 157 in July and 118 postings in August. The summer period marked a move away from the Balkan news items. In August 1999, the first indications of a change of the atmosphere appeared. From an "anonymizer" server, stationed in Trontheim, Norway a short e-mail dialogue was forged, meant to create distrust and confusion.[15] In February/March 2000, the list fell into a loop several times, repeatedly sending out dozens of copies of same message. By April 2000, anonymous postings and net.art from individuals and groups such as propaganda@0100101110101101.org, net_CALLBOY, [brad brace], Dr. RTMark, iatsu.pavu.com and data[h!]bleede began to increase. Approaching 500 subscribers and still open and unfiltered, Syndicate was an easy outlet for e-mail art, varying from low-tech ASCII hoaxes to anonymous personal attacks. While announcements had been important to the social network early on, they now further increased the feeling of anonymity, which in turn encouraged net.artists to fill the gap created by the disappearing Kosovo exchange with more e-mail experiments. In May 2000, there were over 200 postings.

During the days of the "Serbian revolution" (early October 2000), when large demonstrations forced the fall of the Milosevic regime, Syndicate was revived as a p2p communication channel. For a brief moment, Slobodan Markovic, Dejan Strenovic and Michael Benson reappeared on the list, but their thoughts were quickly overrun by an ever-increasing number of announcements from the global new media arts sector. Postings no longer triggered responses. The last action by the Syndicate network was a spontaneous support campaign for the Albanian curator Edi Muka, who had been fired from his post as director of the Pyramid cultural center in Tirana.[16] While throughout 2001 Melentie Pandilovski regularly forwarded news updates from Skopje related to the crisis in Macedonia between Albanian (KLA) fighters and the army, the Syndicate list was de facto silent on this topic. Syndicate was a window to the world and provided useful information about the region, but it could not be considered a close and homogeneous community.

Machinetalk

In January 2001, "Netochka Nezvanova" (NN), named after Dostoyevsky's first full-length novel, began sending hundreds of messages to Syndicate, most often random responses to anything sent to the list. The posting were a mixture of replies, cryptic political analyses, machine talk[17] and personal attacks.[18] NN used a blend of software and Internet-specific styles of writing such as Europanto[19] and B1FF,[20] combined with an agitated Übermensch attitude (perhaps inspired by the Extropians), flaunting a machinic-futuristic "post-human" superiority over the all-too-human fellow subscribers with their petty and corrupt, dirty-dubious intentions. NN had been posting to nettime and other lists before and was a well known phenomena. NN's aim has been not just to dominate a channel but to destroy the online community as such.

In the larger social context, this phenomenon was known as "trolling." First used on the Usenet group alt.folklore.urban, a troll sends out messages designed to attract predictable responses or flames. The jargon file at tuxedo.org defines the troll as "an individual who chronically regularly posts specious arguments, flames or personal attacks to a newsgroup, discussion list, or in email for no other purpose than to annoy someone or disrupt a discussion. Trolls are recognizable by the fact that they have no real interest in learning about the topic at hand—they simply want to utter flame bait. Like the ugly creatures they are named after, they exhibit no redeeming characteristics, and as such, they are recognized as a lower form of life on the net."[21] One often sees the warning "Do not feed the troll" as part of a follow-up to troll postings, but this was exactly what happened on Syndicate.

Unfamiliar with the troll phenomenon, Syndicalists jumped into a dialogue with NN, thereby unwittingly becoming complicit with the troll's goal of becoming the center of the conversation. The strategy to hijack the list and become the central online personality worked. Because the core community had eroded, the list got entangled in the constant stream of NN/integer postings. Some called to filter the NN/integer postings; others tried to challenge the troll.[22] Others such as Diana McCarty took the liberal stand and defended the democracy of the delete button: "It takes 1–2 minutes of your time and you can file or delete and forget. Noise is sometimes music and sometimes incredibly intelligent."[23] Because of the lack of internal electronic democracy (there were no voting systems in place on lists such as Syndicate), there was no way of ascertaining what subscribers wanted to do. It took another seven months before Syndicate exploded over the integer case.

Hijacking Lists

Faced with the conflict between a desire to be noticed and the fear of being humiliated by taking sides in this conflict, most of the Syndicalists remained silent. The community lacked any armor with which to defend itself. The fear of being labeled a totalitarian advocate of censorship was omnipresent and handicapped participants at this crucial hour. Laissez-faire liberalism showed its brutal face. The choice was an impossible one. There was going to be violence in one way or the other: either a handful of posters would be excluded or the community would self-destruct. On August 7, Inke Arns unsubscribed integer, causing protest from a loud minority, while receiving praise from others. The mood on the list was deeply divided.

Inke and Andreas seemed to have hoped that the Syndicate community, as a living entity, would defend itself against the ongoing integer humiliations. Inke Arns wrote: "If you don't take care of your list, and voice your opinion, the list will be taken care of by others. And you won't necessarily like it."[24] Andreas Broeckmann defended the removal: "I don't like filters. I like this list because it makes sense for me to listen to all the different voices. I don't want to censor what comes through. At the same time, I ask for some sort of respect for my position as somebody who is also on this list. This implies not being shouted at all the time. It more importantly implies not being spat on and insulted for writing this message. It implies not seeing messages that call me a criminal."[25] Annick Bureaud also detested filtering and defended unsubscribing integer. "What I really disliked with NN postings was the flood. Once in a while, why not, but minimum 10 per day, as in the last week, come on! This is just a highjack of the list. S/he knew the rules, s/he didn't play by it. Too bad."[26] Instead of relief over the disappearance of integer, the mood on the list became increasingly tense with Andrej Tisma crying censorship and complaining of a conspiracy by Soros swastika people. Or Brad Brace, who equated NN with the martyr Mata Hari. At the "moment supreme" the Australian net.artist][mez][started systematically forwarding integer messages, stacked with personal attacks.[27] This was the sign for Andreas and Inke to step down. The moderators made sure the list was handed over "in a proper and friendly manner." While a small group (mainly net. artists) kept on arguing and kept on defending the anti-censorship case, Syndicate fell apart in a matter of days.

In retrospect, Honor Harger writes regarding the cynical NN/integer/antiorp strategies: "I find it deeply ironic that an entrepreneur so well known for revoking licenses to use her software (nato) when she encounters even the smallest criticisms of her programming—effectively censoring nato users—would react with such petulance when she herself is asked to minimize the 'noise' of her postings. Considering the NN_construct is so intolerant of others views on her work and ideas, I find it rather galling that so many have tried to defend her in the name of 'free speech.' This is something laughably alien to the NN_construct's philosophy of doing business, and it is unfortunate that Syndicate has collapsed based on this issue. Not that this would be of any interest to the NN_construct, who has little concern for this discussion space, absolutely no awareness of the network which has formed around this list for the past 5 years, and no care if her incessant flood of posts destroys the character of this list."[28] Martha Rosler also emphasized that keeping or removing NN from the list was not about free speech: "Freedom of speech is not the primary issue, and threats to sic the correctness police on the list is an ironic reversal of other authoritarian tropes... a list is neither society nor the public sphere in toto. I am not advocating asking NN to leave, for the decision is not mine, but ask yourself, when you play a game, what happens when the bully insists that it is always his/her turn at the bat; at a forum, what if she/he jumps up for the microphone after every remark someone else has made, simply to snipe, and not actually engage their points?... This well-known tactic has a name: disruption."[29]

The Death of a Community

NN's strategy of disruption had proven successful. By mid-August 2001, the Syndicate list had effectively spit in two. After Inke and Andreas resigned, the group that defended integer became the new list owners and, keeping the name Syndicate, moved the list to a server in Norway.[30] Johan Sjerpstra: "During mid-2001, when the new type of aggressive rhetoric appeared, the Syndicate founders/owners left the list without too much hesitation. They seemed to have lost their interest. Maybe their motivation was revenge, but for what? The broad membership could not handle the attack and basically no one wanted to defend the list. The new owners, which took over in August 2001, had no better agenda either. They repeated the old East/West dichotomy plus the info exchange function, but this is no longer of importance because there are a lot well organized sites."

No genuine information appeared on Syndicate, most of the art info was forwarded from other sources. The main traffic had become small talk, internal, nonsensical, repetitive, and redundant textual content, very often with simple ("small is beautiful") messages, often with no more then a URL. Johan Sjerpstra: "The minimal e-mails can be a seen as a new movement in the quickly changing net/web art scene, like a counter reaction to the earlier socially engaged and/or conceptual type of net/web art. We could call it a sort of Dadaist answer to the seriousness and tech orientation of the late 1990s. The significant difference with Dada is that instead of humor they use an aggressive and threatening (hacking) tone. Hate speak, targeted at those they dislike—a sign of an emerging new extremity."[31]

In early September, Inke and Andreas created a follow up to the Syndicate list—"Spectre." Spectre had been prepared on a cc: list during the turbulent weeks in August when it had become clear to Inke, Andreas and a few others who had left the list in protest that Syndicate could no longer be saved. The Spectre announcement included the following "netiquette" rules: "No HTML, no attachments, messages < 40K; meaningful discussions require mutual respect; self-advertise with care!"[32] Soon Spectre had 250 subscribers and continued the Syndicate focus on announcements related to new media culture. Spectre would no longer explicitly focus on the East-West dynamics, but would still refer to "Deep Europe." From the beginning of 2001, Syndicate had already been transforming into a communication community for a few insiders—a small circle of friends squatting in the past, the tradition, mimicking a community, only capable of being a parasite on the past of a dead project. During the NN/integer debates, Igor Markovic had written that Syndicate was pretty much dead anyway, even if you filter out integer and all the announcements.[33] Spectre, the follow-up to Syndicate, also proved to have no relevance, caught in the pragmatics of a redundant, no-nonsense list.

Throughout its existence, Syndicate had the feel of a somewhat safe project, struggling with the obsolete East-West dichotomy it had imposed upon itself. Unconsciously, the project had been built on the Cold War strategy of cultural subversion without ever naming its adversaries. "Being a (potentially interesting) international artist group, Syndicate lacked consistency to push its agenda (if there was any). Beyond the communication paradigm (not a particular Eastern approach anyway) there wasn't much else. No authority was explicitly questioned. The common denominator, working with (networked) computers in an arts environment, did not translate into a specific group aesthetics."[34] Indeed, Syndicate did not end up as a movement, school, style or tendency. Still, this inability was a general problem and did not only affect Syndicate. The impoverished new media arts sector, clustered around the Syndicate node, remained on a boutique level. It was neither "cool" marketwise, nor did it create inspiring and controversial expressions of dissent.

The 1998–99 period around the Kosovo crisis were Syndicate's heydays. While elsewhere on the Net, dotcommania dominated the Internet agenda, the Syndicate network, symbolic of the new media arts sector as a whole, tried—and failed—to claim the moral high ground over war and ethnic tensions on the one hand, and over corporate greed on the other. There simply was no cultural high ground to escape to. The twist of mid-2001 can only be read as a hostile takeover, covered up by lies and a massive abuse of democratic tolerance. The unspoken consensus of mediated communication, based on tolerance, democracy and credibility fell apart, torn apart by fussy controversies. Antiorp/Integer's (efficient) usage of anti-globalization rhetoric ("corporate fascists") with its roots in Stalinism and totalitarianism managed to destroy an already minimal sense of belonging.

The Syndicate list-takeover showed how an aggressive strategy of information warfare could overcome tolerance, a form of weakness. The incident marked the end of the romantic concept of open, unmoderated exchange. Extreme strategies can penetrate existing structures with virtually no resistance. As mediocre viruses are capable to bringing down millions of computers, so too can the net artist increase its impact dramatically, using aggressive memes. This is the age of total information warfare. The "war zone" is no longer a distinct battlefield, but stretches out deep into society. It not only affects the physical civil infrastructure, but also penetrates the civilian mindset.

The strategies of tension, disinformation and uncertainty are now common practices amongst and between social groups. In the case of Syndicate, the naïve East-West communication model turned into a dangerous, manipulative, unreliable network of abuse. This turning point reflects and further accelerates the collapse of hippie dreams of the Net as a utopian, parallel world.

Notes

1. See http://www.n5m.org.
2. Concerns of a Western-led takeover of the East may have been fueled by historical precedents. The stalking horse role played by abstract expressionism during the Cold War and revelations of CIA funding for U.S. exhibitions were more than just wild rumors. Articles reconstructing such cases appeared in *Artforum*, written by Max Kozloff and Eva Cockcroft. See also: Serge Guilbaut, *How New York Stole the Idea of Modern Art* (Chicago: University of Chicago Press, 1983).
3. For instance, an "activist" during Communist times was a low-rank Party member, spreading propaganda on the work floor, spying on others, always ready for betrayal, if necessary. It was therefore not a surprise that Western "media activism" in the East was met with a certain disdain.
4. http://www.medialounge.net/lounge/workspace/deep_europe/. First announcement: Andreas Broeckmann, concept "Deep Europe," *Syndicate*, May 19, 1997. Reports: Dimitri Pilikin, Syndicate, August 2, 1997; Kit Blake, Deep Europe Visa Department, *Syndicate*, August 5, 1997; Andreas Broeckmann, Discussion about a European Media Policy, *Syndicate*, August, 5, 1997; Inke Arns, Report from Deep Europe, *Syndicate*, August 12, 1997; Lisa Haskel, Tunneling to Deep Europe, a letter from my island home, *Syndicate*, August 15, 1997.
5. Andreas Broeckmann: "The war in Yugoslavia and Kosovo is the most current, pressing scenario with which cultural practitioners in Europe are faced. Other scenarios, like the slow-motion disintegration of the Russian empire, the re-emergence of the Baltic region, the hazy reality of Mitteleuropa, the precarious role of Albanian, Hungarian, German, Turkish, Basque, Roma, and other minorities in different parts of the continent, are equally precarious, both potentially productive and destructive. The site of these scenarios is Deep Europe, a continent which has its own mental topography and which is neither East nor West, North nor South, but which is made up of the multi-layeredness of identities: the more overlapping identities, the deeper the region." "Changing Faces, or Proto-Balkanian Dis-Identifications," *Syndicate*, July 7, 1999, beta version, written for Stephen Kovats, ed., *Media Revolution* (Frankfurt am Main/New York: Campus Verlag, 1999).
6. Inke Arns, "Beyond the Surfaces: Media Culture versus Media Art or How we learned to love tunnel metaphors," *Syndicate*, August 23, 1999, written for Stephen Kovats, ed., *Media Revolution* (Frankfurt am Main/New York: Campus Verlag, 1999).
7. Lisa Haskel, "Tunnelling to Deep Europe. A letter from my Island home," *Syndicate*, August 15, 1997, quoted in Inke Arns, "Beyond the Surfaces," *Syndicate*, August 23, 1999.
8. The three Syndicate publications: 1. *Reader of the V2_East / Syndicate Meeting on Documentation and Archiving Media Art in Eastern, Central and South-Eastern* Europe (Rotterdam: V2_Organisatie / DEAF96, 1996), http://colossus.v2.nl/syndicate/synr0.html. 2. *Deep Europe: The 1996–97 edition. Selected texts from the V2_East/Syndicate mailing list* (143 S.), (Berlin, October 1997), http://colossus.v2.nl/syndicate/synr1.html. 3. *Junction Skopje, selected texts from the V2_East/Syndicate mailing list 1997–98*, Syndicate Publication Series 002 (Skopje: SCCA Skopje, 1998), http://colossus.v2.nl/syndicate/synr2.html.
9. Andrej Tisma, U.S.A. Questionnaire, *Syndicate*, March 22, 1999.
10. Drazen Pantic, Radio "B92 closed, Veran Matic arrested," nettime, March 24, 1999. See also Katarina's e-mail, posted on the same day, for more details about the police raid on the B92 offices.
11. Nikos Vittes, "RE: Bombings," *Syndicate* March 25, 1999. See also his posting on March 26 in which he states that the whole region is under U.S. control. Nikos denies that his analysis is ethnically biased. "I don't say that you have to agree but I don't accept the ironic message about disliking the neighbors. I personally come from the Greek Macedonia. I grew up playing in the summer from kids from Serbia and Macedonia."
12. Andreas Broeckmann, "Berlin news," *Syndicate*, March 25, 1999.
13. Melentie Pandilovski, "Developments in Macedonia, part 1 and 2," *Syndicate*, March 25, 1999.
14. See Lisa Haskel's "Radio Deep Europe report," *Syndicate*, March 28, 1999.
15. See Atle Barcley, "re: enough (message from the provider)," *Syndicate*, August 11, 1999.
16. The campaign started with the message from Edi Muka in which he announced that he got sacked (*Syndicate*, November 29, 2000). A petition was written, signed by Syndicate members.
17. An example: "ue = g!vn d!sz g!ft dont u knou. dze ab!l!t! 2 knou. + through uz mattr kan knou !tzelv. out ov dze uomb ov t!me + dze vaztnesz ov zpasz dzat u!ch = uz - h!drogen. karbon. n!trogen. ox!gen. 16-21 elmntz dze uatr. dze zunl!ght - all hav!ng bkom uz kan bg!n 2 undrztnd uat dze! r + hou dze! kame 2 b." integer@www.god-emil.dk, *Syndicate*, January 18,2001. More on the ideas behind NN can be found in the Syndicate postings of January 28 and 29, February 2 and 11, 2001.
18. A closer look at postings shows that the NN vocabulary is rather limited. Someone would either be a "inkompetent male fascist" or a "korporate male fascist". Whereas some contributions contain traces of brilliant poetry, the personal attacks often show signs of repetition.
19. See Diego Mariani, Europanto, fwd. by Ted Byfield, nettime, April 25, 1997.
20. "B1FF /bif/ [Usenet] (alt. `BIFF') /n./ The most famous pseudo, and the prototypical newbie. Articles from B1FF feature all uppercase letters sprinkled liberally with bangs, typos, 'cute' misspellings (EVRY BUDY LUVS GOOD OLD

BIFF CUZ HE"S A K00L DOOD AN HE RITES REEL AWESUM THINGZ IN CAPITULL LETTRS LIKE THIS!!!), use (and often misuse) of fragments of talk mode abbreviations, a long sig block (sometimes even a doubled sig), and unbounded naiveté. B1FF posts articles using his elder brother's VIC-20. B1FF's location is a mystery, as his articles appear to come from a variety of sites. However, BITNET seems to be the most frequent origin. B1FF was originally created by Joe Talmadge <jat@cup.hp.com>, also the author of the infamous and much-plagiarized 'Flamer's Bible.'" http://jargon.net/jargonfile/b/B1FF.html.

21. http://www.tuxedo.org/~esr/jargon/html/entry/troll.html.
22. An example would be Darko Fritz: "oh, I love women media activists in uniforms... please please spam me. spank me. oh yes. more. even more. spam me. (...) what a pity that there is no moderator here so you can masturbate only with no resistance to a father figure, as in good old years. what unforgettable incest scenes... and everyone can watch..." *Syndicate*, February 1, 2001.
23. Diana McCarty, "re: another small syndicalist," *Syndicate*, February 3, 2001.
24. Inke Arns, "re: what happened," *Syndicate*, August 14, 2001.
25. Andreas Broeckmann, "re: Jaka Zeleznikar: NN - what happened?" *Syndicate*, August 13, 2001. The essence of the *Syndicate* project is well summarized in a posting by Eric Kluitenberg, "A short comment on the identity of the syndicate list," *Syndicate*, August 13, 2001 ("In the last year or so I saw the essence of the list get lost in a cloud of confused autistic ascii experiments that had really nothing to do with the initial character of the list."). See also Patrick Lichty, "Bans & Free Speech," *Syndicate*, August 14, 2001.
26. Annick Bureaud, "re: what happened," *Syndicate*, August 14, 2001.
27.]mez][, explaining her decision to forward integer messages: "y* I'm sending NN's replies to the list... as NN has been uns*bbed without a list consensus, I'll continue 2 4ward her replies as I assume a rite-of-reply should be allowed under the paradigm the syndicate list has adopted." : "][mez][" 4warding of NN's mails, *Syndicate*, August 16, 2001.
28. "Re: future," *Syndicate*, August 18, 2001.
29. Martha Rosler, "banning," *Syndicate*, August 14, 2001.
30. Among those who took the name Syndicate with them to restart the list elsewhere were yaNN@x-arn.org, Claudia Westermann, Clemens Thomas, Atle Barkley, Frederic Madre, Jaka Zeleznikar. The administration team changed and since August 27, 2001 the Syndicate mailing list and Web space was hosted by Atelier Nord in Norway. The homepage of the renewed Syndicate provided information about mail filters and how to use them. "Sometimes it might seem to be necessary to set mail filters, in order to avoid getting your in-box stuffed with mails, that are not of interest to you" (http://anart.no/~syndicate/subpages/filter.html).
31. Johan Sjerpstra, in a private e-mail to the author, March 3, 2002.
32. Andreas Broeckmann, "new mailing list: SPECTRE: info," Spectre, September 14, 2001. It also stated: "Requests for subscription have to be approved by hosts. Subscriptions may be terminated or suspended in the case of persistent violation of netiquette. Should this happen, the list will be informed. The list archives are publicly available, so SPECTRE can also be consulted and followed by people who are not subscribed." URL of the Spectre list archive: http://coredump.buug.de/pipermail/spectre/.
33. Igor Markovic, "re: [ot] [!nt] \n2+0\," *Syndicate*, August 7, 2001.
34. Johan Sjerpstra, in a private e-mail to the author, February 25, 2002.

21

The Cell Phone and the Crowd

Messianic Politics in the Contemporary Philippines

Vicente L. Rafael

This essay explores a set of telecommunicative fantasies among middle-class Filipinos within the context of a recent historical event: the civilian-backed coup that overthrew President Joseph Estrada in January 2001. It does so with reference to two distinct media, the cell phone and the crowd. Various accounts of what has come to be known as "People Power II" (distinguished from the populist coup that unseated Ferdinand and Imelda Marcos in 1986) reveal certain pervasive beliefs of the middle classes. They believed, for example, in the power of communication technologies to transmit messages at a distance and in their own ability to possess that power. In the same vein, they believed they could master their relationship to the masses of people with whom they regularly shared Manila's crowded streets, and utilize the power of crowds to speak to the state. Thus they imagined themselves able to communicate beyond the crowd, but also with it, transcending the sheer physical density of the masses through technology while at the same time ordering its movements and using its energy to transmit middle-class demands. At its most utopian, the fetish of communication suggested the possibility of dissolving, however provisionally, existing class divisions. From this perspective communication held the messianic promise of refashioning the heterogeneous crowd into a people addressing and addressed by the promise of justice. But as we will see, these telecommunicative fantasies were predicated on the putative "voicelessness" of the masses. For once heard, the masses called attention to the fragility of bourgeois claims to shape the transmission of messages about the proper practice of politics in the nation-state. In this context, media politics (understood in both senses of the phrase: the politics of media systems but also the inescapable mediation of the political) reveal the unstable workings of Filipino middle-class sentiments. Unsettled in their relationship to social hierarchy, these sentiments at times redrew class divisions, anticipated their abolition, or called for their reinstatement and consolidation.[1]

I. Calling

Telephones were introduced in the Philippines as early as 1885, during the last decade and a half of Spanish colonial rule.[2] Like telegraphy before it, telephony provoked fantasies of direct communication among the colonial bourgeoisie. They imagined that these new technologies would afford them access to colonial leaders, enabling them to hear and be heard directly by the colonial state. We can see this telecommunicative ideal, for example, in a satirical piece written by Filipino national hero Jose Rizal in 1889. Entitled *Por Telefono,* it situates the narrator as an eavesdropper. He listens intently to the sounds and voices that travel between the Spanish friars in Manila—regarded as the real power in the colony—and their superiors in Madrid.[3] The nationalist writer wire-taps his way, as it were, into the walls of the clerical residences, exposing their hypocrisy and excesses. In this sense, the telephone shares the capacity of that other telecommunicative technology, print, to reveal what was once hidden, to repeat what was meant to be secret, and to pass on messages intended for a particular circle.[4] It is this history of tapping into and forwarding messages—often in the form of ironic commentaries, jokes, and rumors—that figured recently in the civilian-led coup known as "People Power II." From January 16 to 20, 2001, more than a million people assembled at one of Metro Manila's major highways, Epifanio de los Santos Avenue, commonly called Edsa, site of the original People Power revolt in 1986. A large cross-section of Philippine society gathered there to demand the resignation of President Joseph "Erap" Estrada after his impeachment trial was suddenly aborted by the eleven senators widely believed to be under his influence. The senators had refused to include key evidence that purportedly showed Estrada had amassed a fortune from illegal number games while in office. The impeachment proceedings were avidly followed on national TV and the radio. Most viewers and listeners were keenly aware of the evidence of theft and corruption on the part of Estrada and his family; once the pro-Estrada senators put an abrupt end to the hearing, however, hundreds of thousands of viewers and listeners were moved to protest in the streets.[5] Television and radio had kept them in their homes and offices to follow the court proceedings, but at a critical moment, these media also drew them away from their seats. Relinquishing their position as spectators, they now became part of a crowd that had formed around a common wish: the resignation of the president.

Aside from TV and radio, another communications medium was given credit for spurring the coup: the cell phone. Nearly all the accounts of People Power II available to us come from middle-class writers or by way of a middle-class controlled media with strong nationalist sentiments. And nearly all point to the crucial importance of the cell phone in the rapid mobilization of demonstrators. "The phone is our weapon now," we hear from an unemployed construction worker quoted in a newspaper article. A college student in Manila testified that "the power of our cell phones and computers were among the things which lit the fuse which set off the second uprising, or People Power Revolution II." And a newspaper columnist advised "would-be foot-soldiers in any future revolution" that "as long as you[r cell phone] is not low on battery, you are in the groove, in a fighting *mood.*"[6] A technological thing was thus idealized as an agent of change, invested with the power to bring forth new forms of sociality.

Introduced in the latter half of the 1990s, cell phones in the Philippines had become remarkably popular by around 1999.[7] There are a number of reasons for their ubiquity. First, there is the perennial difficulty and expense of acquiring land line phones in the Philippines and the service provided by the Philippine Long Distance Company (PLDT) and the more recent, smaller Bayan Tel, is erratic. Cell phones offered the promise of satisfying this pent-up need for connectivity. In addition, cell phones cost far less than personal computers, which are owned by less than 1 percent of the population (though a larger proportion has access through Internet cafes). By contrast, there are over ten million cell phone users in a population of about seventy-seven million. The vast majority of users buy pre-paid phone cards that, combined with the relatively low cost of phones (as little as $50 in the open market and half this amount in secondary markets), make wireless communication more accessible and affordable than regular telephones or computers.

Most importantly, cell phones allow users to reach beyond traffic-clogged streets and serve as an alternative to slow, unreliable, and expensive postal services. Like many Third World countries recently opened to more liberal trade policies, the Philippines shares the paradox of being awash with the latest communication technologies, like the cell phone, while being mired in deteriorating infrastructures: roads, postal services, railroads, power generators and land lines. With the cell phone, one seems able to pass beyond these obstacles. And inasmuch as these broken, state-run infrastructures represent government ineptitude, passing beyond them gives one the sense of overcoming a state long beset by corruption.[8] It is not surprising, then, that cell phones could prove literally handy in spreading the rumors, jokes and information that steadily eroded whatever legitimacy President Estrada and his supporters still had during the impeachment hearings. Bypassing the broadcast media, cell phone users themselves became broadcasters, receiving and transmitting both news and gossip and often confounding the two. Indeed, one could imagine each user becoming his or her own broadcaster; a node in a wider network of communication that the state could not possibly even begin to monitor, much less control.[9] Hence, once the call was made for people to mass at Edsa, cell phone users readily forwarded messages they received as they followed the messages' instructions.

Cell phones, then, were not only invested with the power to overcome the crowded conditions and congested surroundings brought about by the state's inability to order everyday life, they were seen to bring about a new kind of crowd that was thoroughly conscious of itself as a movement headed towards a common goal. While telecommunication allows one to escape the crowd, it also opens up the possibility of finding oneself moving in concert with it, filled with its desire and consumed by its energy. In the first case, cell phone users define themselves against a mass of anonymous others. In the second, they become those others, accepting anonymity as a condition of possibility for sociality. To understand how the first is transformed into the second, it is worth noting how, specifically, the vast majority of cell phone messages are transmitted in the Philippines: as text messages.

II. Texting

Text messages are e-mails sent over mobile phones that can also be transferred to the Internet. Recently, the verb "texting" has emerged to designate the act of sending such messages, indicating its popularity in such places as England, Japan and Finland (where text messaging was first available). In the Philippines, texting has been the preferred mode of cell phone use since 1999, when the two major networks, Globe and Smart, introduced free, and then later on, low cost text messaging as part of their regular service. Unlike voice messages, text messages take up less bandwidth and require far less time to convert into digitized packets available for transmission. It thus makes economic sense for service providers to encourage the use of text messaging in order to reserve greater bandwidth space for more expensive—and profitable—voice messages. Calling cards and virtually free texting, as opposed to expensive, long-term contracts, give cell phone service providers a way to attract a broad spectrum of users from different income levels. Thus, from an economic standpoint, texting offers a rare point of convergence between the interests of users and providers.[10] But it is obviously more than low costs that make cell phones popular in the Philippines. In an essay sent over the Internet signed "An Anonymous Filipino," the use of cell phones in Manila is described as a form of "mania." Using Taglish (the urban lingua franca that combines Tagalog, English and Spanish), this writer, a Filipino *balikbayan* (that is, one who resides or works abroad and periodically visits the motherland) remarks:

> HI! WNA B MY TXT PAL? They're everywhere! In the malls, the office, school, the MRT [Manila Railroad Transit], what-have-you, the cell phone mania's on the loose! Why, even Manang Fishball [Mrs. Fishball, a reference to older working class women vendors who sell fishballs by the side of the road], is texting! I even asked my sisters how important they think they are that they should have cells? Even my nephew in high school has a cell phone.

> My mom in fact told me that even in his sleep, my brother's got his cell, and even when they have a PLDT [land line] phone in the house, they still use the cell phone.[11]

According to the *Oxford English Dictionary*, *mania* is a kind of madness characterized "by great excitement, extravagant delusions and hallucinations and in its acute stage, by great violence." The insistence on having cell phones nearby, the fact that they always seem to be on hand, indicates an attachment to them that surpasses the rational and the utilitarian, as the remarks above indicate. The cell phone gives its holder a sense of being someone, even if he or she is only a street vendor or a high school student—someone who can reach and be reached and is thus always in touch. The "manic" relationship to the cell phone is just this ready willingness to identify with it, or more precisely with what the machine is thought capable of doing. One not only has access to it; by virtue of its omnipresence and proximity, one becomes like it. That is to say, one becomes an apparatus for sending and receiving messages at all times. An American journalist writing in the *New York Times* observes as much in an article on Manila society:

> "Texting?" Yes, texting—as in exchanging short typed messages over a cell phone. All over the Philippines, a verb has been born, and Filipinos use it whether they are speaking English or Tagalog.... The difference [between sending e-mail by computers and texting] is that while chat-room denizens sit in contemplative isolation, glued to computer screens, in the Philippines, "texters" are right out in the throng. Malls are infested with shoppers who appear to be navigating by cellular compass. Groups of diners sit ignoring one another, staring down at their phones as if fumbling with rosaries. Commuters, jaywalkers, even mourners—everyone in the Philippines seems to be texting over the phone... Faye Siytangco, a 23-year-old airline sales representative, was not surprised when at the wake for a friend's father she saw people bowing their heads and gazing toward folded hands. But when their hands started beeping and their thumbs began to move, she realized to her astonishment that they were not in fact praying. "People were actually sitting there and texting," Siytangco said. "Filipinos don't see it as rude anymore."[12]

Unlike computer users, cell phone owners are mobile, immersed in the crowd, yet able to communicate beyond it. Texting provides them with a way out of their surroundings. Thanks to the cell phone, they need not be present to others around them. Even when they are part of a socially defined group—say, commuters or mourners—cell phone users are always somewhere else, receiving and transmitting messages from beyond their physical location. It is in this sense that they become other than their socially delineated identity: not only cell phone users but cell phone "maniacs." Because it rarely leaves their side, the phone becomes part of the hand, the digits an extension of the fingers. In certain cases, the hand takes the place of the mouth, the fingers that of the tongue. One Filipino-American contributor to Plaridel, an online discussion group dealing with Philippine politics, referred to a Filipino relative's cell phone as "almost a new limb."[13] It is not surprising then that the consciousness of users assumes the mobility and receptivity of their gadgets. We can see how this assumption of the qualities of the cell phone comes across in the practice of sending and receiving messages:

> The craze for sending text messages by phone started [in 1999] when Globe introduced prepaid cards that enabled students, soldiers [and others] too poor for a long-term subscription to start using cellular phones.... People quickly figured out how to express themselves on the phone's alphanumeric keypad.... "Generation Txt," as the media dubbed it, was born. Sending text messages does not require making a call. People merely type in a message and the recipient's phone number, hit the phone's send key and off it goes to the operator's message center, which forwards it to the recipient.... Sending text messages by phone is an irritating

> skill to master, largely because 26 letters plus punctuation have to be created with only 10 buttons. Typing the letter C, for example, requires pressing the No.2 button three times; an E is the No. 3 button pressed twice; and so on. After the message is composed it can be sent immediately to the phone number of the recipient, who can respond immediately by the same process. People using phones for text messages have developed a shorthand. "Where are you?" becomes "WRU." And "See you tonight" becomes "CU 2NYT." People have different styles of keying in their messages. Some use their index fingers, some one thumb, others both.... [Others] tap away with one hand without even looking at [their] phone.[14]

As with e-mail, conventions of grammar, spelling and punctuation are frequently evaded and rearticulated in texting. The constraints of an alphanumeric keypad require users to type numbers to get to letters. As a result, counting and writing become closely associated. Digital communication requires the use of digits, both one's own and those of the phone pad, as one taps away. But this tapping unfolds not to the rhythm of one's speech, or in tempo with one's thoughts, but in coordination with the numbers by which one reaches letters: three taps on *2* to get a *C*, for example, or two taps on *3* to get an *E*. Texting seems to reduce all speech to writing, and all writing to a kind of mechanical percussion, a drumming that responds to an external constraint rather than an internal source. In addition, as it were, there are no prescribed styles for texting: one finger will do, or one can use a thumb, and skilled typists can text without looking at the screen. Nor are standardized body postures required with texting: one can sit or walk or drive while sending messages. If hand writing in the conventional sense requires classroom instruction in penmanship and posture, texting frees the body, or so it seems, from these old constraints.

Mimicking the mobility of their phones, texters move about, bound to nothing but the technological forms and limits of the medium. The messages they send and receive condense versions of whatever language—English or Tagalog and, more frequently, Taglish—they are using and so are proper to none. This hybrid language follows the demands of the medium rather than reflecting the idiosyncrasies of its users. The phone companies' recent introduction of limits on free text messaging, and their assessment of a fee per character of text, has led to further shortening of words and messages. Instant messaging, along with the mechanical storage and recall of prior messages, requires only highly abbreviated narrative constructions with little semantic deferral or delay. Using the cell phone, one begins to incorporate its logic and its techniques to the extent of becoming identified with an apparently novel social category: *Generation Txt*.

An obvious pun on Generation X, *Generation Txt* first began as an advertising gimmick among cell phone providers in order to attract young users to their products. Defined by its attachment to and ease with the cell phone, Generation Txt has troubled older generations uneasy about the rise of texting. An anthropologist from the University of the Philippines addresses the dangers of texting in terms that are familiar from other countries where the practice has become popular, especially among youth. He cites the cell phone's propensity to stifle literacy by "[wreaking] havoc" on spelling and grammar, and its erosion "in tandem with mindless computer games and Internet chat rooms, [of] young people's ability to communicate in the real world in real time."[15] Rather than promote communication, texting obstructs it; indeed, cell phones cultivate a kind of stupidity. For the anthropologist, this is evident in young people's gullibility for the marketing ploys of cell phone providers; they end up spending more, not less, in sending messages of little or no consequence. He further charges cell phones with leading to "anti-social" behavior: children "retreat to their own cocoons," while the parents, who give them cell phones, evade the responsibility of "interacting" with them in any meaningful way.[16] Other writers report the students' use of texting to cheat on exams, or the use of cell phones in spreading slanderous rumors and gossip that may ruin someone's reputation.[17] As one Filipino online writer put it, cell phones are like "loaded weapons" and their use must be tempered with some caution. Another contributor writes: "If the text [I received] felt like a rumor masquerading as news, I didn't forward it." An office worker from Manila adds, "Sometimes whenever you receive serious msgs (sic), sometimes you have to think twice if it is true or if perhaps someone is fooling you since there is so much joking [that goes on] in txt (sic)."[18]

Part of the anxiety surrounding texting arises from its perceived tendency to disrupt protocols of recognition and accountability. Parents are disconnected from their children while children in turn defy parental authority. Cheating is symptomatic of teachers' inability to monitor students' cell phone use. And the spread of rumors and gossip, along with irreverent jokes, means that the senders of messages readily give in to the compulsion to forward messages without, as the writers above advise, weighing their consequences or veracity. Indeed, it is the power to forward messages almost instantaneously that transforms the cell phone into a "weapon." The urge to retransmit messages is difficult to resist, and under certain conditions, irrepressible, as we learn from the events leading up to People Power II. Actor and writer Bart Guingona, who organized a demonstration at Edsa on 18 January, describes his initial doubts about the effectiveness of cell phones in a post to the Plaridel listserv: "I was certain [texting] would not be taken seriously unless it was backed up by some kind of authority figure to give it some sort of legitimacy. A priest who was with us suggested that [the church-owned broadcasting station] Radio Veritas should get involved in disseminating the particulars... We [then] formulated a test message... and sent it out that night and I turned off my phone.... By the time I turned it on in the morning, the message had come back to me three times... I am now a firm believer in the power of the text!"[19]

The writer is initially hesitant to use texting, reasoning that messages sent in this way would be perceived as groundless rumors. Anonymously circulated from phone to phone, the text seemed unanchored to any particular author who could be held accountable for its content. Only when the church-owned radio station agreed to broadcast the same information did he agree to send a text message. Upon waking up the next day, he saw the effect of this transmission. Not only did his message reach distant others; it returned to him three-fold. He is converted from a doubter to a believer in the "power of the text." Such a power has to do with the capacity to elicit numerous replies.

There are two things worth noting, however, in this notion of the power of texting: first, that it requires, at least in the eyes of this writer and those he sends messages to, another power to legitimate the text's meaning; and second, that such a power is felt precisely in the multiple transmissions of the same text. The power of texting has less to do with the capacity to elicit interpretation and stir public debate than it does with compelling others to keep messages in circulation. Receiving a message, one responds by repeating it. The message is forwarded to others who are expected to do the same. In this way, the message returns, mechanically augmented but semantically unaltered. They crowd one's phone mailbox just as those who believed in the truth of the call they received crowded the streets of Metro Manila. On this account, the formation of crowds answers the repeated call of texts deemed to have legitimacy by virtue of being grounded in an authority outside the text messages themselves: the electronic voice of the Catholic Church. The voice of the church in effect domesticates the dangers associated with texting. Users can then forward texts and likewise feel forwarded by the expectations these texts give rise to. Finding themselves called by the message and its constant repetition, they become "believers," part of Generation Txt.

Generation Txt thus does not so much designate a new social identity as a desire for seeing in messages a meaning guaranteed by an unimpeachable source residing outside the text. In this sense, there is nothing new or different about the technological fantasy. Most of those who gathered at Edsa and marched towards Mendiola—the road leading to the Presidential Palace—were united in their anger at the corrupt regime of President Estrada and their wish to replace him with a more honest leader. This said, the protesters challenged neither the nature of the state nor its class divisions. Indeed, everything I have read by supporters about People Power II emphasizes the constitutional legality of these protests vis-à-vis the Supreme Court and the Catholic Church (as opposed to the army or left-wing groups) for institutional legitimacy. In the end, Estrada's replacement came from within his own circle of power: Gloria Macapagal-Arroyo was his vice-president and the daughter of a previous Philippine president. It would appear then that Generation Txt comes out of what its "believers" claim to be a "technological revolution" that sets the question of social revolution aside.

Texting is thus "revolutionary" in a reformist sense. Its "politics" seek to consolidate and render authority "transparent," whether this is the authority of the state or of text messages. In an exemplary manifesto titled "Voice of Generation Txt" [Tinig ng Generation Txt], which appeared in what was, until recently, one of Manila's more widely read tabloids, *Pinoy Times*, Ederic Peñaflor Eder, a twenty-something University of the Philippines graduate, credits the "power" (*lakas*) of "our cell phones and computers" for contributing to the "explosion" of People Power II. Texting, he declares, became the medium through which "we" responded quickly to the "betrayal" (*kataksilan*) of the pro-Estrada senators who had sought to block the impeachment hearings. Elaborating on the "we" designating Generation Txt, Eder writes in Taglish:

> We are Generation Txt (sic). Free, fun-loving, restless, insistent, hard-working, strong and patriotic.
>
> We warmly receive and embrace with enthusiasm the revolution in new technology. Isn't it said that the Philippines rules Cyberspace and that the Philippines is the text messaging capital of the world? Our response was rapid to the betrayal of the eleven running dogs (*tuta*) of Jose Velarde (a.k.a. Joseph Estrada). The information and calls that reached us by way of text and e-mail were what brought together the organized as well as unorganized protests. From our homes, schools, dormitories, factories, churches, we poured into the streets there to continue the trial—the impeachment trial that had lost its meaning.
>
> ... Our wish is for an honest government, and a step towards this is the resignation of Estrada. We are patriotic and strong and with principles, since our coming together is not merely because we want to hang out with our friends, but rather to attain a truly free and clean society brought by our love for the Philippine nation....
>
> There were those from our generation that have long since before the second uprising chosen to struggle and fight in the hills and take up arms, trekking on the harsh road towards real change. Most of us, before and after the second uprising, can be found in schools, offices, or factories, going about our everyday lives. Dreaming, working hard for a future. Texting, internetting, entertaining ourselves in the present.
>
> But when the times call, we are ready to respond. Again and again, we will use our youth and our gadgets (*gadyet*) to insure the freedom of our Motherland.... After the second uprising, we promise to militantly watch over the administration of Gloria Macapagal Arroyo while we happily push Asiong Salonga (a.k.a. Joseph Estrada) into the doors of prison.
>
> We are Generation Txt.[20]

This statement of identity curiously enough does not specify who this "we" is except as those who "warmly accept and embrace" the "revolution" in new technology. The "we" is established through an identification with technological novelty and the status of the Philippines as the "text messaging" capital of the world. This is perhaps why the message reads as if it were meant to be received then forwarded: it begins and ends with exactly the same lines: *Kami ang Generation Txt* (We are Generation Txt). Instead of ideas or social critique, Generation Txt is characterized here by attitudes and affects: it is *malaya* (free), *masayahin* (fun-loving), *malikot* (restless), *makulit* (insistent), *masipag* (hardworking) and so forth. Its members pride themselves on having principles and courage, and, unlike the rudderless and Westernized Generation X, they have direction. They stand for "transparent" government, and a "free" and "clean" society. In this sense, they do not see themselves as different from their elders for they are patriots (*makabayan*) dedicated to using their "gadgets" for the sake of the motherland (*Inang Bayan*). Such commitment comes in the form of a "militant" readiness to watch over the workings of the new government in order to ensure "justice" (*katarungan*). Unlike those who have chosen to take up arms and go to the mountains, Generation Txt can be found in schools, offices, and factories, ready to respond to the call of the times. They watch, they wait, and they are always ready to receive and forward messages.

Generation Txt is concerned not with challenging the structures of authority but with making sure they function to serve the country's needs. This reformist impetus is spelled out in terms of their demand for accountability and their intention of holding leaders under scrutiny. Through their gadgets, they keep watch over their leaders rather than taking their place or putting forth other notions of leadership. Thus does Generation Txt conceptualize its historical agency: as speedy (*mabilis*) transmitters of calls (*panawagan*) that come from elsewhere and have the effect of calling out to those in their "homes, schools, dormitories, factories, churches" to flood the streets in protest. Rather than originate such calls, they are able to trace them to their destination which, in this case, is the nation of middle-class citizens that seeks to renew and supervise its government. Like the first generation of bourgeois nationalists in the nineteenth century I mentioned earlier, Generation Txt discovers yet again the fetish of technology as that which endows one with the capacity to seek access to and recognition from authority.[21]

III. Crowding

In the Generation Txt fantasy, texting calls into being a new form of social movement—one that is able to bear, in both senses of the term, the hegemony of middle-class intentions. As we have seen, texting is sometimes used to evade the crowd. But as a political technology, it is credited with converting the crowd into the concerted movement of an aggrieved people. In short, the middle class invests the crowd with the power of the cell phones: the power to transmit their wish for a moral community. Indeed, the act of transmission would itself amount to the realization of such a community. The fantasy projects a continuity between the crowd and middle-class texters. Nevertheless, during People Power II, the middle-class interest in ordering the crowd sometimes gave way to its opposite. At times, it was possible to see the materialization of another kind of desire, a desire for the dissolution of class hierarchy altogether. How so?

The contemporary streets of Manila provide some insight into the contradictory middle-class ideas about crowds. The city has a population of over ten million, a large number of whom are rural migrants in search of jobs, education, or other opportunities unavailable in the provinces. Congested conditions—packed commuter trains, traffic-clogged roads, crowded sidewalks, teeming shopping malls—characterize everyday life in the city, slowing travel from one place to another at nearly all hours of the day and night. These conditions affect all social classes. And because there is no way of definitively escaping them, they constitute the most common and widely shared experience of city life.

Just as Manila's roads are clogged with vehicles, its sidewalks seem unable to contain the unending tide of pedestrians who spill out onto the highways, weaving in and out of vehicular traffic. Indeed, among the most anomalous sights on city sidewalks are signs for wheelchair access. Given the uneven surfaces and packed conditions of the sidewalks, these signs are no more than traces of a possibility never realized, a future overlooked and forgotten. It is as if at one point, someone had thought of organizing urban space along the lines of a liberal notion of accommodation. Instead, that thought quickly gave way to what everywhere seems like an inexorable surrender of space to the people who use it—and use it up.

Urban space in Manila thus seems haphazardly planned, as if no central design had been put in place and no rationalizing authority were at work in organizing and coordinating the movement of people and things.[22] Instead, this movement occurs seemingly on its own accord. Pedestrians habitually jaywalk and jump over street barriers. Cars and buses belch smoke, crisscrossing dividing medians—if these exist at all—inching along to their destinations. Drivers and passengers find it difficult to see more than a few feet beyond their vehicles. The windshields and windows of jeepneys, tricycles and cabs are often cluttered with decals, curtains, detachable sun shades, and other ornaments that make it difficult to get a view of the road, in effect obstructing one's vision and further heightening the sense of congestion. Indeed, given Manila's topographical flatness, it is impossible to get a panoramic view of the city except from commuter trains and the tops of tall buildings. In the West, the "view" is understood as the site for evacuating a sense of internal unease and a resource for relieving oneself of pressure, both social and psychic.[23] This panoramic notion of the view is not possible in Manila's streets. Caught in traffic, one sees only more stalled traffic, so that the inside and the outside of vehicles seem to mirror one another.

The overwhelming presence of garbage adds to the sense of congestion. Garbage disposal has long been a problem in Manila owing to the shortage of adequate landfills, among other reasons. As a result, trash seems to be everywhere, dumped indiscriminately on street corners or around telephone poles, some of which bear signs that impotently forbid littering and public urination. What appears are thus scenes of near ruin and rubble. While certainly not exclusive to Manila, these scenes bespeak a city in some sense abandoned to the pressures of a swelling population. Instead of regulating contact and channeling efficient movement of people and things, the city's design—such as it is—seems to be under constant construction from the ground up and from so many different directions. The thought of regulation occurs, but the fact that construction never seems to end—stalled by crowded conditions, periodic typhoons, floods, and the accumulation of garbage—makes it seem as if these sites were in ruins. The sense is that there is no single, overarching authority. Walking or riding around in Manila, then, one is impressed by the power of crowds. Their hold on urban space appears to elude any attempt at centralizing control. This is perhaps why the largest private spaces open to the public in Manila, shopping malls, play what to an outsider might seem to be extremely loud background music. A shopping mall manager once told me that turning the volume up was a way of reminding mall-goers they were not in the streets, that someone was in charge and watching their actions.[24]

The anonymity proper to crowds makes it difficult, if not impossible, to differentiate individuals by precise social categories. Clothing sometimes indicates the social origins of people, but with the exception of beggars, it is difficult to identify class on the basis of looks alone. The sense that one gets from moving in and through crowds is of a relentless and indeterminable mixing of social groups. This pervasive sense of social mixing contrasts sharply with the class-based and linguistic hierarchies that govern political structures and social relations in middle-class homes, schools, churches and other urban spaces.[25] One becomes part of the crowd only by having one's social identity obscured. Estranged, one becomes like everyone else. Social hierarchy certainly does not disappear on the streets. But like the police who are barely visible, appearing mostly to collect payoffs (*tong* or *lagay*) from jeepney drivers and sidewalk vendors, hierarchy feels more arbitrary, its hold loosened by the anonymous sway of the crowd.

The power of the crowd thus comes across in its capacity to overwhelm the physical constraints of urban planning and to blur social distinctions by provoking a sense of estrangement. Its authority rests on an ability to promote restlessness and movement, thereby undermining pressure from state technocrats, church authorities, and corporate interests to regulate and contain such movements. In this sense, the crowd is a sort of medium, if by that word one means a way of gathering and transforming elements, objects, people, and things. As such, the crowd is also a site for the articulation of fantasies and the circulation of messages. It is in this sense that we might think of the crowd as not merely an effect of technological devices, but as a kind of technology itself. It calls incessantly and we find ourselves compelled to respond to it. As a kind of technology, the crowd represents more than a potential instrument of production or an exploitable surplus for the formation of social order. It also delineates the form and content of a technology of engaging the world. The insistent and recurring proximity of anonymous others creates a current of expectation, of something that might arrive, of events that might happen. As a site of potential happenings, it is a kind of place for the generation of the unknown and the unexpected. Centralized urban planning and technologies of policing seek to routinize the sense of contingency generated by crowding. But in cities where planning chronically fails, the routine sometimes gives way to the epochal. At such moments, the crowd, as I hope to show below, takes on a kind of telecommunicative power, sending messages into the distance while bringing distances up close. Enmeshed in a crowd, one feels the potential for reaching out across social, spatial, and temporal divides.[26]

As we saw, middle-class discourses about the cell phone tend to oppose texting to the crowd as a means for overcoming the latter. But in more politically charged moments such as People Power II, cell phones were credited along with radio, television, and the Internet for summoning the crowd and channeling its desire, turning it into a resource for the reformation of social order. Other accounts, however, suggested the crowd's potential for bringing about something else: the transmission of messages, which at times converged with, but at other times diverged from, those emanating from cell phones. For at times, the crowd made possible a different kind of experience for the middle class. This had to do less with representing the masses than with becoming one with them. In so doing, the crowd became a medium for the recurrence of another fantasy that emanates from the utopian side of bourgeois nationalist wishfulness: the abolition of social hierarchy.[27] We can see the recurrence of this fantasy and the desire to do away with hierarchy in one of the more lucid accounts of the crowd's power in a posting by "Flor C." on the Internet discussion group, Plaridel.[28] The text, written in Taglish, is worth following at some length for what it tells us about this other kind of political experience.

"I just want to share my own way of rallying at the Edsa Shrine," Flor C. begins. She invites others do the same, adding, "I am also eager (*sabik*) to see the personal stories of the 'veterans' of Mendiola." The urge to relate her experiences at the protests comes with a desire to hear others tell their own stories. What she transmits is a text specific to her life, not one that comes from somewhere else and which merely passes through her. Yet, by identifying herself as "Flor C.," she makes it difficult for us to locate her narrative beyond its signature. Nor can we determine who authorizes its telling. In this way, she remains anonymous to her readers, the vast majority of whom likewise remain unknown to her.[29] What is the relationship between anonymity and an eagerness to share experiences, one's own as well as those of others?

Flor C. refers to the "buddy-system" used by protest marchers in the 1970s and 1980s to guard against infiltration by fifth columnists and military and police harassment. But, writes Flor C., because "my feet were too itchy so that I could not stay in the place that we agreed to meet," she ends up without a "buddy" at Edsa. Instead, she finds herself swimming in an "undulating river (*ilog na dumadaloy*), without let-up from Edsa and Ortigas Avenue that formed the sea at the Shrine." She can't keep still. She feels compelled to keep moving, allowing herself to be carried away from those who recognize her. At Edsa, she knows no one and no one knows her. Yet the absence of recognition causes neither dismay nor a longing for some sort of identity. Instead, she relishes the loss of place brought about by her absorption into the movement of the crowd. She finds herself in a community outside of any community. It fills her with excitement (*sabik*). But rather than reach for a cell phone, she does something else: she takes out her camera.

> And so I was eager to witness (*kaya nga sabik akong masaksihan*) everything that was happening and took photographs. Walking, aiming the camera here and there, inserted into the thick waves of people who also kept moving and changing places, walked all day until midnight the interiors of the Galleria [shopping mall], around the stage and the whole length of the Edsa-Ortigas flyover. Sometimes stopping to listen for a while to the program on stage, shouting "Erap resign!," and taking close-ups of the angry, cussing placards, T-shirts, and posters and other scenes; "Good Samaritans" giving away mineral water and candy bars, a poor family where the mother and child were lying on a mat while the father watched over, a group of rich folks on their Harley Davidsons, Honda 500s, and Sym scooters that

> sparkled.... And many other different scenes that were vibrant in their similarities but also in their differences.

Immersed in the crowd, Flor C. begins to take photographs. Here, the camera replaces the cell phone as the medium for registering experience. In the passage above, she initially refers to herself as "*ako*," or "I," the first-person singular pronoun in Tagalog. But once she starts to take photographs, the "I" disappears. The sentences that follow do not contain any pronouns at all. It is as if her walking, moving, listening and looking are performed impersonally. While we can certainly imagine these sentences to imply a person carrying out these activities, Flor C.'s narrative suggests some other agency at work here: an "it" rather than an "I." That "it" of course is the camera that Flor C. takes out and begins to aim (*tinutok*). Led by her desire to join the crowd, she begins to act and see like her camera. She stops, then moves on, taking close-ups of "scenes" (*eksenas*) made up of the juxtaposition of various social classes. She is thus drawn to the appearance of sharp "contrasts" (*pagkaiba*) that are thrown together, existing side by side as if in a montage. The juxtaposition of contrasts, the proximity of social distances, the desire to close in on all sorts of expressions and signs, to draw them into a common, though always shifting, visual field; these are what interest Flor C.'s camera. These are also precisely the features of the crowd. It is the crowd that drives Flor C. to take out her camera; and in registering the mixing of differences, the camera reiterates its workings. Identifying with a camera that brings distances up close and holds differences in sharp juxtaposition, Flor C. begins to take on the telecommunicative power of the crowd. Yet, unlike the cell phone, whose political usefulness requires the legitimation of messages by an outside authority, the crowd in Flor C.'s account seems to derive its power from itself. At least in this instance, the crowd does not look beyond itself, precisely because it erodes the boundary between inside and outside. We can further see this blurring of boundaries in Flor C.'s account of entering the Galleria shopping mall next to the center stage of the Edsa protest:

> Many times I entered the Galleria to line up for the restroom and at the juice store. During one of my trips there, I was shocked and thrilled (*kinilabutan ako*) when I heard "Erap resign!" resonating from the food center, cresting up the escalator, aisles and stores. The mall became black from the "advance" of middle-class rallyists wearing the uniform symbolic of the death of justice. But the whole place was happy (*masaya*). Even the security guards at the entrance simply smiled since they could not individually inspect the bags that came before them...

She is thrilled and shocked (*kinilabutan ako)* by a sonic wave making its way up the shopping mall. Middle-class "rallyists" dressed in black surged through the aisles, protesting rather than shopping. Like all modern retail spaces, the shopping mall has been designed to manufacture novelty and surprise only to contain them within the limits of surveillance and commodity consumption. But during People Power II, it is converted into a site for something unexpected and unforeseen. Ordinarily, the mall is meant to keep the streets at bay. Now it suddenly merges with them, creating a kind of uncanny enjoyment that even the security guards cannot resist. Formerly anonymous shoppers, middle-class protestors now come across en masse. As shoppers, they had consumed the products of others' labor, and constituted their identities in relation to the spectacles of commodities. But as demonstrators, they now shed what made them distinct: their identity as consumers. They are instead consumed and transformed by the crowd. While they may still be recognizable as middle class, they simultaneously appear otherwise, advancing in their black shirts and chanting their slogans. To Flor C., their unfamiliar familiarity produces powerful effects. In the mall, Flor C. finds herself to be somewhere else. As in the streets, the intensification of her sense of displacement becomes the basis for a sensation of a fleeting and pleasurable connection with the crowd.

However, this sense of connection can be a source of not only pleasure but, at certain times, anxiety and fear. What is remarkable about Flor C.'s narrative is the way it takes on rather than evades this fear. The result, as we will see in the concluding section of her story, is not a mastery nor an overcoming of the crowd's disorienting pull, but a realization of what she conceives to be the saving power of the crowd. Back on the streets, she wanders onto a flyover, or an on-ramp, at the Edsa highway.

> When I first went to the flyover, I was caught in the thick waves of people far from the center of the rally. I could barely breathe from the weight of the bodies pressing on my back and sides. I started to regret going to this place that was [so packed] that not even a needle could have gone through the spaces between the bodies. After what seemed like an eternity of extremely small movements, slowly, slowly, there appeared a clearing before me (*lumuwag bigla sa harap ko*). I was grateful not because I survived but because I experienced the discipline and respect of one for the other of the people—there was no pushing, no insulting, everyone even helped each other, and a collective patience and giving way ruled (*kolektibong pasensiya at pagbibigayan ang umiral*).
>
> The night deepened. Hungry again. Legs and feet hurting. I bought squid balls and sat on the edge of the sidewalk.... While resting on the sidewalk, I felt such immense pleasure, safe from danger, free, happy in the middle of thousands and thousands of anonymous buddies.

Finding herself amid a particularly dense gathering of bodies, Flor C. momentarily fears for her life. She can barely breathe, overwhelmed by the weight of bodies pressed up against her. Rather than a medium for movement, the crowd is in this instance a kind of trap, fixing her in place. Yet ever so slowly, the crowd moves as if on its own accord. No one says anything, no directives are issued, no leader appears to reposition bodies. Instead a kind of "collective patience and giving way ruled" (*kolektibong pasyensya at pagbibgayan ang umiral*). The crowd gives and takes, taking while giving, giving while taking and so suffers the presence of all those that compose it. It is for this reason "patient," which is to say, forbearing and forgiving while forgetting the identities of those it holds and is held by. Forbearance, forgiveness and forgetting are always slow, so slow in coming. They thus share in, if not constitute, the rhythm of the work of mourning, that in turn always entails the sharing of work.

After what seemed like an eternity of waiting and very little movement, Flor C. suddenly arrives at a clearing. *Lumuwag bigla sa harap ko* (it suddenly cleared in front of me), she says, which can also be glossed as "the clearing came before me." Who or what came before whom or what remains tantalizingly uncertain in the text. Earlier, she regretted being trapped in the crowd. But now, thrown into a sudden clearing by a force simultaneously intimate and radically exterior to her, Flor C. is grateful. She survives, but for her, this is not the most important thing. Rather, what matters is that she was given the chance to experience the "discipline and respect" of a crowd in which no one was pushed or pushing, no one was insulted or insulting, and everyone seemed to help one another, a condition that in Tagalog is referred to as *damayan*, or cooperation, the very same word used to connote the work of mourning.[30] It is a peculiar sort of discipline that Flor C. undergoes, one that does not interpolate subjects through hierarchies of recognition.[31] Instead, it is a kind of discipline borne of mutual restraint and deference that, inasmuch as it does not consolidate identity, lessens the hold of social distinctions.

Crowding gives rise to a sense of forbearance and a general economy of deference. At the same time it does not precipitate social identities. Rather, it gives way to a kind of saving that Flor C. refers to as the experience of "freedom" (*kalayaan*). Far from being a mob, the crowd is an embodiment of freedom and incalculable pleasure. It is where a different sense of collectivity resides, one that does away momentarily with hierarchy and the need for recognition. Constraint gives way to an unexpected clearing, to a giving way that opens the way for the other to be free, the other that now includes the self caught in the crowd. And because it is unexpected, this freeing cannot last—just as it cannot be the last, in the sense of final, experience of freedom. Here, emancipation, however transitory—and perhaps because it is felt to be so—does not depend on submission to a higher authority that guarantees the truth of messages. Rather, it relies on the dense gathering of bodies held in patient anticipation of a clearing and release.

Accounts of People Power II indicate that over a million people gathered in the course of four days at Edsa. These protesters were not all from the middle class. As Flor C.'s earlier remarks show, many who opposed Estrada came from the ranks of the working class and the urban and rural poor. This heterogeneous crowd was not entirely constituted by texting, for obviously not everyone owned cell phones. It emerged primarily, we might imagine, in response to a call for and the call of justice. Put another way, the crowd at Edsa was held together by the promise of justice's arrival. Here, justice is imagined not simply as a redistributive force acting to avenge past wrongs, its violence producing yet more injustice. The non-violent nature of People Power II instead suggests that the crowd formed not to exact revenge but to await justice. In so doing, it dwelt in the expectation of a promise that was always yet to be realized. Like freedom and no doubt inseparable from it, justice is thus always poised to arrive from the future. And it is the unceasing uncertainty of its arrival that constitutes the present waiting of the political crowd. It is a gathering that greets that whose arrival is never fully completed, and which forbears a coming always deferred. Yet, it is precisely because justice comes by not fully coming, and coming in ways unexpected, that it comes across as that which is free from any particular socio-technical determination. This promise of justice is what Flor C.'s experience of the crowd conveys. The promissory nature of justice means that it is an event whose eventfulness occurs in advance of and beyond any given political and social order. Evading reification and exceeding institutional consolidation, such an event entails a telecommunication of sorts. It is what Jacques Derrida might call the "messianic without a messiah." It would be "the opening up to the future or to the coming of the other as the advent of justice.... It follows no determinable revelation.... This messianicity stripped of everything, this faith without dogma.... "[32] In the midst of messianic transmissions, Flor C. along with others around her imagines the dissolution of class differences and feels, at least momentarily, that it is possible to overcome social inequities. She sees in crowding therefore a power that levels the power of the social as such. Past midnight, Flor C. finds herself no longer simply herself. Her body hurting, bearing the traces of the crowd's saving power, she sits on the sidewalk, eating squid balls, happy and safe, free in the midst of countless and anonymous "buddies."

IV. Postscript

Utopias, of course, do not last even if their occasional and unexpected happenings are never the last.

Some three months after People Power II, the newly installed government of President Gloria Macapagal-Arroyo made good on its promise to arrest former President Estrada on charges of graft and corruption. On 25 April, 2001, he was taken from his residence, fingerprinted and photographed, his mug shot displayed for all to see in the media. The sight of Estrada treated as a common criminal infuriated his numerous supporters, many of whom came from the ranks of the urban poor, who helped him win the largest majority ever in a presidential election. Spurred on by the middle-class leaders of Estrada's party, Puwersa ng Masa (Force of the Masses), and swelled by the ranks of the pro-Estrada Protestant sect, Iglesia ni Cristo and the populist Catholic group, El Shaddai, a crowd of perhaps a hundred thousand formed at Edsa and demanded Estarada's release and reinstatement. Unlike those who had gathered there during People Power II, the crowd in what came to be billed as the "Poor People Power" were trucked in by Estrada's political operatives from the slums and nearby provinces, and provided with money, food, and, on at least certain occasions, alcohol. In place of cell phones, many reportedly were armed with sling shots, home-made guns, knives and steel pipes. English-language news reports described this crowd as unruly and uncivilized and castigated protestors for strewing garbage on the Edsa Shrine, harassing reporters, and publicly urinating near the giant statue of the Virgin Mary of Edsa.[33]

Other accounts qualified these depictions by pointing out that many of those in the crowd were not merely hired thugs or demented loyalists but poor people who had legitimate complaints. They had been largely ignored by the elite politicians, the Catholic Church hierarchy, the middle-class dominated left-wing groups and the NGOs. Even though Estrada manipulated them, the protestors saw their ex-president as a patron who had given them hope by way of occasional hand-outs and who addressed them in their vernacular. The middle-class media treated Estrada's supporters as simpletons deficient in moral and political consciousness, but worthy of compassion. The vast majority of middle-class opinion thus shared in the view that the pro-Estrada crowd differed profoundly from the one that gathered in January during People Power II. While the latter was technologically savvy and politically sophisticated, the former was retrograde and reactionary. Generation Txt spoke of democratization, accountability, and civil society; the "*tsinelas* crowd," so-called because of the cheap rubber slippers many protestors wore, was fixated on its "idol," Estrada. In their mystified state, they seemed to the middle class barely articulate, and incapable of formulating anything other than a desire for vengeance on those they deemed responsible for victimizing Estrada. If the crowds of People Power II responded to the circulation of messages sanctioned by a higher authority, and the prospect of justice as the promise of freedom, the *masa* (masses) in People Power III were merely playing out a tragically mistaken identification with Estrada. They sought, or so it was assumed, the crude sort of payback typical of many of the ex-president's movie plots.[34]

Middle-class accounts of this other crowd regularly made mention of the "voicelessness" of the urban poor. At the same time, these accounts showed a relative lack of concern with actually hearing—much less recording—any distinctive voices. By emphasizing this voicelessness, the middle class in effect redoubled the masses' seeming inarticulateness; as if the masses, without anything intelligible to say, could only act irrationally and sometimes violently. "Voiceless," the masses, it was feared, might only riot in the streets. Indeed, in the early morning of 1 May, they marched from the Edsa Shrine to the presidential palace, in the process destroying millions of pesos worth of property, and suffering several deaths and scores of injuries. They finally were dispersed by the police and palace guards. But it is important to note that the protestors were, in fact, not voiceless. While marching to the palace, the masses chanted slogans. Newspaper reports quoted these slogans, and in so doing, give us a rare chance to actually hear the crowd: *Nandito na kami, malapit na ang tagumpay* (We're here, our victory is close at hand!), and *Patalsikin si Gloria! Ibalik si Erap! Nandyan na kami! Maghanda na kayo!* (Get rid of Gloria! Return Erap! We are coming! Get ready!).[35]

Here, the crowd is fueled by the desire to give back to Gloria what they think she's given to them. In return for her unseating of Estrada, they want to unseat her. She took his place, and now they want him to take hers. Through their slogans, the crowd expresses this giving back of a prior taking away. It says: "We are here, our victory is close at hand!"; "We are coming, you'd better be ready!" The crowd thereby takes itself for an apocalyptic power. The "we" referred to here has already arrived even as it continues to come. Certain of their arrival, the protestors ask those who hear to be ready. Having arrived, they will settle their debts, collect what is owed to them and thereby put an end to their—the crowd's and its audience—waiting. While the crowd in People Power II clung to a sense of the messianic without a messiah, this other crowd comes as a messianic specter delivered by resentments whose satisfaction can no longer be deferred. It is perhaps for this reason that middle-class observers repeatedly referred to it (in English) as a "mob," a "rabble," or "horde." These words imply more than savage or disordered speech and appearance. As the use of the word *horde* indicates, the masses were also seen to be irreducibly alien: foreign invaders encroaching upon a place they had no right to occupy.[36]

Eschewing a stance of forbearance, this crowd demanded recognition without delay. "Here we are!" it shouted. "Be prepared!" For many among the middle class, to hear this crowd was to realize that they were not quite ready to hear them; indeed, that they would always have been unprepared to do so. The masses suddenly became visible in a country where the poor are often viewed by the middle class as literally unsightly, spoken about and spoken down to because they are deemed incapable of speaking up for themselves. They are acknowledged only in order to be dismissed. Marching to the palace, however, and chanting their slogans, they assumed an apocalyptic agency. They threatened to bring about a day of reckoning that was simultaneously desired and dreaded by those who saw them. In their uncanny visibility, the masses did not gain a "voice" that corresponded to a new social identity. Instead, they communicated an excess of communication that could neither be summed up nor fully accounted for by those who heard them. Unprepared to hear the crowd's demand that they be prepared, the middle class could only regard it as monstrous. Hence the bourgeois calls for the conversion of the masses and their domestication by means of "pity," "compassion" and some combination of social programs and educational reform. But these calls also demanded that those who made up the crowd, one that was now totally other, be put back in their place, removed like so much garbage from the Edsa Shrine and from the perimeter of the presidential palace.[37] By the late morning of Labor Day, the military, spooked earlier by the specter of Poor People Power, had dispersed the marchers. The crowds' violent outbursts, like their abandoned rubber slippers, were relegated to the memory of injustices left unanswered, fueling the promise of revenge and feeding the anticipation of yet more uprisings in the future.

Notes

My thanks to Pete Lacaba and the contributors to Plaridel, to RayVi Sunico, Tina Cuyugan, Lita Puyat, Karina Bolasco, Jose and David Rafael, Carol Dahl, Chandra Mukerji, Matt Ratto, Paula Chakrabarty, Teresa Caldeira, James Holston, Jean-Paul Dumont, Adi Hastings, and Michael Silverstein for providing me with a variety of sources and insights that proved invaluable for this essay. I am especially grateful to Rosalind Morris and Michael Meeker for offering thoughtful comments on earlier drafts of this essay.

1. The link between telecommunication technologies and the politics of belief that I pursue here is indebted partly to the work of Jacques Derrida, especially in such writings as "Faith and Knowledge: The Two Sources of 'Religion' at the Limits of Reason Alone," trans. Sam Weber. In Jacques Derrida, *Acts of Religion*, ed. Gil Anidjar (New York: Routledge, 2002): 42–101; "Signature Event Context," in *Margins of Philosophy*, trans. Alan Bass (Chicago: University of Chicago Press, 1982), 307–330; and *The Politics of Friendship*, trans. George Collins (London: Verso, 1977).
2. See the bundle entitled "Telefonos, 1885–1891" at the Philippine National Archives, Manila for sketches of a plan to install a telephone system in the city as early as November, 1885. By December 1885, an office of Telephone Communication had been established (*Communicacion Telefonica*) and the first telephone station set up at Santa Lucia, Manila, was operational.
3. Jose Rizal, "Por Telefono" (Barcelona, 1889); reprinted in *Miscellaneous Writings* (Manila: R. Martinez and Sons, 1959), and in various other anthologies of Rizal's writings. For a more extended discussion of telegraphy and the formation of a wish for a lingua franca among the first generation of nationalists, see Vicente L. Rafael, "Translation and Revenge: Castilian and the Origins of Nationalism in the Philippines," in *The Places of History: Regionalism Revisited in Latin America*, edited by Doris Sommer (Durham: Duke University Press, 1999), 214–35.
4. For an elaboration of other modalities of these telecommunicative fantasies and their role in shaping nationalist consciousness, see Vicente L. Rafael, *White Love and Other Events in Philippines History* (Durham: Duke University Press, 2000), especially chapters 4 and 8 on rumor and gossip as populist modes of communication in Philippine history.
5. For a useful collection of documents and newspaper articles relating to the corruption case against Estrada, see Sheila Coronel, ed., *Investigating Estrada: Millions, Mansions and Mistresses* (Quezon City: Philippine Center for Investigative Journalism, 2000).
6. The quotations above come respectively from Uli Schmetzer, "Cell Phones Spurred Filipinos" *Chicago Tribune* (24 January 2001); Ederic Penaflor Eder, "Tinig Ng Genertion Txt" *Pinoy Times* (8 February 2001); Malou Mangahas, "Text Messaging Comes of Age in the Philippines" *Reuters Technology News* (28 January 2001).
7. Much of the information that follows was gathered from Wayne Arnold, "Manila's Talk of the Town is Text Messaging" *New York Times* (5 July 2000): C1; "Text Generation," special issue of I: The Investigative Reporting Magazine 8, no. 2 (April–June 2002), especially 14–21, 28–32; and Elvira Mata, *The Ultimate Text Book* (Quezon City: Philippine Center for Investigative Journalism: 2000), which is especially good for examples of the more common text messages that circulate among Filipino users.
8. For a succinct historical analysis of the Philippine state, see Benedict Anderson, "Cacique Democracy in the Philippines," in *The Specter of Comparisons* (London: Verso 1998), 192–226. See also John Sidel, *Capital, Coercion, and Crime:*

Bossism in the Philippines (Stanford: Stanford University Press, 1999); and Paul D. Hutchcroft, *Booty Capitalism: The Politics of Banking in the Philippines* (Ithaca: Cornell University Press, 1998).

9. The technology for monitoring cell phone use does exist and there is some indication that the Philippine government is beginning to acquire. It is doubtful, however, that cell phone surveillance technology was available to the Estrada administration. It is also not clear whether the current regime of Gloria Macapagal-Arroyo has begun monitoring or intends to monitor cell phone transmissions.
10. See Arnold, "Manila's Talk of the Town is Text Messaging"; Mangahas, "Text Messaging Comes of Age in the Philippines,"; Schmetzer, "Cell Phones Spurred Filipinos' Coup." See also Leah Salterio, "Text Power in Edsa 2001," *Philippine Daily Inquirer* (22 January 2001) (hereafter *PDI*); Conrad de Quiros, "Undiscovered Country," *PDI* (6 February 2001); Michael L. Lim, "Taming the Cell Phone," *PDI* (6 February 2001). However, the economic advantages of texting are limited. For example, any transmission across cell phone networks is expensive, so that calling or texting from a Globe phone to a Smart phone is rarely done. Indeed, the Department of Transportation and Communication (DOTC) had to intervene in late 1999 to get the two companies to improve interconnectivity and service as well as lower their costs.
11. This article was circulated on the listserves of various non-governmental organizations in the Philippines and bore the title "Pinoy Lifestyle." I have no knowledge as to the original source of this piece, so it exists in some ways like a forwarded text message. Thanks to Tina Cuyugan for forwarding this essay to me. All translations are mine unless otherwise indicated.
12. Arnold, "Manila's Talk of the Town."
13. Message posted by rnrsarreal@aol.com, in Plaridel, (plaridel_papers@egroups.com), 25 January 2001.
14. Arnold, "Manila's Talk of the Town"; See also Richard Lloyd Parr's untitled article on People Power II and cell phone use in *The Independent*, London (23 January 2001).
15. Michael Tan, "Taming the Cell Phone" *PDI* (6 February 2001).
16. Tan, "Taming the Cell Phone"; De Quiros, "Undiscovered Country" *PDI* (6 February 2001).
17. Arnold, "Manila's Talk of the Town."
18. These messages were forwarded by rnrsarreal@aol.com, to the Plaridel discussion group (plaridel_papers@egroups.com), 25 January 2001.
19. Bart Guingona, Plaridel, (plaridel_papers@egroups.com), 26 January 2001. Texting is widely credited with bringing about the rapid convergence of crowds at the EDSA Shrine within approximately seventy-five minutes of the abrupt halt of the Estrada impeachment trial on the evening of 16 January. Even prior to Cardinal Sin and former president Cory Aquino's appeal for people to converge at this hollowed site, it has been estimated that over 20,000 people had already arrived there, perhaps drawn by text messages they received. As Danny A. Gozo, an employee at Ayala Corporation, points out in his posting on Plaridel (plaridel_papers@egroups.com), 23 January 2001, during the four days of People Power II Globe Telecom reported an average of 42 million outgoing messages and around an equal number of incoming ones as well, while Smart Telecom reported over 70 million outgoing and incoming messages texted through their system *per day*. He observes enthusiastically that "the interconnectedness of people, both within the country and outside is a phenomenon unheard of before. It is changing the way that we live!"
20. Ederic Penaflor Eder, *Pinoy Times* (8 February 2001). The translation of this text is mine.
21. I owe this term to James T. Siegel, *Fetish Recognition Revolution* (Princeton: Princeton University Press, 1997).
22. My remarks on Manila's streets were gleaned from the notes and observations I made in the 1990s. On Manila's urban forms, see the excellent essay by Neferti X. Tadiar, "Manila's New Metropolitan Forms," in *Discrepant Histories: Translocal Essays on Filipino Cultures,* ed. Vicente L. Rafael (Philadelphia: Temple University Press, 1995), 285–313. For a lucid portrait of Manila's fantastic street life, see the novel by James Hamilton-Paterson, *The Ghosts of Manila* (New York: Vintage, 1995). Contemporary Philippine films, which often traverse the divide between rich and poor and explore their spaces of habitation, are excellent primary source materials for the study of Manila's urban forms. For a recent collection of essays on Philippine cinema, see Roland Tolentino, ed., *Geopolitics of the Visible: Essays on Philippine Film Cultures* (Quezon City: Ateneo de Manila University Press, 2000).
23. See chapter four of Wolfgang Schivelbusch's *The Railway Journey: The Industrialization of Time and Space in the 19th Century* (Berkeley: University of California Press, 1986).
24. I owe this information to Mr. David Rafael, former manager of the Glorietta shopping mall in the Ayala Center in Makati.
25. For a discussion of the historical link between linguistic and social hierarchies, see Vicente L. Rafael, "Taglish, or the Phantom Power of the Lingua Franca." In *White Love and Other Events in Filipino History* (Durham: Duke University Press, 2000), 162–189.
26. Here, I draw from Martin Heidegger, "The Question Concerning Technology," in *The Question Concerning Technology and other Essays,* trans. William Lovitt (New York: Harper and Row, 1977), 3–35. See also the illuminating commentary by Samuel Weber, "Upsetting the Setup: Remarks on Heidegger's 'Questing After Technics." *Mass Mediauras: Form Technics Media* (Stanford: Stanford University Press, 1996), 55–75. My remarks on the crowd are indebted to Walter Benjamin, *Charles Baudelaire: A Lyric Poet in the Era of High Capitalism* (London: Verso, 1977).
27. For a discussion of the history of this nationalist fantasy, see the Introduction to Vicente L. Rafael, *White Love and Other Events in Filipino History,* 1–18. For a comparative approach to the radical potential of nationalist ideas, see Benedict Anderson, *Imagined Communities: Reflections on the Origins and Spread of Nationalism* (London: Verso, rev. ed., 1991).
28. Flor C., Plaridel listserve (plaridel_papers@yahoogroups.com), January 24, 2001.
29. "Flor C." I have subsequently learned, is Flor Caagusan. She was formerly editor of the editorial page of the *Manila Times* and at one point served as the managing editor of *Diliman Review*. I owe this information to the journalist Pete Lacaba. While she would be known to a small group of journalists who are part of the Plaridel discussion group, she would presumably be unknown to the majority of participants in this group.

30. For an elaboration of the notion of *damayan*, see Reynaldo Ileto, *Pasyon and Revolution: Popular Uprisings in the Philippines, 1840–1910* (Quezon City: Ateneo de Manila University Press, 1979). See also the important work of Fenella Cannell on Bikol province, south of Manila, *Power and Intimacy in the Christian Philippines* (Cambridge: Cambridge University Press, 1999).
31. Flor C.'s account also recalls the experience of crowding in certain religious gatherings, notably the all-male procession of the image of Black Nazarene that marks the high point of the fiesta of Quiapo, a district of Manila on the ninth of January. For a description of the 1995 procession that conveys some sense of the dangers and pleasures experienced by onlookers and practitioners alike in the experience of crowding, see Jaime C. Laya, "The Black Nazarene of Quiapo," in *Letras y Figuras: Business in Culture, Culture in Business* (Manila: Anvil, 2001), 86–90.
32. Jacques Derrida, "Faith and Knowledge: The Two Sources of 'Religion' at the Limits of Reason Alone," in *Acts of Religion,* 56–57. The relationship among politics, promise, and technology intimated by Derrida is, of course, a key preoccupation of this essay. Promises arguably lie at the basis of the political and the social. The possibility of making and breaking pledges, of bearing or renouncing obligations, of exchanging vows and taking oaths forges a sense of futurity and chance, allowing for an opening to otherness. It is this possibility of promising that, Derrida has argued, engenders the sense of something to come, of events yet to arrive. But promises can be made and broken only if they can be witnessed and sanctioned, confirmed and reaffirmed. They must, in other words, be repeatable and citable, capable of being performed again and again. Repetition underlies the making of promises and, thus, the practices of politics. We can gloss this iterative necessity as the workings of the technical and the mechanical that inhere in every act of promising. Technology as the elaboration of the technical, including the technics of speech and writing, is then not merely an instrument for engaging in politics. It is that without which the political and the futures it claims to bring forth would simply never emerge, along with the very notion of emergence itself.
33. See for example the news reports and opinion columns of the *Philippine Daily Inquirer* from April 26 to May 5, 2001 for coverage of the "Poor People Power," or as others have referred to it, "People Power III." In particular, see the following, Alcuin Papa, Dave Veridiano, and Michael Lim Ubac, "Estrada Loyalists Overwhelm Cops on Way to Malacanang," *PDI* (2 May 2001): 1; Amando Doronilla, "The State Defends Itself" *PDI* (2 May 2001): 9, "Now the Fight Over Semantics," *PDI* (4 May 2001): 9; "Exchanges on Edsa 3," *PDI* (3 May 2001); Blanche S. Rivera and Christian Esguerra, "Edsa reclaimed by Edsa II Forces," *PDI* (2 May 2001): 1; Blanche Gallardo, "Tears of Joy for Tears of Sadness," *PDI* (6 May 2001): 1. See also Jarius Bondoc, "Gotcha," in *Philippine Star*, 1 May 2001; Howie G. Severino, "The Hand that Rocks the Masa" *Filipinas Magazine* (June 2001): 70–72; Pete Lacaba, "Edsa Puwersa" *Pinoy Times* (29 April 2001).
34. See for example Conrado de Quiros, "Lessons" *Philippine Daily Inquirer* (4 May 2001); Walden Bello, "The May 1st Riot: Birth of Peronism RP Style?" *Philippine Daily Inquirer* (8 May 2001); La Liga Policy Institute (Quezon City), "Poor People Power: Preludes and Prospects," as it appears in filipino-studies@yahoogroups.com , 6 May 2001; Ferdinand Llanes "Edsa at Mendiola ng Masa," filipino-studies@yahoogroups.com, 3 May 2001.
35. Papa et al., "Estrada Loyalists Overwhelm Cops."
36. "Horde" comes from the Turkish *ordi/ordu,* meaning "camp," and originally referred to "troops of Tartar or other nomads dwelling in tents or wagons and moving from place to place for pasturage or for war and plunder," according to the *Oxford English Dictionary*.
37. See "Edsa Reclaimed by Edsa II Forces" *Philippine Daily Inquirer* (May 2, 2001) which reports, among other things, how people involved in People Power II "brought their own towels, sponge and scrubs," to clean the garbage that had been left behind by the pro-Estrada crowd, hosing down "the filth from the ground," and "disinfecting," the Shrine with chlorine. Estrada supporters had "heaped mounds of garbage, sang and danced lustfully over the Edsa Shrine marker, rammed a truck into the landscape and directed huge loudspeakers to the shrine door," according to the Shrine rector, Monsignor Soc Villegas.

Part V

Theorizing "New" Media

22

Cybertyping and the Work of Race in the Age of Digital Reproduction

Lisa Nakamura

Software engineers and academics have something in common: they both like to make up new words. And despite the popular press's glee in mocking both computer-geek and academic jargon, there are several good arguments to be made for the creation of useful neologisms, especially in cases where one of these fields of study is brought to bear on the other. The Internet has spawned a whole new set of vocabulary and specialized terminology because it is a new tool for communicating which has enabled a genuinely new discursive field, a way of generating and consuming language and signs which is distinctively different from other, older media. It is an example of what is dubbed "the new media" (a term refreshingly different from the all purpose "post" prefix so familiar to critical theorists, but destined to date just as badly). Terms such as "cybersex," "online," "file compression," "hypertext link" and "downloading" are now part of Internet's user's everyday vocabulary since they describe practices or virtual objects which lack analogues in either offline life or other media. The new modes of discourse enabled by the Internet requires new terminologies and conceptual frameworks to describe it.

Just as engineers and programmers routinely come up with neologisms to describe new technologies, so too do academics and cultural theorists coin new phrases and terms to describe concepts they wish to introduce to the critical conversation. While these attempts are not always well-advised, and certainly do contribute at times to the impenetrable and unnecessarily confusing nature of high theory's rhetoric, there are some compelling reasons that this move seems peculiarly appropriate in the case of academic studies of the Internet. Lev Manovich and Espen Aarseth both make a persuasive case for the creation and deployment of a distinctively new set of terminologies to describe the new media, in particular the Internet. In *The Language of New Media* Manovich asserts that "comparing new media to print, photography, or television will never tell us the whole story" and that "to understand the logic of new media we need to turn to computer science. It is there that we may expect to find the new terms, categories, and operations which characterize media which became programmable. From media studies, we move to something which can be called software studies; from media theory—to software theory."[1] This statement calls for a radical shift in focus from traditional ways of envisioning media to a new method which takes the indispensability of the computer-machine into account. It truly does call for a reconceptualization of media studies, and constitutes a call for new terms more appropriate to "software studies" to best convey the distinctive features of new media, in particular the use of the computer.

Manovich identifies two "layers" to new media: the cultural layer, which is roughly analogous to "content," and the computer layer, or infrastructure, interface, or other machine-based forms which structure the computer environment. His argument that the computer layer can be expected to have a "significant influence on the cultural logic of media"[2] is in some sense not original; the notion that form influences content (and vice versa) has been around since the early days of literary criticism. It has been conceded for some time now that certain forms allow or disallow the articulation of certain ideas. However, what is original about this argument is its claim that our culture is becoming "computerized" in a wholesale and presumably irrevocable fashion. This is a distinctively different proposition from asserting the importance of, say, electronic *literacy*, a paradigm which is still anchored by its terminology in the world of a very old medium: writing. Manovich calls for a new terminology, native to the computer: he goes on to write that "in new media lingo, to 'transcode' something is to translate it into another format. The computerization of culture gradually accomplishes similar transcoding in relation to all cultural categories and concepts. That is, cultural categories and concepts are substituted, on the level of meaning and/or language, by new ones which derive from the computer's ontology, epistemology, pragmatics. New media thus acts as a forerunner or this more general process of cultural re-conceptualization."[3]

If we follow this proposition, we can see that our culture is in the process of being 'transcoded' by the computer's "ontology, epistemology, pragmatics." While this statement has far-reaching implications, at the least it can be seen as an argument for a new openness in new media studies towards the adoption of terminology which at least acknowledges the indispensable nature of the computer in the study of new media. This would be a "transcoded" kind of terminology, one which borrows from the language of the computer itself rather than from the language of critical theory or old media studies. In his article "The Field of Humanistic Informatics and its Relation to the Humanities," Espen Aarseth argues that the study of new media needs to be a "separate, autonomous field, where the historical, aesthetic, cultural and discursive aspects of the digitalization of our society may be examined.... We cannot leave this new development to existing fields, because they will always privilege their traditional methods, which are based on their own empirical objects."[4]

In an attempt to "transcode" the language of race and racialism that I observed online, I coined the term "cybertype" to describe the distinctive ways that the Internet propagates, disseminates, and commodifies images of race and racism. The study of racial cybertypes brings together the cultural layer and the computer layer; that is to say, cybertyping is the process by which computer/human interfaces, the dynamics and economics of access, and the means by which users are able to express themselves online interacts with the "cultural layer" or ideologies regarding race that they bring with them into cyberspace. Manovich is correct in asserting that we must take into account the ways that the computer determines how ideological constructs such as race get articulated in this new medium.

Critical theory itself is a technology or machine which produces a particular kind of discourse, and I'd like to conduct a discursive experiment by poaching a term from nineteenth century print technology; that term is "stereotype."

The word "stereotype" is itself an example of machine-language, albeit pre-computer; the first stereotype was a mechanical device that could reproduce images relatively cheaply, quickly, and in mass quantities. Now that computer-enabled image-reproducing machines like the Internet are faster, cheaper, and more efficient than ever before, how does that machine language translate into critical terms? Might we call new formulations of machine-linked identity "cybertypes"? This is a clunky term; in hacker-speak it would be called a "kludge" or "hack" because it's an improvised, spontaneous, seat-of-the-pants way of getting something done. (Critical theory, like the software industry, is a machine which is good at manufacturing linguistic kludges and hacks). I'd like to introduce it because it acknowledges that identity online is still "typed," still mired in oppressive roles even if the "body" has been left behind or bracketed. I pose it as a corrective to the disturbingly utopian strain I see embodied in most commercial representations of the Internet in general. Chosen identities enabled by technology such as online avatars, cosmetic and transgender surgery and body modifications, and other cyber-prostheses are not breaking the mold of unitary identity, but rather are shifting identity into the realm of the "virtual," a place not without its own laws and hierarchies. Supposedly "fluid" selves are no less subject to cultural hegemonies, rules of conduct, and regulating cultural norms than are "solid" ones.

While telecommunications and medical technologies can challenge some gender and racial stereotypes, they produce and reflect them as well. Cybertypes of the biotechnologically enhanced or perfected woman and of the Internet's invisible minorities, who can log onto the net and be taken for "white," participate in an ideology of liberation from marginalized and devalued bodies. This kind of technology's greatest promise to us is to eradicate Otherness, to create a kind of better living through chemistry, so to speak. Images of science freeing women from their aging bodies, which make it more difficult to conceive children and ward off cellulite, men from the curse of hair loss, and minorities online from the stigma of their race since no one can see them, reinforces a post-body ideology which reproduces the assumptions of the old one. (In an example of linguistic retrofitting, I've termed this phenomenon an example of the "meet the old boss, same as the new boss" product line). In other words, machines which offer identity prostheses to redress the burdens of physical "handicaps" such as age, gender, and race produce cybertypes which look remarkably like racial and gender stereotypes. My research on cross-racial impersonation in an online community reveals that when users are free to choose their own race, all were assumed to be white. And many of those who adopted non-white personae turned out to be white male users masquerading as exotic samurai and horny geishas.

Of course, this kind of vertiginous identity-play which produces and reveals cybertyping is not the fault of or even primarily an effect of technology. Microsoft's corporate slogan "where do you want to go today?," another example of the discourse of technological liberation, situates the agency directly where it belongs: with the user. Though computer memory modules double in speed every couple of years, users are still running operating systems which reflect phantasmatic visions of race and gender. Moore's Law does not obtain in the "cultural layer." In the end, despite academic and commercial post-identitarian discourses, it does come down to bodies; bodies with or without access to the Internet, telecommunications, and computers and the cultural capital necessary to use them, bodies with or without access to basic healthcare, let alone high-tech pharmaceuticals or expensive forms of elective surgery.

Cybertypes are more than just racial stereotypes "ported" to a new medium. Because the Internet is interactive and collectively authored, cybertypes are created in a peculiarly collaborative way; they reflect the ways that machine-enabled interactivity gives rise to images of race which both stem from a common cultural logic and seek to redress anxieties about the ways that computer-enabled communication can challenge these old logics. They perform a crucial role in the signifying practice of cyberspace; they stabilize a sense of a white self and identity that is threatened by the radical fluidity and disconnect between mind and body that is celebrated in so much cyberpunk fiction. Bodies get tricky in cyberspace. That sense of disembodiment engendered by cyberspace which is both freeing and disorienting creates a profound malaise in the user which stable images of race works to fix in place.

Cybertypes are the images of race that arise when the fears, anxieties, and desires of privileged Western users (the majority of Internet users and content producers are still from the Western nations) are scripted into a textual/graphical environment that is in constant flux and revision. As Rey Chow writes in a chapter entitled "Where Have All the Natives Gone?," images of raced Others become necessary symptoms of the postcolonial condition. She writes, "the production of the native is in part the production of our postcolonial modernity,"[5] and that "we see that in our fascination with the 'authentic native' we are actually engaged in a search for the aura even while our search processes themselves take us farther and farther from that 'original' point of identification."[6] The Internet is certainly a postcolonial discursive practice, originating as it does from both scientific discourses of progress and the Western global capitalistic project. When Chow attributes our need for stabilizing images of the "authentic native" to the "search for the aura," or original and authentic object, she is transcoding Walter Benjamin's formulation from "The Work of Art in the Age of Mechanical Reproduction" into a new paradigm. In the subsection to this chapter entitled "The Native in the Age of Discursive Reproduction," Chow clarifies her use of Benjamin to talk about postcoloniality and the function of the "native." While Benjamin maintained that technology had radically changed the nature of art by making it possible to reproduce infinite copies of it, thus devaluing the "aura" of the original, Chow envisions the "native" himself as the original, with his own aura. When natives stop acting like natives, that is to say, when they deviate from the stereotypes that have been set up to signify their identities, their "aura" is lost: they are no longer "authentic." Thus, a rationale for the existence of racial cybertypes become clear: in a virtual environment like the Internet where *everything* is a copy, so to speak, and nothing has an aura since all cyber-images exist as pure pixellated information, the desire to search for an original is thwarted from the get-go. And hence the need for images of cybertyped "real natives" to assuage that desire. Chow poses a series of questions in this section: "why are we so fascinated with 'history' and with the 'native' in 'modern' times? What do we gain from our labor on these 'endangered authenticities' which are presumed to be from a different time and a different place? What can be said about the juxtaposition of 'us' (our discourse) and 'them'? What kind of *surplus value* is created by this juxtaposition?"[7] The surplus value created by this juxtaposition (between the Western user and the discourses of race and racism in cyberspace) lies precisely within the need for the native in modern times. As machine-induced speed enters our lives—speed of transmission of images and texts, of proliferating information, of dizzying arrays of decision trees and menus—all of these symptoms of modernity create a sense of unease which is remedied by comforting and familiar images of a "history" and a "native" which seems frozen in "a different time and a different place."

This is the paradox: in order to think rigorously, humanely, and imaginatively about virtuality and the post-human, it is absolutely necessary to ground critique in the lived realities of the human, in all their particularity and specificity. The nuanced realities of virtuality—racial, gendered, Othered—live in the body, and though science is producing and encouraging different readings and revisions of the body, it is premature to throw it away just yet, particularly since so much postcolonial, political, and feminist critique stems from it.

The vexed position of women's bodies and raced bodies in feminist and postcolonial theory has been a subject of intense debate for at least the past twenty years. While feminism and postcolonial studies must, to some extent, buy into the notion of there being such a thing as a "woman" or a "person of color" in order to be coherent, there are also ways in which "essentialism is a trap,"[8] to quote Gayatri Spivak. Since definitions of what counts as a woman or a person of color can be shifting and contingent upon hegemonic forces, essentialism can prove to be untenable. Indeed, modern body technologies are partly responsible for this: gender reassignment surgery and cosmetic surgery can make these definitions all the blurrier. In addition, attributing essential qualities to women and people of color can reproduce a kind of totalizing of identity which reproduces the old sexist and racist ideologies. However, theorists such as Donna Haraway, who radically question the critical gains to be gotten from conceptualizing "woman" as anchored to the body, take great pains to emphasize that she does not "know of any time in history when there was greater need for political unity to confront effectively the dominations of 'race,' 'gender,' 'sexuality,' and 'class.'"[9] Though she replaces the formerly-essential concept of "woman" with that of the "cyborg," a hybrid of machine and human, she also acknowledges that feminist politics must continue "through coalition—affinity, not identity."[10] Both she and Spivak write extensively about the kinds of strategic affinities that can and must be built between and among "women" (albeit in quotation marks), racial and other minorities, and other marginalized and oppressed groups.

Is it a coincidence that, just as feminist and subaltern politics—built around affinities as well as identities—are acquiring some legitimacy and power in the academy (note the increasing numbers of courses labeled "multicultural," "ethnic," "feminist," "postcolonial" in university course schedules), MCI and other tele-technology corporations are staking out their positions as forces which will free us from race and gender? Barbara Christian, in her 1989 essay "'The Race for Theory': Gender and Theory: Dialogues on Feminist Criticism," saw a similar kind of "coincidence" in regards to the increasing dominance of literary theory as a required and validated activity for American academics. She asserts that the technology of literary theory was made deliberately mystifying and dense to exclude minority participation; this exclusionary language "surfaced, interestingly enough, just when the literature of peoples of color, of black women, of Latin Americans, of Africans, began to move 'to the center.'"[11] The user-unfriendly language of literary theory, with its poorly designed interfaces, overly elaborate systems, and other difficulties of access happened to arise during the historical moment in which the most vital and vibrant literary work was being produced by formerly "peripheral" minority writers.

Perhaps I am like Christian, who calls herself "slightly paranoid," in this essay (it has been well documented that telecommunications technologies encourage paranoia), but I too wonder whether cyberspace's claims to free us from our limiting bodies are not slightly too well timed. Learning curves for net-literacy are notoriously high; those of us who maintain Listservs and Web sites and MUDs learn that to our rue. Indeed, it took me a few years of consistent effort, some expensive equipment, and much expert assistance to feel anything less than utterly clueless in cyberspace. Rhetorics which claim to remedy and erase gender and racial injustices and imbalances through expensive and difficult to learn technologies such as the Internet entirely gloss over this question of access, which seems to me *the* important question. And it seems unlikely that this glossing over is entirely innocent. Cybertyping and other epiphenomena of high technologies in the age of the Internet is partly the result of people of color's restricted access to the means of production—in this case, the means of production of the "fluid identities" celebrated by so much theory and commerce today.

Increasing numbers of racial minorities and women are acquiring access to the Internet: a hopeful sign indeed. Ideally, this equalizing of access to the dominant form of information technology in our time might result in a more diverse cyberspace, one which doesn't seek to elide or ignore difference as an outmoded souvenir of the body. Indeed, sites such as ivillage.com, Oxygen.com, Salon.com's Hip Mama web pages, and NetNoir which contain content specifically geared to women and African Americans indicate a shift in the Internet's content which reflect a partial bridging of the digital divide. As women of color acquire an increasing presence online, their particular interests which spring directly from gender and racial identifications, that is to say, those identities associated with a physical body off-line, are being addressed.

Unfortunately, as can be seen from the high, and ultimately dashed, feminist hopes that new media such as the Oxygen Network would express women's concerns in a politically progressive and meaningful way, gender and race can just as easily be co-opted by the e-marketplace. Commercial sites such as these tend to view women and minorities primarily as potential markets for advertisers and merchants rather than as "coalitions." Opportunities for political coalition-building between women and people of color are often subverted in favor of e-marketing and commerce. (NetNoir is a notable exception to this trend. It is also the oldest of these identitarian Web sites, and thus was able to form its mission, content, and "look and feel" prior to the gold rush of dot.com commerce which brought an influx of investment capital, and consequent pressure to conform to corporate interests, to the Web).[12] Nonetheless, this shift in content which specifically addresses women and minorities, either as markets[13] or as political entities, does acknowledge that body-related identities such as race and gender are not yet as fluid and thus disposable as much cybertheory and commercial discourse would like to see them.

However, such is the stubborn power of cybertyping that even when substantial numbers of racial minorities do have the necessary computer hardware and Internet access to deploy themselves "fluidly" online they are often rudely yanked back to the realities of racial discrimination and prejudice. For example, on March 13, 2000 "in what its lawyers called 'the first civil rights class action litigation against an Internet company,' the Equal Rights Center, a Washington-based civil rights group, and two African-American plaintiffs are suing Kozmo for racial 'redlining' because of what they believe is a pattern of those neighborhoods not being served."[14] Kozmo.com, an online service that delivers convenience foods and products, claims to deliver only to "zip codes that have the highest rates of Internet penetration and usage"[15]; however, the company's judgment of what constitutes an Internet-penetrated zip code follows racial lines as well. African-American Washingtonians such as James Warren and Winona Lake used their Internet access to order goods from Kozmo, only to be told that their zip codes aren't served by the company. Kozmo.com also refuses to deliver to a neighborhood of Washington D.C. occupied primarily by upper-class African Americans with equal "Internet penetration" as white neighborhoods.[16] It seems that these African-American Internet users possessed identities online too firmly moored to their raced bodies to participate in the utopian ideal of the Internet as a democratizing disembodied space. Unfortunately, it would appear that online identities can never be truly fluid if you live in the wrong zip code.

As the Kozmo.com example shows, actual hardware access is a necessary but not sufficient component of online citizenship. All of the things that citizenship implies—freedom to participate in community on an equal basis, access to national and local infrastructures, the ability to engage in discourse and commerce, cyber and otherwise, with other citizens—are abrogated by racist politics disguised as corporate market-research. This example of online "redlining" or "refusing to sell something to someone due to age, race or location" puts a new spin on cybertyping. Rather than being left behind, bracketed, or "radically questioned," the body—the raced, gendered, classed body—gets "outed" in cyberspace just as soon commerce and discourse come into play. Fluid identities aren't much use to those whose problems exist strictly (or even mostly) in the real world if they lose all their currency in the realm of the real.

It is common to see terms such as "the body," "woman," and "race" in quotation marks in much academic writing today. The after/images of identity which the Internet shows us similarly attempt to bracket off the gendered and raced body in the name of creating a democratic utopia in cyberspace. However, postmortems pronounced over "the body" are premature, as the Kozmo.com lawsuit shows. My hope is that these discourses of cyber-enabled fluidity and liberation do not grow so insular and self-absorbed as to forget this.

In the mechanical age, technology was viewed as instrumental, a means to an end; users were figured as already-formed subjects who approach it, rather than contingent subjects who are approached and altered by it. However, this view has been radically challenged in recent years, in particular by the Internet and other telecommunications technologies, which claim to eradicate the notion of physical distance and firm boundaries not only between users and their bodies but between topoi of identity as well.

The Internet generates both images of identity and after-images. The word "after/image" implies two things to me in the context of contemporary technoscience and cyberculture.

The first is its rhetorical position as a "y2K-ism," part of the millennial drive to categorize social and cultural phenomena as "Post" and "After." It puts pressure on the formerly solid and anchoring notion of "identity" as something we in the digital age are fast on our way to becoming "after." This notion of the post-human has evolved in other critical discourses of technology and the body, and is often presented in a celebratory way.[17]

The second is this: the image which you see when you close your eyes after gazing at a bright light: the phantasmatic spectacle or private image-gallery which bears but a tenuous relationship to "reality." Cyberspace and the images of identity that it produces can be seen as an interior, mind's eye projection of the "real." I'm thinking especially of screen fatigue—the crawling characters or flickering squiggles you see inside your eyelids after a lot of screen-time in front of the television, CRT terminal, movie screen, any of the sources of virtual light to which we are exposed every day. How have the blinding changes and dazzlingly rapid developments of technology in recent years served to project an altered image or projection of identity upon our collective consciousness? This visual metaphor of the after-image describes a particular kind of historically and culturally grounded seeing or mis-seeing, and this is important. Ideally, it has a critical valence and can represent a way of seeing differently, of claiming the right to possess agency in our ways of seeing; of being a subject rather than an object of technology. In the bright light of contemporary technology, identity is revealed to be phantasmatic, a projection of culture and ideology. It is the product of a reflection or a deflection of prior, as opposed to after/ images of identity. When we look at these rhetorics and images of cyberspace we are seeing an after/image—both post-human and projectionary—meaning it is the product of a vision re-arranged and deranged by the virtual light of virtual things and people.

Similarly, the sign-systems associated with advertisements for reproductive and "gendered" technologies reveal, in Valerie Hartouni's words, "The fierce and frantic iteration of conventional meanings and identities in the context of technologies and techniques that render them virtually unintelligible."[18] According to this logic, stable images of identity have been replaced by "after/images." When we look at cyberspace, we see a phantasm which says more about our fantasies and structures of desire than it does about the "reality" to which it is compared by the term "virtual reality." Many of cyberspace's commercial discourses such as those seen in television and print advertisements work on a semiotic level which establishes a sense of a national self. However, in a radically disruptive move they simultaneously deconstruct the notion of a corporeal self anchored in familiar categories of identity. Indeed, this example of "screen fatigue" (commercials are great examples of screen fatigue because they're so fatiguing) projects a very particular kind of after-image of identity.

The discourse of many commercials for the Internet includes gender as only one of a series of outmoded "body categories" like race and age. The ungendered, deracinated self promised to us by these commercials is freed of these troublesome categories, which have been done away with in the name of a "progressive" politics. The goal of "honoring diversity" seen on so many bumper stickers in Northern California will be accomplished by eliminating diversity.

It's not just commercials that are making these post-identitarian claims. Indeed, one could say that they're following the lead or at least running in tandem with some of the growing numbers of academics who devote themselves to the cultural study of technology. For example, in *Life on the Screen* Sherry Turkle writes: "When identity was defined as unitary and solid it was relatively easy to recognize and censure deviation from a norm. A more fluid sense of self allows for a greater capacity for acknowledging diversity. It makes it easier to accept the array of our (and others') inconsistent personae—perhaps with humor, perhaps with irony. We do not feel compelled to rank or judge the elements of our multiplicity. We do not feel compelled to exclude what does not fit."[19] According to this way of thinking, regulatory and oppressive social norms such as racism and sexism are linked to users' "unitary and solid" identities off-screen. Supposedly, leaving the body behind in the service of gaining more "fluid identities" means acquiring the ability to carve out new, less oppressive norms, and gaining the capacity to "acknowledge diversity" in ever more effective ways. However, is this really happening in cyberspace?

I answer this question with an emphatic "no." I coined the term "identity tourism" to describe a disturbing thing that I was noticing in an Internet chat community. During my fieldwork I discovered that the "after/images" of identity that users were creating by adopting personae other than their own online as often as not participated in stereotyped notions of gender and race. Rather than "honoring diversity," their performances online used race and gender as amusing prostheses which could be donned and shed without "real life" consequences. Like tourists who become convinced that their travels have shown them real "native" life, these identity tourists often took their virtual experiences as other-gendered and other-raced avatars as a kind of lived truth. Not only does this practice provide titillation and a bit of spice—as bell hooks writes, "one desires a 'bit of the Other' to enhance the blank landscape of whiteness"[20]—it also provides a new theater in cyberspace for "eating the Other." For hooks, "the overriding fear is that cultural, ethnic, and racial differences will be commodified and offered up as new dishes to enhance the white palate—that the Other will be eaten, consumed, and forgotten."[21] Certainly, the performances of identity tourists exemplify the consumption and commodification of racial difference; the fact that so many users are willing to pay monthly service fees to put their racially-stereotyped avatars in chatrooms attests to this.

Remastering the Internet

The racial stereotype, a distinctive and ongoing feature of media generally, can be envisioned in archaeological terms. If we conceive of multimedia, in particular what's been termed the "new media" engendered by the Internet, as possessing strata, layers of accretions and amplifications of imageries and taxonomies of identity, then it is possible and indeed, for reasons I will show shortly, strategic to examine the structure of these layerings. Old media provide the foundation for the "new," and its means of putting race to "work" in the service of particular ideologies are re-invoked, with a twist, in the new landscape of race in the digital age. Visions of a "post-racial democracy" evident in much discourse surrounding the Internet, in particular print and television advertisements, are symptomatic of the desire for a cosmetic cosmopolitanism which works to conceal the problem of racism in the American context.

I could put this another way: "where's the multi(culturalism) in multimedia?" or "where is race in new media?" What is the "work" that race does in cyberspace, our most currently privileged example of the technology of digital reproduction? What boundaries does it police? What "modes of digital identification" or disidentification are enabled, permitted, foreclosed vis à vis race? Has the notion of the "authentic" been destroyed permanently, a process that Benjamin predicted had begun at the turn of the century with the advent of new means of mechanical reproduction of images? How do we begin to understand the place of authenticity, in particular racial and cultural authenticity, in the landscape of new media? Digital reproduction produces new iterations of race and racialism, iterations with roots in those produced by mechanical reproduction. Images of race from older media are the analog signal which the Internet optimizes for digital reproduction and transmission.

On the one hand, Internet use can be seen as part of the complex of multimedia globalization, a foisting of a Western (as yet) cultural practice upon Third World, minority, and marginalized populations. Recent protests in the Western world against the IMF critique global capitalism and globalization as not only economically exploitative of the Third World, but also culturally exploitative as well, essentially creating a "Monoculture of the Mind."

A recent full page advertisement from the *New York Times* (June 19, 2000) entitled "Megatechnology" uses this term and superimposes it with an image of a television being carried on a top of an African woman's head. The subtitle reads "Ours is the first culture in history to have moved inside media—to have largely replaced direct contact with people and nature for simulated versions on TV, sponsored by corporations. Now it's happening globally, with grave effects on cultural diversity and democracy." This advertisement, produced for the Turning Point Project, a coalition of more than 80 non-profit organizations including Adbusters, Media Alliance, the International Center for Technology Assessment, and the International Forum on Globalization, includes AOL/Time Warner among the "biggest three global media giants," and explains that cultural diversity cannot survive "virtual reality," of which television is cited as the "earliest form."

It claims that global media, including and especially the Internet, produce a kind of "mental retraining; the cloning of all cultures to be alike." The positioning of this advertisement in a mainstream mass media publication could seem to a cynical reader as an exercise in bad faith, since the *New York Times* is itself a part of the global media complex the ad is critiquing. Nonetheless, the situatedness of this argument within a non-academic publication demonstrates that concerns about "virtual reality" or cyberspace as a culturally imperialistic practice exist outside of the academy as well as inside of it.

Monocultures are posed here as the opposite of diversity. Ziauddin Sardar characterizes cyberspace itself as a monoculture, the West's "dark side" and thus a powerful continuation of the imperalist project.[22] The discourse of agribusiness and bioengineering of crops is central here: monocultures are economies of scale, an erasure of diversity under current attack by the fashionable as offering little resistance to disease. But where does the hybrid, specifically the hyphenated American of color, stand in relation to this?

In this ad, the image of the African woman in native dress walking a dusty road with a television balanced skillfully on her head is meant to be jarring, to operate as part of the argument against globalization and television watching in native cultures. Viewers are supposed to react with horror at the evil box contaminating her culture and the landscape. Yet, ought we (or do we) experience a similar horror when seeing a Filipino youth in Monterey Park carrying a boom box, break dancing, and eating a McDonald's hamburger? Or when we see a Chinese rock group, performing in Britney Spears-type outfits? In the first example, vegetarians may well take offense, but the fact is that such sights are common, and are examples of what could be seen as resistant practices.

In the first example with the African woman, the tourist gaze would like to see her outside of time, protected from the incursions of digital "culture" (or monoculture) by Western intervention. The authenticity of the timeless primitive is threatened by the television set. In the second example, cultural appropriations and borrowings are commonly celebrated as hybridity and assimilation. In the culture of popular music, the productive samplings, mixings, and re-masterings of hip-hop are envisioned as vital signs of a flourishing youth culture. The technologies of contemporary music create a space for these cultural mixings, scratchings, and bricolage.

How do these paradigms from music fit the Internet? Does the Internet indeed create a monoculture? Is there space for the subaltern to speak in it? How do representations of the subaltern in reference to the Internet preserve or deny diversity? How is the paradigm of tourism invoked to stabilize threatened ideas of the authentic native post-Internet?

The Internet has a global sweep, a hype (hysteria?) attached to it; it makes distinctive claims to a radical post-racial democracy that other media have failed to employ effectively. Racial cybertyping is at work on the Internet today, and its implications both *for* its "objects" and for the cultural matrix it is embedded in generally are far-reaching. Groups such as racial and ethnic minorities who are prone to being stereotyped in older media are now being "remastered" to use more digital terminology, ported to cybertyping. Remastering, the practice of converting an analog signal, say from a vinyl record, to a digital one, like a DVD, CD, or HTML, preserves the "content" of the original piece while optimizing it for a new format. Remastering fiddles sound levels, timbre, erases scratchy silences, smoothes roughnesses, alters signal to noise ratios in such a way that the same song is made infinitely available for reproduction, replay, and re-transmission. But with a difference: variations in tone, timbre, and nuance are detectable; while the song remains the same, some of its qualities are altered, as are the possibilities for different audiences, different occasions for capture, replay, and transmission. The web-like media complex of images of the racialized Other as primitive, exotic, irremediably different and fixed in time are an old song, which the Internet has remastered or retrofitted in digitally reproducible ways. I wish to get back in the studio, so to speak, and to see how this remastering happens, and what its effects are upon social formations and readings of race in the age of digital reproduction. When you feed racism into this machine, what you get are images of "exotic" non-American racial minorities using technology, not American minorities.

The Internet is the fastest, most effective image-reproduction machine this world has yet seen. Just as the stereotype machine, that clumsy mechanical device which produced multiple but imperfect copies of an original image, has been replaced by more efficient and clearer, cleaner modes of image reproduction, so too are racial stereotypes being replaced by cybertypes. While racial stereotypes can now be perceived by our ever more discerning eyes as crude and obvious, and thus have been appropriated as camp (as in Bill Cosby's collection of racist black memorabilia), parody (black humor, like Chris Rock's, turns upon this), or incorporated into a history of oppression, cybertypes have as yet managed to sneak under the radar of critical and popular scrutiny. The digital images of natives, others, and the "raced" which proliferate on and around the Internet are clean, non-mechanical, carried upon a beam of fiberoptic light. Cybertyping's phantom track can be traced in a Cisco television advertisement, produced as a series entitled "The Internet Generation," which participates in a subtle blend of racism and racialism. Rather than stereotyping different races, it cybertypes them. The children in the first ad, "Out of the Mouths of Babes" repeat statistics about the Internet's improvements on older media, i.e., "The Web has more users in the first five years than television did in the first thirty," in distinctively accented voices whose speakers are depicted in "native" dress in "native" settings, such as a temple pool, a mosque, and a rural schoolyard. In addition, their dialogue is fractured, as each sentence is continued or repeated by a different child in a different locale. Thus, the ad tries to literalize the smaller world which Benjamin predicted audiences accustomed to proliferating mechanical images, and, by extension, digital images, would come to desire and expect. One child tells us that "a population the size of the United Kingdom joins the Internet every six months. Internet traffic doubles every 100 days." This depiction of the Internet as a population one joins, rather than a service one purchases and consumes or a practice one engages in, significantly uses the un-imperial nation, the United Kingdom, as the yardstick of measurement here. This language of a "united kingdom" of multiracial "generations" seems utopian, yet polices the racial and ethnic boundaries of this world very clearly. Global capitalism is envisioned as a United Nations of users from different countries united in their praise of the Internet, yet still preserved in their different ethnic dress, languages, and "look and feel."[23] Despite the fact that international Internet users are likely to be city-dwellers, these ads depict them in picturesque and idealized "native" practices uncommon even in rural areas.

Cybertyping's purpose is to bracket off racial difference representatively to assuage fears that the Internet is indeed producing a monoculture. The greater fear, however, which cybertyping actively works to conceal, is the West's reluctance to acknowledge its colonization of global media, and ongoing racist practices within its own borders. The ad's claims that "soon, all of our ideas will be free of borders" tries to stake out the notion that America's responsibility for its own problems with race, the greatest problem of our age in DuBois's terms, will be erased when "borders" (between nations, between the mind and the raced body) are figuratively erased. The subtlety of this argument is necessary in our postcolonial, postmodern age: scenarios that invoke the Scramble for Africa, an emblematic episode of the West's division and exploitation of the non-Western world, just will not "play" anymore. However, re-porting the imperialist impulse to a commercial like Cisco's "Generations" series, which cybertypes race as useful rather than divisive, sneaks it under the surveillance cameras.

This commercial re-masters race. Remastering implies subjugation, the re-colonization of Otherness in a "postcolonial" world, and its method rests upon the ideological rock of cultural "authenticity." On the contrary, rather than destroying authenticity, cybertyping wants to preserve it. Just as intellectuals in ethnic studies and women's studies are starting to question radically the efficacy of "authenticity" as a flag to rally around, a way to gain solidarity, the commercial discourse of the Internet (that is, the way it figures itself *to* itself) scrambles to pick up that dropped flag.

The Internet must contain images of authentic natives, in the service of militating against particular images of cultural hybridity. The Internet functions as a tourism machine; it reproduces digital images of race as Other. Missing from this picture is any depiction of race in the American context. The vexed question of racism here, racism now, is elided. Racism is recuperated in this ad as cosmetic multiculturalism, or cosmetic cosmoplitanism. In this ad and others like it, American minorities are discursively fixed, or cybertyped, in particular ways to stabilize a sense of a cosmopolitan, digerati-privileged self, which is white and Western.

Post-Racial Cosmopolitanism

In "The Unbearable Whiteness of Being: African American Critical Theory and Cyberculture," Kali Tal writes that "in cyberspace, it is possible to completely and utterly disappear people of color," and that the elision of questions of race in cyberspace has led to its "whitinizing." On the contrary, race is far from elided in these narratives; instead it is repurposed and remastered, made to do new work. The following passage by James Fallows, taken from "The Invisible Poor," an article written for the *New York Times Magazine,* elucidates this:

> The tech establishment has solved, in a fashion, a problem that vexes the rest of America—and therefore thinks about it in a way that seems to prefigure a larger shift. The hallway traffic in any major technology firm is more racially varied than in other institutions in the country. (It is also overwhelmingly male). But the very numerous black and brown faces belong overwhelmingly to immigrants, notably from India, rather than to members of American minority groups. The percentage of African-Americans and Latinos in professional positions in booming tech businesses is extremely low, nearing zero at many firms.[24]

Fallows goes on to write:

> People in the tech world inhabit what they know to be a basically post-racial meritocracy. I would sit at a lunch table in the software firm with an ethnic Chinese from Malaysia on one side of me, a Pole on the other side, a man from Colombia across the table and a man born in India but reared in America next to him. This seems, to those inside it, the way the rest of the world should work, and makes the entrenched racial problems of black-and-white American seem like some Balkan rivalry one is grateful to know is on the other side of the world.[25]

The above quote refers to the technologically (and in this case Internet) driven diaspora of brown, black, and yellow foreign high tech workers into America's technology industry. This contributes to a cosmetic multiculturalism, a false sense of racial equality—or post-racial cyber-meritocracy—which I would term "cosmetic multiculturalism." As Fallows notes, this cosmetic multiculturalism actively works to conceal "the entrenched racial problems of black and white America." The presence of black and brown faces from other countries, notably Asian ones, encourages white workers to inhabit a *virtually* diverse world, one where *local* racial problems are shuffled aside by a *global* and disaporic diversity created by talented immigrants as opposed to "hyphenated Americans." This is a form of tourism, benefiting from difference in order to make the American/Western self feel well-rounded, cosmopolitan, "post-racial." This is not "digital identification," but digital disidentification, disavowal of the recognition of race in local contexts in favor of comfortably distant global ones. In the new landscape of cyberspace, other countries (i.e., markets, and sources of cheap expert immigrant labor in information fields) exist, but not American minorities. It only seems commonsensical, as Reed Koch, a manager at Microsoft, puts it, that "if you go 10 years [in the high tech corporate world] and extremely rarely in your daily life ever encounter an American black person, I think they disappear from your awareness."[26] One of the symptoms of cybertyping is this convenient "disappearance from awareness" of American racial minorities, a symptom that "multiculturalist" Internet advertising and the discourse of technology work hard to produce.

Cybertyping and the American Scene

In Vijay Prashad's important work *The Karma of Brown Folk*, he poses a question to Asian readers: "How does it feel to be the solution"? In this volume, Prashad invokes DuBois's rhetorical question to African Americans: "How does it feel to be a problem?" and repurposes it in order to trace the construction of the Asian, in particular the South Asian, as a model minority. The figure of the Asian as model worker is inextricably tied to this stereotype, which has been reiterated as a particular cybertype of the Asian as an exemplary information worker. If one sees race as a major "problem" of American digital culture, an examination of these cybertypes reveals the ways in which Asians prove to be the "solution." Different minorities have different functions in the cultural landscape of digital technologies. They are good for different kinds of ideological work. And in fact, this taxonomy of work and identity has been remastered: seeing Asians as the solution and blacks as the problem is and has always been a drastic and damaging formulation which pits minorities against each other and is evident in the culture at large.

On the contrary, in a fascinating twist, cybertyping figures both Asians and blacks as the solution, but for different problems. While Asians are constructed as anonymous workers, an undifferentiated pool of skilled (and grateful) labor, African Americans serve as a semiotic marker for the "real," the vanishing point of cyberspace in particular and technology in general.[27]

The New New Thing: Headhunting the S. Asian Cyborg

The issue of the *New York Times Magazine* which contains Michael Lewis's article "The Search Engine" features a cover graphic which repeats the words "The New New Thing" hundreds of times. The subtitle is "How Jim Clark taught America what the techno-economy was all about." Clark, the founder of Netscape, Silicon Graphics, and Healtheon is described as "not so much an Internet entrepreneur as the embodiment of a new kind of economic man." This article reveals that the "new kind of economic man," specifically an American man, attains pre-eminence partly by his ability to repurpose the discourse of racism, to create new cybertypes of Asian technology workers, in ways which at first seem unobjectionable because they have become so common.

Clark spent a great deal of energy recruiting Indian engineers from Silicon Graphics (like engineer Pavan Nigam) to work for his new start-up Healtheon. As Lewis writes: "Jim Clarke [of Netscape] had a thing for Indians. 'The Indian outcasts of Silicon Valley,' he usually called them, 'my Indian hordes' in less sober moments. 'As a concentrated group,' he said, 'they were the most talented engineers in the valley... *And they work their butts off.*"[28]

These "less sober moments" reveal cybertyping in action. This idea of Indians as constituting a horde devoid of individuality, a headless mob, reveals both a fear of their numbers and a desire to become the head of the horde, their leader.[29] These "Indian outcasts" are seen as a natural resource to be exploited, valuable workers, like Chinese railroad workers. What's more, they're a racial group characterized as "naturally" or always-already digital, like Asians as a whole. In 1997, Bill Gates indulged in a moment of foot-in-mouth cybertyping when he declared during a visit to India that "South Indians are the second-smartest people on the planet (for those who are guessing, he rated the Chinese as the smartest; those who continue to guess should note that white people, like Gates, do not get classified, since it is the white gaze, in this incarnation, that is transcendental and able to do the classifying!)."[30] Asian technology workers are thought not to need a "personal life," just as Chinese railroad workers were thought to have nerves farther away from the skin. This characterization of Asians as being superior workers because of inherent, near-physiological differences, seeing them as impervious to pain, in their butts or elsewhere, places them squarely in a new, digital "different caste": the Outcasts of Silicon Valley. This phrase repurposes the old language of "caste," an ancient system which preserves hierarchical distributions of privilege and oppression, for use in the digital age. Keeping to this logic, no amount of work can make them a part of the digital economy as "entrepreneurs" or "new economic men"; they are figured as permanent outcasts and outsiders.[31] Yet, such is the power of cybertyping that Clark's and Gates's comments are not viewed as racist, but rather as strategic, a canny recognition of the rightful work of race in the digital age: this is what makes Clark the "new economic man."

As Lisa Lowe writes, "stereotypes that construct Asians as the threatening 'yellow peril,' or alternatively, that pose Asians as the domesticated 'model minority,' are each equally indicative of these national anxieties."[32] Clarke's figuration of South Indian engineers, his "thing," cybertypes them as simultaneously, rather than alternatively, the threatening horde *and* the model minority: both threatening as a quasi-conspiratorial "concentrated group" and enticing because of their engineering talents. This cybertype of the South Asian seeks to fix the "unfixed liminality of the Asian immigrant—geographically, linguistically, and racially at odds with the context of the 'national'—that has given rise to the necessity of endlessly fixing and repeating such stereotypes."[33]

Indeed, the discourse of Internet technology has a "thing" for Asians. In the article quoted above, Jim Clarke describes himself as a headhunter, in at least two senses of the word. A headhunter, in the language of the cultural digerati, is an entrepreneur who locates professional "talent" and lures it away from one job to another. Much of the tension in this story has to do with Clark's quest to acquire Asian engineers he'd worked with previously for his new venture. A high-tech headhunter facilitates the flow of human capital and labor, often across national borders.[34] The term has roots in colonial discourse: a headhunter is a mythologized figure, like the cannibal, constructed by colonists to embody their notions of the "native" as a savage, a creature so uncivilized and unredeemable that he cannot be broken of his habit of collecting humans as if they were trophies, thus he must be exterminated or civilized. The figure of the headhunter was a justification for colonization. Envisioning South Asians as if they were trophies, outcasts, or hordes, having a "thing for Indians," is a form of cybertyping; it homogenizes South Asians as a group in such a way that they constitute both the familiar model minority paradigm as well as a resource for global capital. And what's more, cybertyping permits this kind of speech, even allows it to signify as "cool," or "new" in a way that Jimmy the Greek's better-intentioned comments about the superiority of black athletes could not be.

Lewis goes on to write: "By 1996 nearly half of the 55,000 temporary visas issued by the United States government to high-tech workers went to Indians. The definitive smell inside a Silicon Valley start up was of curry."[35] This insistence upon the smell of curry in the context of global commerce and capitalism works to discursively fix Asians as irredeemably foreign in order to stablilize a sense of a national self. This smell, here invoked as a stereotyped sign of South Asian identity, is figured as a benefit of sorts to white workers, a kind of virtual tourism: they need never leave their start-up offices (a frowned-upon practice in any event) yet can conveniently enjoy the exotic cuisine and odors of "another" world and culture.

At the dawn of the 21st century, cultural digerati live lives composed of these "less sober moments"; culturally and economically, Americans are living in intoxicated times, a Gold Rush of sorts. The fever of acquisition, creation, and entrepreneurship engendered by dot.com culture licenses specific forms of racialism, if not overt racism, which are no more descriptive of the lived realities of Asian immigrants or Asian Americans than earlier colonialist or racist ways of speaking were. Just as the Gold Rush depended upon the exploited labor of Chinese immigrants, black slaves, and Mexican workers, and consequently created racial stereotypes to justify and explain their exploitation as "Western expansion," so too does our current digital gold rush create mythologies of race which are nostalgic. That is, they hearken back to earlier narratives of race and racialism which were always-already "virtual," in the sense that they too were constructed narratives, the product of representational labor and work. As Susan Stewart defines nostalgia, it is a "sadness without an object." Nostalgia is "always ideological: the past it seeks has never existed except as narrative, and hence, always absent, that past continually threatens to reproduce itself as a felt lack."[36] The construction of post-racial utopias enabled by the Internet, and so prominently troped in television advertising for the Internet, seek to fill that "lack" by supplying us with new narratives of race which affirm its solidity in the face of global culture, multiracialism, and new patterns of migration. Cybertyping keeps race "real" using the discourse of the virtual. The object of digital nostalgia is precisely the idea of race itself. As Renato Rosaldo defines it, nostalgia is "often found under imperialism, where people mourn the passing of what they themselves have transformed," and is "a process of yearning for what one has destroyed that is a form of mystification."[37] Cybertyping works to rescue the vision of the authentic raced "native" which, firstly, never existed except as part of an imperialist set of narratives and, secondly, is already gone, or "destroyed" by technologies such as the Internet.

African American Digital Divides: Bamboozled by the Myth of Access

Two thousand was a banner year, for "Web use became balanced between the sexes for the first time year with 31.1 million men and 30.2 million women online in April, according to Media Metrix. In some months this year...female users have significantly outnumbered their male counterparts."[38] The digital divide between the genders is shrinking, which is not to say that there isn't gender cybertyping occurring online. (This contradicts prior predictions from the early and mid-nineties that a masculinist web would repel women from logging on: on the contrary, as in television, sexism didn't repel women from the medium). The hegemony of the Web is still emphatically male. However, the article from which these statistics came, entitled "Studies Reveal a Rush of Older Women to the Web," also notes that "lost in the rush to use the Web, however, are the nation's poor."

While the article provides graphs and statistics to track Web use by gender, nationality, income, and whether users log on from home or work, it neglects to mention race as a factor at any point. This elision of race in favor of gender and class is symptomatic of what Radhika Gajjala sees as the tendency of "this upwardly mobile digiterati class to celebrate a romanticized 'multiculturalism' and diversity in cyberspace."[39] It is widely assumed that the digital divide is created by inequities in access; indeed, institutional efforts to address this divide seem solely focused on getting everyone online as quickly as possible. African Americans are cybertyped as information "have nots," occupying the "wrong" side of the digital divide; it tropes them as the "problem." This fallacy that access equals fair representation in terms of race and gender can be traced by examining the ways that race has worked in other media.

No sane person would contend that once everyone has cable, television will become a truly democratic and racially diverse medium, for we can see that this has not come to pass. Mainstream film and television depicts African Americans in consistently negative ways despite extremely high usage rates of television by African Americans.[40] Hence, the dubious goal of 100% "penetration" of African American communities by Internet technologies cannot, by and of itself, result in more parity or even accuracy in representations of African Americans. How does the Internet perpetuate this myth of access-as-ultimate-equalizer? Cyberspace's rhetorics make claims which are distinctively different from those of other media: its claims to "erase borders" and magically produce equality simply via access can be seen nowhere else. However, Internet usage by racial minorities is a necessary, but not sufficient, condition of a meaningfully democratic Internet. As Spike Lee's brilliant film parody *Bamboozled* (2000) makes all too clear, even the presence of black writers or content producers in a popular medium such as television fails to guarantee programming which depicts "dignified black people" if audiences are unwilling to support the show in large numbers. In *Bamboozled*, the Harvard-educated black television writer Pierre Delacroix produces the most offensive, racist, "ignorant" variety show he can come up as a form of revenge against his white boss. He fully intends that the show, which depicts blacks as Topsys, Aunt Jemimas, Sambos, and Little Nigger Jims will be a resounding flop: he entitles it the "Man Tan New Millenium Minstrel

Show" and requires the African American performers to appear in authentic blackface made of burnt cork. Of course, it is a major hit with the networks and the audience. This can be seen as an object-lesson to people interested in the Internet's potential as a space for activism and anti-racist education: what needs to happen on the Internet to ensure that it doesn't become the newest of the new millennium minstrel shows? The film contains a clip from Lee's earlier film, *Malcolm X*, in which the protagonist addresses a crowd of African Americans, crying out, "you been hoodwinked, bamboozled." Until we acquire some insight into racial cybertypes on the Internet, we are quite likely to be hoodwinked and bamboozled by the images of race we see on the net, images which bear no more relation to real people of color than minstrel shows do to dignified black people.

Due to the efforts of black activists and scholars working in older media studies, we can better see what's at stake in this limited range of representations of racial minorities. Studies of race and the Internet are just now beginning to catch up (which is not surprising, considering the familiar lag time in media criticism when it comes to critical readings of race).

We should wish Internet access for the betterment of material and educational conditions of African Americans, but ought not expect that the medium itself is going to represent them fairly without any strategies or plans put into place to encourage this direction.

Post-Racial Digerati?: Cybertyping the Other

Some studies claim that the Internet causes depression. The 1998 Carnegie Mellon study posits that this is so because the Internet reduces the number of "strong social ties" that users maintain IRL (In Real Life) and replaces them with "weak" or virtual ties, which don't have the same beneficial psychological effects as face to face social interactions.[41] The Internet's ability to produce depression in its users (at least in me), can be traced at least in part to cybertyping, a kind of virtual social interaction which constructs people of color as "good" workers or bad, on the "right" side of the digital divide or the wrong side. The Internet's claims to erase borders, such as gender, class, and racial divisions, and the ways in which public policy makers' attentions to bridging the "digital divide" that is erroneously seen as being the source of these problems in representation, overshadow these more subtle varieties of cybertyping. This dynamic is indeed depressing, all the more so because it remains largely silent and undiscussed.

Radhika Gajjala writes that

> Race, gender, age, sexuality, geographical location and other signifiers of "Otherness" interact with this class-based construction of "whiteness" to produce complex hierarchies and contradictions within the Digital Economy. While we can continue to call this "whiteness" because the status quo is still based upon a cultural hegemony that privileges a "white" race, it might be more appropriate to refer to this upwardly mobile subject as a "privileged hybrid transnational subject" who is a member of the "digiterati" class.[42]

Here, Gajjala posits that "privileged hybrid transnational subjects" such as Clark's coveted South Asian programmers can be read, for all intents and purposes, as "white" since they participate in the "cultural hegemony that privileges a white race."

While they are no doubt part of that hegemony, as is every person of color who consumes, produces, and becomes the object of representation of information technologies, I contend that they are put to work in that hegemony in distinctively raced ways. The "work" that they do in this hegemony, their value-added labor in the system of information practices dubbed "global capitalism," is this: their cybertypes work to preserve taxonomies of racial difference. The nostalgia for race, or visions of racial "authenticity" invoked by the Cisco advertisements, assuages a longing. The espoused public desire for technological uplift, in the discourse of science fiction narratives, the desire to create a new class of "digiterati" which is in some sense post-racial, is matched by a corresponding longing for "race" as a spectacle of difference, a marker to function as the horizon to the vanishing point of postmodern identities.

Contemporary debates about the digital divide tend to be divided roughly into two camps. The first of these maintains that the master's tools can never dismantle the master's house, to paraphrase Audre Lorde's formulation.[43] In other words, if people of color rush to assimilate themselves into computer culture, to bridge the digital divide, they are simply adopting the role of the docile consumer of Microsoft, Intel, and other products, and are not likely to transform the cyberspace they encounter. Like feminists who adopt the values of the patriarchy, they may succeed as isolated individuals in what has thus far been a privileged white male's domain—technology and the Internet—but they cannot bring about the kind of change that would bring about true equality. As Lorde writes, taking up the master's tools "may allow us to temporarily beat him at his own game, but they will never allow us to bring about genuine change. And this fact is only threatening to those women who still define the master's house as their only source of support."[44]

The second camp maintains that people of color can only bring about "genuine change" in the often-imperialistic images of race which exist online by *getting* online. Envisioning cyber-technologies as less the master's tools than tools for discourse which can take any shape is an optimistic ways of seeing things.

While it is impossible to say, definitively, which path is correct, there is no question that the digital divide is both a result of and a contributor to the practice of racial cybertyping. It is crucial that we continue to scrutinize the deployment of race online as well as the ways that Internet use figures as a racialized practice if we are to realize the medium's potential as a vector for social change. There is no ignoring that the Internet can and does enable new and insidious forms of racism. Whether the master's tools present the best way to address this state of affairs has yet to be seen.

Notes

1. Lev Manovich, *The Language of New Media* (Cambridge, MIT Press: 2001): 65.
2. Manovich, *New Media*, 63.
3. Manovich, *New Media*, 64.
4. Espen Aarseth, "The Field of Humanistic Informatics and its Relation to the Humanities." Online: http://www.hf.uib.no/hi/espen/HI.html.
5. Rey Chow, *Writing Diaspora: Tactics of Intervention in Contemporary Cultural Studies* (Bloomington: Indiana University Press, 1993): 30
6. Chow, *Writing Diaspora*, 46.
7. Chow, *Writing Diaspora*, 42.
8. Gayatri Spivak, *In Other Worlds: Essays in Cultural Politics* (New York and London: Routledge, 1988): 89.
9. Donna Haraway, *Simians, Cyborgs, and Women: The Reinvention of Nature* (New York and London: Routledge, 1991): 157.
10. Haraway, *Simians*, 155.
11. Barbara Christian, "The Race for Theory." In *Feminist Literary Theory: A Reader*, edited by Mary Eagleton (London: Blackwell, 1986): 278.
12. In an article entitled "Survivor: As Internet Industry Plays Survival of the Fittest, Netnoir.com celebrates 5th Anniversary," which appeared in Netnoir.com's online newsletter in 2000, the company announced that San Francisco Mayor Willie Brown had proclaimed June 22 as "Netnoir.com day in the city and county of san Francisco." In 1994, Netnoir.com's E. David Ellington received an award from the AOL Greenhouse Project to fund information technology entrepreneurs, and "soon after, AOL backed NetNoir with a 19.9% equity stake." Currently, Netnoir has partnered with AOL, Syncom Ventures, and Radio One. Netnoir's slogan, "taking you there. Wherever there is" stands as an interesting contrast to Microsoft's "where do you want to go today?" in the sense that it is far more open ended about the Web's topography and structure (Netnoir.com Newsletter. Online. Mailing List. July 7, 2000).
13. Since the incredible dominance of the Internet by the World Wide Web in the mid-nineties, it has consistently supported this construction of women *as* bodies. The saying that the Internet is 90% pornography and advertising, while it may be a slight exaggeration, gestures towards the Internet's role as an extremely efficient purveyor of exploitative images of women. Similarly, the Internet's current bent towards merchandising and selling online constructs women as either "markets" or more commonly as scantily-clad figures in commercials for products.
14. Frances Katz, "Racial-Bias Suit Filed Against Online Delivery Service Kozmo.com." *KRTBN Knight-Ridder Tribune Business News: The Atlanta Journal and Constitution-Georgia*. April 14, 2000: C9.

15. Martha Hamilton, "Web Retailer Kozmo Accused of Redlining: Exclusion of D.C. Minority Areas Cited." *The Washington Post.* April 14, 2000. Online. http://www.washingtonpost.com/wp-dyn/articles/A9719-2000Apr13.html
16. Snigdka Prakash, "All Things Considered." NPR broadcast. May 2, 2000. Kozmo.com has since gone out of business, for reasons unrelated to this lawsuit.
17. See *Posthuman Bodies*, edited by Judith Halberstam and Ira Livingston (Indiana University Press, 1995), as well as Scott Bukatman's *Terminal Identity: The Virtual Subject in Postmodern Fiction* (Duke University Press, 1998).
18. Valerie Hartouni, "Containing Women: Reproductive Discourse in the 1980s." In *Technoculture*, edited by Constance Penley and Andrew Ross (Minneapolis: University of Minnesota Press, 1991): 51.
19. Sherry Turkle, *Life on the Screen: Identity in the Age of the Internet* (New York: Simon and Schuster, 1995): 51.
20. bell hooks, *Black Looks: Race and Representation* (Boston: South End Press, 1992): 29.
21. hooks, *Black Looks*, 39.
22. Ziauddin Sardar, "Alt.Civilizations.FAQ: Cyberspace as the Darker Side of the West." In *The Cybercultures Reader*, edited by David Bell and Barbara Kennedy (New York: Routledge, 2000).
23. Just as computer users become accustomed to the "look and feel" of particular interfaces (the loyalty of Macintosh users to the desktop metaphor is legendary), so too do consumers of popular discourse become strongly attached to particular images of race. As software designers and webmasters have learned, users are quick to protest when familiar websites, such as Amazon's, are redesigned, and have often responded to consumer protests by changing them back to their original appearance. This is also the case for the ways that the "native" is portrayed in popular culture.
24. James Fallows, "The Invisible Poor." *New York Times Magazine* (March 19, 2000): 95.
25. Ibid.
26. Ibid.
27. See Alondra Nelson et. al.'s essay collection *Technicolor: Race, Technology and Everyday Life* (New York: New York Univeristy Press, 2001), for a critique of this formulation; their work posits a re-framing and redefinition of the "technical" to include sampling and sound technologies and communications technologies such as the beeper, cellphone, and pager in ways that would "count" African Americans as innovators and users of note.
28. Michael Lewis, "The Search Engine." *New York Times Magazine.* (This issue entitled "The New New Thing") October 10, 1999. Section 6. pgs. 77–83, 100, 108, 112–113.
29. As in Neal Stephenson's cyberpunk novel *The Diamond Age*, which represents Chinese girls as members of a faceless "horde" of model minorities.
30. Vijay Prashad, *Karma of Brown Folk* (Minneapolis: University of Minnesota Press, 2000): 70.
31. Growing attention has been paid to the existence of a "glass ceiling" for Asian engineers in the high technology industry, particularly in Asian-American publications and newspapers. However, despite this glass ceiling, Mutthuswami asserts that "highly educated Indians…serve as CEOs of 25 percent of the companies in Silicon Valley" (Amitava Kumar, "Temporary Access: the Indian H1-B Visa Worker in the United States." In *Technicolor:* 81).
32. Lisa Lowe, *Immigrant Acts: On Asian American Cultural Politics* (Durham: Duke University Press, 1996): 82.
33. Ibid., 19.
34. One can see the headhunter's analogue in the more down-market image of the "coyote." The coyote, as Mexican "smugglers of workers and goods are locally known [supply workers] for the farms of South Texas, the hotels of Las Vegas and the sweatshops of Los Angeles" (Mike Davis, *Magical Urbanism: Latinos Reinvent the U.S. Big City.* London: Verso, 2000): 27). They guide people across the U.S.-Mexican border, and there are often casualities along the way.
35. Lewis, "The Search Engine," 82.
36. Susan Stewart, *On Longing: Narratives of the Miniature, the Gigantic, the Souvenir, the Collection* (Durham and London: Duke University Press, 1993): 23.
37. Quoted in hooks, *Black Looks*, 25.
38. Ian Austen, "Studies Reveal a Rush of Older Women to the Web." *New York Times*, Circuits (Thursday, June 29, 2000): D7.
39. Radhika Gajjala, "Transnational Digital Subjects: Constructs of Identity and Ignorance in a Digital Economy." Conference Talk.
40. Despite the existence of black-oriented programming on smaller cable networks such as the WB and UPN, the majority of African Americans, much less Asians and Latinos, groups even less depicted on television as primary characters, understandably feel that their lived realities are entirely unrepresented on television. Of course the same is true for whites: few possess the limitless leisure and privilege of characters enjoyed by characters on the show *Friends*. But at least they might aspire to these roles. What African American woman wants to be the "hoochy mama" depicted on Rikki Lake's "reality" programming or the noble black mammy "Oracle" in the film *The Matrix*?
41. Robert Kraut, et al. "Internet Paradox: A Social Technology That Reduces Social Involvement and Psychological Well-Being?" *American Psychologist*, Vol. 53, No. 9 (September 1998): 1029.
42. Gajjala, "Transnational Digital Subjects," 6.
43. Audre Lorde, "The Master's Tools Will Never Dismantle the Master's House." In *This Bridge Called My Back: Writing by Radical Women of Color*, edited by Cherrie Moraga and Gloria Anzaldua (New York: Kitchen Table Press, 1981).
44. Lorde, "Master's Tools," 99.

23

Network Subjects
or, The Ghost is the Message

Nicholas Mirzoeff

This is a piece from another time, a message that is itself now a ghost, even as it speaks of and to the ghost. Whatever globalization is now, it is remote from this moment of digital utopianism. The piece remains a fragment that can be used as by archaeologists of the contemporary.

It has widely been argued that there is not much new about the new media. It was in 1971, for example, that Roy Ascott established "electric media" as a degree program at the Ontario College of Art, Toronto. By 1984, Ascott was promoting the interactive "electronic space" as being of "evolutionary significance." Writing in the same year in William Gibson first introduced the term cyberspace, Ascott asserted: "The true consequence of the combination of art and electronic information technology will not properly be seen until there is universal availability at very low cost of the means of transmission of digital information within a planetary interactive network embracing the audio, visual, and data/text modes. Even at this stage of development we can sense the emergence of a planetary consciousness which I call 'network consciousness.'"[1] While Ascott's rather odd mix of Darwin and Hegel is representative of the intellectual moment in which he was writing, his technological forecast seems accurate, if not quite fulfilled. The network, like the ghost, is from the past but is still yet to come. It might be said that digital criticism is developing an awareness of past, present and future networked subjects. The question is precisely how those modes of subjectivity are related and who may inhabit them, both now and in the past.

Take a medium, perhaps any medium, but for the sake of argument, photography. Photography is more than ordinarily important to answering these questions. Photography's invention has long been taken as a critical step in the formation of modernity and modernism. Despite repeated attempts at debunking, the myths of its origins and recent death have been very durable. Let us once more recall the French painter Paul Delaroche declaring on first seeing a daguerreotype in 1839: "From today painting is dead!" Similarly, from the 1982 announcement by Lucasfilm that "photography is no longer evidence for anything" to William J. Mitchell's academic obituary for photography in 1992,[2] the digital assassination of photography has been so widely announced that it is now the subject of bottom-feeding advertising campaigns by the likes of Circuit City. But as my grandmother would say, who died? These myths of photography both reinforce the idea that photography's essence is its indexical relationship to reality: that is to say, whatever is shown in the photograph must really look like that. Photography was never a purely indexical medium, depicting only what was really there. Rather, as Ruth Iskin has argued, "the 'real' became structured like its photograph."[3] In other words, photography was not a passive mirroring of an unquestionable exterior reality but rather the formation of a "real" that conformed to a positivist notion of indexicality. The desiring subject that formed photography operated across the medium itself, creating an inside/outside distinction that, as Diana Fuss has argued: "cannot be easily or ever finally dispensed with."[4] This subjectivity was not a philosophical monad but was always already connected. Today, the panoptic gaze within which this photography was constituted has become indifferent to what it surveys. The digital subject's desire to photograph has reconfigured itself accordingly, denoting a self-reflexive space of inside/inside in which the medium is the object and subject of its own desire.

These thoughts depart from Geoffrey Batchen's remarkable work that has undone the neat packaging of photography as a once and for all historical event. By showing that there was no single historical instant in which photography was invented, or one single inventor of the process, he has highlighted what he intriguingly calls the desire to photograph.[5] What is under examination here, then, is not the much-debated essence of photography but the contingent relationship in which a camera becomes a medium. The camera is itself understood as a space, a room as the word itself means, or even a closet. But this space is not simply empty. It is the space of the medium and consequently of the ghost. Kaja Silverman's definition of the camera as being "less a machine, or the representation of a machine, than a complex field of relations"[6] opens up a new field of investigation before and after photography. These relations were those between the apparatus, the user, the viewer and the medium. The medium is not simply the light-sensitive surface but the ensemble of conditions that make it possible for a light-sensitive film to be created, used and understood.

Here I can supply only some snapshots and screen grabs towards the wider project of understanding networked visual subjectivity. In what follows, I shall suggest that the Enlightenment embodied a new desire for clarity of vision, which enabled both photography and what Foucault called surveillance. These devices, among others, provided a means to restrain the proliferation of the network that has now broken down. In what we call photography, there was a new and different configuration of desire motivating the representational circuit between the apparatus, the medium and the viewer that marked a transition from a theological subject to an existential subject. This dyadic subject was bounded by what Freud called equilibrium that is now out of control, beyond the economy principle.

Snapshot 1: Flickering Enlightenment

In recent years a broad periodization of modern visual culture has emerged. Beginning with Descartes' re-evaluation of classical philosophy in the mid-seventeenth century is what Jonathan Crary has called the camera obscura model. It is derived from the camera obscura itself, a darkened room or box into which light is admitted through a small hole fitted with a lens, forming a reversed picture of the exterior on the far wall of the room. In Descartes' view, the judgment (what we might call the subject) examines external sense perceptions in this interior chamber of the mind. Descartes' famous thinker is certain of his existence because of his self-awareness of thought, a self-awareness that is guaranteed by the presence of God. In representations of such Classical space, like Velazquez's *Las Meninas*, we see figures in the interior, representing the judgment in the space of the mind. The accuracy of representation, and hence of judgment, is given by the presence of the quasi-divine state in the person of the Absolute monarch Philip IV, who is both the presumed viewer of the painting and its subject. The point of view is singular, reinforced by the use of one-point perspective. While every sense perception is subject to doubt, in the last instance knowledge of the self is Absolute, confirmed by King and God.

During the eighteenth century, the camera obscura became less of a philosophical obsession and more a fairground amusement. While the Cartesian world-view was broadly accepted, Enlightenment thinkers were concerned with trying to eliminate the uncertainties of vision. For French critics like Diderot, this mean above all rejecting the aesthetics of *papillotage*, that is to say, blinking or flickering. Marion Hobson defines papillotage as expressing "both the gaze, the acceptance of the object seen, and the blink which cuts off the eye from contact with the world and, in so doing, brings the self back to self."[7] This singular sense of the self and vision was very much in line with the Cartesian or camera obscura model. But as the century advanced, critics were increasingly scathing about *papillotage*. The Comte de Caylus huffed that: "It is difficult to conceive how one can abandon oneself to colors that do not go together in the least; or which distract the eye from the principal figure or the dominant object and prevent it from being led without revolt or obstacle... These highly inappropriate false awakenings are like so many instruments at a concert interrupting the beautiful effect, which cause the listener to despair."[8] In his first Salon review of 1759, Diderot opened his remarks with an attack on *papillotage*, looking at a portrait of Madame de Pompadour wearing a floral print dress: "I do not like floral prints in painting at all. They have neither simplicity nor nobility. Inevitably, the flowers flicker [*papillotent*] against the background, which, above all if it is white, forms a multitude of little scattered lights. However skilful an artist is, he will never make a beautiful painting of the *parterre*, or a beautiful garment from a flowered dress."[9] Diderot combines here a formal dislike for the flickering effect of Rococo painting with a social disdain for the *parterre*, the standing audience at the theater. Neither has what he calls "nobility," which should be taken to mean a nobility of spirit rather than hereditary aristocracy. This is no egalitarian criticism, however, as Caylus's concern to avoid revolt makes clear. Diderot wants to be sure of what he sees. In 1765, he claimed that "imperceptible cords" should tie the painting to its object.[10] This new desire for indexicality replaced an aesthetics of uncertainty that had been encapsulated in the Abbé Dubos's hunt for what he called the "je ne sais quoi" in art. Diderot's concern that the painted object be an adequate representation of its subject led him to revert to a much older theory of vision. In this view an *eidolon* or likeness of the object was emitted and physically received into the eye via the iris. This ghost object was therefore less subject to doubt because it was in a sense part of the object. Philosophy, science and aesthetics in fact refuse to line up here to create a single "scopic regime." Rather, a social and critical contestation over the mediation of vision was opened up that placed indexical media in opposition to their flickering counterparts. This new direction in criticism had its counterpart in the social practice of street lighting. In 1760, Paris was newly lit by advanced oil lanterns called *réverbères*. By using multiple wicks whose light was amplified by the use of two reflectors, the new lights seemed to contemporaries to have conquered the demons of the night. Louis-Sebastien Mercier, in his well-known *Tableaux de Paris*, disparaged the old lamps as offering "only a weak, flickering and uncertain light, which was, moreover, impaired by dangerous shadows." He was much taken with the new ones" "the city is extremely brightly lit. The combined force of 1,200 *réverbères* creates an even, lively and lasting light."[11]

Snapshot 2: Vision in Prison

In the photographic era (1795–1982), the subject desires certainty regarding not just sense perceptions but its very existence as a body in space.[12] This anxiety is the product of being subjected to the gaze of the Panopticon, the now legendary device for social organization created by Jeremy Bentham in 1791 and adopted by Michel Foucault as a symbol for the modern surveillance of the body. In panopticism, the actual surveillance of authority in disciplinary institutions like schools, factories, prisons and hospitals gradually becomes internalized. By the twentieth century, the visual subject, in Lacan's famous phrase, sees itself seeing itself. In the digital era, desire is uncertain whether it can see itself at all. Although Foucault stressed the role of surveillance in the panopticon, he took visibility for granted as a self-evident term. More precisely, a new form of visuality came into being in the period between the first musings on panopticism in the 1780s and the first actual construction of panoptic social structures in the 1830s. This period was a chaotic transition in many arenas ranging from the birth of the prison to the new medicalized regimes of modernity and the political revolutions of the period. One of its products would be the term visuality itself.

It might seem, then, that a direct line of sight opens up between the indexical philosophizing of the Enlightenment and the panoptic institutions of the 1840s. While there is clearly a connection, the intervening revolutionary era produced a proliferation of visual subjectivities, including the first experiments with photography. Diderot's call for an orderly and indexical mode of painting quickly became official orthodoxy, represented most clearly in the work of Jacques-Louis David. Paintings like his *Oath of the Horatii* (1785) use a limpid, crisp visual field that does everything possible to convince the viewer of its materiality. However, the *Horatii* and David's subsequent masterpiece *Brutus* (1789) also make use of a tightly controlled interior space that could be understood as a camera obscura. This format persisted in the revolutionary period in works like *The Death of Marat* (1793). But after Thermidor, when the Jacobin regime with which he had been closely associated was overthrown, David found himself in prison, on the inside one might say. Among his prison paintings at this time was this *View of the Luxembourg Gardens* (1795), a view from inside the disciplinary institution out. Its very ordinariness speaks to an intense desire for a return to everyday existence, away from the high drama of revolution.

At the same moment, another former revolutionary was meditating on the collapse of his dreams in a brief interlude between bouts of imprisonment. The Marquis de Sade, although an aristocrat, was an enthusiastic supporter of the Revolution. As secretary of the radical Section des Piques, Sade presented a petition to the Convention at the height of the Year II, demanding the suppression of religion: "The reign of philosophy is finally about to destroy that of imposture; finally man is enlightening himself, and, destroying with one hand the frivolous playthings of an absurd religion, he elevates with the other an altar to the most cherished divinity of his heart. Reason replaces Mary in our temples, and the incense which was burnt on bended knees to an adulterous woman will now be lit only at the feet of the goddess who broke our chains."[13] Philosophy was now in the open light, judging only by the precepts of Reason. These sentiments were not uncommon in the period. Sade's distinctive contribution was to continue this line of thought in defeat, when philosophy had retreated from the outside world to the private space of the boudoir. Sade's 1795 *Philosophy in the Boudoir* was explicitly framed as the response of still convinced republicans to the change in political climate. It is a retreat to an interior world, the private space of the boudoir or closet. This space was what one might call a panopticon of perversity, if perversity were taken to mean a refusal to conform. The space is described as being covered with mirrors: "This is so that, repeating these attitudes in a thousand different senses, they multiply these same pleasures to infinity to the eyes of those who enjoy them on this ottoman. In this way no part of any body can be hidden. Everything must be in full view; there are so many groups assembled round those enchained by love, so many imitators of their pleasures, so many delicious pictures, whose lubricity is intoxicating."[14] Sade at once parodied the famous Hall of Mirrors at Versailles, the Rococo passion for mirrors in painting and in life and the new disciplinary gaze where "visibility is a trap." In Sade's boudoir, visibility enchains its subjects to the maximum of sexual and sensual practice.[15]

Yet this is still a revolutionary project. The book includes a lengthy pamphlet, entitled "People of France. One more effort if you wish to be republicans" that advances Sade's hostility to religion and his pursuit of post-Cartesian reason. Of his three guiding principles—"sodomy, sacrilegious fantasies and cruel tastes"—the middle term, or medium, has received less attention.[16] Dispensing with religion is key to Sade's philosophy as the underpinning of his other tastes. In a secular world, no regeneration was possible without destruction: "destruction is therefore one of the laws of nature, like creation."[17] Regeneration was a key term of French revolutionary discourse that Sade chained—to use his term of choice—to cruelty. The Sadeian world was one in which "movement is inherent to matter."[18] Using evidence from Enlightenment philosophy and the discoveries of European expansion,[19] Sade asserted that this uncontrollable natural energy was the guiding force of life. This force was both creative and destructive without regard to moral principles. Sexual energies were forces in themselves without regard to questions of reproduction. From these arguments, Sade concluded that: "All our ideas are representations of objects that strike us; what could possibly represent to us the idea of God, which is evidently an idea without an object? . . . All principles are judgments, all judgments are the effect of experience, and experience is only acquired by the evidence of the senses."[20] This argument would not have been unfamiliar to Enlightenment thinkers like Condillac but Sade linked it with the historical experience of 1789 to suggest, without expectation of success, that all conventions should be rejected. By rejecting the possibility that the judgment had divine authority, Sade made the camera obscura model impossible. As if to emphasize the modernity of his philosophy, at the end of each sexual adventure, the practice of his theory, Sade's protagonists declare: "I discharge!" to announce orgasm, using the terminology of Volta's new battery. The metaphor of the electric circuit was newly available as a result of the discovery of current electricity in the 1790s that influenced the photographic usage of the terms positive and negative.[21] Here philosophy was explicitly in the closet or camera, theorizing nature as a flow that established boundless connections, a euphemism in the period for sex. Jeremy Bentham extolled the benefits of his inspection house in disciplinary terms: "morals reformed, health preserved, industry invigorated, instruction diffused, public burthens lightened, economy seated as it were upon a rock, the gordion knot of the poor laws not cut but untied—all by a simple idea in architecture."[22] Panopticism did not have to be directed toward these ends, as we are in the process of discovering and as Sade had anticipated.

The political reaction against what came to be seen as the sanguinary excesses of the revolution was aware of the challenge. Thermidor generated a new theory of the relationship between mind and body.[23] In this view, the mind was far from being secure in its camera obscura, with or without little cords. Instead it was closely interactive with the body, as Sade suggested. But rather than give into the unrestrained flow of desire, the new regime required an orderly regimen to maintain stability that could be upset by excesses whether political or of the imagination. The alienist Esquirol described a wave of insanity that he saw as being directly caused by the French Revolution. In the new asylums for the insane, devices were created to convince a sufferer that he or she was in fact sane. One such *appareil* or machine sought to persuade a paranoid tailor that he was not liable to guillotined for lack of revolutionary fervor by staging a mock trial for him. He was triumphantly acquitted and his patriotism was hailed by the judges, leading to an immediate recovery that lasted until the unfortunate tailor learned that the event was a fake. *Appareil* is now the French word for camera, a device that seeks to reassure the objects of the disciplinary institution that their existence is not under threat. It was not a radical device but a construct of post-revolutionary reason, literally reactionary discourse. This overlap is more than a coincidence of vocabulary. It highlights the photographic construction of reality as a discursive network of the psychic and the social whose seriousness can be undermined by something as small as a joke.

Rather than being a transhistorical expression of the subject's lack, desire here represents the need to connect from inside to outside, a desire that was both a cause of, and then constrained by, the disciplinary institution. When Nicephore Niépce succeeded in fixing an image in 1823 it was logically a view from inside his chambers to the outside. When Louis Daguerre heard of this, he wrote to Niépce that he was "burning with desire" to see the results, a phrase powerfully appropriated by Geoffrey Batchen. It is not surprising, though, that Daguerre's wife worried instead that he was going mad.[24]

Snapshot 3: Arresting the Shadows

Photography's representation of the desire to configure a stable relation between inside and outside was not universal but rather the expression of a European male existential condition at the beginning of modernity. Both race and gender disrupt the simple inside/out formula with their own inside/out. Schematically, I would argue that the abolition of British colonial slavery in 1838 was an essential precondition for Talbot's invention of a photographic process patented under English law in 1839. In his first description of his new device Talbot wrote in January 1839 that it could "receive on paper the fleeting shadow, arrest it there and in the space of a single minute fix it there so firmly as to be no more capable of change."[25] Photography calls a halt to the dynamic flow of nature evoked by Sade and other revolutionaries. In fact, it put it under arrest, as if evoking the process of recapturing fugitive, or fleeting, slaves. In the moment of abolition a group of people went from the shadow of what Orlando Patterson has called the "social death" of slavery to subjectivity as members of the British Empire. It is not surprising in this context to read of the photographic pioneer Hercule Florence experimenting on the veranda of his Brazilian plantation and later taking pictures of the town jail.

As visuality came to represent the exterior of the disciplinary institution, it came to be a means of social control in itself. That process was not immediate. It is interesting to observe how post-emancipation societies negotiated this transition. On the island of Jamaica, British slavery came to an end in 1838, leaving a minority of Europeans, "coloureds" and Jews much outnumbered by the newly enfranchised Africans. Adolphe Duperly (1801–64), a French painter and lithographer, picked up the daguerreotype camera as a medium for the creation of new subjects in the emergent black public sphere. Although his biography is uncertain, Duperly seems to have arrived in the Caribbean around 1822, while slavery was still in force. He next appears teaching lithography in Haiti, the first free black nation. This instruction may have taken place in the Lycée in Port-au-Prince, which offered classes in drawing at this time.[26] He also visited Cuba and then arrived in Jamaica in 1824. Between 1840 and 1850 he took a series of daguerreotypes of Jamaica that were published as a volume of lithographs made by Jacoltelt for Thierry Brothers under the title *Daguerrian Excursion In Jamaica* in London and Paris.[27] His scenes showed the new African subjects of Jamaica creating social, commercial and political space. The daguerreotype, which by its nature could not be reproduced, was intended to serve as a guarantee that these scenes were not simply fantasies. The second image in the book showed newly enfranchised Africans lining up to vote at the Court House, in front of the offices of the formerly pro-slavery *Jamaica Gazette*.[28] As the book progressed, more images of everyday life after slavery asserted a new social order as Africans were seen chatting in the main streets of town, at church and in the Falmouth Market, while a sugar plantation was shown with the former slave quarters in ruin. Duperly, a circum-Atlantic artist and intellectual, was using the new apparatus of photography to legitimate his vision of an emancipated, free Jamaica.

Photography reasserted discipline by racializing the body within space, as Allan Sekula, John Tagg and others have shown.[29] Nineteenth-century racism ought properly to be known as photographic not scientific. This racializing of the camera transformed the device from a modest reaffirmation of the self in the panoptic gaze into a powerful expression of racial hierarchy. The white subject finds reassurance from within his panopticon that he is not black and has photographs to prove it. In Jamaica, a process unfolded from the abolition of slavery to the reassertion of direct rule from London following the abolition of the Jamaica Assembly in 1866 that one historian has called "the Jamaican panopticon."[30] The transformation can clearly be seen in a volume published under the name of Adolphe Duperly and Son in 1905, entitled *Picturesque Jamaica*. The African population appeared in very different roles here. While there are many views of Jamaica's natural beauty, when Africans are seen, it is usually in the context of the banana plantation. The photographer depicted himself in *Me and My Family*, a middle-aged man of perhaps fifty or sixty, with one foot resting on a donkey's stirrup. He stands slightly to the side of his wife and six children in what seems to be the temporary quarters constructed for workers on the plantations. All appear to be of African descent. Their Jamaica is clearly under the disciplinary gaze that led the scientist Joseph Hooker to declare in 1868: "We do not hold an Englishman and a Jamaican Negro to be convertible terms, nor do we think that the cause of human liberty will be promoted by any attempt to make them so."[31] Duperly's pictures include a scene entitled *Off to the Jail*, showing two women on donkeys. Without the caption its disciplinary content would not have been clear. By contrast, *Loading Bananas, Port Antonio* shows a line of workers filing into a cargo ship, under the watchful eye of three supervisors, one armed with a machete. It might have been a scene from the days of slavery but for the fact that at least one overseer was also African. The supervisors are themselves being watched by various whites from the deck of the ship. Only one worker turned their head to look at the camera, an indication perhaps that they were used to working under visual supervision. Overhead a large electric light hangs, making it possible to work at night or to guard the ship. The men are visibly disciplined, but the women offer a point of disorder within the otherwise intensely regulated plantation economy. As they are not under restraint, why are they going to jail? As visitors? Or with more subversive aims in mind? Compared to the public efforts to create an emancipated polity in the 1840s, this aporia is small comfort.

Screen Grab 1: Acting Out Indexicality

Let's take a jump[32] via hyperlink to 1982 when the panoptical regime has been itself made visible in its reconfiguration as the society of control. In that year, as mentioned above, the special effects studio Lucasfilm asserted that photography was meaningless because they could now digitally alter the photograph in unrecognizable ways or create a photograph that showed a scene which had never existed. In the science-fiction film *Bladerunner* released that same year, 1982, this scenario was played out as a key part of the action. *Bladerunner* depicts a dangerous future version of Los Angeles in which the lead character Deckard (Harrison Ford) hunts down and destroys renegade replicants, the artificial humans that colonize "off-world" planets for the benefit of their human employers. In a key scene, Deckard is confronted in his apartment by Rachel, who has just learned from Deckard's tests that she herself is a replicant. Replicants, as their leader Roy often remarks, are slaves, living under the erasure of physical as well as social death because they are programmed to die after a few years of active life. Rachel offers her childhood photographs to prove that she is really human. But Deckard is able to show that her most "private" memories are known to him as an expert on replicant manufacture. Her photographs replicate nothing and she herself is literally a simulacrum, a copy with no original. She is a digitized photographic effect, devoid of photography's until then taken for granted connection to the "real." Deckard's apartment serves as a camera in which the dark room—literally a camera obscura—is flooded with light from outside and as a result, something develops: only what develops isn't the truth but the emergence of a manipulation.[33] The numerous accounts of this scene often overlook the ending. Rachel does not disintegrate in the face of this challenge to indexicality. She literally acts out: that is to say, she exits. By the end of the film, especially in the director's cut where it is suggested that Deckard himself is a replicant, Rachel is the last "person" standing. Her machine-made desire wins.

A decade later the artist John Dugdale used photography as a machine for his own desire to see. In his 1992 picture *Life's Evening Hour*, the photographer is seen next to the gravestone of photographic pioneer Henry Fox Talbot and his wife. The story seems clear: at photography's end, we go back to its origins. But all is not as it seems. Dugdale has used the cyanotype process, a light-sensitive solution of cyanide that was popular in the nineteenth century for cheap reproductions but was not used by Talbot.[34] In a sense, the photograph itself is deadly, being made from a poison much favored by suicides. The deliberate anachronism speaks to a by-now familiar postmodern sense of irony. But there is another dimension to the photograph. Dugdale is living with AIDS and had lost over 80 percent of his sight by the time that this photograph was taken. The image represents a desire to see, haunted by the ghost of Talbot's creation of an artificial prosthesis to sight that is now unavailable to the image-maker. Dugdale takes a blind walk with the ghost and records it in what must be called a spirit photograph. The photographer's condition is of course not visible. In a certain sense, his desire in this photograph remains closeted until revealed by means external to the image. No criticism of Dugdale, this is to assert that the photographic desire to represent is always in a certain sense closeted. Pure indexicality was the fantasy of panopticism whether in representations of stones or spirits.[35]

Screen Grab 2: In-Different Gazes

With the rise of digital culture and the Internet, desire no longer tries to escape its interiority but celebrates and consumes it. In his *Travel Diaries* (1999–2000), Fred Cray creates a mediascape out of found images from film, photography, the Internet and animation that are rendered in filmstrip format arranged to form pages in a book. The journey that is recorded in these diaries is not physical but virtual, a journey from one medium to the next and in between that refuses to be a photography of exteriority. The omission in these diaries is the self, usually the key point of a diary entry. They are his counterpoint to a series of self-portraits that a number of critics have compared to spirit photographs. Rather than try and show his intimate interior self as so many photographers have done, Cray wears extravagant make-up and exotic hairstyles. Using a long exposure, Cray deliberately moves while the shutter is open to create a blurred, moving image that he further enhances with the use of color. Cray's diaries and self-portraits reveal that there is no one home but the ghost; which is not to say that he is insane but that the endless repetition of visual selves leads to an indifference, a loss of difference, that can end in the loss of the self. The ubiquitous surveillance cameras of today's society in no way seek to prevent crime or other breaches of social norms, nor do they claim any moralizing effect on the individual. The video record of the abduction of the toddler James Bulger from a Liverpool shopping center was, in retrospect, simply a well-publicized example of the indifference of contemporary surveillance to the individual.

This is in no way to join in the disapproving chorus that blames identity politics for the collapse of the "left" or to assert that identity has been displaced once again by the economic in globalization. Rather it is to try and place the epoch of identity politics into a discursive network that may shed light on the seemingly intransigent categories of visualized identity. For the hypervisuality of the present defeats even deliberate efforts to erase and displace it. Paul Pfeiffer's video piece *Fragment of a Crucifixion (After Francis Bacon)* (1999) creates an endless loop from a short clip of an African-American basketball player celebrating a score, while being photographed from all around the arena. In this repetition, his shout of triumph seems like a tortured scream, or a moment of crucifixion for the Christian artist. The viewer becomes aware of the incessant flicker of flashlights from individual cameras in the audience. Television films people photographing and in the loop it comes to seem like an act of violence. At the same time, the tiny Sony video projectors used by Pfeiffer create a counterpoint of consumer desire and fascination for the technology. The title seeks to place the clip in the already self-referential loop of high art by referencing the work of British painter Francis Bacon, known for his queering of the painterly canon. Pfeiffer tried to strip the indexical references out of his image by digitally erasing the team name from the player's shirt as well as the other players, score, advertising and so on. The only detail left unaltered on the court was the ethnicity of the player, giving a polemical edge to the notion that the piece represents a crucifixion, rather than a lynching. But this clip was so often used by the NBA to promote itself that we know it anyway. It shows Larry Johnson after making a dunk in his All-Star days at the Charlotte Hornets. This clip is already part of our media memory and resists being reframed as art. It recalls other NBA commercials, such as one featuring a young man playing imaginary basketball with his laundry in a nondescript basement washroom under the slogan "I love this game." The ad works—just as Pfeiffer's piece does—because it recognizes the extent to which daily life in advanced capitalist society is now lived and imagined within the mediascape.

The reconfigured spatialization of digital desire finds a metonym in the intensely popular web cam format. Web cams come in two distinct types. First, as a gaze out on a particular view or geographic location, ranging from skyline views, to wilderness sites and traffic stops. These can be seen, as Bolter and Grusin have argued, as a remediation of television.[36] The second, more popular variety, turns the gaze inwards on itself. Where Niépce pointed his camera out of his bedroom window, web cam users make the bedroom interior the scene of the action. On popular sites like jennicam or annacam, the viewer sees the ostensibly private space of the photographer. In the digital age where your tracks are always visible to those who know how to look, privacy is scarcely a concern for these young people. Rather the camera serves as a device to validate desire itself. Desire wants to see itself desiring, using the closeted camera to reveal and conceal at once. It is not surprising that women have most quickly adopted the web cam format both because of the hypervisibility of the female body in consumer culture and because women since Lady Hawarden have queered photography by not looking out of its closet.[37] For Jennifer Ringley of jennicam, "I am doing jennicam not because I want other people to watch but because I don't care if people watch."[38] What matters, then, is the interiorized sensation of being monitored by a digital other that is controlled by the self. For this is self-surveillance and self-display that leads to a digitizing of desire. Natacha Merrit, author of the *Digital Diaries*, a collection of images of sexual encounters in hotel rooms, claims that "my photo needs and my sexual needs are one and the same." Merrit uses only digital cameras in her practice, creating a digital desire that dissolves the self at the heart of the subject. This erasure was predicted by Foucault in the famous conclusion to his 1967 *Order of Things*. It is the unnerving task of the contemporary critic to find out what comes next.

In advanced capitalist societies across the planet, people are teaching themselves to be media. They attach digital camcorders to their eyes at any event of public or private importance and make endless overlapping records of their memories, which, like those of the replicants, are given out in advance. As a new wave of hyperreal digital animated features comes to dominate North American multiplexes in the summer of 2001, it seems that audiences are teaching themselves to see like computers. That is to say, given that we are all cyborgs, we need to know how the computer sees, to learn how to recognize it and then to imitate it. In *Final Fantasy: The Spirits Within* (2001), the heroes battle the aliens for spirits. In short, can humans still be media? As this is still "Hollywood," the answer is never in doubt. Elsewhere things are less certain.

In June 2000, the Royal Court Jerwood Theatre Upstairs staged Sarah Kane's piece *4.48 Psychosis*. To call *4.48* a play would be to miss its extraordinary power as a visualized text exploring whether it is possible for desire to see itself when mind and body are not just separated but unrelated. The piece takes its title from the notion that at 4.48 in the morning the body is at its lowest ebb, the most likely time for a person to kill themselves. In a long meditation on the possibility of self-killing that is written in different voices but not as separately named characters, Kane mixes Artaud and Plato, a mix that can only be called performed deconstruction.[39] Three actors perform on a stage whose emptiness was broken only by a table. The mise-en-scène, created by director James McDonald and designer Jeremy Herbert, placed a mirror the length of the stage at a forty-five degree angle facing the audience. The mirror made it possible for the actors to perform lying down and still be seen by the audience but at the same time it converted the entire performance space into a camera, mirroring the reflex lens. Within this camera space, a video was played at frequent intervals, showing the view from a London window, as traffic and pedestrians passed by. It was in effect a web cam. The web cam was projected onto the table, forming a screen that was visible in the mirror. The speech of the actors was broken at intervals by the white noise of a pixilated screen without a picture, like a television set that has lost reception. In short, *4.48 Psychosis* played out in the contested space of the contemporary camera, a dark room in which digital, performative and photographic renditions of exteriority were explored, compared and analyzed. In Kane's view, Cartesian reason was a barrier to understanding existence: "And I am deadlocked by that smooth psychiatric voice of reason which tells me there is an objective reality in which my body and mind are one. But I am not here and never have been."[40] Despite the endless assertions of identity over the past two decades, Kane simply asserts its non-existence in the hypervisual digital world.

From time to time the performers simply write numbers on the table in a series they invest with significance that cannot be understood by others. There is an uncanny echo here of de Sade's numerological musings in his journal written in the insane asylum of Charenton. For Kane body and soul do not form a unit or even a schizophrenic network: they simply do not belong together: "Do you think it's possible for a person to be born in the wrong body? (*Silence*) Do you think it's possible for a person to be born in the wrong era?... [F]uck you god for making me love a person who does not exist."[41] Kane explores how metaphysical reason, personal love, and pharmacological psychiatry all attempt to close the gap in which the mind is a camera admitting light all too infrequently and with uncertain results:

> a consolidated consciousness resides in a darkened banqueting hall near the ceiling of a mind whose floor shifts as ten thousand cockroaches when a shaft of light enters as all thoughts unite in an instant of accord body no longer expellant as the cockroaches comprise a truth which no one ever utters[42]

The camera of the mind is deserted now, inhabited only by parasitic insects. Confronted by the indifferent surveillance of late capitalist society and an absent god, the subject disintegrates. At 4.48 "sanity visits/ for one hour and twelve minutes" and as the performing voices suicide themselves, "it is done." The piece ends with a final aphorism:

> It is myself I have never met
> whose face is pasted
> on the underside of my mind.

It is as if there was no mirror stage for Kane to identify herself as an image, only the indifferent reflection of the all-encompassing mirror of the mass media. The stage-long mirror is a catachresis for this loss of identity. There is a long pause and then an actor says: "Please open the curtains." The three performers silently move to the sides of the space and pull back black-painted shutters, opening the camera to the quiet West London light. There is no stage direction to indicate this anti-Platonic gesture which may read as a banal coup de théatre but the audience of which I was a part experienced it as shock. In 1839 Hipployte Bayard performed a mimicry of mimesis when he photographed his *Self-Portrait as a Drowned Man*, a knowing play on photography and death. On February 20, 1999 at the age of twenty-eight Sarah Kane had killed herself in a small room adjoining her hospital bedroom, her camera, her closet. The networked subject is everywhere on screen but no one is watching, least of all herself.

Notes

1. Roy Ascott, artist's note on his collaborative writing piece *La Plissure du Texte (A Planetary Fairy Story)*, in Frank Popper (ed.), *Electra: L'électricité et l'art électronique au Xxe siècle* (Paris: Musée d'Art Moderne de la ville de Paris, 1984), 398.
2. William J. Mitchell, *The Reconfigured Eye* (Cambridge MA: MIT Press, 1992).
3. Ruth E. Iskin, "In the Light of Images and the Shadow of Technology: Lacan, Photography and Subjectivity," *Discourse* 19: 3 (Spring 1997): 46.
4. Diana Fuss, "Inside/Out," in *Inside/Out: Lesbian Theories, Gay Theories* (New York: Routledge, 1991), 1.
5. Geoffrey Batchen, *Burning With Desire* (Cambridge MA and London: MIT Press, 1998).
6. Kaja Silverman, *The Threshold of the Visible World* (Routledge: New York and London, 1996), 136.
7. Marion Hobson, *The Object of Art: The Theory of Illusion in Eighteenth-Century France* (Cambridge: Cambridge University Press, 1982), 52
8. Cited by Hobson, ibid.
9. Denis Diderot, "Salon de 1759," in Jean Seznec (ed.), *Salons de 1759–1761–1763* (Paris: Flammarion, 1967), 9.

10. Hobson, *Object of Art*, 53.
11. Wolfgang Schivelbusch, *Disenchanted Night: The Industrialization of Light in the Nineteenth Century*, trans. Angela Davies (Berkeley and London: University of California, 1988), 93.
12. Talbot's notion of photography as "the pencil of nature" indicates that it is not light which is to be represented but form. Lacan's description of perspective suggests that it represents space rather than sight. Perspective is not a constant, however, but like all visual technologies a discursive practice that constantly renews itself while retaining traces of its older incarnations. In this sense, it recalls both early nineteenth-century definitions of the mind as soft wax and Freud's later metaphors like that of the writing-pad. Most exactly, it is like Freud's 1937 reworking of his earlier figures of the mind in *Civilization and Its Discontents* as an archaeological site, like Rome, with various sedimented layers underneath, surviving fragments peeking through the surface among a contemporary city. You'll recall that Freud's image was formed as Mussolini was rebuilding Rome in the shape of his disturbed imagination.
13. Sade, "Pétition de la Section des Piques aux Représentants du Peuple Français," [25 Brumaire, an II], in Annie Le Brun and Jacques Pauvert (eds.), *Oeuvres complètes du Marquis de Sade*, tome 13 (Paris: Pauvert, 1986), 363.
14. Sade, *La Philosophie dans le boudoir* in Le Brun and Pauvert (eds.), *Oeuvres complètes*, tome 13
15. Satish Padiyar in his "Sade/David," *Art History*, Vol. 23 no. 2 (September 2000), 365–395.
16. Sade, *La Philosophie*, 443.
17. Sade, *La Philosophie*, 470.
18. Sade, *La Philosophie*, 407.
19. Examples include Buffon's assertion that some races have become extinct and the global reach of sodomy: "If we discover a hemisphere, we will find sodomy there. Cook sailed into a new world: there it was king. If our balloons floated to the moon, we would find it there as well" (*La Philosophie*, 472).
20. Sade, *La Philosophie*, 499.
21. Batchen, quoting Reece Jenkins, 152.
22. Quoted in Janet Semple, *Bentham's Prison: A Study of the Panopticon Penitentiary* (Oxford: Clarendon Press, 1993), 100.
23. Ewa Lajer-Burcharth, *Necklines* (Cambridge MA: MIT Press, 2000).
24. Batchen, 33.
25. Quoted by Geoffrey Batchen, *Each Wild Idea: Writing Photography History* (Cambridge, MA: MIT Press, 2001), 11.
26. Charles Mackenzie, *Notes on Haiti: Made during a residence in that Republic* (London: Frank Cass 1971 [1831]), 121.
27. The firm of Adolphe Duperly (later and Son) was established in 1840 at 85 King Street, Kingston, Jamaica. Most libraries suggest 1844 as the publication date for the book, although some say 1850. In each case there is an intriguing question as to how Duperly obtained his apparatus and what machine he was in fact using.
28. Adolphe Duperly, *Daguerrian Excursion in Jamaica* (Paris, c.1845), n.p.
29. Allan Sekula, "The Body and the Archive," October, John Tagg, *The Burden of Representation*.
30. Thomas C. Holt, *The Problem of Freedom: Race, Labor and Politics in Jamaica and Britain, 1832–1938* (Baltimore and London: Johns Hopkins University Press, 1992), 105–15.
31. Cited by Holt, *The Problem of Freedom*, 306. It should be noted that Charles Darwin and Herbert Spencer opposed this position (n.105).
32. Note the queer connotation to the jump as in Isaac Asimov's novel *The Naked Sun*: "There was a queer momentary sensation of being turned inside out. It lasted an instant and Bailey knew that it was a jump. That oddly incomprehensible, almost mystical, momentary transition through hyperspace that transferred a ship and all it contained from one point in space to another light years away." The trope of inversion is of course another figure for homosexuality. Quoted by Steven Holzmann, *Digital Mosaics: The Aesthetics of Cyberspace* (New York: Simon and Schuster 1997), 172.
33. This failure to represent also challenges the theory of cinematic spectatorship as "regressive to some imaginary anterior moment," as Anne Friedberg has put it. Friedberg instead argues for a mobilized virtual gaze. Anne Freidberg, *Window Shopping: Cinema and the Postmodern* (Berkeley and Los Angeles: University of California Press, 1993), 132.
34. Carol Armstrong, *Scenes in a Library: Reading the Photograph in the Book, 1843–75* (Cambridge, MA and London: MIT Press, 1998), 184.
35. Diana Fuss, "Inside/Out," in *Inside/Out: Lesbian Theories, Gay Theories* (New York: Routledge, 1991), 1.
36. Jay David Boulter and Richard Grusin, *Remediation: Understanding New Media* (Cambridge MA and London: MIT Press, 1999), 208.
37. See Carol Mavor, *Becoming* (Durham: Duke University Press, 1999)
38. Compare Ana of Anacam.com, whose artwork—made with Paint Shop Pro—in her gallery declares: "I'll be your mother, mirroring back 2 U."
39. In one monologue a character describes:
    "abstraction to the point of…
    dislike
    dislocate
    disembody
    deconstruct."
    Sarah Kane, *4.48 Psychosis* (London: Methuen, 2000), 20.
40. Kane, 6.
41. Kane, 13.
42. Kane, 3.

24

Modes of Digital Identification

Virtual Technologies and Webcam Cultures

Ken Hillis

For the Victorians, one of archeology's principal activities was displaying old structures and buried relics of the remote past. My research on immersive Virtual Reality (VR)[1] borrows from this understanding to incorporate histories of vision, space and light that address how "old structures and buried relics" are exhumed and embodied within the technology's logics of vision and sight. These "relics" include age-old metaphysical desires expressed in the belief that images allow access to "direct unmediated perception" and thus might allow us to "see what we mean"; an empiricist privileging of sight; and more recent cultural instruction, largely of corporate origin, that relentlessly encourages people to identify as commodities and images. These understandings, together with an accelerating shift towards an image culture supported by visual technologies, have significant implications for how subjectivity and self identity are reconceptualized and practiced.

Virtual Reality is an assemblage of technical developments and social practices. The technology and what it allows users to do inflects their grasp of embodied perceptual processes, what we mean by experience, and how users make meaning of the world around them within cultural contexts that necessarily impinge upon sense-making and its component physiological processes. As Hubert Dreyfus has noted, the West builds its philosophies as technologies.[2] Yet, given the West's preference for empiricist approaches, VR's very materiality renders the debatable philosophies of perception built into it—such as the implication that a person and her or his human physicality might constitute different entities—more resistant to critique.

At the scale of user experience, VR suggests that the question "How do we know what we perceive is real?"—a central question haunting theories of perception and epistemology since early modernity—might no longer be worth asking. This implicit dismissal of history and of the value of distinguishing between simulacra and reality today adopts other forms too, from postmodern critiques of "the real," to televisual "reality television" as different as *Survivor* and *The West Wing*, or to the U.S. Department of Justice's use of the TV drama *Law and Order* to educate Russian lawmakers about the American criminal justice system, or to EyeVision, an instant replay technology that debuted in the 2001 American SuperBowl and which uses more than thirty revolving robot cameras to synthesize what its inventors at Carnegie Mellon University have termed and trademarked "Virtualized Reality™."[3] This cultural dismissal or blurring of distinctions between the real and the virtual parallels the increasing commodification, branding, and mediation of experience. Within such contexts, though immersive VR remains largely an elite device within the mediascape, the technology constructs and therefore works to confirm, if not exactly prove, the experiential equivalency between virtuality and reality. This blurring of reality and virtuality at the level of appearance—an idea interrogated in popular films such as *The Truman Show*, *Being John Malkovitch*, and *Pleasantville*—underlies my research into perception and reality, if only so that, to borrow a phrase from *The Matrix* (which borrows it from Jean Baudrillard), I might make some small contribution to our not coming to *actually* reside with/in "the desert of the real."

In this essay I first examine aspects of the form of immersive VR along with certain identity issues it raises. Because the current range of potential Web identity formations is vastly more varied than with VR,[4] I then discuss how websites and webcams further complicate subjectivity claims and political identities. The Web's promise has been more attainable than VR's within the mandates of consumer electronics, and in looking at Web technologies I extend relevant findings from my research on form and subjectivity in VR. The virtual technologies I discuss exhibit two overarching desires at play: (1) to depict reality as a vision drawn in light, the vividness of which interpellates directly to sensation and an immateriality which reifies contemporary Gnostic and Cartesian inflected desires to transcend bodily-centered limitations; and (2) to fabricate illuminating technologies that confirm and promote the desirability and utility of globalized, circulating, commodifiable and multiple identities that also trade in creation myths and the divine.

Light, Space, and Sensory Perception in VR

The modes of digital identification immersive VR makes possible depend strongly upon the interplay among light, users' sensory perception, and the conceptions of those who design and build the technology. Elsewhere, I have probed the intermingled conceptions of space, light and the body that guide the theorization and construction of VR technologies, arguing that VR's ability to question what constitutes "the real" is central to its cultural appeal, and that the technology operates at the level of sensation to blur distinctions between perception and conception.[5] In brief, immersive virtual environments reduce real three-dimensional "space" to the flat plane of the computer screen. Compared to TV, film or video, immersive VR technology collapses the distance, or space, between the subject's eyes and the computer screen to almost nothing. In part, this then allows the technology to suggest to users' perceptive faculty of sight that its interface has no frame or screen. In sophisticated applications, images display across more of the human visual field's 150° vertical span and 180° horizontal span than other visual technologies to date. This "extreme close-up" fosters users' sense that they have experientially entered the virtual environment, and, by extension, become part of the panoramic landscape constituted by light itself.

The technology's ability to suggest that users can "enter the light" draws on numerous theories of light, from antiquity onward. In VR metaphors of light organize spatial relationships between the subject/viewer/seeker and light itself positioned as a source of truth. The technology synthesizes archaic and Enlightenment ideas about light to suggest that users might experience an exalted transcendent status and divine illumination. Early classical thought understood humans as being *in* the light. Though shining from on high, light was not yet conceived as having an original source spatially discrete from the earth. Light was also accorded transcendent status because it illuminates matter but is not of the matter it reveals. For Platonists, light took on the status of a metaphysical truth and over time was theorized, along with the truth light carries, as conceptually withdrawn from the world to a more supernatural, Ideal realm. Having distinguished the earth as a place of suffering and illusion in contradistinction to the broader cosmos, philosophers reconceptualized light as otherworldly and pure so as not to scandalize its now transcendent and divine nature.[6]

Subsequently, and I am jumping centuries here, Christian theology, drawing from Neoplatonic philosophies, identified humanity as having fallen from grace and as requiring Christian instruction, therefore, on how to step back *into* the light. Christian Neoplatonists eliminated the articulation that light once had been seen to provide between God and humans and replaced it with a distance that light now had to traverse in the fashion of a message from a unitary heavenly Sender to multitudes of earth-bound receivers. This distance or space was partially bridged by the Enlightenment metaphor of a secondary light within the self, a light that reflected the emergent god-fearing bourgeois subject's progressive self-illumination through reason and education.[7]

VR applications trade on positioning virtual environments as a light into which users step (one dons the display), and also on confirming that subjective interiority will be augmented and enhanced by the technology. Promotional print material for VR consumer applications mine this connection. For example, a late 1990s advertisement for I-O Display Systems' "Virtual I-glasses" depicts a young man who wears the company's lightweight Head Mounted Display technology. While the technology actually emits light *from* the display directly *into* the user's eyes, the ad depicts the glasses along with parts of the man's head as glowing with a suffused golden light that appears to radiate *from within him*. The overall effect suggests the cyborg's transcendence from the mundane realities of the world and conceptual relocation to the supposedly more desirable state of "enlightenment" that VR makes possible. As if taking its cues from the human potential movement the ad further suggests that if individuated subjectivity first ideated and desired the immersive product on offer, then the product itself now enhances and completes this individual, today aglow with a power emanating from within the prosthetic self who seems to have merged with the display. Or, in the words of VR researchers Richard Held and Nathaniel Durlach: "[t]aking liberties with Shakespeare, we might say that 'all the world's a display and all the individuals in it are operators *in and on* the display'"[8]; and articulating their comment to light and VR reveals the implicit proposal that subjective interiority might best be relocated within virtual worlds that also model a prelapsarian state and which draw together competing metaphors of light so that these environments, and the enhanced agency they implicitly claim to situate, seem constituted *in* and *of* the immaterial and purifying light *into* which one enters. Stated otherwise, with respect to VR's privileging of sight, and its reliance on the ambivalent combination of the West's acceptance of empiricism coupled to an ongoing complex desire for transcendence, the technology functions according to the hybrid logic of what I term enlightened magical empiricism.

Earlier I equated virtual space with light. The conceptions of light from which immersive virtual reality draws have a complex history; and so too do its built-in conceptions of space. There is a widespread belief that space (understood variously as distance, extension, or orientation) constitutes something elemental, and VR reflects support for a belief that because light illuminates space it may therefore *produce* space a priori. As a result, VR users may experience desire or even something akin to a moral imperative to enter into virtuality where space and light, following on early classical theories, have become one immaterial "wherein." The ability to experience a sense of entry into the image and illumination enabled by VR's design, coupled with both esoteric and pragmatic desires to view the technology as a "transcendence machine" or subjectivity enhancer, works to collapse distinctions between the conceptions built into virtual environments by their developers and the perceptive faculties of users.

Compare, therefore, concepts of space deployed in VR to one's experience of spatial relationships in the everyday embodied world: In the room in which I was seated as I typed this passage, if I looked up from the desk, I saw the screen upon which were displayed these words you now read as well as the wall paneling behind it. Turning my head slightly, I saw the black bookshelf, then the window and the garden beyond, together with the small rip in the sofa upholstery, et cetera. And if I stepped back through the archway linking this room with the adjacent one, I could see these "discrete objects in space" in one fell swoop. In virtual environments, such as those encountered in

VR gaming arcades, when I turn my head to look around the virtual space, the technology reads my proprioceptive movements and everything in the space reconfigures (or "refreshes") before my eyes. Space itself can seem to acquire magical agency and updates itself depending upon how the technology monitors and interprets my own body's movement. Advanced applications, many of which are funded by military agencies, read me as an "other" in relation to the computer's own perceptual abilities, so that it can then offer me a "perspective," or point of view, from which my perception will be saturated with its imagery. In this way virtual technology achieves a new form of dazzling spatialized power based on unseen computational abilities with which a user's body is rendered complicit.[9]

Looking around any environment in which we find ourselves we perceive objects arrayed in space differently as we move our embodied capacity for sight amongst them. In immersive Head Mounted Display-dependent virtual environments, however, I depend upon "the kindness of strangers" as it were, who in their software designs conceive, represent and continually refresh this array for me in order that I might, after the fact, perceive its simulation. Consequently, as a process taking place at any one moment of experience, conception precedes perception in immersive VR. A book or novel also performs such a task—the author's conceptions precede the reader's perceptions—but a book does not collapse the actual space between itself and its reader in the same wrap-around fashion as do immersive virtual environments. In these ersatz worlds, the space users perceive is not the non-human parts of the natural world or their reorganization for human occupation or consumption; rather, "space" is other people's encoded conceptions or uses of language. In so doing, VR also depicts aspects of theoretical arguments asserting there is no world beyond the text.

Virtual environments propose that users interact with someone else's conceptions materialized for users as highly vivid sensations and experienced by them through a process of immersion. This is a new phenomenon that shrinks the space between object and subject, though I would note that immersion has a complex history in the camera obscura, the panorama, or even the stereoscopic View-Master. Immersion works to conflate users' perceptual experience with the conceptions coded into software. "Everything in the field of view is presented to the senses . . . VR is a literal enactment of Cartesian ontology, cocooning a person as an isolated subject within a field of sensations and claiming that everything is there, presented to the subject"[10] in a way that reformulates the saliency of Barthes' suggestion that a photograph is violent because it fills sight by force. The extent of any "violence" done in VR—in the formal mechanisms of the technology's flooding the eye with light and reducing bodies to their spatial coordinates—depends on one's philosophical stance. Nevertheless, the machine becomes the primary sender and the naturalized location for the cosmographic map into which we are invited to insert ourselves. Sensation aligns more closely to the conceptual orbit of the technology and its producers, leaving the viewer with biological perception stripped of its more active meanings. This suggests a reduction of the primacy of a user's perception and, by extension, a loss of self-reflexive abilities supported, for example, by the decoding process abstract print demands.

Finally, while in both real and virtual environments my experience of space takes place primarily in my sensory perceptions of it, in the latter I assent to a kind of double recursivity—the world designed *by* humans in such a fashion as to authorize identifying the world as designed *for* humans. A world designed *for* humans accepts as a moral good a reduction of the sensory interplay between people and their lived worlds to a concept of "world picture" from which the non-human natural world has been excluded, save for the degree to which it has been depicted or rendered via software applications.

Marketing Light

VR's ability to seize the 1990s technical imaginary was linked to a massive amount of hype. People were encouraged to believe, for example, they might actually travel through cyberspace to exotic places without leaving their homes. And while the inability of the technology to live up fully to its exaggerated press has led to a post-millennial waning of media coverage, this in no way diminishes hype's centrality within the political economy of technology diffusion and therefore its *necessary* value to VR proponents, or even how hype inflects sensory perceptions and conception. Staying home while on the road would give users a sense of cultural knowledge disengaged from the material realities of distant places. In several late 1990s' print advertising fantasies of immersive VR's "near future"—and which predate the explosion of Web-based access to data bases—individuals who seek a natural world or information about it are depicted as immersed in a variety of landscapes and ecologies so that they appear to join with technology in a kind of ecstatic, "out of body," new-age reunion even as they equally appear to have donned a "thinking cap" (the Head Mounted Display) that allows them access to a vast "records machine" along the lines of Borges' library. Further, these individuals do so in a fashion that also retains traces of the American pastoral tradition as discussed by Leo Marx in *The Machine in the Garden*—a tradition which seeks to reconcile tensions between nature and art by virtue of a symbolic "middle landscape" earlier described in poetry and prose and rendered in landscape paintings and, with VR advertising, depicted as the subject's access to data banks whereby, for example, he or she "walks through" picture-languages and grasps visual images that represent, and are also "portals" or links to, the information he or she seeks to *experience* as well as retrieve. The discursive positioning of the *idea* of VR, therefore—one where hype renders the actual powers of the technology secondary to the mythic powers it invokes—was also critical in constructing and sustaining a neo-cosmopolitan globalism attached to the transcendent belief that the so-called "biased" materiality of "limited" human bodies and material places and their politics might finally be "set aside." And this positioning also reflects the West's ongoing projection of utopian desires onto the most current but uncharted technologies.

At the increasingly seamless interface where popular culture meets commercially inflected desire, VR's status in the 1990s as "the next thing" was critical in sparking consumer desire for interactive technologies—a desire finally commercialized less as immersive VR and more so in Web formats and portable communication technologies. Interactivity operates as a strategy that increasingly directs people to speak through technologies, and it sutures users more firmly to the economic interests that interactive technologies represent.[11] The ideology underlying VR hype can be summarized as, "you ain't seen nothing yet but in the meantime there's Instant Messenger, Limewire and, now, 3G T-Mobile GSM networks." It remains the case, however, that military and academic VR research and development continues, and so too does the promise of immersive environments, which still occupies that necessary place in the technical imaginary—just around the corner, at the intersection of Progress Boulevard, Consumer Crescent and Disembodied Drive. VR remains a keystone of information technology's advance brigade—a locus of utopian dreaming. Hence speech technology pioneer and futurist Ray Kurzweil, interviewed in the October 2000 issue of *Fortune* magazine, can predict that "by 2030 you'll see full-immersion, shared virtual-reality environments, or spaces, involving all the senses, where we can actually go inside our brains and tap into the flow of signals coming from our senses."[12] If the time line has been extended a bit, the cyborg's dreams have been even more so. Kurzweil's vision updates the longstanding desire for a post-linguistic, post-symbolic communication on the part of those who seek to escape the isolation imposed by Cartesian epistemology—hence Kurzweil's hope to actually get inside "our brains"—and his vision also anticipates a convergence of subjective interior states and material facts, of virtual and real spaces and, therefore, of the private and the public—a vision centered on a spatial metaphor with the potential to authorize the logic of schizophrenia as a model of social relations.

In an increasingly privatized public sphere, promotional media conflate marketing and education, and the meanings of experience with those of sensation, as captured in such advertising slogans as "an experience you will remember" or "one you'll never forget." Implicitly, in this scenario, consumers are Locke's *tabula rasa*. Blank slates without imprint, we are positioned as seeking the learning that experience brings and, in the case of sensation, to have etched in memory some sense of the truly vivid and outstanding. Interactivity contributes to sensing vividness and this idea of experience dovetails perfectly with the commodity form's need to stand out from the clutter and then to mask its origins by means of an interrelated appeal to, and reliance upon, experiential intensity's ability to temporarily overwhelm memory's embodied role in qualitatively ranking experience. For the moment of, or just after, an individual's most recent intense experience often feels to him or her as the most intense experience she or he has had, even though comparison with earlier and later experiences becomes easier over time. To wit, commodities to which an experience of intensity is linked, such as consumer applications of VR, are frequently promoted and often received by consumers as "Sensational!" Commodification, however, has not entirely expunged experience's more complex and over-determined associations from its purviews. The roots of the word's first two syllables—ex-peril—suggest an exit from or outcome of risk, of putting oneself on the line, and this speaks to the role of testing in experimentation, whether this be a test of self, others, or objects.

We are trained by experience and the OED defines the term as a state of having been occupied in some way or "knowledge resulting from actual observation or from what one has undergone." Now, if we think of the tourist versus the traveler, in the context of modes of digital identification, does the VR user tour or travel or both? Any answer comments directly on the nested interplay between our sensorium and experience (and experience's meaning of "having been occupied in some way"). A touristic consumption of distant places as sensations organized by others contrasts to the traveler who today often seeks to confirm social and geographic realities not entirely manipulated in advance for her viewing pleasure. The traveler (somewhat like the archeologist) assumes some risk to gain this knowledge. The tourist eschews this risk for sensation yet with the pre-packaged tour (seen either from the screen of the tour bus or the screen of the VR display) courts the risk of having her imaginative materiality overly colonized by the conceptions of others.

Web Technologies

If immersive VR may never fully live up to its hype, Web sites and webcams incorporating icon-animating programs, Quick Time VR,[13] IM functionalities and pop-up frames within frames have realized in 2-D formats some of VR's promise of interactive immersion. And like immersive VR, therefore, Web-based sites and webcams and the ways by which they are used also complicate subjectivity claims and political identities. As complex and important contemporary sites for research, Web technologies suggest at least four intersecting sites of enquiry: ritualistic uses of visual information technologies, the accelerating value of celebrity status, the aestheticization and fetishization of online performances of the self, and visual technologies' convergence with commodity culture. Web practices indicate the emergence of a new form of ritual, one in which users "construct" then "inhabit" virtual environments in an invocation of the power of celebrity and technology itself. Much of the literature on celebrity treats the activity of celebrities as exceptional, in a class by itself; yet online communication technologies are increasingly akin to churches in which the culture of individuality and niche celebrity is practiced. Online webcams that allow individuals to craft themselves as "24/7 stars" of their own Web sites democratize (without fully localizing) the scale of celebrity dynamics, and users actively embrace, to the point of seemingly desiring to be subsumed by, a technologically constructed reality that increasingly subtends the social.

Space precludes discussing each of these areas. I will, therefore, focus on fetishization of online performances of the self in a way that also addresses issues of ritual and celebrity. I do so by reference to a user group I have studied for several years. Flowing from the complex interplay of several cultural factors, some of which I will touch on shortly, certain English-speaking, first-world, gay men (largely young adults but not exclusively so) who are versed in Web technologies—"digital queers"—were at the forefront of developing specific forms of online cultural practices, and they offer a prototype for new forms of mediated social relations important to interrogate as models and cautionaries in an increasingly wired and wireless world.

Though any one user's stated intention for using Web technologies may be pleasure and communication across distance, webcam site operators reveal themselves in images and text via idealized performances ranging from the confessional mode to the sexual and often both at once. Operators who perform themselves as "moving pictures" suggest that enacting a fantasy self allows for an enhanced feeling of being expressed in the claim that "the online projection of my fantasy self *is* the true inner me." This outing of the inner, this revealing of the light within, is rendered seemingly more concrete by virtue of the truth status afforded depiction, photography and the metaphysical status of the icon. Moreover, digital queers' use of Web technologies in performing identity claims online resonates with archaic meanings of the fetish and a pre-modern sense of time and the present such as those described by early modern European commercial traders, anthropologists, and Christian missionaries. I find the 'neither within me nor without me yet both at once' dynamic that articulates archaic fetishes and their users, together with a cyclical sense of time, productive in thinking through aspects of Web-based identity claims.

The OED notes fetish's etymological connection to the modern English "factitious," as in technical and arising from custom or habit and not natural or spontaneous, and also to the Latin *facticius*, as in "made by art, artificial, [and] skillfully contrived." A subsequent entry summarizes C. de Brosses' 1760 account (*Le Culte des Dieux Fétiches*) of the fetish as "an inanimate object worshiped by savages on account of its supposed inherent magical powers, or as being animated by a spirit," and subsequent subentries note that "a *fetish*... differs from an *idol* in that it is worshiped in its own character, not as an image, symbol, or occasional residence of a deity." Furthermore, a pre-modern fetish is a "visible object," the power of which "resides in its own right as a dwelling for a deity," yet early European traders, missionaries and anthropologists also noted that the fetish was "thrown away as useless when the consecrating nostrum [was] discovered to be inoperative."[14]

Digital queers renovate a similar set of understandings when using information technologies to fabricate online personae experienced both as a component of personal identity and as a set of discrete images. In text-based Internet Relay Chat (IRC) or in Web-based discussion environments or chat rooms, for example, men assert a transparency through naming strategies, yet over time these names exhaust themselves and are discarded, along with the identities they connote to self and others, when no longer useful or when found to be inoperative. Incorporating strategies of recycling and discarding bears similarity to the continual updating of Web sites, and to webcams that "refresh" at ever more frequent intervals so as to approximate more closely the ideal of an interactive, *transparent* and *lively* interface.

In a sense this is a radically old way of making sense of the world: people replacing talismans according to the rhythm of temporal magic. What is contemporized or updated through creative queer uses of digital technologies, however, is an active engagement with magical thinking by marginalized yet well-connected men who understand that screen identities are not fully "real" even as they hope they might be, if only for "the [actual] moment." The increasing power of these technologies to suggest telepresence—briefly, the ability to remain here while seeming also to be there, courtesy of networked digital technologies of vision—coupled with the desire of gay men to fabricate online personae that seem experientially real or even quasi-embodied to others, together suggest the emergence of what I call the *telefetish*: an online, interactive fetish image experienced as the seemingly alive projection of a visualizable and desirable aspect of an individual's identity. These fantasy images garner unto themselves an ironic power: as in the definitions of the fetish provided above, they are skillfully contrived, made by art, animated by the owners' spirit, and worshiped in their own character to the extent that webcams suggest individuals and their embodied selves are fully interpolated within the conceptual apparatus of the technology. Webcam operators who face the online screen image of themselves both embody *and* depict Sandy Stone's argument that computer communication technologies form part of ourselves because we know that "inside the little box are other people."[15]

Telefetishism, the activity and practice of the telefetish, is a powerful means for making sense of the late capitalist commodity-body. A webcam owner is both a body and his iconic telefetish: material and virtual, either both at once, or, in a now-here, then-there fashion. As a site owner, telepresence allows the images of my body as a commoditized "work of art" on the Web to be telepresent to me. I am the camera that Christopher Isherwood and later Baudrillard[16] understand as already inside my head, but having interpellated the technology's logic into my sense of self, with a webcam I can also interact with myself-the-camera. Telepresence allows this commodified self-image (and potentially solipsistic dynamic) to haunt me even as telepresence also gives the sense of distance between the iconography and my embodied presence that modernity requires.

And like a fetish object discarded when found to be inoperative, the telefetish identity is never stable. Personal webcam owners are overwhelmingly concerned that visitors not find their sites "boring." The hypertextual apology—"I hope you don't find my site boring, email me if you do"—is a strategy for dealing with an overarching anxiety of being discarded as a dated commodity. This need for confirmation articulated to the dynamics of webcam niche celebrity culture subtends the voting function written into many personal gay webcams so that thousands of sites operate, as it were, as a vast self-organizing Web site with the owners of the most popular sites competing to achieve a "top of the charts" niche celebrity status. Owners encourage fans to vote and results are tallied daily on Web sites such as Jasbits, which features "Top5 CAM Current Guy and GirlCam Rankings."[17] Becoming a telefetish, therefore, is an ambivalent mastery, and the voting game suggests the precarious status of the commodity form experienced by these individuals "inside" the cams. Kant argues that fetishism is a *relation* based on minimizing the fear that springs from human powerlessness,[18] and while webcams allow making oneself into a niche celebrity, complete with an attentive fan base in order to acquire enhanced identity status and stave off the commodity's fate to be discarded, these practices do so only so long as a Web site charms its visitors. The advance apology also speaks to an ongoing belief that the power of photographic images flows from "the romantic metaphysics of inner, individual truth."[19] In other words, the apology announces a fear that viewers who find a site's pictorial content boring also confirm owners' fears that their inner self might be "boring," without aura, insufficiently illuminated, a poorly packaged Enlightenment product. Further, the apology suggests an individual who has become a "working fetish" even to himself. He may be an interactive experience for some, an object to others, or, as is the case with such minor webcam stars as SeanPatrick,[20] who, when his site was operational, was accorded by fans the status of minor celebrity who could, potentially, be most anything fans wished him to be.[20] Others may acknowledge a webcam operator's subjectivity and labor. But regardless of his over-determined status, he seeks to keep the magic alive and avoid being discarded. This an object cannot do.

The act of producing one's digital self image is a first step in becoming a special kind of commodity fabricated (if not destined) for exchange. As a person being viewed, the webcam owner stars as a willingly commoditized and interactive telefetish. He is alive; both a pictured object and not; frozen in time yet discardable by users; an individual who decides when he will no longer be a fetish, yet desiring the email feedback that confirms both the exchange and use values of his existence as a communicatory act and worthwhile commodity. All of this transpires within a networked environment which serves both as a "place" *and* the means to get to this "place," and within which each person may be a plural fetish to others. Commodity culture suggests an ever-expanding present linked to a *seemingly infinite* world of goods and multiple identity formations in the here and now. Such an expansive sense of presence that occludes the past and future bears theoretical consonance with the ancient meanings of the Hydra. The Hydra was a poisonous water snake with numerous, conceivably infinite, heads. When one was severed another grew in its place. I raise the Hydra to suggest linkages among: (1) the commodity, with its discardability and subsequent replacement by the factory/consumer apparatus with the new that sooner or later becomes the same (exhausted of meaning); (2) a pre-modern fetish belief that temporal experience is eternally cyclical; and (3) the fear of many webcam owners of being discarded online. Applied science is modernity's powerful answer to the conceived weakness of the archaic magic of the fetish against fate and the vicissitudes of the natural world. The applied science of information technologies, coupled with their central role in extending and diffusing the progress myth, diverts attention away from considering how the archaic logic of the eternal return of the same still circulates, and onto the infinite virtual future located within information technologies and their iconographic displays. The commodification of everyday experience increasingly directs the Western notion of transcendence to cut a deal with the eternal. Eternity and infinity, the temporal and the spatial, set apart by theorists of antiquity and modernity, seem to converge within the virtual realm of Web technologies. Webcam visitors may see performed online the making actual of a repressed or heretofore unseen inner self.

The State of Disconnect

Might not the above arguments apply to webcam ownership generally? I suggest that this depends on the contexts and social reception of specific identity claims and the ways that they are advanced. Why, then, do certain gay/queer men desire online augmentation or reconceptualization of identity—to mount a fetish site that seems to stand alone yet at the same time to achieve a quality of persona that combines the representational qualities of the idol and the telepresent status of the online fetish. In his 1904 book, *Fetichism in West Africa*, the Reverend Robert Nassau offers an account of indigenous conceptions of the future useful in theorizing the need to continually update the telefetish. He writes, "The future is so vague that in the thought of most tribes it contains neither heaven or hell... The future life is to each native largely a reproduction, on shadowy and intangible lines, of... this earthly life."[21] The need for digital queers to update their telefetish suggests something similar may be at play to what Nassau describes as an absent Western sense of a future or heaven: fetish strategies operate within a kind of hyper-present; they have a shelf life, and fetish powers require ongoing renewal in direct proportion to an ambivalent or absent sense of the future.

A diffuse loss of faith in, or perhaps indifference to, a sense of the future not articulated to technological progress experienced by many first world individuals suggests the resurgence and renovation of the aboriginal view noted by Nassau that the future will be "the same as today." This is a cyclical understanding of temporality—the eternal return of the same. I am not suggesting that gay men who trade in fetishized Web identities lack any sense of a meaningful future or that somehow these men constitute a pre-modern culture. Rather, for many gay men who have always lived with what have become the conjoined realities of HIV and homophobia, "the future," as a metaphor for achieving fuller acknowledgment of their embodied citizenship within the heteronormative contexts which necessarily imbricate the gay/queer lived world, seems a *tantalizing* prospect continually denied by a regressive application of the logic of "the eternal return of the same," one that is experienced by many of these men as working to keep them "in their place." That is to say, "not here, not yet." Don't ask, don't tell. The OED defines tantalization as "torment by the sight, show or promise of the desired thing which is kept out of reach on the point of being grasped." Many self-defined middle-class American gay men buy into commodity culture's inflation of desire and its seduction of consumers with power fantasies; and they also participate within this culture's logic of filling the present with objects and distractions so that the past and future recede from view. However, these men also frequently comprehend, at an iterative level, that while political struggle has augmented their own gay/queer subjectivities, they live in a post-Matthew Sheppard, post-Brandon Teena world fueled by a persistent cultural othering and an attendant withholding by the State (the 1996 U.S. Defense of Marriage Act and subsequent proposed Constitutional Amendment to define marriage solely as the union of a biological male and female) of a fuller acknowledgment of their subjectivities that they already understand themselves to possess. Hence, perhaps, the number of U.S. Web sites whose owners are "only out on the net, closet in real life." Men seeking to achieve online commodification as a means to control the fragmentation they experience on *and* offline therefore embrace risk to autonomy yet also hold it at bay. Fleshiness is celebrated but never touched, and a panoply of utopian ideas of queerness coupled with a mediatized belief that the truth is that which sells supersede the ambivalent here-and-now realities of queer flesh.

Extending Vaid's thesis, the political and cultural class promises such men a "virtual equality"—a hegemonic dynamic that takes the form of a veneer of acceptance in exchange for conforming to a heteronormative standard of "good taste."[22] This is performed through donning the limited image of what heteronormativity prescribes the appearance of gay/queer identity to be. If online performances promise both connection and separation so too does the promise of virtual equality demand segregating gay experiential and embodied realities into sanitized public fronts acceptable for heteronormativized and capitalized standards of consumption. "Unpalatable" queer politics and sexualities are relegated, in Goffmanesque fashion, to "back stage" places which take on increasingly virtual forms, thereby reifying spatial distinctions between secrecy and exhibitionism that also inhere in fetish practices. The neither public nor private but both at once character of online environments permits marginalized aspects of queer cultures to attain something like a holistic "return to the center"—yet only virtually, in the telepresent.

While the considerable success of a webcam star such as SeanPatrick indicates the possibility for site operators to accrue a transient measure of power, the "so near yet so far" tantalization of viewers' desires this may entail mirrors the kinds of tantalization of gay/queer men and lesbians promised by liberal social discourse. With respect to queer telefetish strategies, if certain digital queers and other marginalized individuals and groups are at the forefront in deploying virtual technologies as symbolic actions, most of my virtual correspondents and owners of sites I have canvassed also buy into, to varying degrees, the dominant cultural equation that information technology = progress = a hopeful future made real. And this is so because Web technologies are accorded mythic status by virtue of their ability to depict a fantasy of embodiment—a fantasy in that webcam performances, in the context of the gay/queer lived worlds I have just sketched, become or depict an alternative to the seemingly endless experience of disavowal of embodied queerness or difference that does not dovetail with an instrumental definition of equality that, in the words of one of my virtual correspondents, perpetuates the politics of Christianity's "great chain of being where fags are on the bottom." Even as several postmodern theorists have critiqued the progress myth and transcendence as Western magical cultural technologies, information technologies, part of the discourse of a progressive definition of the future and the ambivalence it reveals about the present, are increasingly fetishized as new and utopian syntheses of identity and commodity forms.

In a sense, emerging gay webcam cultures also reveal a quasi-diasporic dynamic. Though there is no specific homeland per se from which these men have emigrated, many gay/queer Web sites assert that the material territory upon which everyday gay political battles and lives are waged and lived is too small—both the homophobic places (and closeted lives) from which many younger gay men tend to move on and the ghettoes of large cities. If Wyoming and Nebraska are "the heartland," they are also battlegrounds—the deserts of the real—to which most of these men do not seek even imaginative or telepresent returns for their own sake. And I suggest this is part of the reason why gay webcam sites frequently assert identity claims that transcend or ignore locality and the State. Telepresence—the "site" where technology and cultural practices conjoin—permits conceiving aspects of the self as an array of negotiations between the place where one is (imaginative, experiential, and material) and the webcams that allow a sense of movement among spatially discrete individual users engaged in a continual practice of moving in and out of commodity status and appropriation.

If telepresence proposes multiple identities so too does the Christian doctrine of the Trinity—a single God can be three persons. And the dogma of incarnation allows for a single person to be both God and human. These Web sites, as both telepresent locations *for* and new forms *of* ritual practice, are the crossroads where the carnival and play meet the shrine and worship as a new civil religion that blurs distinctions between acolyte and priest, worship and pleasure, history and archive, carnival and shrine. If Dr. Frankenstein's monster revealed a Western bourgeois fear of technology's dark potential as science's child spurned, with immersion, light and interactivity a way has been opened to imagine virtual technologies as a terrestrial landscape of the gods with space enough for us to digitally identify.

Notes

1. Ken Hillis, *Digital Sensations* (Minneapolis: University of Minnesota Press, 1999).
2. Hubert Dreyfus, *What Computers Still Can't Do* (Cambridge MA: MIT Press, 1992).
3. http://www-2.cs.cmu.edu/afs/cs/project/VirtualizedR/www/VirtualizedR.html (5 November 2003).
4. An earlier version of this paper's focus on telefetishism and webcams is published as "So Near, So Far and Both at Once: Telefetishism and Performing Identities in Web Technologies," *Space and Culture 8* (2000): 107–125.
5. Hillis, *Digital Sensations*.
6. See Hans Blumenberg, "Light as a Metaphore for Truth: At the Preliminary Stage of Philosophical Concept Formation," trans. Joel Anderson, in *Modernity and the Hegemony of Vision*, ed. David Michael Levin (Berkeley: University of California Press, 1993).
7. See Charles Taylor, *Sources of the Self* (Cambridge MA: Harvard University Press, 1989).
8. Richard Held and Nathaniel Durlach, "Telepresence, Time Delay and Adaptation," in *Pictorial Communication in Virtual and Real Environments*, edited by Stephen Ellis (New York: Taylor and Francis, 1991): 232. Emphasis added.
9. For an extended discussion of space seeming to acquire magical agency in virtual environments see Hillis, *Digital Sensations*, Ch.3, "The Sensation of Ritual Space."
10. Richard Coyne, "Heidegger and Virtual Reality," *Leonardo* 27, 1 (1994): 68.
11. Jean Baudrillard, "Aesthetic Illusion and Virtual Reality," In *Jean Baudrillard, Art and Artefact*, edited by Nicholas Zurbrugg (London: Sage 1997): 22.
12. David Kirkpatrick, "14 Minds Look (Way) Out," *Fortune* 142, 8 (October, 2000):253.
13. Quick Time Virtual Reality (QTVR) software allows for the virtual recreation of the Panorama online. See, for example, http://panoramas.dk where site owner Hans Nyberg has organized a range of virtual experiences where you, the viewer, become the camera at the center of a 360° virtual space/experience.
14. Since the early 1980s there has been an upsurge of academic interest in fetishism, much of it organized around or devolving from an interest in eroticism, and the sexual practices of sexual sub-groups; see, for example, Danae Clark, "Commodity Lesbianism," *Camera Obscura* 25/26 (1991):181–201. A growing body of work extends Freudian and Lacanian psychoanalytic theories of fetishism; see, for example, Henry Krips, *Fetishism: An Erotics of Culture* (Ithaca: Cornell University Press, 1999), which, broadly stated, position the fetish as a substitution mechanism for displacing psychic distress or lack through projecting certain attributes onto a particular external object. Karl Marx's original writings on commodity fetishism, informed by a mid-Victorian positioning of fetishism as the practice of "dark Others," in part to confirm the Enlightenment premise of the unitary rational subject, continue to inform research. And Anthropology's long history of interest in fetish practices of aboriginal peoples also provides a resource base for current work on the meaning and use of fetishes and fetishism, and also on the relationship between people and objects often theorized to be at the base of the fetishist-fetish relationship. The Victorian missionaries and archeologists whose work I touch on here of necessity "produce" highly contestable "knowledge" of the fetish. But, so, too, do the more recent approaches noted above. I have identified a quality of "magical empiricism" at work in the ways that people use web technologies to "produce" digital identifications and I am interested here in the ways by which certain Victorian observations of fetish practices uncannily anticipate certain contemporary first world Web practices.
15. Alluquere Rosanne Stone, *The War of Desire and Technology at The Close of The Mechanical Age* (Cambridge: MIT Press, 1997): 16.
16. Baudrillard, "Aesthetic Illusion,"19.
17. http://www.jasbits.com (24 January 2004).
18. Oded Balaban, *Politics and Ideology* (Aldershot UK: Avebury, 1995): 80.
19. Anne McClintock, *Imperial Leather* (New York: Routledge, 1995): 124.
20. At its peak of popularity, www.seanpatricklive.com received between 25,000 and 80,000 hits per day. In 2001, its owner closed the site but information about (and certain portions of mirrored pages from) this "ghost site" remains accessible through internet search engines.
21. Robert Hamill Nassau, *Fetichism in West Africa* (New York: C. Scribner's Sons, 1904): 77.
22. Urvashi Vaid, *Virtual Equality: The Mainstreaming of Gay and Lesbian Liberation* (New York: Anchor Books, 1995).

25

Hypertext *Avant La Lettre*

Peter Krapp

The transition from analog to digital media is perhaps too readily understood as a shift from continuity to fragmentation, from narration to archeology. One might instead view it as a process of translation, since what is completely untranslatable into new media will disappear as fast as what is utterly translatable.[1] Such threats of disappearance tend to lead to symptomatic cultural formations.[2] The implications of digitalization for learning and pedagogy are the topic of numerous scholarly efforts; the most widely used hypertextual systems seemed to bear witness to the creation of a "new economy." But while some saw the Internet conquering the world, others formed their neo-Luddite resistance.[3] Their discontent concerned not so much the machine as its purported effects. Both positions pivot on the same unquestioned assumption: that something irreversibly, incontrovertibly new is intruding on the turf of textual production and reception.

Hypertext is the popular form of computer-mediated communication that has raised perhaps the highest expectations for a transformation of culture.[4] It has been hailed as a new form of literature, a new encyclopedia, a universal library, and as a meta-medium that would ingest and replace all older media. Theodor Nelson proposed to consider hypertext a "generalized footnote," and other media theorists like Jacob Nielsen, Norbert Bolz, and Friedrich Kittler have followed him in this respect.[5] However, the footnote is still for the most part coextensive with the technology of the printing press, even as it expresses a certain strain against the linearity of narrative conventions.[6] More than constituting an extension of annotation and gloss, hypertext draws on processes of subverting, inverting, and exploding the apparent linearity of the page, in self-referential ways modern literature had already exploited.[7] At the same time, broader acceptance of hypertext in and as culture will only partly be achieved by way of improved technical concepts.[8] Required, therefore, is an attentive reading both of the promises that throw historical caution to the winds of mass distraction, and of the quick assimilations that tend to reduce the complexity of any new situation to something already known. Thus if one were to maintain a truly innovative character of hypertext, a more promising model is actually the relational database.[9] Indeed, new media art no longer presents itself as narrative, its forms have no beginning or end, no predetermined sequence. These and related observations about the symbolic form of computer-age fiction, cinema, games, art, and literature may or may not carry the full weight of the hype with which an absolute innovation was heralded; the point of the present argument will be to test, as a selective probe in the genealogy of media, whether claims of an absolute departure are justified. If the following paragraphs focus mostly on hypertext, it is because the widespread aestheticization of digital forms of expression, distinguishing between hyper- and inter-media, separating fiction from interactive art, and so forth, in the end invariably fails to account for the fundamental question raised exemplarily by hypertext: namely, how to explain the anachronism of claiming precursors and forefathers while by the same token presenting a radical departure. It is a curious side-effect of positing such a paradigm shift that the logic of the break is applied to itself, and suddenly, with hindsight, it appears as if everyone knew it all along: as hypertext is hyped, much of what it supposedly superseded turns into hypertext *avant la lettre.*

To be sure, a text that would contain its own exhaustive index would already be nothing *but* its own index, and therefore the end of what it indexes: thus, the computer explodes the boundaries of the book. Hypertext makes relational references within the textual machine available, while their exact manner of connection remains open. The factors that affect and transform culture are less a matter of the media achievements that challenge the capacity of cultural memory than indeed of the conditions that question the functioning of memory as such.[10] However, it is not enough to counter the promise of new media with the oldest critique on the books, that they scatter knowledge, undermine memory, and expose thinking to its deterioration. It is feasible to see hypermedia as little more than an improved means to an old end, as Thoreau said of the telegraph—but with hindsight, we know that technologies not only change the institutions of learning, they also transform the juridical and political milieu of culture.[11]

To arrive at an appreciation of the relational database, one may look back at the development of the card index. Nevertheless, the point is not to historicize what goes beyond the book by pointing out that what first took shape as a bound sheaf recently has begun to fall apart again. Certainly in the sixteenth century, one knew to generate and copy excerpts and to summarize them in a register, but the loose pages were invariably threaded together, not handled individually.[12] For rhetorical memory it was imperative not to work with loose sheets; since such excerpts were to be re-read and committed to memory, it would imperil the entire project if their position in the collection were variable.[13] The ability to sort and shift entries in varying correlations was long perceived not as a strength, a valued feature of knowledge management, but indeed as a dangerous weakness of the system. At the end of the seventeenth century, a historical comparison of different techniques for excerpting and indexing led to the development of a "learned box" which would enable the relational manipulation of notes.[14] This repository was soon adapted and adopted by writers, lawyers, historians, and philosophers: while John Locke had published the description of his card index in 1686 anonymously, by 1796 Jean Paul could publish a novel called *The Life of Quintus Fixlein, pulled from 15 card indexes.* Whatever occurred to Leibniz while reading or even on his walks, he scribbled onto slips for which he had a special cabinet constructed.[15] The search for a page norm was easily settled: playing cards were in use for indexing at least since the French revolution. On May 15, 1791, the French government decreed that a list of confiscated books was needed to decide their fate: sell the libraries of noble families and monasteries, or make them accessible to the public. Local authorities resisted the scheme, since they had good reason to fear that after a book index went to Paris, the books themselves would not be far behind. Thus the National Assembly recommended quick new ways of indexing. Instructions were issued to inexperienced aides who would take stock where the intractable librarians seemed to procrastinate. Regardless of local library customs, they were to go and copy each book's publishing information on a numbered playing card. These cards would later be more easily handled and sorted than a number of incongruent lists from the 83 departments; sure enough, the operation netted the commission 1.2 million cards, to be used for a national library.[16] As contemporaries of Hegel describe in detail, he systematically hoarded ideas and excerpts on note cards, and carried them with him from school days, when he started at age 15, to his death.[17] Gerhart Hauptmann "wrote his nocturnal ideas on the wallpaper near his bed," then cut it up to paste it into his daily output.[18] Raymond Carver taped citations and fragments on three-by-five cards to the wall beside his desk; Georges Perec, who had worked as an archivist in a scientific laboratory, likewise yielded to the "temptation towards an individual bureaucracy" and developed a complex filing system, using his index cards for most of his literary publications.[19]

Despite this respectable lineage (itself reconstructed from excerpts of excerpts), the card index figures only as an anonymous, furtive factor in text generation, acknowledged—all the way into the 20th century—merely as a memory crutch.[20] Since the enlightened scholar is expected not just to reproduce knowledge but to produce innovative thought (figured not just as a recombination of good quotations but opening new arguments and lines of investigation), knowledge management became and remained a private matter.[21] But then as now, the question remains whether there is indeed a departure from the "neolithic mind" Claude Lévi-Strauss glosses over in an interview, when he admits that his own memory "is a self-destructive thief" counter-balanced only by his extensive use of a card index:

> I get by when I work by accumulating notes—a bit about everything, ideas captured on the fly, summaries of what I have read, references, quotations... And when I want to start a project, I pull a packet of notes out of their pigeonhole and deal them out like a deck of cards. This kind of operation, where chance plays a role, helps me revive my failing memory.[22]

In his subversion of the rigorous constraints of memorial order by dint of chance and play, Lévi-Strauss seems to allow that the notes may either restore memory—or else restore the possibility of contingency which gives thinking a chance under the conditions of modernity. That hypertext may instantiate such an epistemology of chance and play on-screen is therefore no innovation; the encoding and deciphering practices of computer-linked textuality merely recapture what had been possible already with the relatively primitive means of note cards—or playing cards. Hence the temptation to claim them for hypertextual ancestry.

Suggesting encyclopedic fulfillment and yet accessible only in constant dispersion, it has been suggested that hypertext has the potential to radicalize literary production. Writing was never simply a means of data storage; as it inscribes and erases traces of textual work, of memory and anticipation, it seemed as if literalizing this structure as hypertext could approach the most exalted hopes of literature. The bulk of critical commentary tends to focus on the question of hypertext-reception, but insight into textual production complicates a careful archeology of the self-reflective poetics of literature written under the conditions of the personal computer. Just as early cinema lagged behind the aesthetic possibilities of theater when it imitated its devices, hyper-fiction tends to lag behind the poetics of pre-screen literature. As with many technological innovations, at first hypertext appeared to spell the end of the book, the end of literature, the end of the humanistic constraints of perception. But instead of an immense extension of aesthetics, as media optimists envisioned, computing technologies soon turned out to have an *anesthetic* effect, threatening to turn the user of a tool into a mere consumer of anachronisms. Despite the widespread digitalization of all media, most attempts to put computers to literary use restrict themselves to hypertext, and the result more often than not falls back behind much modern prose. To be sure, hypertext can pose significant challenges to the conventions of canon, author, reader, and text. That does not prevent philologists from using hypertext for their analyses.[23] Even the most skeptical media critics demonstrate increasing technical competence.[24] On the other hand, numerous cultural commentators who seek to establish the renewed relevance of their particular intellectual lineage claim prescience when it comes to this knowledge system and interface. Vilém Flusser called Champollion a computer *avant la lettre,* since he cracked the hieroglyphic code.[25] Friedrich Kittler considers Hegel's notebooks "hypertextual" and Babbage a "precursor of the computer," and with Lacan, he identifies the "first machine" based on empty placeholders as Pascal's invention of the arithmetic triangle in the year 1654.[26] Lacan called cybernetics and psychoanalysis parallel instances of the same thought experiment.[27] With hindsight, everybody knew all along. Recollection becomes oblivion, the interface-principle WYSIWYG becomes WYSIWYF: what you see is what you (for)get. Such parapraxis slips into the discussion of hypertext and the Internet wherever you look. One might say that the symptom of new media studies is this screen memory. As long as we remain blind to the texture of this symptom, we seem to get over it simply enough, beheading hypertext and arriving at psycho-biographic significance: hypertext will have been nothing but the metalanguage which never presents itself and remains folded in.[28] In the age of digital modification and insufficient version control, the screen is the horizon of memory.[29] Context hides directly beneath the surface, always a click away; there is no world before the machine.

By far the most enthusiastic reception of hypertext in all its dimensions was extended by cultural theorists: at long last, all the promises of their approaches seemed to have come into their own, be they hybridity, nomadism, polyphony, intertextuality, or discourse analysis. Hypertext was going to prove Foucault, Iser, Barthes or Deleuze right.[30] Whether the attention paid to hypertext is seen as confirmation of rhizomatics, actualization of semiotic theory, or a return with a vengeance of reception aesthetics, all of these modes fail to recognize a basic and pivotal fact about the precarious status of hypertext: programs can be called writing, but in order to run, in order for text to be displayed properly, to be distributed and received, they need to be translated into other codes. Despite the obvious misgivings that a grand narrative of textual and theoretical innovation might smuggle traditional hermeneutics back in through the back door of technological determinism, it has been claimed as belated support for a certain poststructuralist and semiotic claims. George Landow was among the first academics to claim a "convergence" of hypertext and the theoretical micrologies of the last three decades.[31] He identified the key feature of hypertext as the link, and presented it as a kind of parodic hypertrophy of the footnote. Landow's identification of Derrida's writing as hypertextual *avant la lettre* itself exhibits this sort of drift, if we follow the notes: Landow cites Ulmer, who refers to an interview with Derrida regarding one passage from Derrida's *Glas*, in which citations from the French *Littré* dictionary are listed... Across the Atlantic, Norbert Bolz agreed—calling both Wittgenstein's *Philosophical Investigations* and Derrida's *Glas* hypertext *avant la lettre*.[32] I have written about this tendency to avoid reading *Glas* elsewhere; mention of Wittgenstein invites scrutiny of another aspect of such contagious retrospective anachronism. His papers, dispersed between Britain, Norway, Austria, and elsewhere, presented the executors of his estate with a conundrum when they found a box labeled ZETTEL, containing numerous loose pages and fragments. Anscombe and von Wright numbered no fewer than 717 such "scraps," the earliest dating from 1929, the latest from 1948 (the bulk was dictated between 1945 and 1948). Were they excess material, occasional ideas, sources and excerpts? Should the typescripts and hand-written notes be published, destroyed, classified—and according to which criteria? A closer look demonstrated that they constituted a card index, and offered clues on the ways in which Wittgenstein's writing relied on fine-tuning and copying; version control after his death proved to be an extremely difficult, but on rare occasions very informative, task. Though far from presuming to reconstruct what Wittgenstein had "meant" to say in unfinished works, the editors simply ordered and published what they deemed the significant finds from this card catalog. Throughout, Wittgenstein's practice of cut-and-paste was integral to his writing method to an extent that puts the avant-garde claims of hyperfiction to shame: "Usually he continued to work with the typescripts. A method which he often used was to cut up the typed text into fragments ('Zettel') and to rearrange the order of the remarks."[33] As von Wright reports of the Wittgenstein papers, some cuts of longer texts are still extant, others were destroyed, and yet other fragments never made it into print. A typescript of 768 pages (called simply *The Big Typescript*) was dated to 1933, and it had been in the estate's control since 1951, but only in 1967 did they discover the "Zettel" from which it was made. Despite extensive cut-and-paste, the end-product was always a linear argument, not a multi-dimensional arrangement.

Above all other unwitting forefathers, Landow and other adopters of the convergence hypothesis claim that Roland Barthes anticipated hypertext.[34] Be it Proust, the daily newspaper, or the television screen—to Barthes, it was all text, and in the age of the Internet, it was going to be Barthes who had always already anticipated its structures and strictures. Admittedly Barthes' writing lends itself to such pretexts, because he often read in a manner that generated, despite all categorical, classificatory zest, a kind of constant *déjà vu* effect.[35] In *S/Z,* Roland Barthes goes so far as to claim that, faced with the impure communication or "intentional cacophony" that is literature, one must accept "the freedom of reading the text as if it had already been read"—and he goes further in asserting that faced with the plural text, there is no such thing as forgetting its meaning. Indeed Barthes believes that one truly reads only in such quasi-forgetting.[36] Reading would be a certain kind of constructively modified forgetting; inversely, it might mean that one only ever reads as if one had already read. Here, click theorists and critics of digitextuality find themselves in agreement with the impresario of the *Desktop Theatre of Amnesia.*[37] Interestingly, reading Barthes is to experience *déjà lu*, too: the distinctions Barthes made in 1960 between writerly and readerly texts return in 1968, and his semiological definition of text crops up in his arguments from 1963 through 1976. "Though most of Barthes' now 'canonical' formulations on textuality occur in the period from 1968 to 1975, the issues that pushed him toward it were organizing his writing much earlier," observed John Mowitt, "in essence adumbrating the move that directed his attention to the work's status."[38] Mowitt notices how 'articulation', Barthes' term in "The Structuralist Activity" of 1963, "reappears eight years later in the Preface to *Sade/Fourier/Loyola*"—and such continuities abound:

> Though I might be accused of stretching the point, it is also worth noting that in order to exemplify the procedural category of "dissection" (articulation's twin) Barthes has recourse in this essay to the sonoric distinction between s and z—precisely the distinction that Barthes later exploited in his most ambitious demonstration of how one might read "textually," namely, *S/Z.*[39]

Faced with such textual echo, Mowitt concludes "it becomes difficult to dismiss this tangle of associations as merely fortuitous." The reason became evident to the public when the *Centre Pompidou* opened an exhibition on Barthes' work: he had worked, daily throughout his intellectual life, with an extensive card index. In an interview, Barthes described his method:

> I'm content to read the text in question, in a rather fetishistic way writing down certain passages, moments, even words which have the power to move me. As I go along, I use my cards to write down quotations, or ideas which come to me, and they do, curiously, already in the rhythm of a sentence, so that from that moment on, things are already taking on an existence as writing.[40]

From 1942 to his death, Barthes amassed 12,250 note cards, constantly rewritten and re-ordered. He had given an outline of this intellectual tool in an interview, but it was only upon opening his papers to the manuscript researchers of IMEC that the scope of his card index could be studied. "There is a kind of censorship," Barthes said, "which considers this topic taboo, under the pretext that it would be futile for a writer to talk about his writing, his daily schedule, or his desk."[41] Almost all of these cards, a quarter of letter-size paper, were written in pencil or blue ink; sometimes words or phrases are (partially) crossed out or corrected. Barthes marked a group of cards simply by noting the category on an upright card, and the rectangular cards that followed it would contain quotes, observations, or diagrams. In the left or right top corner, he sometimes noted the date, and often the page numbers of his publications where he used the information contained on the card (e.g., a *fiche* on "acting out" refers to *S/Z* pages 71–72). Several of the cards exhibited showed more than one use—including the passages noted by Mowitt.[42] There are no obvious techniques Barthes used to refer from one card to another beyond underlining, or sometimes circling, a word, term, or topic taken up on another card (some cards list up to three such links). For Barthes, outing his card catalog as co-author of his texts was "an anti-mythological action," as he said: "it contributes to the overturning of that old myth which continues to present language as an instant of thought, inwardness, passion, or whatever." As one of the editors of the exhibition catalog concluded, Barthes' *fiches* were not the carcass of an unfinished project—there are no missing works by Roland Barthes, despite his sudden death in 1980.[43] "I know that everything I read will somehow find its inevitable way into my work," he had said confidently. The last course Barthes taught, however, was called *La préparation du roman*, preparing the novel. Spread over two years, it simulates the exercises leading up to a novel; a week after the last class, Barthes was run over by a bus. On the one hand, his death may have prevented him from actually writing his novel—on the other hand, the entire course, now published as a notebook, marks the novel as a lost object from the start. These notes are quite condensed and fragmented, just as the short sections of his *Lover's Discourse* were; Barthes had planned to include a postscript to that book, discussing his card index and method of writing. But that plan was abandoned, and the postscript was found only later among his papers.[44] All of Barthes' papers are now available at IMEC; the one thing that can be learned from the manuscripts is his tendency to pare, to erase and efface certain words, especially pronouns, pruning his writing of autobiographical and self-referential elements while retaining a novelistic propensity.[45] However, we must not confound the exposition of text design with what makes up the core of the card database: the so-called content.

If Landow's convergence hypothesis is to be tested in its reliance on Barthes as a model, the question is to what extent the card index, not the footnote, constitutes the precursor and technical model for hypertext in general and hyperfiction in particular.[46] Admittedly, some experimental story-tellers mimicked the gloss of Talmudic annotation; Queneau and Calvino made their mark with the quasi-formalist poetics of Oulipo; and some novelists and even a few poets intersperse their texts with the occasional footnote. Yet while annotation remains crucial for the documentation of philological or bibliographical accuracy, or for the demonstration of philosophical or pedantic veracity, it is only rarely a poetic model. There is, however, a poetics of erudition and concealment around reading and writing, as long as there remains a vested interest in the appearance of originality or creativity, in preparing a novel or other literary form as well as in new media art. One need only think of Chris Marker's IMMEMORY or Olia Lialina's *Anna Karenina Goes to Paradise* for intelligent use of the database form; George Legrady's art makes the structure even more obvious.[47] This poetics of intellectual capital was first embodied in the card index, and perhaps hypertext goes no further than to make it more explicit than before. Yet already in 1951, the Prussian writer Ernst von Salomon had published a novel that takes its shape as a questionnaire; to read it is to construe it as text-generator in following commands to jump recursively from questions to answers and page to page.[48]

It was Walter Benjamin who announced that "the card index marks the conquest of three-dimensional writing, and so presents an astonishing counterpoint to the three-dimensionality of script in its original form as rune or knot notation."[49] Arguably, the true forefather of the web is not the footnote of yore, but the vision of the Belgian bibliographer Paul Otlet, whose fantastic project of a Universal Book was to manifest the connections each document has with all others, and to open this referential structure to further annotation and restructuring by each user. Since 1895, Otlet had envisioned a master bibliography of the world's libraries, but found one fatal flaw all systems shared: they stopped at book titles. Otlet wanted his system to penetrate that boundary, to link up the substance, sources and conclusions of all books. Long before Vannevar Bush or Ted Nelson laid claim to radicalizing knowledge management with memex or hypertext, Otlet developed a scholar's workstation that was, in essence, a database using millions of index cards.[50] He imagined the *réseau* would eventually be accessible by telephone lines, retrieving facsimiles projected onto a flat screen. Today as in Otlet's vision, hypertext foregrounds one feature: it tends to present itself as the sum of its links. However, the defining trait of hyperlinks is not just a web of self-annotation—they set in motion the three-dimensionality of letters that Benjamin saw mainly in the typographic innovations of advertising. It is important to note that under the efficiencies of the networked computer, hyperlinks in effect may also result in a poetics of the relational database. With this realization, new perspectives have been opened for the presentation and production of meaning. Few commentators accept this, however, surmising, again with Benjamin, that the new media spell the end of narration. As the limits and combinations of the new machines were tried and applied, the conventions of time-space perception are challenged and transformed. While film still maintains an affinity to linear narration, it also marks a significant departure from its conventions, by dint of cut and montage, fast-forward and slow-motion. In a note for his storyteller essay, Benjamin articulated the fear that

> Everything is repudiated: narration by television, the hero's words by the gramophone, the moral by the next statistics, the storyteller by what one knows about him. [...] *Tant mieux.* Don't cry. The nonsense of critical prognoses. Film instead of narration.[51]

Perhaps under the conditions of computerized society, the assumption that literature is the highest form of human language may seem obsolete. There is no Turing-test for literature.[52] But before we hasten to the conclusion that the introduction of computers turns "even the most intelligent poetry into myth or anecdote," as Kittler mocks, the fact remains that the new systems are used not only for the technical documentation of airplane construction and open-heart surgery, but also for the writing of poetry.[53] Of course historically (and systematically), the first electronic texts were computer programs, and without them there could be no hypertext. But there is also plenty of serious work on literary software. In 1962, the software "Auto-Beatnik" was introduced by R.M. Worthy in *Horizon Magazine,* "Auto-Poet" and "Scansion Machine" followed, and in 1984, *Scientific American* reported on "Racter," the first prose generator.[54] It uses a vocabulary database to generate complex, grammatically correct sentences. By now, numerous such programs are available on the Internet; among the best known are "Eliza," imitating a psychiatric conversation, and sentence generators like "Prose."[55] Many commercial websites now use customer service bots that interact with visitors handling standard queries and complaints. Search engines parse natural language to better determine the exact nature of your question. A program, it turns out, is just a text that generates text. With this development, the task of the critic seems impossible. How can the reader recognize an object as belonging to a class of objects, such as poetry, in such a way that it does not resemble the other members of that class too closely, as in plagiarism or direct imitation? One solution would be to distinguish between dissimulation and membership in the class. Twenty years ago, the literary critic Hugh Kenner collaborated in the development of a "travesty generator," a software that would imitate literary texts. He concluded that all texts already followed his travesty principles, and language itself follows the rules of his software.[56] But impossible anteriority leads into paradox. One way to address the issue is to remind ourselves that not every text about literature is literature; not every text generated under the conditions of the machine is machine-generated text.

Of course computers have no need to distinguish between a poem, a portrait, a video file, or a chunk of Unix code—sounds, images, texts all disappear into binary states and are only simulated on screen. The readability of hyperfiction relies on HTML and its extensions like Javascript, on the server software and its integral and occasional components that make the Internet possible, and on the operating software the computers run. Thus in the final analysis, literature on the computer is simulated literature; strictly speaking, there is no hyperfiction, there is no net literature. But before this is seen as belated confirmation of the again and again greatly exaggerated news of literature's death, informed hypertext criticism requires competence both in the technologies of literary form and in the arsenals of code.[57] The true challenge of multi- or hyper-mediality and interactivity is that the integration of sound and image tends to distract from the fact that ultimately, they are all code—and integrated only to the extent they are compatible on that level. As for hyperlinks, they challenge policies covering citation and fair use only to the extent that they go beyond the confines of a web or net of references internal to a text; rather than radicalize the poetic possibilities of creation, the whole tangle of questions is reduced to a matter of user interface design. What few commentators care to address is how the practice, for instance, of Proust, Joyce, or Arno Schmidt demonstrates the transition from an extensive card index to a complex textual montage. The next step would be to recognize what lessons their exploration of the frontiers of textual production may yield for writing and reading under the conditions of the computer. On either side of this equation, the technologies of data processing and poetics surely go back further than to Modernism. Nevertheless, it is against the yardstick of twentieth century writing that digitextuality is mostly measured.

One twentieth century German writer often claimed as forefather of hypertextual literature is Arno Schmidt.[58] Voraciously citing, inveterately punning, Schmidt distilled his card index into literary texts, published as complex typescripts, photo-mechanically reproducing his montages without editing. Between 1963 and 1969, Schmidt worked on his 130,000 cards for up to 16 hours per day, producing a text of 1,130 pages, 13 by 17.5" large, and managed to publish it as *Zettel's Traum* (in a Shakespearean allusion, *Bottom's Dream*) in the following year. But he sought recognition not only as creative writer, but also as a theorist of linguistic and stylistic elements of modern prose. According to Schmidt, only diaries constitute a serious attempt at dealing with internal human processes—they help recollect, just as a photo album does, and Schmidt calculated the graphic dimensions of his textual arrangements so as to assist you in following certain associations and connections. Critics even speak of Schmidt's guidance "luring the reader into identification, into the *déjà vu* conviction that these recollections are his own."[59] Joining impulses from Joyce and Freud, among others, Schmidt documents how literature springs from less than divine sources. *Zettel's Traum* is an extended essay on E. A. Poe; over the course of 24 hours, the four protagonists discuss Poe's works, and Schmidt arranged his text in three parallel columns: the center column contains the action, the left one the Poe discussion, and the right column is made up of comments, footnotes, and auctorial opinions. Page (or card) 914 of this proto-hypertext contains the passage most critics view as the key to this gigantic structure.[60] Each of the four characters in this card index fiction is spaced out on Schmidt's pages in a collective score, and here, the book is allegorized as a quartet of voices—the voluptuous unconscious, the mean super-ego, the observant ego, and a fourth instance —something which, according to Schmidt, accrues to men in their fifties, when the sex drive wanes and gives way to what the detached, smiling alter ego of the author represents. Like Derrida's *Glas* or Joyce's *Finnegans Wake*, often claimed as proto-hypertexts that court unreadability, Schmidt's book earns its inclusion here not by virtue of any such purported or real difficulties, but simply because it dares to declare itself made, not always already fully formed.[61] Such unforgivable artifice stands in the way of naïve investments in make-believe, auctorial inspiration, or genius.[62] Similar textures are also evident in Benjamin's *Passagenwerk*, in Butor's *Mobile*, or in Nabokov's *Pale Fire*, a self-declared novel that falls into four parts—a preface, a poem, a lengthy annotation, and an index focusing almost exclusively on the notes.[63] In the preface, Nabokov recommends that readers start with the annotations, then return to them after cursorily picking the poem apart; he even goes so far as to suggest taking the book apart in order to cut and paste pages together at will, or at least buying a second copy to read them side by side. The poem itself is said to be written on 80 index cards of 14 lines each, as the preface dryly describes.[64] Over the moon, Jules Verne's writing is equally illuminated by the reflective fire of a card index, since the source code for his science fiction output was a box of some 20,000 excerpts and notes on scientific journals and books.

The palimpsestic structure of such cosmic writing presents itself differently, again, in the 24 books of *A*, by Louis Zukovsky: "A/child learns on blank paper,/an old man rewrites palimpsest."[65] In this self-interpreting long poem, lines here gloss other lines there, allusions there become references here, and the whole successfully stages what many experimental hypertexts aspire to: a fascinating textual machine that explodes the pages of a book and yet holds together aesthetically. Zukovsky's poetry implicitly uses both Wittgenstein and Benjamin, whom he had read carefully;

but when he was working on *A*, as he records in *Bottom: On Shakespeare,* he had also acquired the habit of performing, for himself, Shakespearean texts.[66] As for Schmidt, Bottom is for Zukovsky the performative weaver, the character who wants to play all the parts out of a fear that the audience might take the play for reality. In their craft, both Schmidt and Zukovsky hone a Shakespearean attention to particulars, scraps, contingencies. But unlike Schmidt, who in his punning ways always sought out vernacular spellings and colloquial phrases, Zukovsky's writing is not an imitation of speech but written to be performed. If some of Schmidt's spellings disgorge their single and double entendres only when read out loud, Zukovsky's poetry has other ways of straining against typographic convention. Despite such partial confirmation of a convergence hypothesis, it is certainly not satisfying to offer Joyce or Schmidt, Zukovsky or Nabokov as advanced hyperfiction writers if by the same stroke their writing is rendered (virtually) illegible under the burden of theoretical proof. At the same time, it remains questionable whether even the most accomplished new media art could or should be measured against high modernism.

Finally, the hypothesis of convergence must be tested inversely: if hypertext instantiates what cultural theory knew all along, can a theorist's work be presented hypertextually? This has been tried with the silicon sociology of Niklas Luhmann's recombinant excerpts from an archive of excerpts.[67] His card index, Luhmann confessed, cost him more time than the writing of his numerous books: little surprise, then, that they demonstrate a certain amount of systematic redundancy.[68] Shortly after Luhmann's death in 1998, a dictionary and a glossary appeared to facilitate access to his thought, and an interactive database is offered on disk, marketed as "Luhmann on your computer." To be sure, nearly everything Luhmann read and wrote was part of his extensive card index, and his theory is incorporated in it perhaps more even than in his numerous books. The question, as in the case of Barthes, would be whether from the depths of such a memory bank, further texts could have been generated, or still can be. Users of the Luhmann CD-ROM may try their hand at emulating his arguments within the recursive parameters of his systems theory.[69] The assumption of such an introductory multi-media tool, even without Luhmann's examples or a decent full-text search function, is that the theory comes alive, lives on, in its card index. Exploring the referential complexities of observation and differentiation, of circularity, structure, method, contingency, of communication and autopoiesis, the user navigating the database is held to make distinctions of increasing complexity while exploring the concepts and questions along their converging paths and definitions. Luckily for the uninitiated, the CD-ROM offers more than just continuous jumps—at the bottom level, one finds an introductory essay on the historical development of Luhmann's systems theory, and most screens also display an alphabetical menu, thus firmly anchoring the hyper-theoretical drift in an encyclopedic project.

A different approach to associative indexing is explored in another collaborative database tool, developed by a Swiss team of programmers.[70] Called nic-las in homage to the great late sociologist ("nowledge integrating communication-based labeling and access system"), and billed as a "software prototype of an *autopoietic* knowledge landscape for social systems," it is basically a cooperative digital space for research groups, made up of textual components and java objects. Shielded and organized by a multi-user access portal, each team can decide to what extent their collaboration is visible also to outsiders, and to what extent their notes, citations, exchanges, and other documents are made available to search engines. Anonymous use is possible at least in principle, but experience has shown that the thirty or so research collaboratives currently using *nic-las* tend to express themselves in the idiosyncratic ways of a typical academic gathering, with concerns over attribution, credit, and accreditation still extant. New entries or modifications of existing entries are recognized and dynamically linked to relevant other notes in the system. An intriguing feature is that deleted elements end up, for a while, in a digital unconscious; they remain accessible to certain search operations, and can even return in unforeseen ways. The system distinguishes between a Freudian and a Deleuzian unconscious; while the former pushes some deleted objects back onto the documentation surface, the latter generates a random selection of deleted and undeleted objects in the form of new virtual index cards. Whether this is seen as new media art or as a software tool for academic work, and more importantly whether or not this succeeds in inscribing theory in software or vice versa, is ultimately a matter of the user experience, not just of the user interface. Here, as in other single- or multi-user hypersystems, if the archive is intricately linked to the institution which authorizes it, then the law of selection, inclusion or exclusion would appear to be a dark outside. Although this law is itself implied in the archive, it decides what is represented, and what is not. Yet hypertext's champions still claim that it accomplishes a virtually universal memory as envisioned by its pioneers Bush and Nelson.[71] Claiming to have foreseen in 1960 the development of personal computing, word-processing, hypermedia, and desktop publishing, Nelson protests that nobody had yet understood how this structure can organize every connection and use of information, beyond inclusion or exclusion: hence his neologism, transclusion.[72] Transclusion would enable one to re-use information with its identity and context intact.[73] However, just what the identity of context would be is the question: arguably, such a limitless memory of "intertwingularity" would not be a memory at all, but infinite self-presence, while memory constantly revives the aposemiological corpse of the sign in referential paraphrases to recall its necessary relation with the non-present.[74] This "diadeictic" relationship presupposes, as Lyotard writes, "the empty gap, the depth separating shower and shown, and even if this gap is referred onto the table of what is shown, it will there be open to a possible index, in a distance which language can never signify without a remainder."[75] Hyperlinks alone do not allow one to surmount this obstacle. If every word were its own index, referring to something else—another word, another meaning—it does not follow that the word index, even when it appears in an index, is already that index.

That the exclusionary meaning of the word index, in the sense of an instrument of censorship, can never be excluded, even in the most efficient file management, is illustrated amply by the computer art installation *The File Room* (1994) by Antonio Muntadas, which indexes cases of governmentally suppressed speech from classical Greek drama to contemporary journalism.[76] It includes works censored throughout the history of art because of their sexual content, and directly addresses freedom of speech; when the project opened in Chicago in May 1994, it contained 400 cases spanning 25 centuries, from Aristophanes to Salman Rushdie. Viewers could ponder Diego Rivera's dispute with the Rockefeller Center over his depiction of Lenin, or TV moderator Ed Sullivan's request to The Doors to change one line of their lyrics in "Light My Fire." The architectural refinement of the installation belies the immense amount of information compressed into its representation of censorship; in its dark chambers of bureaucratic compartmentalization, containing black file cabinets and low lamps, viewers browse case histories—or indeed add their own case to the archive.[77] Chicago high school students reported the confiscation of pamphlets about teen sexuality; entries were also made possible via the Internet. Hypertextual case management allowed the integration of images and other data from the Internet into *The File Room*—hundreds of users logged on daily and explored notorious or half-forgotten incursions into private or public lives. Thus *The File Room* earned its reputation as pioneering "net art."[78] But while such computer-mediated extension seems to explode the frame of the project, the installation remained site-specific in another sense: Muntadas had chosen the Cultural Center in Chicago because it had originally been built as a city library in 1897. Foregrounding the precarious and unfinished nature of archival processes, *The File Room* attempts a re-integration of the exclusions of the archive into the institution that has been shaped by censorship as much as by preservation. In the final analysis, *The File Room* can never be closed, its promise to render invisible images and make unreadable texts legible must remain in permanent deferral. By the same token, with the inclusion of formerly censored art and literature now widely available online, the specificity of Muntadas' hypertext project is in peril of paling into the grand nowhere of the Internet, an unremarked irony for an art installation which despite (or because) of the intentionally claustrophobic atmosphere of its physical setting sought to transcend certain limitations of time and space. Muntadas' *The File Room* is clearly indebted to the conceptual works of the Art & Language collective, particularly to card index systems such as *Index 01* (1972), consisting of eight tall file cabinets of variable dimensions (appearing like columns topped with drawers) and photostats; *Index 2* (1972), consisting of a similar installation and surrounded by a wallpaper of index cards, plus file boxes on a table; and *Index 5* (1973), offering "Instructions for reading the index."[79] While net art may disregard the modernist ideal of the artist who originates or perfects a single skill or style, it still differs from conceptual art in that it often suffers a separation of interface and content; projects such as the *I/O/D Webstalker* (1997) strive to make that gap of digital representation the main theme.[80] Full comprehension of the influence new technologies have on literature and literary studies in particular, and on our culture and its self-representation in general, may seem to recede perpetually into the distance. But while popular views of distance remain cathected with forgetting and repression, distance is arguably nothing but the medium of appearing—as long as simultaneity equals noise, distortion, incomprehensibility, the delays and processing cycles of human or machine intelligence remain necessary. Information lies dormant until it is accessed through an interface; yet that same interface may be distorting the information, obscure its sources, and perhaps even its crucial processes. This kind of information hiding is at work in every machine, and in the recesses of the very code that carries hypertext; it is what database art tries to tease out and foreground.[81]

Since Hegel, writing and calculating machines are understood as a threat, because they interrupt and disperse the cultural fabric of sublation, recollection, idealization, and the history of spirit; the mechanical prevents any recuperation into complete and infinite self-presence. Neo-Luddites and technophiles share the assumption, enthusiastically or apocalyptically, that machines are omnivores, imploding all referentiality and excluding humans by means of their illegibility. Fredric Jameson worries that no society has ever been as oversaturated with information as ours.[82] On the other hand, qualified net-critique beyond mere consumerism requires new competencies and access for all; one can learn Fortran, C++, Unix, and Java—and still concede that most programming is a synthetic group effort, not a critical analysis. And it is somewhat anticlimactic for new media studies to beat a retreat to interface design if it means giving up the crucial access to what interfaces only cover over. At times, this retreat is even dressed up as progress, as in the demand that a filmmaker, for instance, "needs to become an interface designer," as Lev Manovich urges: "Only then will cinema truly become new media."[83] Surely the political, technological, or economic impulses of new media will have aimed higher than at generating mere screen memories for the bureaucratic entertainment of an interface culture.[84] In the end, preserving access beyond user interface design is a necessity, as the index card demonstrated many times over since the French revolution. While it is clear that computer programs and hypertexts by themselves will not revolutionize textual production or digestion, the archeology of multimedia reminds us that fiction and technology "converge" long before the age of the personal computer, which turned their convergence into an ever more technologized fiction. To observe the issues at stake is to observe how literature and the human sciences observe themselves and each other. This mutual second-order observation of information hiding becomes legible only if you are able to access systems such as that which Barthes, as well as the collector Nabokov or the accountant Schmidt, the lawyer Luhmann or the philosopher Wittgenstein, all knew as a reliable tradition of archiving and handling the knowledge they would use as writers. Thus to study media is often if not always to study the political economy of an open secret.[85] Discussing the documentary system of police surveillance, Foucault points to a "partly official, partly secret hierarchy" in Paris that had been using a card index since 1833 to manage data on suspects and criminals. In a note, he dryly remarks:

> Appearance of the card index and constitution of the human sciences: another invention the historians have celebrated little.[86]

Notes

1. This text is a companion piece to Peter Krapp, *Déjà Vu: Aberrations of Cultural Memory* (Minneapolis: University of Minnesota Press, 2004). My argument here develops a line of investigation indicated, but not explored in chapter 6 of that book; it was presented and discussed at Brown University in 2003, and at the University of California, Irvine, in 2004. In its present shape, it repeats neither that presentation nor the argument in my book—but it was produced from the same computerized card index, so one may consider it a DJ Vu re-mix.
2. Nicholson Baker, "Discards", in: *The Size of Thoughts and Other Lumber* (London: Picador, 1996), 125-181; Richard A. Lanham, *The Electronic Word* (Chicago: University of Chicago Press. 1993); Richard J. Finneran, ed. *The Literary Text in the Digital Age* (Ann Arbor: University of Michigan Press, 1996).
3. Curiously, the news media are constantly abuzz with reports of cyber-slacking and other forms of corporate-employee ambivalence, but without much critical reflection along the lines of what Michel de Certeau analyzed as "la perruque"—see Michel de Certeau, *The Practice Of Everyday Life* (Berkeley: University of California Press, 2002), 24–30.
4. A different concept of hypertext was proposed in Gérard Genette, *Palimpseste* (Paris: Gallimard, 1982), who opposes it to hypotext as defining transtextual relations.
5. Theodor Holm Nelson, "Opening Hypertext: A Memoir." In *Literacy Online. The Promise (and Peril) of Reading and Writing with Computers,* edited by Myron C. Tuman (Philadelphia: University of Pennsylvania Press, 1992), 43–57. See Jacob Nielsen, *Multimedia and Hypertext* (Boston: AP Professional, 1996), 2; Norbert Bolz, "Zur Theorie der Hypermedien," *Raum und Verfahren* (Basel: Stroemfeld/Roter Stern, 1993), 17–27, and Friedrich A. Kittler, "Bewegliche Lettern. Ein Rückblick auf das Buch," *Kursbuch* 133 (1998), 195–200.
6. Not to forget a footnote on some recent books about the footnote, above all Anthony Grafton, *The Footnote. A Curious History* (Cambridge: Harvard University Press, 1997), who refuses to date the origin of footnoting but wants to "connect scattered threads of research," and Chuck Zerby, *The Devil's Details. A History of Footnotes* (Montpelier: Invisible Cities, 2002), a book that invites being read as an extended note to Grafton's ambivalent defense of pedantry.
7. The radicalization of general self-annotation is playfully illustrated by Heath Bunting's *readme.html* (1998), where every word is a hypertext link; see http://www.irational.org/heath/_readme.html.
8. George P. Landow, "Changing Texts, Changing Readers: Hypertext in Literary Education, Criticism, and Scholarship," in *Reorientations: Critical Theories & Pedagogies,* edited by Bruce Henricksen and Thais E. Morgan (Urbana: Illinois University Press, 1990), 133–161; compare Paul Edwards, "Hypertext and Hypertension: Post-Structuralist Critical Theory, Social Studies of Science, and Software," *Social Studies of Science* 24:2 (May 1994), 229–278.
9. For the claim that new media art presents a divorce of database interface and database content, see Lev Manovich, *The Language of New Media* (Cambridge: MIT Press, 2001), 226–227.
10. Aleida and Jan Assmann, "Schrift und Gedächtnis." *Schrift und Gedächtnis* (Munich: Fink, 1983), 277 and 281.
11. Joseph Tabbi, "Review of Books in the Age of their Technological Obsolescence," *American Bookreview* 17:2 (Dec 1995–Jan 1996), 31. Compare Jacques Derrida, who wondered "whether the digressive, complicated, parenthetical, sophisticated structure of this discourse, which seems to include notes within notes, infinitely *en abyme*, derives from the fact that I wrote it on a computer." Jacques Derrida, "This is not an Oral Footnote," in *Annotation and its Texts,* edited by Stephen A. Barney (Oxford: Oxford University Press, 1991), 199.
12. Conrad Gessner, *Pandectarum sive partitionum universalium libri XXI* (Zurich 1548); compare H. Wellisch, "How to Make an Index—16th Century Style: Conrad Gessner on Indexes and Catalogs," *International Classification* 8 (1981), 10–15. Despite Gessner's recommendations, most libraries worked with printed and bound catalogs all the way into the mid-twentieth century.
13. See Christoph Meinel, "Enzyklopädie der Welt und Verzettelung des Wissens: Aporien der Empirie bei Joachim Jungius," in *Enzyklopädien der frühen Neuzeit. Beiträge zu ihrer Erforschung,* edited by Franz Eybl, Wolfgang Harms, Hans-Henrik Krummacher, Werner Welzig (Tübingen: Niemeyer, 1995), 162–187.
14. Vincent Placcius, "De scrinio litterato," *De arte excerpendi* (Stockholm and Hamburg, 1689), 121–159.
15. John Locke, "Méthode nouvelle de dresser des Recueils communiquée par l'Auteur," *Bibliothèque universelle et Historique* (Amsterdam, 1668), vol. 2, 315–340; Jean Paul, *Das Leben des Quintus Fixlein* (Stuttgart: Reclam, 1987) and Jean Paul, "Die Taschenbibliothek." In *Sämtliche Werke* II:3 (Frankfurt: Zweitauseneins, 1996), 772; Ch. G. von Murr, "Von Leibnizens Excerpirschrank," *Journal zur Kunstgeschichte und allgemeinen Litteratur* VII (1779), 211, here cited after Markus Krajewski, "Zitatzuträger. Aus der Geschichte der Zettel/Daten/Bank." In *Anführen—Vorführen—Aufführen. Das Zitat in Literatur und Theorie,* edited by Nils Plath and Volker Pantenburg (Bielefeld: Aisthesis, 2002), 177–195.
16. "Only a historian of playing cards might find this relevant," cautioned Jean-Baptiste Labiche, *Notices sur les depôts littéraires et la révolution bibliographique* (Paris: Parent, 1880), 64. But see the commentary by Hans Petschar, "Einige Bemerkungen, die sorgfältige Verfertigung eines Bibliothekskatalogs für das allgemeine Lesepublikum betreffend." In *Der Zettelkatalog. Ein historisches System geistiger Ordnung,* edited by Hans Petschar, Ernst Strouhal, Heimo Zobernig (Vienna: Springer 1999), 17.
17. Johann Jacob Moser, "Einige Vortheile für Cantzley-Verwandte und Gelehrte in Absicht auf Acten-Verzeichnisse, Auszüge und Register," *Lebensgeschichte, von ihm selbst geschrieben* (Frankfurt and Leipzig, 1777), vol. 3; Karl Rosenkranz, *Georg Friedrich Wilhelm Hegels Leben* (Berlin, 1844), 12, and Hermann Schmitz, "Hegels Begriff der Erinnerung," *Archiv für Begriffsgeschichte* 9 (1964), 37–44; compare Friedrich Kittler, *Die Nacht der Substanz* (Bern: Benteli, 1989), 18.
18. Günter Kunert, "Zettel," *Akzente* 33:5 (1986), 391–394. See already Francesco Sacchini, *Über die Lektüre, ihren Nutzen und die Vortheile sie gehörig anzuwenden* (Karlsruhe, 1832), 101–102.
19. Raymond Carver, "On Writing," *Fires. Essays, Poems, Stories* (New York: Vintage 1968), 22-27. Georges Perec, "Notes Concerning the Objects that are on my Work-Table," *Species of Places and Other Pieces* (New York: Penguin 1999), 145 and 152. Perec's novel *Life: A User's Manual* (London: Harvill 1987) features characters who share his obsession with indexing; see also David Bellos, *Georges Perec: A Life in Words* (London: Harvill 1999), 207 and passim.

20. In 1981, when the Internet consisted of only 256 computers, Bob Kahn – co-designer of the TCP/IP networking protocol – was in charge of issuing Internet addresses and carried around index cards in his shirt pocket to keep track of newly issued addresses.
21. Harold Innis is the rare exception; the eighteen inches of index cards in his idea file were themselves indexed by means of another five inches of cards (Innis Papers, Archives of the University of Toronto, Thomas Fisher Library, Box 8). The cards themselves seem to be lost, but a typescript based on them was published as *The Idea File of Harold Innis* (Toronto: University of Toronto Press 1980)
22. Didier Eribon, *Conversations with Claude Lévi-Strauss* (Chicago: University of Chicago Press, 1991), vii–viii.
23. Daniel Ferrer, "Hypertextual Representation of Literary Working Papers," *Journal of the Association for Literary and Linguistic Computing*, 10/2 (1995), 143–45; Tim William Machan, "Chaucer's Poetry, Versioning, and Hypertext," *Philological Quarterly* 73/3 (1994), 299–316.
24. Edward Barrett, ed. *The Society of Text* (Cambridge: MIT Press, 1989); Charles Platt, "Why Hypertext Doesn't Really Work," *The New York Review of Science Fiction* 72 (August 1994), 1–5; Stuart Moulthrop, "You Say You Want a Revolution? Hypertext and the Laws of Media." In *Essays in Postmodern Culture,* edited by Eyal Amiran und John Unworth (Oxford: Oxford University Press, 1993), 69–97; Robert Markley, ed. *Virtual Reality and its Discontents* (Baltimore: Johns Hopkins University Press, 1996).
25. Vilém Flusser, *Schrift* (Düsseldorf: Bollmann 1995), 79.
26. Friedrich Kittler, "Geschichte der Kommunikationmedien," *Raum und Verfahren* (Basel: Stroemfeld/Roter Stern 1993), 169–188, here: 183 and 186.
27. Jacques Lacan, "Psychanalyse et cybernétique, ou de la nature du langage", *Le Seminaire, Livre II: Le moi dans la théorie de Freud et dans la technique de la psychanalyse* (Paris: Seuil, 1978), 339–354; see Laurence Rickels, "Cyber-Lacan," *Nazi Psychoanalysis*, vol. 2 (Minneapolis: University of Minnesota Press, 2003), 60–62.
28. Jacques Derrida, *Glas* (Paris: Galilée, 1974). The fold of so-called "metalanguage" is irreducible like a pocket or cyst that incessantly forms anew; Derrida suggests that for this theoretical question, no other word is possible.
29. Jacques Derrida maintained that the horizon is the "*toujours-déjà-là* of a future that keeps the indeterminacy of infinite openness intact." *Introduction à 'L'Origine de la géométrie de Husserl'* (Paris: PUF, 1962), 123.
30. E.g., Darryl Laferte, "Hypertext and Hypermedia: Toward a Rhizorhetorical Investigation of Communication," *Readerly/Writerly Texts: Essays on Literature, Literary/Textual Criticism, and Pedagogy*, 3/1 (Fall-Winter 1995), 51–68; Julian Stallabrass, *Internet Art. The Online Clash of Culture and Commerce* (London: Tate Publishing, 2003).
31. George P. Landow, *Hypertext: The Convergence of Contemporary Critical Theory and Technology* (Baltimore: Johns Hopkins University Press 1992); and George P. Landow, ed. *Hyper/Text/Theory* (Baltimore: Johns Hopkins University Press, 1994).
32. Norbert Bolz, "Zur Theorie der Hypermedien," *Raum und Verfahren* (Basel: Stroemfeld/Roter Stern 1993), 17–27, here: 17. See my own efforts at http://www.hydra.umn.edu/derrida.
33. Georg Henrik von Wright, "The Wittgenstein Papers," *The Philosophical Review* 78:4 (1969), 483–563, here: 487.
34. George P. Landow, "Hypertext, Metatext, and the Electronic Canon," in *Literacy Online: The Promise (and Peril) of Reading and Writing with Computers,* edited by Myron C. Tuman (Pittsburgh: University of Pennsylvania Press, 1992), 67–94.
35. See Paul de Man's attack on Barthes' literary-historical assumptions: "You distort history because you need a historical myth to justify a method which is not yet able to justify itself by its results," in *The Structuralist Controversy: The Languages of Criticism and the Sciences of Man,* edited by Richard Macksey and Eugenio Donato (Baltimore: Johns Hopkins University Press, 1972), 150.
36. Roland Barthes, *S/Z* (Paris: Plon, 1970), 9-28, sections iv, v, ix. Compare Theodor Adorno, "Skoteinos, oder Wie zu lessen sei," *Drei Studien zu Hegel* (Frankfurt: Suhrkamp, 1969), 105–173, e.g., 154: the retroactive injunction *already to have read* takes the form of a reprise.
37. Anna Everett, "Digitextuality and Click Theory: Theses on Convergence Media in the Digital Age," in *New Media. Theories and Practices of Digitextuality,* edited by Anna Everett and John T. Caldwell (New York: Routledge/AFI, 2003), 3–28, referring to Barthes' "narrative within a narrative"-see Roland Barthes, *S/Z* (New York: Hill and Wang 1974, 90). Jon Dovey, "Notes Toward a Hypertexual Theory of Narrative," in *New Screen Media: Cinema/Art/Narrative,* edited by Martin Rieser and Andrea Zapp (London: BFI, 2002), 19–20, and Jon Dovey, *Desktop Theatre of Amnesia* (Liverpool: Moviola, 1995).
38. John Mowitt, *Text. The Genealogy of an Antidisciplinary Object* (Durham: Duke University Press, 1992), 117.
39. Mowitt, *Text,* 118. See Mowitt, "What is a Text Today?" *PMLA* 117:5 (2002), 1217–1221.
40. "An almost obsessive relation to writing instruments" (interview with Jean-Louis de Rambures of *Le Monde,* September 27, 1973), in Roland Barthes, *The Grain of the Voice* (Berkeley: University of California Press, 1985), 177–182.
41. Barthes, *The Grain of the Voice,* 182: "I have my index-card system, and the slips have an equally strict format: one quarter the size of my usual sheet of paper. At least that's how they were until the day standards were readjusted within the framework of European unification." But Barthes found solace about his mental health in this unwelcome change: "Luckily, I'm not completely obsessive. Otherwise, I would have had to redo all my cards from the time I first started writing."
42. Barthes' note card titled "fiches" reads: "D'origine érudite, la fiche devient le coin vengeur que le désir insère dans la loi compacte du travail. Principe poétique: ce carré savant ira dans le tableau de l'écriture, non dans celui du savoir."
43. "Le fichier n'est pas le livre à venir: il n'y a pas d'oeuvre manquante que quelques milliers de fiches inédites viendraient constituer. Barthes a écrit tout ce qu'il avait à écrire." Nathalie Leger, "Immensément et en detail," *R/B* (Paris: Centre Pompidou/Seuil/IMEC, 2002), 94. However, her co-editor Marianne Alphant thinks the notes for his last course limn the ichnographic *moi-poisson* book he was working towards: Marianne Alphant, "Presque un roman," *R/B*, 125–128. And the executor of Barthes' unpublished papers also believes "these courses revolve around the idea of a possible

novel, a novel that death prevented him from writing." Eric Marty, "Interview with Jacques Henric," *Art Press* 285 (Decembre 2002), 51.

44. Roland Barthes, "Comment est fait ce livre," *Art Press* 285 (Decembre 2002), 55. Interestingly, Daniel Ferrer contends that at several points in his career, Barthes seemed to stop short of embracing genetic criticism: "Genetic Criticism in the Wake of Barthes." In *Writing the Image: After Roland Barthes*, edited by Jean-Michel Rabaté (Philadelphia: University of Pennsylvania Press 1997), 217–227.
45. As Anne Herschberg Pierrot writes, *Roland Barthes par Roland Barthes* "manifests the pleasure of auto-commentary and of reflexivity which includes the relation of the author to his manuscript." Anne Herschberg Pierrot, "Les manuscripts de *Roland Barthes par Roland Barthes.* Style et genèse," *Genesis* 19 (2002), 195.
46. Another recent proponent of the convergence theory along the lines of a generalized footnote is Joe Amato, "Endnotes for a Theory of Convergence," in *New Media. Theories and Practices of Digitextuality*, 255–264. This piece is written as if it contained only endnotes, but then nevertheless augmented by two supplemental "ending notes" after the main body of fake "endnotes" ends.
47. Chris Marker, IMMEMORY (Paris: Centre Pompidou, 1998); Olia Lialina, *Anna Karenina Goes to Paradise*, http://www.teleportacia.org/anna/. For an overview of George Legrady's work, see *Catalogue George Legrady: From Analogue to Digital* (Ottawa, Ontario: National Gallery of Canada, 1998).
48. Ernst von Salomon, *Der Fragebogen* (Reinbek: Rowohlt, 1951); translated, often inadequately but always valiantly, as *The Questionnaire* (Garden City, NY: Doubleday, 1955). Von Salomon takes as his poetic program the de-Nazification questionnaire handed to Germans after the end of World War II; while he answers some questions at epistolary length, others are merely marked with cross-references to later or earlier Q&A.
49. Walter Benjamin, "Vereidigter Bücherrevisor," *Gesammelte Schriften* vol. IV.1 (Frankfurt: Suhrkamp, 1991), 102–104.
50. Boyd Rayward, "The Case of Paul Otlet, Pioneer of Information Science, Internationalist, Visionary," *Journal of Librarianship and Information Science* 23 (Sept. 1991), 135–145, and Boyd Rayward, "Visions of Xanadu: Paul Otlet (1868–1944) and Hypertext," *Journal of the American Society of Information Science* 45 (1994), 235–250. See Paul Otlet, *Traité de Documentation* (Brussels: Editiones Mundaneum, 1934).
51. "One might consider these things eternal (e.g. storytelling), but one can also see them as temporal and problematic, dubious. Eternal things in narration. But probably totally new forms. Television, gramophone and so forth make all these things dubious." Walter Benjamin, "Vorstufen zum Erzähler-Essay," *Gesammelte Schriften* vol. II.3 (Frankfurt: Suhrkamp, 1990), 1282.
52. The mathematician Alan Turing became famous for the unsolved test which was to show statistically that the distinction between human language and computer-generated language is beyond human capacity. (Turing usually referred to this as a "game," only twice does he call it a "test." See Alan Turing, "Computing Machinery and Intelligence," *Mind* vol. LIX, n. 236/1950, 433–460.) Arguably, if this game of imitation is to be decided this side of eternity, it must be stopped by someone who occupies the position of external observer. (See Jean Lassegue, "What Kind of Turing Test did Turing have in Mind?" *Tekhnema* 3/1996, 37–58.) Turing himself became the literary material, for instance in Ian McEwan, "The Imitation Game," *Three Plays for Television* (London: Picador, 1981), or in Alan Hodges, *The Enigma of Intelligence* (London: Allen Unwin, 1983). The artificial intelligence advocate Minsky even published a science fiction novel about Turing: Marvin Minsky, *The Turing Option* (New York: Warner Books, 1992).
53. Friedrich Kittler, "Die künstliche Intelligenz des Weltkriegs: Alan Turing." In *Arsenale der Seele*, edited by F. Kittler and Georg Christoph Tholen (Munich: Fink, 1989), 198. Yet poets have been trying for a few decades to generate experimental computer poetry, referring to William Carlos Williams, who seemed to grant them permission when he wrote: "a poem is a small (or large)/machine made of words."
54. Robert Pinsky, "The Muse in the Machine, or: The Poetics of Zork," *New York Times Book Review*, March 19, 1995.
55. Charles Hartman's program "Prose," somewhat unstable in DOS, but satisfying in its Apple OS version, is found at http://www.conncoll.edu/ccother/cohar/programs/
56. Hugh Kenner und Joseph O'Rourke, "A Travesty Generator for Micros," *Byte* 9/12 (November 1984), 129–131, 449–469. Their fundamental insight is that material is limited; the challenges are posed by technical and economical iteration of connection.
57. Alvin Kiernan, *The Death of Literature* (New Haven: Yale University Press, 1990); see the critique of Sven Birkerts' *Gutenberg Elegies* in Friedrich Kittler, "Computeranalphabetismus," in *Literatur im Informationszeitalter*, edited by Dirk Matejowski and F. Kittler (Frankfurt: Springer, 1997), 237–251.
58. See for instance Jochen Meißner, "Von der Schrift zum Hypertext. Typographie in der *Schule der Atheisten,*" in *'Alles=Gewendet!' Zu Arno Schmidts 'Die Schule der Atheisten,* edited by Horst Denkler/Carsten Würmann (Bielefeld: Aisthesis, 2000), 219–252.
59. F. Peter Ott, "Tradition and Innovation: an introduction to the prose theory and practice of Arno Schmidt," *German Quarterly* 51:1 (1978), 26.
60. See the contributions to the special issue of *Text & Kritik* 20 (1971), as well as Siegbert Prawer, "Bless Thee Bottom! Thou Art Translated," in *Essays in German and Dutch Literature,* edited by WD Scott-Robson (London: Institute of Germanic Studies 1973), 156–191, and Heinrich Vormweg, "Traum eines Babylonikers," *Merkur* 25 (1971), 354–361.
61. For the popular comparison of hypertext and Joyce, see Landow, *Hypertext*, 10: "implicit hypertext in nonelectronic form. Again, take Joyce's *Ulysses* as an example." Others draw Schmidt and Derrida's *Glas* into the mix. H.C. Lucas, "Zwischen Antigone und Christiane. Die Rolle der Schwester in Hegels Biographie und Philosophie und in Derridas *Glas,*" *Hegel-Jahrbuch 1984–1985* (1988), 409–442, here 433: "ein Leseerlebnis, das wohl nur dem von Arno Schmidts *Zettels Traum* oder von James Joyces *Finnegans Wake* vergleichbar ist." More recently, Volker Langbehn considers *Zettels Traum* an "anti-classical work" like Joyce's *Finnegans Wake* and Derrida's *Glas.* See Volker Langbehn, *Arno Schmidt's Zettels Traum: An Analysis* (Rochester: Camden House, 2003), 6; compare Söke Dinkla, "The Art of Narrative—Towards the Floating Work of Art." In *New Screen Media: Cinema/Art/Narrative*, 31–32.

62. Arno Schmidt, "Der Platz, an dem ich schreibe," *Essays und Aufsätze* vol. 2 (Zurich: Haffmanns Verlag, 1995), 28–31. For the history of library technology, see Markus Krajewski, *Zettelwirtschaft. Die Geburt der Kartei aus dem Geiste der Bibliothek* (Berlin: Kadmos, 2002).
63. Vladimir Nabokov, *Pale Fire* (New York: Putnam, 1962).
64. For further information, see Brian Boyd, *Nabokov's Pale Fire. The Magic of Artistic Discovery* (Princeton: Princeton University Press, 1999), and Markus Krajewski, "Ver(b)rannt im Fahlen Feuer. Ein Karteikartenkommentar," *Kunstforum International* 155 (June–July 2001), 288-292. As Krajewski notes, there is at least one book structured as a card game: Marc Saporta, *Composition numéro 1. Roman* (Paris: Seuil, 1962); see also Reinhold Grimm, "Marc Saporta oder der Roman als Kartenspiel," *Sprache im technischen Zeitalter* 14 (1965), 1172–1184.
65. Louis Zukovsky, *A* (Berkeley: City Lights, 1978), 525. For an introduction, see Don Byrd, "Getting Ready to Read 'A,'" *Boundary 2* 10:2 (Winter 1982), 291–308.
66. Louis Zukovsky, *Bottom: On Shakespeare* (Austin: University of Texas Press, 1963); Hugh Kenner, "Bottom on Zukovsky," *Modern Language Notes* 90:6 (1975), 921–922.
67. Niklas Luhmann, "Kommunikation mit Zettelkästen. Ein Erfahrungsbericht," *Universität als Milieu*, ed. André Kieserling (Bielefeld: Haux, 1993), 53–61.
68. Niklas Luhmann, *Archimedes und wir. Interviews* (Berlin: Merve, 1987), 142–149.
69. Detlev Krause, *Luhmann-Lexikon* (Stuttgart: UTB, 2001), as well as Claudio Baraldi, Giancarlo Corsi, Elena Esposito, *GLU. Glossar zu Niklas Luhmanns Theorie sozialer Systeme* (Frankfurt: Suhrkamp, 1997); and Theodor M. Bardmann and Alexander Lambrecht, *Systemtheorie verstehen. Eine multimediale Einführung in systemisches Denken* (Wiesbaden: Westdeutscher Verlag, 1999).
70. Joachim Maier and René Bauer, www.nic-las.com; on collaborative authorship, see also Martha Woodmansee, "On the Author Effect. Recovering Collectivity," *Cardozo Art and Entertainment Law Journal* 2 (1992), 279–292.
71. *From Memex to Hypertext: Vannevar Bush and the Mind's Machine*, ed. James M. Nyce and Paul Kahn (Boston: Academic Press, 1991). See Hilmar Schmundt, "Autor ex machina. Electronic Hyperfictions: Utopian Poststructuralism and the Romanticism of the Computer Age," *Arbeiten aus Anglistik und Amerikanistik* 19:2 (1994), 223–246.
72. Theodor Holm Nelson, "The Transclusion Paradigm," *Project Xanadu* (Sapporo Hyperlab, 1995), d8; compare Nelson, "A File Structure for the Complex, the Changing and the Indeterminate," in *Proceedings of the ACM 20th National Conference*, ed. Lewis Winner (ACM 1965); and Nelson, "What is Literature?" in *Literary Machines: The Report on, and of, Project Xanadu* (Swartmore, PA: Nelson, 1981).
73. Theodor Nelson, "Hypertext is Ready: HTML for Home and Office," *New Media* 5/8 (August 1995), 17. A working model of Nelson's Xanadu is found at www.udanax.com.
74. See also Peter Krapp, "Derrida Online," *Oxford Literary Review* 18 (1996), 159–173.
75. Jean-François Lyotard, "Dialectique, index, forme," *Discours, figure* (Paris: Klincksieck, 1971), 27–52, here: 41. Compare Geoffrey Bennington, *Legislations* (London: Verso, 1994), 274-294.
76. http://www.thefileroom.org; see Susan Snodgrass, "Antonio Muntadas: The file room," *New Art Examiner* 22 (October 1994), 48–49, and Judith Russi Kirshner, "The File Room," *Ars Electronica* 1995, http://www.aec.at/festival1995/catalog/muntadas.html.
77. For the presentation in Chicago, 138 file cabinets of four drawers each were placed around the ground floor of the Cultural Center; seven color monitors installed in various networked file cabinets invited the audience to point and click through case histories, sorted by location, time, medium, and reasons for censorship. A centrally located table with another computer allowed the entering of new cases.
78. Robert Atkins, "The Art World and I Go On Line," *Art in America* 83:12 (December 1995), 60, and Miriam Rosen, "Web-specific works: the Internet as a space for public art," *Art & Design* 11 (January-February 1996), 86–96.
79. Art & Language, *Index 01* (1972). Installation at P.S.1, New York, 1999 (Collection Daros, Switzerland); *Index 02* (1972), Installation Lisson Gallery, London, 1978 (Collection Herbert); *Index 05: Instructions for reading the index* (1973) (Collection Carine and Philippe Meaille, France). See the catalog to the PS1 retrospective *The Artist Out of Work: Art & Language 1972–1981*; some archived images can be found at http://www.lissongallery.com/theArtists/Art&Language/artlanguage.html.
80. The most recent version of *Webstalker* can be found at http://bak.spc.org/iod/
81. Software engineering is "information hiding," where each module hides its function: see Jörg Pflüger, "Distributed intelligence agencies," in *Hyperkult: Geschichte, Theorie und Kontext Digitaler Medien,* edited by M. Warnke, W. Coy, G.C. Tholen (Frankfurt: Stroemfeld/Nexus, 1997), 433–460. See Laura U. Marks, "Invisible Media," in *New Media. Theories and Practices of Digitextuality*, 40.
82. Fredric Jameson, *The Political Unconscious* (Ithaca: Cornell University Press, 1980), 60–61.
83. Lev Manovich, "Old Media as New Media: Cinema." In *The New Media Book,* edited by Dan Harries (London: BFI, 2002), 209-218, here: 217.
84. See Steven Johnson, *Interface Culture* (New York: Basic Books, 1997).
85. Although secretaries in 17th-century France or Italy were forbidden to speak of their work in public, their confiscated speech never dampened their drive to express the master-medium dialectic of their employment. And as Foucault demonstrates, doctors (not unlike confessors) were figured as stenographers of the client's secrets, until the birth of the clinic forced them out of their secretarial role.
86. "Apparition de la fiche et constitution des sciences humaines: encore une invention que les historiens célèbrent peu." Michel Foucault, *Surveillir et punir. Naissance de la prison* (Paris: Gallimard, 1975), 287, referring to A. Bonneville, *De la recidive* (Paris, 1844), 92–93.

26

Network Fever

Mark Wigley

We are constantly surrounded by talk of networks. Every third message, article, and advertisement seems to be about one network or another. We are surrounded, that is, by talk on networks about networks. It is as if our technologies feed on a kind of narcissistic self-reflection. Everyone has become a kind of expert, ready to discuss the different types of nets (computer, television, telephone, airline, radio, beeper, bank . . .) or scales (global, national, infra, local, home . . .) or modes (cable, wireless, digital, optical . . .). And where would we be without our opinions about *the* Internet, a net of nets against which all others are now referenced? How many ways do we have to express our amazement at such a vast space in which any address is just a few clicks away from all the others? Attaching oneself to a seemingly marginal thread soon accesses an endlessly dense weave, as if a walk down a quiet country lane would suddenly bring one to the heart of a metropolis of unprecedented dimensions. In celebrating this new kind of territory, we recast questions of individual identity in terms of unimaginable levels of connectivity, ignoring the equally dramatic rise of new forms of inaccessibility to stage an institutionalized simulation of euphoria in which discourse about openness, democracy, free exchange, and speed dominates over that of control, surveillance, blockage, sedation, and crime.[1]

The message is clear. Nowhere escapes the net. A map of all the webs passing through any particular space would be impossibly dense. Invisible networks seemingly threaten visible means of defining space, dissolving the walls of buildings. The architecture of borders, walls, doors, and locks gives way to that of passwords, fire walls, public key encryption, and security certificates. Indeed, the idea of a space occupied by networks or superimposed by them has been replaced by that of overlapping networks within which physical space only appears as a fragile artifact or effect. Space itself can only be seen when caught in the net. It is as if the modern perforation and lightening up of architecture in the face of speed, industrialized technology, and mass production at the turn of the twentieth century has gone a step farther as buildings dissolve into information flow, to be either discarded as a relic of a previous time or nostalgically preserved as a quaint memento.

The Internet is relatively new, emerging out of ARPANET, a 1969 cold war operation of the U.S. Defense Department that combined the computers of four universities. It grew exponentially ever since and now bounces from school to house to car to plane to beach. But what if we are actually at the end point of the network logic? What if contemporary discourse about the net simply realizes nineteenth-century fantasies that were acted out throughout most of the last century? What if the much-advertised dissolving of architecture occurred long ago? What if much of our net talk is just an echo? An echo of an echo?

Dancing Gurus

The radical confusion of architecture and networks can be marked by the July 6, 1963, meeting of a short man in dark pants, close-fitting white jacket, crisp shirt, and tie with a tall man in light pants and a loose-fitting summer shirt covered with a geometric pattern. The shorter man was born in the last decade of the nineteenth century and has been using communication networks as a model for architecture since the late 1920s. The taller one is a forty-year-old expert in communication who has just published a book on networks in which architecture plays a decisive but less obvious role. Photographed together on the deck of a boat, the architect is clearly no stranger to the sun but stands a little defensively, holding a text with two hands in front of him, as if about to deliver an important statement somewhere else. The communications expert is pale but leans casually toward the camera with his hands tucked behind his back, smiling openly as if he has nothing much to say and will stay on deck for as long as possible. Yet the odd couple take an instant liking to each other and quickly become a kind of intellectual tag wrestling team, tormenting colleagues and audiences around the globe until the younger man's death at the end of 1980. Both were regarded as entertaining but crazy in their respective fields. Both regarded their fields as crazy. But their mark is everywhere. They voiced so much of what is said today. They wrote a lot of our script.

Buckminster Fuller and Marshall McLuhan met for the first time after boarding the *New Hellas* in Athens for an eight-day boat trip around the Greek Islands. The two gurus of the electronic age had been invited on the trip, along with thirty-two other leading intellectuals from fourteen countries, by Constantinos Doxiadis, a Greek architect and urban planner. The idea was to have a

Figure 26.1 On board the *New Hellas,* in the Aegean, July 1963. Photograph courtesy Corrine McLuhan.

"symposion," a radical mixing of intellectual activity and sensual pleasure as the boat traveled from island to island. Each morning, the group would have informal but intense discussions onboard about "the evolution of human settlements." In the afternoon and evening, they would leave the boat to go swimming, visit famous historic sites, eat in restaurants, see performances, go dancing, and shop. High-level theoretical discourse was well lubricated with retsina and ouzo.

McLuhan and Fuller admired each other's eccentricity. McLuhan liked to speak in aphoristic punch lines thrown as grenades into the morning discussions. A pun was as likely as a formal statement. Fuller surprised the group by seeming uncomfortable with the rapid exchanges. Having difficulty following the conversation because of his bad hearing, he preferred to give speeches. He would talk for hours on end, continuing his line of thought during meals, while drinking, and while changing in the cabin—enthralling yet ultimately exhausting everyone.[2] He moved wildly when speaking but said McLuhan's moves were more extreme:

> After dinner on the Doxiadis ship we used to dance and Marshall would dance with his wife all over the place, so much so that he took up the whole dance floor. He thought we had all stopped to marvel at his and his wife's performance, but that wasn't it; the way he was dancing there wasn't room for the rest of us and we had to leave the floor.[3]

Even if the others on the boat regarded McLuhan as "outlandish," as he later wrote to a friend, his arguments had a marked effect. The group included some prominent architects and planners, but most came from outside the traditional limits of architectural discourse. Led by superstars like Margaret Mead and Barbara Ward, there were representatives of psychiatry, engineering, economics, sociology, anthropology, political science, language, law, metallurgy, animal genetics, meteorology, biotechnology, aesthetics, physics, history, philosophy, literature, agricultural science, and geography. Each field was seen to have an important contribution to make to architectural discourse. When Doxiadis sent his letter of invitation to McLuhan just seven weeks before the event, for example, he said that he had just read *The Gutenberg Galaxy* of the year before and saw ideas in it that are "essential" to a reconsideration of human settlements.[4] McLuhan had no problem seeing his work in that light. He wrote an unsuccessful fund-raising letter to another Canadian who had been invited to the event, citing the letter of Doxiadis and explaining that he was currently completing a book "which includes matters of immediate concern in housing and town planning." Since the extension of the human nervous system in an electric age "confuses the problems of living space," his own participation in the event "could be of very real importance to the study of changing problems of our national housing."[5]

Once onboard, McLuhan used the event to explore the architectural implications of his work. The boat became an amplifier for his argument that electronics is actually biological, an organic system with particular effects. The evolution of technology is the evolution of the human body. Networks of communication, like any technology, are prosthetic extensions of the body. They are new body parts and constitute a new organism, a new spatial system, a new architecture. This image of prosthetics—which McLuhan had first presented a year earlier in *The Gutenberg Galaxy* and was busy elaborating for *Understanding Media: The Extensions of Man*, which would launch him to superstardom when it came out a year later—was now reframed as an architectural image. McLuhan only waited until the second morning of the boat trip to get up and present his work as a question of urban planning, insisting, in a paradoxical twist, that the latest technologies have expanded the body so far that they have shrunk the planet to the size of a village, creating a "tremendous opportunity" for planners.[6]

This was all too familiar to Fuller, who had been describing technology as an extension of the body ever since his first, but not well known, book, *Nine Chains to the Moon* of 1938, and had been insisting that traditional architecture had to give way to a "world wide dwelling services network" modeled on the telephone network. Indeed, Fuller had visualized global electronic networks long before they arrived. Unsurprisingly, he felt that his ideas, including the concept of the global village with which McLuhan would soon become famous, had been taken without acknowledgment. Yet a strong friendship was immediately established. This was greatly assisted by the fact that, as Fuller recalls it, McLuhan was carrying copies of his *Nine Chains to the Moon* (which had just been republished) and *No More Second Hand God* when they first met on the boat, declaring, "I am your disciple. . . . I have joined your conspiracy."[7] McLuhan, who had denied getting the idea of prosthetic extension from anyone until he met Fuller, later told his friends that Fuller was too much a "linear" thinker.[8] Fuller told his friends that McLuhan never had original ideas, nor claimed to.[9] He simply remixed available material in an original way. Yet a firm bond was established, and from then on they defended each other's work, seeking out any opportunity to be together and pursuing the global implications of prosthetics and networks to the limit.

Animate Nets

Doxiadis was ready for such sport. Like Fuller and McLuhan, he always thought at the scale of the planet. To say the least, he was a global architect. The design office that he started in Athens in 1951 had already completed major buildings, complexes, infrastructures, urban plans, and regional studies in Greece, Pakistan, India, Ghana, Spain, Denmark, Sudan, Libya, Syria, Venezuela, Lebanon, the United States, Australia, Iran, Jordan, and Iraq. After just a decade of work, he was able to publish a world map dotted by all his projects as if by a spreading virus.[10] Indeed, global spread was his obsession. Doxiadis was an expert in growth. His starting point was that cities were expanding out of control, as marked by the Tokyo Taxation Department's decision that it could only keep up with the spread of buildings by using aerial photography. Insisting that the speed and scale of such growth defied traditional analysis, Doxiadis launched the field of "Ekistics" in the mid-fifties and founded the Athens Technological Institute in 1958 as a research center and architecture school based on the idea of global statistics. The idea was to think at the largest possible scale by domesticating vast amounts of global information. If the data could be controlled, cities could be controlled. Courses in statistical analysis became "indispensable" for architectural training. Spatial patterns would follow from detecting patterns in the flow of information. Design would begin with precisely calibrated charts rather than artistic sketches.

For Doxiadis, a settlement is a continually evolving "organism," at once biological and technological, a technology with a biology. On the one hand, he keeps referring to the city as a body with nerves, arteries, and heart and uses the growth and multiplication of organic cells as a model—presenting images from biology textbooks to clarify the behavior of urban form. On the other hand, he represents the evolution of cities with sequences of "electromagnetic maps" and computerized "cartographatrons" showing shifting patterns and hidden force fields through time. The combination of biological and technological images creates the impression of a dynamic biotechnological organism, ever widening its scale of operation until it becomes dysfunctional or extinct.

Doxiadis never tired of insisting that the real dimension of cities is not space, but time. What counts is a city's trajectory of development rather than its form. If a city simply grows radially outward from its center, as usually happens, pressure increases on the center until the organism collapses. "Surgery" on its "heart," like feeding new "arteries" as highways into the core, will only speed up its death. Doxiadis's prescription, as worked out in most of his projects since the mid-fifties, is that a city should grow in one direction. The core itself needs to move sideways and expand as the scale of the city increases. Such a city doesn't simply grow; it moves across the landscape. Growth becomes movement.

Settlements become a mobile species, and their movements are further accelerated by the multiple patterns of mobility made available by numerous overlapping networks. Despite designing many fixed buildings, complexes, and neighborhoods, Doxiadis rejected the traditional conception of architecture as a static self-contained object in favor of nomadic organizations animated by circulation patterns. The internal life of each building is extended by ever-larger-scaled networks, from the pedestrian journey to a neighbor's house to an airline flight to the opposite side of the planet. Doxiadis's basic image of a building is a minimal form, a single thick semicircular line defining a shelter containing a dense internal life that is extended out by the wandering tentacles of different forms of circulation. Buildings are but "shells" for movement patterns that reach out far beyond them. Whereas buildings house function, networks are pure function, function without shell. If modern architects are serious in their commitment to function, they will have to reduce their fixation on shells and become responsible for networks.

This concern for networks became clearest in the "City of the Future" project that Doxiadis launched in 1960 and kept working on until his death. First published just a month before the Delos event, it predicts the emergence of a single city covering the whole earth like a lava lamp network, a fluid biomorphic growth extending itself everywhere. The modern architects' fantasy of free-floating generic forms that could be dispersed anywhere on the planet gives way to a single planetary scaled dwelling: "a continuous network of centers and lines of communication" in which "all parts of the settlement and all lines of communication will be interwoven into a meaningful organism."[11]

And it is not just architectural form that turns into a network. Doxiadis draws the discipline of architecture in the same way as he draws the city. To survive the global explosion, architects must be as networked as the spaces they produce. The discipline must take the form of an efficiently webbed biotechnical organism capable of new forms of growth. Architects can only conquer the planet by becoming an animate global net.

Floating Amplifier

The 1963 boat trip was intended to be such a networking operation. Experts from heterogeneous countries and disciplines were linked together in a tight web. Lines of communication were effectively drawn between every participant. Yet this web of global figures took the form of a withdrawal. To engage with the global networks whose key feature is that no point has any more value than any other, the group disconnected from those networks and returned to a very singular point, the ruins of the mythical source of western philosophy in the Greek islands.[12] In withdrawing to archaic origins, they withdrew from the media. All participants were warned before going that there would be no telephone, newspaper, or mail.[13] Once onboard, there were "no formal minutes or records, no stenographers and no tapes,"[14] just a set of handwritten notes. The body was used to record the group's analysis of the displacement of the body by new technologies of communication. The only concession was a mimeograph machine that was used to convert typed statements by participants into documents distributed to all the cabins, establishing a local net that only reached as far as the sides of the ship. The implied fantasy is that the boat is a pre- or postdisciplinary space, drifting freely between islands, unaffected by the explosive global growth it so earnestly addresses.

Yet the trip was ultimately a media event. The isolation was staged as such in the very networks supposedly left behind. Edited notes on the discussions, photographs, the boat's itinerary, and biographies of the participants were circulated to the international press and specialist journals in the represented fields. Each participant was sent the same material and encouraged to send further information to journals—which many of them did.[15] The event gathered information, accumulating expert opinion from diverse fields and countries, only to reorganize it and send it back out in a unified form. What was retransmitted around the world was the media image of a premedia event—reinforced by photographs of metropolitan experts in relaxed vacation clothes, basking in the sun, in restaurants, ruins, and the water. It is as if the technological expansion of the body could only be faced by returning it to its original state.

This strategic primitivism was exemplified in the closing event, when everyone signed a collective "declaration" in the ancient theater on the island of Delos. A group of globe-trotting intellectuals gathered at sunset in a ruined amphitheater to solemnly endorse a manifesto for a global makeover. They sat in the first row, a semicircle of close packed expertise facing a piece of paper that rested at the center of the stage on a rustic altar improvised out of stacked stones. The sun dropped while they listened to speeches and, in a torchlight ceremony, solemnly stepped forward to leave their mark.

Unsurprisingly, the Delos Declaration ends by reaffirming Doxiadis's vision of a single global city growing out of control, with the human species portrayed as the victim of the uncontrolled growth of architecture.[16] The "Delians" had withdrawn from this destructive exploding organism to reaffirm their physical, intellectual, and emotional humanity in a symbolic display. Yet the whole point of their radical disconnection from the modern world was to set up a better reconnection, as became clear in Fuller's speech at the beginning of the signing ceremony. An ancient amplifier was being used for a global broadcast:

> The acoustics of the Greek theatre are phenomenal, and I believe that our voice here, relaying the voice of every man, will be heard around the world and that it will catalyze the efforts to prevent man from eliminating himself from his extraordinary role in the universe.[17]

This fantasy seemed to be realized when each participant became a kind of missionary, spreading the word "Ekistics" around the globe, and the Delos Declaration was reported extensively in newspapers and journals, was cited in discussions on housing at the United Nations, and was entered into the official records of the U.S. Congress—events that were eagerly monitored by the monthly in-house magazine of Doxiadis's design office.[18]

The model for all this was the fourth meeting of CIAM (The Congrès Internationaux d'Architecture Moderne) in 1933, the boat trip from Marseilles to Athens and back, out of which the famous Athens Charter on the future of the city emerged. Sigfried Giedion, the longtime secretary general of CIAM, was symbolically invited to the Delos event and was asked to give the last speech at the signing ceremony affirming the fundamental "continuity" between the Athens Charter and the Delos Declaration. He noted that the collegial atmosphere of the two meetings was very similar, insisting that "Greece has done it again!"[19] Ekistics had officially picked up the legacy of CIAM. There would be twelve annual Delos meetings to match the ten CIAM congresses held between 1928 and 1956. But Doxiadis tried to go farther than his role model, networking a wide range of disciplines rather than just architects. Networks had to be taken to the next level.[20] In fact, the starting point of the Delos meetings was the call made at the end of the Athens Charter to improve the condition of transportation networks.

With each Delos meeting, more and more time was devoted to networks, and a collective attitude toward them evolved. In the fourth Delos in 1966, the participants accepted Doxiadis's claim that networks are historically the youngest element of settlements and will therefore change the most radically in the future, while other speakers emphasized that many of the key networks are invisible. By the eighth Delos meeting in 1970, networks had become the official theme, and Doxiadis was arguing that they are the single most important element in settlements: "the foundations of everyday life and the most decisive element for man's well-being."[21] He was even starting to describe buildings as networks. At one point in the morning discussions, he called a house "a network of walls," and at another point he referred to a theater as a "physical network between actor and audience."[22] Other speakers added the idea of social networks to the physical ones, arguing that there had to be an "interface" between them, and Doxiadis agreed that "every non-physical network requires a physical network for its delivery."[23] The final report of the meeting confirmed that networks "proliferate and interlock, crossing every barrier, physical and political, that has previously divided man.... Networks are the key to the making or breaking of cities."[24]

By the tenth Delos in 1972, networks had become the central focus of all urban design, with the final report insisting that their configuration determines the growth patterns of cities. Networks were now the beginning rather than the end point of city form. Everyone now agreed that many of the most decisive networks are invisible and that designers unwisely focus on the dense visible form rather than the diffuse communication patterns that extend that form to constitute the real settlement. Doxiadis argued that cities are simply the product of networks used to minimize effort to maximize contacts, yet typically it is only the shells that are designed. To demonstrate a necessarily wider role for the architect, he redesigned the invisible global airline network to match the latest version of his visible global network on the ground, hanging an immaterial triangulated web over the planet that is linked to the physical web on the ground at a series of strategic nodes.[25]

Such an image of the invisible extension of the physical was always the central goal of the Delos meetings. Instead of sitting in front of finished drawings of projects, as in the CIAM meetings, the participants always sat around a blackboard, drawing and discussing diagrams of network flows. No matter what discipline the speakers represented, they all drew and energetically criticized one another's diagrams. Everyone was treated as a kind of architect—or, rather, the whole group tried to act as a single architect. The boat was a collaborative design studio. Following Doxiadis's lead, the Delos events were all about making a certain kind of drawing, trying to visualize the invisible by conjuring up a coherent picture of an unseen order.

From Scan to Plan

The major vehicle for disseminating these new kinds of pictures of invisible architecture was *Ekistics,* the journal that Doxiadis started toward the end of 1955. The latest attempt to come up with a network pattern often appeared on the cover—as a kind of hidden architecture of the month. Each Delos meeting was given a special issue, and the content of other issues often responded to developments at Delos or inspired them. The lines scribbled onto the onboard blackboard in response to the international scene, and repeatedly modified during the morning debates, were quickly sent back out to an international audience.

The special issue on the first meeting begins with a glossy foldout text of the Delos Declaration with all the signatures, followed by detailed notes on the daily debates interspersed with photographs of the morning onboard discussions, afternoon tourism, and evening entertainment. Intense discourse about the architectural implications of the latest technologies appears against the background of ancient ruins. There are photographs and short bios of each "Delian" and a collage of newspaper reports of the event. The issue closes with a copy of the original Athens Charter. The result is a carefully constructed image of strategic networking, a particular network trying to draw the very principles of networks.

Ekistics is itself a networking instrument. Indeed, it explicitly exaggerates the networking operations of all magazines. It only publishes abstracts of already published texts, repackaging and rebroadcasting existing data. The magazine is a scanning device, constantly monitoring information flow in other magazines. Whenever original material appears, like the annual special issues on the Delos meetings, a special explanation has to be offered, reversing the usual pattern in which journals have to explain the republication of a text. Everything that is picked up in the scan is filtered: abstracted and reframed by editorials and introductions to individual articles. If all magazines are prosthetic extensions of their readers, far-reaching eyes monitoring a distant world for a particular community, *Ekistics* is a precise and efficient instrument.

This relentless networking logic is most evident in the Ekistics Grid, a classification system used by the magazine since January 1965. Everything that is republished in the magazine, discussed in conferences, studied in research projects, taught at the Athens Institute, and even the character of students, is codified as a visual pattern within the grid, a generic frame through which all planetary activity can be monitored. Once again, this is an extension of modernist ambitions, as it is based on the CIAM Grid that Le Corbusier introduced in 1949.[26] And again it is used to further intensify the obsession with networks. Within six months of its introduction, "Networks" was added to the four basic "elements" of settlements ("Nature," "Man," "Society," and "Shells") that it monitors. A year later, even "Networks" was placed into a network with all the other elements. Networks were no longer discrete. Everything was seen to be networked. Even the monitoring grid evolved into a tightly woven network—so tight that by 1971 it was almost unusable. Likewise, the diagrams of Ekistics itself as a networking of diverse disciplines became denser and denser. The field had been overwhelmed by network fever.

In 1972, Doxiadis presented photographs of a spider's web before and after the animal had been drugged with amphetamines. The distorted organization of the doped spider was compared to a map showing "the chaos of networks" in the urban Detroit area.[27] Doxiadis's own design for a neatly geometric system of underground networks of transportation and utilities for the region tried to negotiate a compromise between the existing arrangement and that of an ideal spider. At this point, network fever had him firmly in its grip. As with the latest version of the grid, the central role of the architect was no longer just the form of networks but the connections between them: "We must coordinate *all* of our Networks *now*. All networks, from roads to telephones."[28] The architect is seen as a networked animal that networks networks that are themselves animate. In extending the body, networks have to extend its organic logic. Doxiadis bases design decisions for regional and global systems on the internal operations of the body. The architect elaborates the human body rather than houses it. Designing networks has become a biological necessity.

These associations are classical. The ancient forms of the word "network" were applied at once to the work of humans and that of animals—as in fishing nets and spiders' webs. In the eighteenth century, it was common to use the word to describe the inside of the body itself, as in the organization of veins, muscle bundles, etc., and in the nineteenth century it was a standard label for systems of rivers, canals, railways, cables, electricity, sewers, etc. Finally, it gets applied to organizations of immaterial things like property and groups of people. The word slides seamlessly from biology to technology to society. Any appeal to new networks in the organization of space or society carries some of the original biotechnical association.

Yet it is precisely for this reason that it remains significant that modern architects like Le Corbusier only used the word "network" to describe the old street pattern and the new ones that they proposed. The full biological argument was not used beyond physical form. The key move of the Athens congress of CIAM was precisely to place greater emphasis on the idea of networks. At the first congress in 1928, the key functions of cities were identified as "Dwelling," "Working," and "Recreation." CIAM 4 added "Traffic" (*circuler*—circulation) and gave it a special coordinating relationship to the first three: "The fourth, that of traffic, should have only one objective: to bring the other three into effective communication with one another."[29] Transport networks become an organizing concept. In picking up where CIAM left off, Ekistics simply exponentially increased the role of networks.

The key figure in this escalation of network thinking was the urban planner Jacqueline Tyrwhitt, the editor of *Ekistics* since its first issue and a member of the planning committee of all the Delos meetings. In addition to playing a key role in the selection of the participants, Tyrwhitt was responsible for all organizational details during the events, and she attended every single session, taking the official notes of the discussions and editing them for publication. She typically sat to one side of the lead speaker but rarely spoke. While all heads are up in animated debate, hers is usually down. As "secretary-general," she is at the very center of the Delos events yet maintains a low profile, facilitating the interactions of others rather than displaying her remarkable expertise. Despite being the only person who attended all the meetings other than Fuller, she only reluctantly accepts the role of full participant in the tenth Delos, almost always adopting the stereotypical role of the ostensibly subordinate woman as secretary. Yet Tyrwhitt had a major effect on Doxiadis. It was not just that she was the one who chose and summarized all the articles in *Ekistics* and produced most of his books. Much of his position is actually coming through her, along with many of the key organizational strategies he deployed. The ever-public Doxiadis is unthinkable outside the ever-private Tyrwhitt.

In fact, it is Tyrwhitt who provided the key link with CIAM. Giedion was happy to come to Delos because it was Tyrwhitt who invited him. They had first met in 1947 at the sixth CIAM congress at Bridgewater. Tyrwhitt immediately became an integral part of the CIAM operations, being secretary to the Council of CIAM from 1948 and maintaining a tight circuit of communication between the ring leaders. Giedion, Le Corbusier, José Luis Sert, Walter Gropius, and Tyrwhitt constituted the "committee of five" at the heart of CIAM. Tyrwhitt played the same role at each CIAM congress that she would later play in their Delos descendants, being responsible for organization, communication, notes, and the editing of all the proceedings.[30] She maintained a particularly close working relationship with Giedion, collaborating, corresponding, and translating all the books he published from 1951 onward. For Giedion, coming to the first and third Delos was continuing the project with Tyrwhitt rather than signing on to a new venture with Doxiadis.

Tyrwhitt was a professor in planning at the University of Toronto when Doxiadis first met her in 1954 at a United Nations seminar on housing in Delhi that she was directing. Once again, it was Tyrwhitt making the invitation. And again, the sense of a shared venture was immediate.[31] Shortly afterward, Doxiadis asked if she could help put together a set of relevant readings on third world housing and planning for the use of the branches of his office. She started to do so in 1955, after taking a position at Harvard, and eventually the monthly set of mimeographed abstracts became a full-blown magazine.[32]

The collaboration with Ekistics literally picked up where CIAM left off. Tyrwhitt traveled directly from the final CIAM congress at Dubrovnik in August 1956 for the first of her annual summer-long working sessions with Doxiadis. The Athens Technical Institute was founded in 1958, a year before CIAM formally dissolved itself, and in 1960 Doxiadis was announcing all the major moves he had schemed up with Tyrwhitt. She played a key role in the teaching, research, and publication operations at the institute (especially the Athens Center of Ekistics, which was formally established within it in 1963), and she eventually left her position in Harvard in 1969 to take up permanent residence in Greece. Symbolically, she lived in a hillside house designed by her Harvard and CIAM colleague Jerzy Soltan, with a dome by Fuller and an extraordinary garden of her own design that would eventually be the subject of her last, posthumous, book.

If Doxiadis picked up the CIAM mentality, it was Tyrwhitt who affected the form of that pickup and maintained its trajectory. After all, she was really the supreme networking figure. The majority of the key people invited to Delos were from her own circles of London, CIAM, the United Nations, Toronto, Harvard, and MIT, and they shared her particular concerns. She was invaluable to Doxiadis because she had done it all before: launching teaching programs, conferences, proceedings, and books. Their collaboration was extremely fruitful for both because Tyrwhitt was using Doxiadis to continue her long-standing project just as much as he was using her for his.

Digital Traffic

The key move at Delos was to take the CIAM argument in the direction of electronics—starting with McLuhan's announcement on the second morning of the first Delos boat trip that electronics presents new challenges to planners because this latest prosthetic extension of the body defines an entirely new form of space. Tyrwhitt was yet again the link, having been a member of McLuhan's inner circle in Toronto since the end of 1952 (after Giedion had written a letter of recommendation to McLuhan).[33] From 1953 to 1955, Tyrwhitt was one of McLuhan's four colleagues working under a Ford Foundation grant to carry out an interdisciplinary study of the effects of the new media, the key project out of which McLuhan's famous arguments would emerge. It was while she was in the middle of that project and was one of the associate editors of *Explorations,* the group's magazine, that she met Doxiadis. The radical position that McLuhan brought to Delos in 1963, and would not become internationally renowned until the following year, had been very familiar to Tyrwhitt for a long time. They had been discussing the media's transformation of the world into "one city in space" since the mid-fifties. In December 1960, McLuhan's letter responding to Tyrwhitt's suggestion that he take a position at Harvard spells out the view that the traditional city has been displaced by the electronic extensions of the body that have constructed a "global village" in which traditional conceptions of space have been overturned. This already existing electronic village calls for the construction of a new form of physical world city by planners—"the job is to create a global *city,* as center for the village margins"—and McLuhan speculates that this post-Euclidean city will have to be assembled by computer in the same way that airports use computers to coordinate flights.[34]

When making these same points three years later at Delos, McLuhan was supported by Fuller and two former associates of Tyrwhitt, the planners Edmund Bacon and J. Gorynski. They agreed that the "electronic scale" had to be integrated with the human scale and that the latter could actually be maintained within man's "electronic extensions."[35] CIAM thinking had to be retooled for an electronic world. The Delos meetings would turn the CIAM idea of settlements held together by transportation networks into the idea of inhabitable information networks. But it took a while. The final point of the first draft of the *Delos Declaration* did include a reference to the concept of electronic extensions and its effect on the emergence of urban form.[36] But it did not survive the editing process. It is not until the fourth Delos of 1966 that the whole meeting accepts the basic point of Fuller, McLuhan, and Mead that communication networks have produced a single planetary society—that it is no longer possible to research the city without discussing electronics.

The trajectory from the physical city to the electronic one was even more evident in *Ekistics,* through the strategic selections of Tyrwhitt and her unsigned editorials. Again, the first Delos appears to have acted as a catalyst. Momentum builds through the gradual accumulation of individual articles and then special issues on communication. Diverse media, including telephone, radio, television, telex, cable, closed circuit, and satellites, are analyzed. Particular attention is paid to their influence on the third world, where their transformative effect is most pronounced. The journal steadily and increasingly radically explored the displacement of the physical. There was a continuous feedback loop between the journal and the Delos events. On the one hand, the journal presented detailed studies into the questions raised each year at Delos. On the other hand, most of the participants appeared in the journal before being invited to participate. Between the conferences and the journal's monthly scan, an extraordinary discourse about the architecture of electronics developed.

In fact, the journal's concern with electronics precedes the first Delos. The computer, for example, had been a theme in the magazine since the late fifties, starting with the editorial of the August 1959 issue that describes the usefulness of computerized analysis of data on punch cards and the graphic representation of that analysis on computer monitors, when introducing two articles on the "science fiction"-type machines that do this.[37] Doxiadis started using computers in 1962 to develop mathematical models of settlements, and after the first Delos, he wanted everything computerized and proudly published photographs of each new computer installed in his office. At the beginning of 1964, the report on the research projects at the Athens Institute said that the main emphasis was now on computer programming, data processing, and methodology. An "Electronic Computer Center" was set up in the office, and the ground floor of the building was actually used as a computer training center once Doxiadis discovered that no one in Athens was qualified to run the machines. The following year, *Ekistics* featured a special issue on "Architecture and the Computer," discussing computer design of buildings, computer conferencing, and so on.[38] The journal itself was soon computerized, with ever more detailed indexes becoming computer printouts, and in 1969 a special issue was needed on "Computers in the Service of Ekistics." By then, three shifts of workers were employed twenty-four hours a day to type in statistical data on cities in a room filled with punch card machines.[39] The more philosophical discussions of the architecture of electronics at Delos were paralleled by their ongoing practical use.

The appeal of the computer is that it offered a new viewing point to survey the explosive rise of ever larger and less visible networks. If the uncontrolled growth of the city had first demanded the surveillance view from the airplane and then the view from outer space, the growth of invisible networks demanded new scanning instruments. The computer was the ideal mechanism to negotiate between the visible and the invisible. The computer is both a means of diagnosis and a symptom, both a mechanism that reveals hidden patterns in an overwhelming conglomeration and one of the forces that dematerializes or transforms the occupation of that physical organization. Ekistics oscillated between using electronics to expose hidden circulation patterns and producing images of hidden electronic patterns.

Basically, Ekistics radicalized the logic of traffic and moved it into the world of electronics. At the fourth Delos trip, it was argued that "modern transport networks extend far beyond the visual horizon."[40] This shift became clear in "From Man's Movements to His Communications," the May 1970 special issue of *Ekistics,* whose foreword speaks of the "more complete move from a mechanical to an electronic environment." At the Delos meeting of that year, it was argued that physical transportation might go away when moving ideas replaces moving bodies. This had become a mantra by the June 1973 special issue on "Networks: Information, Communication and Transportation." The editorial insists that they are inseparable, and the cover conveys McLuhan's basic point by showing the progressive exponential shrinkage of the world to a small point. Electronics is the new form of traffic and therefore the new form of the city.

In the end, this was the key move with which Ekistics transformed the CIAM mentality. Le Corbusier, for example, was acutely aware of the role played by new systems of communication like the telephone and often refers to it.[41] He even had images of the dense weave of the international telephone network in his files in the twenties, but *Le Ville Radieuse* of 1930 only uses the word "network" to refer to the visible traffic patterns of the city.[42] When the Athens Charter ends by referring to the inadequacy of the "existing network of urban communications," the issue is likewise only physical traffic. Doxiadis, Tyrwhitt, and their friends set out to multiply the concepts of traffic. Traffic was conceived as information flow. Symptomatically, drawings of cities, continents, disciplines, and computers tended to be the same.

The Biology of Information

The new traffic of electronic exchange was seen as biological, once again hyperextending parts of the established discourse of modern architecture rather than abandoning it. A blurring of biology and information occurred throughout the Delos meetings. At the signing of the first *Delos Declaration,* Conrad Waddington, a renowned animal geneticist who would be a central figure in most of the meetings, said that the type of world city being envisioned by the group would be "a new level of organization of the living material of the universe."[43] This echoed the discussions of the second morning when, in response to McLuhan's first statements about media extensions of the body, he joined with geographer Walter Christaller and psychiatrist Leonard Duhl in inventing urban schemes by comparing the evolution of electronic networks to that of animals:

> As animals become more complex they develop increasingly differentiated limbs and organs and a highly efficient communications center. Should we move towards a newly constructed type of organization with highly differentiated centers tied together by a complex communications network, each center having special functions and a special location?[44]

Waddington, whose books Fuller had followed closely since the late forties, read evolution in cybernetic terms. The growth of electronic systems of communication is biological in exactly the same terms that biological growth is itself an evolution of systems of communication. Waddington relentlessly applied this model at most of the Delos meetings, often using spiders' webs as a model system of organization.[45] He was supported by Margaret Mead, who had been a pioneering member of the key group that had established cybernetics after the war. At the second Delos, they got further support in connecting this model to architectural form from Richard Meier, who had published *A Communication Theory of Urban Growth* in 1962, a cybernetic account of the city as a living organism and information system to be analyzed in biological terms.[46] Meier had come to Tyrwhitt's attention when he was writing the book in 1959–60 at the Joint Center for Urban Studies at MIT and Harvard, of which she was a member. He became one of the leading researchers in the City of the Future project. Already in a January 1962 discussion of the project, Doxiadis said cybernetics and information theory would be necessary,[47] a point repeated at the very end of his 1963 report on the project just before the first Delos, but apart from referring to the city as an "organism" he was not yet talking that much about either biology or information.

In fact, the biological argument rose very slowly in Doxiadis's writing. This is most obvious in the City of the Future project, which gradually becomes a vast prosthetic. Working closely with Meier, Doxiadis rationalizes the new city in 1964 with Halleresque charts showing the exponential increase in "the average speed of man's displacement since 10,000 BC" and "the extension of man's vision through mechanical means."[48] He eventually starts talking of the possibility of developing new organs for the city. The biological organism is capable of improvement. The genetics can be rearranged. A new body can and should be developed. Cities can be helped to reach an ever higher biological order. Networks, particularly electronic ones, are the means of this upgrade. Doxiadis used computerized traffic control as a model for cities to be higher-order biological individuals than plants, animals, and humans.[49] With each year, he went deeper into the prosthetic logic.

The Delos discussions clearly had an effect. McLuhan's initial image of prosthetic growth was elaborated in more and more detail as the annual boat trips gradually embraced the centrality of electronics. At the second Delos, for example, Fuller reasserted his old line about prosthetic networks in a lengthy argument about the way computers augment the human brain, before rolling around on the floor to make his point about synergy. He concluded that the human's "externalized organics (the world industrial network)" will eventually become as unconscious as the automated operation of internal organs.[50] In immediate agreement was sociologist Edward Hall, who had been invited to the second Delos after McLuhan suggested it to Tyrwhitt and had located architecture as one of the tools within the array of bodily extensions in his 1959 book, *The Silent Language*. McLuhan was citing the passages in his latest book and identifying them as the inspiration for his use of the prosthetic argument. But Hall in turn had been inspired by Fuller, having been a close friend of the Fullers ever since he became a college teacher of their daughter Allegra. Margaret Mead was likewise no stranger to the prosthetic argument from her years in the cybernetics debates. This was a very persuasive group, and by the eighth Delos everyone was able to agree with them that "information systems today have more power in social systems than ever because computers have magnified the capabilities of the human senses." And at that point, the discussion of prosthetics had become extreme, embracing genetically grown limbs, "brain extension" systems, and so on.

Tyrwhitt's journal relentlessly pursued this more radical view of prosthetics, embracing engineering psychologist J. C. R. Licklider's theory of future "symbiosis" of human and computer and his drawings of the possible interface between the two organisms, with the human network entangled with the electronic.[51] In this line of argument, it is not so much that the latest technology has constructed a new world for us to inhabit. The global city is the global body. We inhabit our own hyperextended body. When Ekistics calls for a redesign of networks, it is calling for a redesign of the human body—network, city, and body being the same thing.

This equivalence of prosthetics and architecture is exemplified in a 1969 essay in *Ekistics* by engineer Koichi Tonuma analyzing contemporary flows of information, with charts of telephone and telex communication within the main island of Japan, to predict the transformation of the island into a single continuous "living space," a vast urban network that looks like a nervous system. The invisible electronic lines connecting people become the matrix on which a visible biomorphic form emerges. The necessary correlate of this biological vision of electronics and buildings is a technological vision of the human body, when "our organs are being replaced with artificial tools."[52] It is the confusion of the body and its extensions that explodes a single biotechnical infrastructure across the landscape. Tonuma presents a sequence of drawings, showing settlements gathering size and complexity from single cells to complete biological organisms that are similar to those of Doxiadis. In fact, Tonuma spent two years in Athens doing research at the center. Japan played a key role in Ekistics, as a model for contemporary statistical trends and as a site for imagining radical futures, featuring prominently in the City of the Future project from the beginning.

Yet this biotechnical vision is not simply projected onto Japan as a unique experiment of the Athens laboratory. Tonuma's whole argument, even its prosthetic aspect, is coming from specific architectural proposals made in Japan in the early sixties. All it does is to add particular statistical readouts of information flow and particular drawings of biological cells to an existing scheme that had its own flow readouts and cell drawings. It is crucial to remember that Ekistics is only ever a networking operation, a scanning mechanism coordinating and editing already existing ideas, not just in the sense of design as statistical analysis of given information rather than artistic innovation, but also design as the recirculation of tested strategies. The network fever in Ekistics can be found in the work of numerous architects. Indeed, the fever was endemic to architectural discourse during those years. The specific contribution of Ekistics was simply to relentlessly monitor it and thereby feed it.

Nerve Design

A major accelerant of the fever was Kenzo Tange, the preeminent Japanese architect. As one of the passengers on the fourth Delos boat trip in 1966, for example, he addressed the "tentacles" of the communication network in biological and evolutionary terms. To inhabit the modern city is to inhabit the information system of an artificial brain:

> Society is evolving into a more advanced state, as plants evolved into animals, and animals into men. We have begun to create a new nervous system in society using the advanced communication technology that will enable the social brain to function more effectively. In large contemporary urban complexes, communications networks twist and interlink into a complex which must be something like the nervous system of the brain.... whirling around in these brains are the people and the information. The citizens are like electrons flowing in an electronic brain.[53]

Tange drew on cybernetics to discuss the influence of all the contemporary systems of communications—arguing, in McLuhanesque fashion, that there has been a second industrial revolution, an information revolution that prosthetically extends the nervous system in the same way that the first one physically extended the body.[54] He predicts that Japan can only maintain its "organic life" by eventually turning into a single colossal city through the linkup of physical, social, and information networks into a single "central nervous system."

Tange's spoken statements were actually taken from an article he had published the year before, entitled "Tokaido-Megalopolis: The Japanese Archipelago in the Future," which presented charts of the "evolution" of media use (telegraph, telephone, radio, and computer) and maps of flow in each "communication network" of Tokaido (rail, car, mail, and telephone) before drawing the island as one colossal biomorphic city.[55] His drawing of the shape of this new organism is similar to those that Doxiadis had been publishing since 1962, but Doxiadis's idea of network form had itself been informed by the earlier work of Tange. The July 1961 *Ekistics* had devoted an unprecedented ten pages to Tange's renowned Tokyo Bay project of 1960, in which a vast floating linear "network of elevated lattices" blurring traffic and building is grafted onto the radial network of the traditional city, and the whole organization is "tied together by the invisible cords of a communication system" (telephone, radio, portable telephone, video telephone). Alongside images from biology textbooks of the growth of spines, Tange describes the project as an "organism" precisely because "communication is the factor that gives organic life to the organization." The paradoxical rationale of the network is that the possibility of infinite extension actually produces density. In an argument that resembles that which McLuhan will present three years later in the first Delos, Tange insists that the capacity of networks to extend anywhere actually produces the need for concentration: "People say that organization man is alone, but even more alone is the man who is separated from this network. It is in order to connect themselves to this network that people gather in the cities."[56] Tange was an important reference point in taking the discourse about urban networks toward electronics.

Tyrwhitt was very familiar with this discourse; Tange's work had been shown by others at CIAM 8 in 1951, and he had taken part in the 1959 meeting in Otterloo where CIAM dissolved. Her editorial on the Tokyo Bay scheme said that the project was the direct outcome of experimental plans that he developed with MIT students in 1959, which she had seen firsthand and had already published, along with his Boston Harbor project, the scheme based on the growth of plants that is generally accepted as the first move in the so-called megastructure movement.[57] At the third Delos in 1965, Giedion singled out the Tokyo Bay project when embracing "the youngest generation" for two concepts that handle variable density: "megastructure" and "group form."[58] Tange exemplified the former, while the latter was introduced in the first Metabolist manifesto of 1960 by Fumihiko Maki, the architect who was also the first to publish the term "megastructure," with Tange as the central example, in a little-known 1962 publication that was immediately republished in *Ekistics* because Maki was teaching with Tyrwhitt at Harvard from 1962 to 1965.[59] As a parallel line of research into networks, all the Japanese experiments were closely monitored by *Ekistics* and had their effect on the field's evolving doctrine.

Another parallel trajectory monitored by the journal was that of Team 10, the dissident younger faction of CIAM. Alison and Peter Smithson, for example, had been designing webbed urban projects since the late fifties, including the extremely influential Hapstaudt Berlin scheme of 1957.[60] Already at CIAM 9 in 1953 they had been insisting against their elders that "the street and the network of streets has to be seen as the arena in which social relationships were played out" rather than a mode of efficient connection.[61] Georges Candilis and Shadrach Woods likewise developed huge "mat buildings" as infrastructural weaves of movement patterns and wrote key articles on the principle of "the web" in 1961.[62] Aldo van Eyk saw the role of the architect to provide a "network of crevices." And so on. A different attitude toward the network—both in terms of physical form, social structure, communication system, and analytical concept—was precisely what differentiated the young group from CIAM and led to their separation from it and the subsequent dissolution of the old organization. As Alison Smithson (who seemingly took over Tyrwhitt's role) puts it at the very beginning of her account of the separation, it was decided by the older generation at CIAM 9 that "life falls through the net of the four functions" but "we wanted a more delicate, responsive, net."[63] When the group became independent, much of the talk was about networks. The 1962 meeting of Team 10 at Royamount, for example, is dominated by it, as exemplified by Stefan Wewerka's description of cities as "compact bundles of overlaid net-structures."[64] A major influence in this discourse, as it was for the parallel Japanese experiments, was Louis Kahn and Anne Tyng's 1953 traffic scheme for Philadelphia, which dematerializes the physical form of the city in favor of pure flow, like an electrical circuit. Streaming arrows become more solid than buildings. The image was a key reference point. The new generation of architects was under the spell of its radicalization of CIAM's long-standing commitment to traffic. But the next step of blending physical network and information flow would only start to become evident in the work of an even younger generation.

The bio-informational language that Tange had used since the Tokyo Bay project had actually come from his assistants on the project: Arata Isosaki, Kisho Kurokawa, and Sadao Watanabe, some of whom were part of the group setting up the Metabolist movement, presenting their founding manifesto with images of organic cell development in the same year at the World Design conference in Tokyo. The group would write extensively on biology, symbiosis, cyborgs, cybernetics, and prosthetics throughout the sixties.[65] Their projects were drawn as delicate systems of intersecting fibers—architecture as biological circuitry.

In fact, all these groups were networked together. Tange had shown the first Metabolist projects (Kiyonori Kikutake's 1959 Marine City and Cell City, "a complete network of living facilities," based on a "move-net" in which fixed structures allow building units to "grow and die and grow again") alongside his own megastructure at the Otterloo meeting where Team 10 began its independent life. The Smithsons included Tange in their Team 10 survey for *Architectural Design* in May 1960; and the conference in Tokyo of the same year, in which the Metabolists launched themselves with Tange speaking as the father figure, was also attended by Kahn and the Smithsons. The younger architects became linked to the evolving Team 10 discourse. Maki was invited to the Team 10 meeting in 1960, and Kurokawa would talk about "nets" at the 1962 and 1966 meetings. A global network of experimental architects devoted to networks was established.

Most of the network organizations dreamed up by these groups were periodically scanned by Tyrwhitt in *Ekistics* as a kindred post-CIAM research.[66] When Team X had started its assault on the older generation at the last CIAMs, Tyrwhitt had initially resisted on behalf of the rest of the central committee, but she eventually embraced the idea of handing over to the next generation. When doing so, she circulated a text to Giedion by her close colleague McLuhan on the need for interdisciplinary research to open up unknown horizons through a cubist multiplicity of viewpoints.[67] The impact of electronics represented a shared threshold. The Smithsons, for example, had been in the middle of intense discussions of the new systems of communication in the early fifties with their Independent Group colleagues in London, very much under the influence of McLuhan's first book, *The Mechanical Bride*. Yet electronics was never an overt feature of their projects, nor those of their colleagues. This would be the task of their respectfully rebellious students.

Invisible Pictures

It was only with the post-1963 work of the young Archigram group that information flow became visible as such. Where the Metabolists emphasized the biological side of the biotechnological equation, Archigram emphasized the technological. Architecture became indistinguishable from communication. Warren Chalk and Ron Herron's City Interchange project of 1963 is just a "net" of intersecting forms of traffic, including invisible traffic: "electronic data transmission, traffic control and administration, radio-telephone tower, communication and news service relay station, intercommercial closed circuit television hook ups, public television and telstar rediffusion center."[68] This principle underlies all the subsequent Archigram work and starts to take a particular form, as can be seen in the 1964 projects that *Ekistics* scanned in 1965. What counts in Ron Herron's Walking City, Peter Cook's Plug-in City, and Warren Chalk's Underwater City is movement in a diagonal net. In Walking City, it is the usually overlooked network of diagonal links between the huge mobile animals that makes the system possible. Plug-in City is likewise a "giant network-structure... with diagonals of lifts making up the grid," and in Underwater City, to leave the diagonal structure/movement system is of course to drown. In each project, the diagonal weave becomes the main event. Activity occurs within the net itself.

Even dense, blurry, psychedelic events like the Instant City "traveling metropolis" of 1968 are actually based on a triangulated network plan covering England. Each of its intense explosions of sound, smell, and color occurs on the node of a net or constructs such a node. A series of six drawings showing an Instant City descending on a "sleeping town" concludes with one entitled "Network Takes Over." The apparatus has moved on, but the infiltrated town has become hooked up to all the others by landline and wireless transmitters. Already in 1963, the group raised the possibility that expendable and flexible communication networks would invalidate fixed physical ones,[69] and projects like Peter Cook and David Greene's Ideas Circus, published by *Ekistics* in August 1969, were "offered as a tool for the interim phase until we have a really working all-way information network."[70] When Archigram folds soon afterward, it is not by chance that Peter Cook starts the Art Net forum in London and names its journal *Net*—unconsciously echoing McLuhan's 1951 proposal to start a newsletter called *Network*.

In all these projects, the grid gives way to the web. Movement in the spaces defined between intersecting lines gives way to flow within lines. Triangulation rules. Of all of Doxiadis's hundreds of charts, he kept presenting one from 1962 that showed that grids were the most efficient form of network at the smallest scales of rooms and neighborhoods, that hexagons made sense in the local region around a center, and that triangulation is best at the largest scales. This combination of orthogonal grid and triangular net can be seen in all his schemes. But the parallel research by other architects had moved triangulation into individual buildings. The 1952–58 project for Tomorrow's Town Hall by Kahn and Tyng was particularly influential in this. Well known since its publication in *Perspecta* in 1953 and *L'architecture d'aujourd'hui* in 1954, the building was conceived as a series of triangulated structural systems operating at different scales. Appearing from a distance, in Kahn's words, as "a lacey network of metal,"[71] it even stood in the center of a plaza marked with a triangulated networked pattern. Another key source of inspiration was A. and J. Pollack and A. Waterkeyn's 1958 Atomium for the Brussels World Fair, which appeared in *Archigram 4* alongside the City Interchange. The Atomium, in which people move up inside the diagonal links to occupy the spherical nodes, has exactly the shape of Doxiadis's drawing of the basic principle of networks. Triangulation is at once identified with the micro scale of atoms and the supermacro scale of transplanetary connections.

The application of a physical image of a global network at all scales became polemically clear in the schemes and systems in which there is no difference between building and extended web, notably in a series of projects published at the end of the fifties: Fuller's demonstration of an Octet-truss at MoMA; Constant Nieuwenhuys's "wide world web" New Babylon; and Eckhard Schultze-Fielitz's Space City. Each of these was influenced by Konrad Wachsmann's enormous Airline Hanger for the U.S. Air Force that was first published in 1954 (and immediately made its way into a section on space frames in Giedion's *A Decade of New Architecture* that was edited by Tyrwhitt in the same year).[72] It became a major influence on the Japanese architects, as all the young architects who would later form the Metabolist group attended Wachsmann's lecture on the project in Japan in 1955, a lecture organized by Tange. Yet all these practical schemes by an international community of architects, including Wachsmann's, can be seen as a compromise of the more radical web that Wachsmann had developed in 1953 with students at Chicago. It envisions a structural system that refuses any difference between the horizontal and vertical strands of the web and the intersections between them. Structural threads are simply twisted together in an endless fabric.

It is symptomatic that information flow is crucial for all these web designers. Wachsmann and Fuller speak of the way communications can activate any point in their systems. Constant and Schultze-Fielitz argue that electronics will change the shape of their spaces, with computers continuously rearranging the forms on the basis of constant feedback from the occupants, an idea that Archigram's Dennis Crompton would take to the extreme with his Computer City project of 1964, in which the city itself is nothing more than a computer, a hardwired "sensitized net" with "local net feedback." When it was published a year later in the November 1965 *Ekistics*, the short explanatory text asserted, "The activities of an organized society occur within a balanced network of forces which naturally interact to form a continuous chain of change."[73] The mesh of links activated or modified by electronics has turned into an endless mesh of self-adjusting information channels.

Indeed, it can be argued that the whole obsession with triangulated space frames in the sixties, and even the concern with building systems as such, was just an attempt to make poetic images of the invisible communication infrastructure whose influence had grown throughout the century—a visible aesthetics for the invisible net. In 1966, Tange, who would polemically have himself photographed against a dense weave of triangulated scaffolding to open his monograph, said exactly that: "Creating an architecture and a city may be called a process of making the communication network visible in a space."[74]

It was not by chance, then, that the first triangulated space frame was actually produced in 1902 by the same person who invented the telephone: Alexander Graham Bell.[75] Fuller's more famous webbed space frames, hovering as light as they could be, as little in the world as possible, likewise came after his fascination with nonphysical technologies of communication. Already in his first book of 1938, Fuller, who would eventually be given Bell's original tetrahedral models by his great-grandchildren, was using the "inter-communicating web" of the telephone as a model for housing and was describing the house as an apparatus for receiving and broadcasting. Fuller always referred to his structural systems as "nets," understood not as systems of physical interconnections but as networks of energy flow, information systems—an association between visible and invisible net that would finally become literal in his dome for Expo '67, the first computer-controlled structure, with each metal link in the net carrying the wiring for a continuous adjustment of the color, opacity, and porousness of the building's surface.

All the webs that proliferated in the sixties, including Doxiadis's City of the Future, were likewise an attempt to establish a physical image of the invisible space of electronics, even if electronics itself is not discussed. All the projects by Tange, the Metabolists, Team X, Archigram, Constant, and others were practical and even took their character from their engagement with the pragmatics of construction, but they were first and foremost polemical images—and were presented as such. It matters little that virtually nothing from all those experiments was built. Or, to be more precise, what was carefully built was a set of images that remain polemical today, a commentary on the networks we already inhabit rather than a dream of a future world.

It took decades to forget such experiments so that a new generation could present itself as the first to engage seriously with the architecture of electronics. Much of what we hear today is an echo—but so delayed that it sounds fresh. It is as if the discourse forgets its own history precisely because it is too afraid to leave those earlier positions behind. Supposedly avant-garde visions manifest the discipline's greatest fears.

Unsettlement

The point here is that McLuhan's influential discourse about networks during the sixties was exactly paralleled by that of experimental architects during the same years. And the architects did not simply follow the communication expert. Rather, they all followed an even earlier generation of designers. Indeed, McLuhan's work begins as a kind of rethinking of architecture. It was his close alliance with Giedion and Tyrwhitt that opened up a new way of reading the space of technology, one that leaned heavily on Fuller.

As the key link between the twenties and the Internet, Fuller played a crucial role. His work was first monitored by *Ekistics* in 1957, but his Dymaxion Map showing the planet as a single network was on every cover from the beginning until mid-1959 and occasionally reappeared. Fuller was president of the World Society for Ekistics and, remarkably, was the only person other than Tyrwhitt to attend all the Delos conferences. Not by chance did another version of his map return for the cover of the issue of the Delos on "Networks" in 1970. Delos was now positioned at the center of the triangulated map, and radiating lines show where all the participants had come from. It is as if Ekistics occupied Fuller's world. His vision of a fundamental continuity between visible and invisible architecture had always lurked in the background and slowly took over. Making the same argument from the side of communication, McLuhan was a crucial ally. The two hovered over the discourse in the same way that they hovered over all the experimental architects obsessing about networks in the sixties.

Yet Fuller and McLuhan pursued such a radical line that even those deeply infected by network fever could not handle it. Already by the fifth morning of the first Delos, Fuller was saying that the idea of permanent settlements and neighborhoods is obsolete in the contemporary hypermobile age. The very idea of settlement so treasured by Ekistics is challenged in a time characterized by "stirring up rather than settling down."[76] McLuhan was quick to agree and even suggested that the whole framework had to change: "Are we selecting as key problems things that are possibly about to disappear with the rise of information levels, such as congestion and confusion?"[77] Fuller both mentored the group and criticized it. In 1966, he wrote "the longest letter I have ever written" to Doxiadis outlining his "general strategy." Published in 1969 as "Letter to Doxiadis," it repeats his prosthetic account of architecture and insists that electronics will lead the way. The latest computer techniques will render conventional architecture and planning obsolete, substituting static urban planning with "Instant city!" New world networks foster a hypermobility of bodies and spaces—technologically upgraded and endlessly circulating bodies being the new spaces.[78] The new biology of technology doesn't leave the kind of fixed trace Doxiadis tried to establish.

After all, there had always been a fundamental conflict between Fuller's original call in the late twenties for a physical disconnection from infrastructural networks and the Ekistics obsession with establishing vast physical nets. Fuller shares the commitment to a single world city. After all, his very first project in 1927 was for a "one world town," but everything in it is mobile and physically disconnected. Buildings are dropped by airplane, have autonomous service systems of plumbing and electricity, and are only interconnected by invisible air and radio links. Fuller rejected physical infrastructure, no matter how flexible, preferring atomized nomadic systems. For him, the capacity to disconnect from a system was as important as the capacity to connect. He was fatally attracted to Ekistics, attending every event and publishing articles like "Why I Am Interested in Ekistics,"[79] but ultimately he had to go beyond it.

At the tenth Delos in 1972, Fuller pointed out that the group was not yet ready to deal with the fact that the boat in which they were talking was actually filled with the signals of over a million radio stations.[80] At the same meeting, McLuhan also acted as the dissident by presenting James Joyce's *Ulysses* as "the greatest piece of city planning and building in this century" and rock music as "an enormous world-wide network of culture which directly relates to the health of human settlements" since it processes the sounds of the city.[81] He insisted that books and music fossilize buildings. Cities should be designed less for occupancy than for performance, a "global theatre" built out of the hidden electronic networks. Tyrwhitt kept her distance from her old teammate, publishing his intervention in a section entitled "Communication via Humor." The earnest reconfiguration of the global city as a physical image of an invisible order was not to be held up by the thought that physical configurations were already redundant. In letters to friends, McLuhan concluded that the other Delos participants were "earnest men, rather all 19th-century types, still preoccupied with bricks and mortar" and that people already existed electronically in "a new kind of world city far outside the keen of Doxiadis."[82]

Fuller had meanwhile become excited by the idea of strategic unsettlement and had elaborated an even more radical defense of it at a conference in the Bahamas organized by McLuhan.[83] The odd couple was once again in the sun, entertaining each other by taking every argument to the next level. Ekistics was losing its appeal. At the beginning of the last Delos in 1975, Fuller mourned the death of Doxiadis just two weeks earlier. When the meeting was finished, he went dancing with "young" people in his dome at Jacky Tyrwhitt's house for the last time.[84] A year later, he did attend a United Nations conference on habitat as the representative of Ekistics, but he published his contribution under the title "Accommodating Human Unsettlement," arguing that it is precisely the stability of unseen infrastructural networks that makes global physical instability possible and desirable.[85] The global village supports a hypermobility of people and architecture. Designers are to aim for "formless" systems of unsettlement rather than overcome them.

Such thoughts rarely appear in the otherwise faithful but unwitting echo of the sixties that occupies so much architectural discourse today. As the computer is rediscovered on saliva-drenched glossy pages featuring the excited commentary of breathless critics, networks are portrayed as playgrounds of the future. Young designers are persuaded that they are pioneer explorers, shockingly oblivious of how well traveled are their paths and how many architects went so much farther. In the face of destabilizing forces, the romantic figure of the architect as stabilizer is reasserted. Now digital architects have moved into housing, with competitions on the virtual house, house forms inspired by information flows, mass-production techniques for infinite variations of housing forms within generic parameters, and so on. Electronic space is being settled. The architect is yet again a figure of order, of pattern within chaos, of comfort. The architectural species has survived by ignoring a century of intense discourse about networks. In a kind of Warholian dream, every echo has become an original artwork.

Notes

This essay is part of a research project on the prehistory of virtual space, which has been supported by the Graham Foundation. It was first presented as the Myriam Bellazoug Memorial Lecture at Yale University, February 12, 2000.

1. The proliferation of electronic forms of public space has been matched by the proliferation of highly restricted private networks. New forms of inaccessibility breed under the cover of our utopian image of infinite mobility. Half aware of this, since by definition one cannot be fully aware of the spaces one cannot enter, we all too eagerly celebrate the new forms of connection.
2. "On a certain day, Bucky gave an afternoon presentation. [Norman] Cousins says: 'Then they adjourned for dinner, but Bucky kept right on talking, not eating himself, and resumed in the saloon after dinner. Then he walked with the individual participants to their cabins. Finally he ended up with Doxiadis and Jim Perkins in the former's cabin long after midnight.... Perkins and Doxiadis, totally exhausted, looked at this man, who by now had been on this talking marathon for nine hours. You can picture it—Doxiadis getting undressed, Jim Perkins slumped down in a chair, and Bucky sounding off his thoughts, exhilarated, fresh, energetic. And there was no doubt in either Doxiadis's or Perkins's minds that he could have kept on through the entire night, and he would still be fresh.'" Albert Hatch, *Buckminster Fuller: At Home in the Universe* (New York: Crown, 1974): 234.
3. Letter from Buckminster Fuller to E. J. Applewhite, July 10, 1973, in *Synergetics Dictionary: The Mind of Buckminster Fuller*, ed.. E. J. Applewhite (New York: Garland, 1986): 592.
4. "I realize of course that this is a rather belated time to invite someone on such short notice but I want to say frankly that I have just finished reading your wonderful book 'Gutenberg Galaxy,' in which I found so many of the things that we also believe in and so many of the ideas which I think are relevant and essential to human settlements and their problems." Letter from Constantinos Doxiadis to Marshall McLuhan, May 20, 1973, cited in letter from Marshall McLuhan to Stewart Bates, June 17, 1973, in Matie Molinaro et al., eds., *Letters of Marshall McLuhan* (Toronto: Oxford University Press, 1987): 289.
5. McLuhan to Bates, June 17, 1973.
6. McLuhan noted that, "The electronic age with its tremendous speedup of communications has created a situation of 'implosion' rather than explosion. The technological age extended man's physical senses enormously, the electronic age is now extending the nervous system. Technology separated our different functions and distributed them widely in space; electronics fuses them together and overlays them. We have all become totally involved and—in terms of communication—the whole globe has been compressed to the dimensions of a village. This global extension of the human brain is as involuntary as seeing when one's eyes are open. It represents a new kind of continuous learning and an enormous upgrading of man. The task of the planner is to prepare the environment for the exploitation of this new tremendous opportunity." Notes from the second meeting, July 8, 1963, "Delos Documents," Avery Classics Collection, Columbia University. McLuhan's statement was reproduced, without identifying the date of the meeting, in "The

Delos Symposium," *Ekistics* 16 (October 1963): 206. McLuhan continued in a similar vein: "Electronic technology has extended the brain to embrace the globe; previous technology had only extended the bodily servants of the brain. The result now is a speedup of information that reduces the planet to the scale of a village—a global consciousness thus becomes the new human scale." Ibid., 257.

7. Letter from Fuller to Applewhite, July 10, 1973. The prosthetic argument also appears in *No More Second Hand God,* as when Fuller refers to "irreversible physical evolution technologically extrapolated as extra-corporeal simplex or complex of tooled man-process extension and augmentation." Buckminster Fuller, *No More Second Hand God and Other Writings* (Carbondale: Southern Illinois University Press, 1963): 77.
8. "Have you encountered the work of Ed. T. Hall? He says he got the idea of our technologies as outerings of sense and function from Buckminster Fuller. I got it from nobody." Marshall McLuhan, *Sheet* (privately circulated newsletter), February 27, 1962, in Molinaro et al., eds., *Letters of Marshall McLuhan,* 287.
9. "McLuhan has never made any bones about his indebtedness to me as the original source of most of his ideas. The 'Global Village' indeed was my concept. I don't think he has an original idea. Not one. McLuhan says so himself. He's really a very great enthusiast, a marvelous populariser and teacher. He has an irrepressible sense of the histrionic, like no one I've known other than Frank Lloyd Wright.... My concept of the 'Mechanical extensions of man' is the basis for his talk of the 'Electrical Extensions' of man.... McLuhan has always been the first to say 'Bucky is my master. I am only his disciple.'" Letter from Fuller to Applewhite, July 10, 1973. "Regarding McLuhan, I have known him for five years. He acknowledges use of my concept and phrasing of the 'Mechanical' and other 'Extensions of Man' which was first published in the 'predictions' in my preface to the *Nine Chains to the Moon,* Lippincott, 1938, and also in my charts in 1938 and republished in my book *The Epic of Industrialization,* written in 1940. I speak about such phenomena as a scientist, McLuhan speaks as a Professor of Literature. He is well read and has good insights... [and] he is skilled in verbal dueling.... I greatly enjoy his foot and rapier work. I have been present when hostile audiences thought they had him on the run only to discover themselves chasing themselves up dead-end alleys as he himself reappeared far down another highway. I like him, personally, respect him and appreciate the respect and friendliness he shows toward my own work." Letter from Buckminster Fuller to John Ragsdale, editor of the *Biophilist,* November 7, 1966, in Molinaro et al., eds., *Letters of Marshall McLuhan,* 308.
10. The map appears on the cover of *Ekistics* 12 (July 1961). By 1956, Doxiadis had five hundred colleagues working in six countries and two continents. By 1961, the branches of his office were working in every continent. By 1966, "our responsibilities, commissions and colleagues had increased even more, to the point that growth had to be controlled in order to avoid creating a mammoth organization." C. A. Doxiadis, "Fifteen Years of Life," *DA Review: House Organ of Doxiadis Associates, Consultants on Development and Ekistics* 3 (January 1, 1967): 1.
11. C. A. Doxiadis, *Ecumenopolis: Towards a Universal Settlement,* Document R-GA 305 (Athens: Athens Technological Institute, June 1963): 116.
12. "When the group visited ruins of the ancient cities of Miletus and Priene, in Ionia, they were reminded that they were in the birth place of western philosophy, the place 'where the first rational myth was born, the myth which gave wings to man's mind and power to his hands to conquer and transform the world.'" E. Papanoutsos, *Ekistics* 16 (October 1963): 205.
13. "For the period of our cruise we shall be to some extent cut off from the outside world. Cables can be received and sent at rates available from the Information Desk, but there are no telephone connections from the boat. We shall not receive newspaper or mail on board, but letters for mailing can be handed in at the Information Desk and will be stamped and sent off at the next convenient port." Delos Symposion Document 9, "General Information, July 1, 1963," 2.
14. C. A. Doxiadis, "Comment on the Delos Symposium," *Ekistics* 16 (October 1963): 204.
15. All participants were sent a copy of the package of information about the meeting that was sent to "the most important national and international technical and scientific journals of all fields represented in the Symposion," along with a list of journals that "we thought might be of special interest to you, just in case you wish to provide them with additional information." Delos Symposion Document 17, "Delos Documents," Avery Classics Collection, Columbia University.
16. "We are citizens of a worldwide city, threatened by its own torrential expansion and... at this level our concern and commitment is for man himself." "Delos Declaration," foldout insert, *Ekistics* 16 (October 1963).
17. Buckminster Fuller, cited in "The Delos Symposium," *Ekistics* 16 (October 1963): 205.
18. The introduction of the Delos Declaration into the *Congressional Record,* for example, is reported in *DA Newsletter* 4 (January 1964): 2. Press clippings on the Delos meetings appear in *DA Newsletter* 4 (January 1964): 2–3.
19. "When I recall the congress at which we wrote the Charte d'Athens I can only think that Greece has done it again! There must be something in the air to induce a peaceful working together and loosen normally constrained behavior." Sigfried Giedion, quoted in "Ninth Meeting—July 12, 1963. The Declaration of Delos: Statements and Comments," *Ekistics* 16 (October 1963): 254.
20. Doxiadis was so committed to the idea of the boat that when a meeting of the Delians was held in Washington in May 1968, the so-called Delos 5 1/2, it took place on a barge for an excursion and a picnic lunch. The event was reported in "Delians Stage Special USA Meeting," *DA Magazine* (July 1968): 15.
21. "Points Made in Discussions," *Ekistics* 30 (October 1970): 261.
22. C. A. Doxiadis, "The Networks We Build and the Networks We Need to Build," *Ekistics* 30 (October 1970): 263; and C. A. Doxiadis, "A Methodological Approach to Networks," *Ekistics* 30 (October 1970): 331.
23. "Points Made in Discussions," 317.
24. "Report of Delos Eight," *Ekistics* 30 (October 1970): 245.
25. Doxiadis had a long-standing interest in airline systems. In a 1959 project for Pakistan, he had made a detailed study of airline and sea links. *Ekistics* had presented an analysis of global airline connections in July 1966. Again, precedent for this can be found with Le Corbusier, who, in his last book, published a global map showing an airline network linking every continent into one accessible space. "Nations, religions, principalities, powers, going to sleep, waking

up, everything is different, changing, moving, flexible. A prodigious new broom has swept through the world order." Le Corbusier, *My Work* (London: Architectural Press, 1960): 152.

26. Le Corbusier, "Description of the CIAM Grid, Bergamo, 1949," appendix to J. Tyrwhitt, J. L. Sert, and E. N. Rogers, eds., *CIAM 8: The Heart of the City: Towards the Humanisation of Urban Life* (London: Lund Humphries, 1952): 171–76.
27. C. A. Doxiadis, "The Two-Headed Eagle: From the Past to the Future of Human Settlements," *Ekistics* 33 (May 1972): 406–20.
28. Doxiadis, "The Two-Headed Eagle," 418.
29. Le Corbusier, *The Athens Charter,* trans., Anthony Eardley (1943; reprint, New York: Grossman, 1973): 98.
30. As in Delos, Tyrwhitt rarely took center stage, but at CIAM 8 at Hoddeston in 1951, again organized by the MARS group (of which she had been the assistant director since 1949), she played a central role in the discussions, put together the proceedings, and edited the resulting book with José Luis Sert and Ernesto Rogers.
31. Tyrwhitt refers to Doxiadis for the first time in a February 1955 article. J. Tyrwhitt, "The Moving Eye," *Explorations,* no. 4 (February 1955): 115–19.
32. For the first issues, Tyrwhitt was assisted in the selection of material by the architecture librarian at Harvard, a role taken over the following year by the architecture librarian at MIT. The journal was distributed to a private list of people under the title *Tropical Housing and Planning Monthly Bulletin* until October 1957, when the name became *Ekistics: Housing and Planning Abstracts*. The subtitle kept evolving until the January 1975 version, *Ekistics: Problems in Human Settlements,* which has been the title ever since.
33. Giedion recommended Tyrwhitt to McLuhan in a letter thanking him for sending a copy of his first book, *The Mechanical Bride*. Tyrwhitt had arrived in Toronto in July 1951. In 1950, she had become a partner with Wells Coates (one of the original founders of the MARS group in 1933) in London and went to Toronto after a short stint in Yale at the beginning of 1951. She got together with McLuhan for the first time in November 1952.
34. Letter from Marshall McLuhan to Jacqueline Tyrwhitt, December 23, 1960, in Molinaro et al., eds., *Letters of Marshall McLuhan,* 277.
35. "This means that we are today dealing with two widely different dimensions: the eternal human scale—man with his five natural senses—our continuity with the ancient world, and the electronic scale—multi-dimensional integration—our emergence into the new world. Man of the renaissance and technological man no longer directly concern us. The human scale needs to develop the totality of the electronic scale, and the electronic scale the discipline of the human scale. Perhaps the latter can be done by maintaining the proportions of the human scale in man's electronic extensions." Edmund Bacon and Marshall McLuhan, "The Delos Symposium," *Ekistics* 16 (October 1963): 208. For Bacon, the most important thing about the conference was establishing the need for a "new science of human settlement for the purpose of comprehending city and regional growth as a total organic process within the framework of the emerging concepts of the electronic age." Edmund Bacon, cited in "Edmund N. Bacon, USA," *Ekistics* 16 (October 1963): 218.
36. "But beyond his senses, man has his 'electronic extensions.' Do these imply that, in addition to his immediate neighborhood, he can also relate to a wide variety of other centers within a single urban region or on a wider scale?" "Conclusions: Preliminary to a First Draft by Barbara Ward (Lady Jackson), Delos Meeting Document B17," July 11, 1963, "Delos Documents," Avery Classics Collection, Columbia University.
37. The editorial of the August 1959 issue describes the usefulness of analyzing the "mass" of data by punched cards, as shown in the two articles abstracted from recent issues of the English *Journal of the Town Planning Institute* and the *Journal of the American Institute of Planners*. The first article describes "a complete handling and mapping of data," and the second, on "Data Processing for City Planning," covers the role of computers, highlighting the cartographatron, an electronic device for displaying "desire lines" on the face of a cathode ray tube as the result of an analysis of over 700,000 punched cards.
38. The article by Jonathan Barrett, "Will the Computer Change the Practice of Architecture," for example, explores the new kind of shapes that could be modeled digitally and have retained the fascination of designers in recent years. *Ekistics* 19 (April 1965): 247–49.
39. The report on the "Human Community" research project says that the main emphasis in January was on computer programming, data processing, and methodology (IBM punched cards) using IBM 1620 machines. *DA Newsletter* 4 (January 1964): 41. The May 1964 issue published pictures of the new computer—organization of a new "Electronic Computer Center" in the office. *DA Newsletter* 4 (May 1964):29. *DA Review* 7 (January 1971) shows all the punched card machines "busy around the clock. The August 1972 issue shows pictures of students with the latest computer system.
40. Diana Rowntree, "The Science of the City," *Ekistics* 22 (October 1966): 243.
41. See Beatriz Colomina, *Privacy and Publicity: Modern Architecture as Mass Media* (Cambridge: MIT Press, 1994).
42. Le Corbusier, *Le Ville Radieuse* [1933], trans. Pamela Knight as *The Radiant City* (New York: Orion Press, 1967). Much of the book is devoted to the coordination of traffic networks handling different speeds.
43. "The Delos Symposion," *Ekistics*16 (October 1963): 205. On other connections between Waddington and other interdisciplinary communities, see Rheinhold Martin, "Crystal Balls," *Any* 17 (1997): 35–39.
44. "Delos Symposion, Notes from 2nd meeting," "Delos Documents," Avery Classics Collection, Columbia University.
45. Particularly in Delos 3 of 1966, an argument repeated after a discussion of "foodwebs." See C. H. Waddington, "Biology and Human Environment," *Ekistics* 21 (February 1966): 90–94. The spider's web is a model of the resilience of biological systems to change because it is not affected by removing links.
46. "A city is a complex living system. Its anatomy and composition can be studied and analyzed like any other living system.... A comprehensive analysis of the interactions within an urban population could be conducted in a manner equivalent to that used by biologists in their studies of living systems were it not for the presence of shields that have been created to fend off messages. This shield is called *privacy.* The small groups clustered inside these screens for communications (homes) are known to exchange messages to which the stranger is not granted immediate or complete access." Richard Meier, *A Communication Theory of Urban Growth* (Cambridge: MIT Press, 1962): 1.

47. C. A. Doxiadis, "Ecumenopolis," *Ekistics* (January 1962): 12.
48. The charts are first described in a report on the City of the Future project in *DA Newsletter* 4 (March 1964): 40. The speed drawing first appears in *DA Newsletter* 4 (April 1964): 39–41.
49. C. A. Doxiadis, *Ekistics: An Introduction to the Science of Human Settlements* (New York: Oxford University Press): 43.
50. Buckminster Fuller, "The Prospects of Humanity: 1965–1985," *Ekistics* 18 (October 1964): 232–42, special issue on Delos 2.
51. This was exemplified in the September 1965 issue, which featured Licklider's article on the "Man-Computer Partnership" and one of his drawings on the cover. J. C. R. Licklider, "Man-Computer Partnership," *Ekistics* 20 (September 1965): 165–69. The article was written when Licklider moved from the Air Force research laboratories to the research center of IBM. Other articles on planning buildings by computer cite his increasingly influential idea of a "symbiosis" between man and computer, and one of them argues that evolution culminates in "computer systems or automata, prostheses which will extend the brain's ability to manipulate symbols just as the bulldozer is a prosthesis for muscles and the microscope, telescope, and laser are for the eye…the upgrading of human capacity on a large scale.…Men can be upgraded. At the strictly biological level, there is no question that the principles of genetics and selection, used widely by man in the plant and animal world, could be used to enhance desired traits in man." R. W. Gerard, "Intelligence, Information and Education," *Ekistics* 20 (September 1965): 162–64. A similar issue in 1967 begins with John McHale's speculations about the "New Symbiosis," which also cites Licklider and weaves together prosthetics, cybernetics, McLuhan, and Fuller; then a series of other articles explores the roles that computers can play in design and analysis.
52. Koichi Tonuma, "Network City," *Ekistics* 29 (June 1970): 458. This essay was first published in *Journal of High Speed Transportation* 3 (May 1969): 203–219.
53. Kenzo Tange, cited in "Kenzo Tange," *Ekistics* 22 (October 1966): 259.
54. "In the First Industrial Revolution, men learned how to extend the functions of their hands and bodies through the use of tools or machinery. The Second Industrial Revolution, which has begun only recently, is a revolution created by information theory and communications techniques, a revolution in which man is learning to extend the functions of his nervous system." Tange, "Kenzo Tange," *Ekistics* 22 (October 1966): 275.
55. Kenzo Tange, "Tokaido-Megalopolis: The Japanese Archipelago in the Future," in *Kenzo Tange: Architecture and Urban Design 1946–1969,* ed., Udo Kultermann (New York: Praeger, 1970): 150–67.
56. Kenzo Tange, "A Plan for Tokyo, 1960," *Ekistics* 12 (July 1961): 9–19.
57. Kenzo Tange, "A Building and a Project: On the Kurashiki City Hall and a Project at MIT," *Ekistics* 11 (June 1961): 469–72.
58. S. Giedion, "Density and Urbanism," *Ekistics* 20 (October 1965): 208–9.
59. Fumihiko Maki, "Linkage in Collective Form," *Ekistics* 14 (August–September 1962): 100–103. When Maki put together a more formal publication, photographs of Tange's Tokyo Bay and Boston Harbor projects were added as key examples. Fumihiko Maki, *Investigations in Collective Form* (St. Louis: Washington University, 1964).
60. Alongside an "ideogram of a net of human relations," Peter Smithson describes it as "a constellation with different values of different parts in an immensely complicated web crossing and recrossing. Brubeck! A pattern can emerge." Peter Smithson, in *Team Ten Primer,* ed., Alison Smithson (Cambridge: MIT Press, 1968): 79, caption.
61. Peter Smithson, "Recollection by Peter Smithson, September 20, 1990," *Team 10 Meetings 1953–1984*, ed. Alison Smithson (New York: Rizzoli, 1991): 60.
62. See Shadrach Woods, "Web," *Le Carrér Bleu,* no 3 (1962). The article was reported in *Ekistics* two years later. The web is "highly flexible," "non-centric," "open-ended," and "can be plugged-into at any point and can itself plug-in to greater systems at any point." See also Shadrach Woods, "Urban Environment: The Search for System," in John Donat, ed., *World Architecture,* no. 1 (London: Studio Vista, 1964): 151–54.
63. Alison Smithson, in *Team 10 Meetings,* ed., Alison Smithson, 9.
64. Stefan Wewerka, in Smithson, ed., *Team 10 Meetings,* 75.
65. "The capsule is cyborg architecture. Man, machine and space build a new organic body which transcends confrontation. As a human being equipped with a man-made internal organ becomes a new species which is neither machine nor human, so the capsule transcends man and equipment.…The capsule is a feedback mechanism in an information-oriented, a 'technetronic,' society." Kisho Kurokawa, "Capsule Declaration," *Space Design* (March 1969), reprinted in Kisho Kurokawa, *Metabolism in Architecture* (London: Studio Vista, 1977): 75–85. On the "Cybernetic Environment," see also Arata Isosaki, "Invisible City," *Tenbou* (November 1967), trans. in *Architecture Culture, 1943–1968,* ed. Joan Ockman (New York: Rizzoli, 1993): 403–7.
66. Frequent articles appear in *Ekistics* from Alison and Peter Smithson, Jaap Bakema, Aldo Van Eyck, George Candallis and Shadrach Woods, Hermann Hertsberger, Fumihiko Maki, Kisho Kurokawa, and others. Tyrwhitt's editorial for the June 1963 issue describes Team 10 as "an intelligent group of thoughtful and highly intelligent architects." In a July 1963 review of a book by Doxiadis, she compares the work of Doxiadis to that of Team 10 and the Japanese architects. Jacqueline Tyrwhitt, "Architecture in Transition, Doxiadis, C.A.: A Review," *Ekistics* 16 (July 1963): 60.
67. Jos Bosman, "CIAM after the War: A Balance of the Modern Movement," *Rassegna* 52 (December 1992): 6–21.
68. Warren Chalk and Ron Herron, "City Interchange—Project," *Living Arts* 2 (1963): 73.
69. "The foreseeable rapid rate of change in transportation method may eventually make invalid the concept of a rigid mobile communications network as the main urban structure. A whole area of study is open for experiment of expendable systems and more flexible technology in terms of communication networks.…Large organizations will control their own visual communications network, allowing for a city center control with satellites dispersed in constant touch with the communication center, no longer dependent on physical communication." Archigram, "Living City," *Living Arts,* no 2 (June 1963).
70. Peter Cook and David Greene, "Metamorphosis," *Ekistics* 28 (August 1969): 104–6.

71. Louis Kahn, "Order in Architecture," *Perspecta* 4 (1957): 64.
72. Sigfried Giedion, *A Decade of New Architecture* (Zurich: Girsberger, 1954): 262. The project was the centerpiece of the special issue on structure of *L'architecture d'aujourd'hui*, no. 55 (September 1954).
73. "Archigram Metropolis Issue," *Ekistics* 20 (November 1965): 282. Crompton had already elaborated the idea of the city as a self-regulating bio-technological computer a year earlier: "The city is a living organism—*pulsating*—expanding and contracting, dividing and multiplying. The complete functioning of the city is integrated by is *natural* computer mechanism. This mechanism is at once digital and biological... The overall network is formed from this information and then absorbs it, processes it, and throws out the subsequent stages... The network is modified and amplified and the substance of the city created... The *City Scene* enveloped in a net of inter-relationships, ultimately controlled by the *Natural Computer*." Dennis Crompton, "City Synthesis," *Living Arts* 2 (1963): 86.
74. "The other factor is the rapid advancement of organizations in modern civilized society brought about by the modern communication system, informational technology, and the sharp reflection of this phenomenon on spatial organization. In modern civilized society, space is a communication field, and it is becoming more and more organic with the development of the communication system.... Creating an architecture and a city may be called a process of making the communication network visible in a space.... We can say that the spatial organization is a network of energy and communication." Kenzo Tange, "Function, Structure and Symbol," in Kultermann, ed., *Kenzo Tange*, 240.
75. A picture of Bell's tower is featured in Wachsmann's book and Archigram's Living City of 1963.
76. "The notion of self-contained permanent settlements is obsolete. We live under conditions of mobility which result in continual stirring up rather than settling down. It is indeed doubtful whether the notion of a neighborhood of people who fulfill the internal functions of a village community makes sense for the future." "Notes from the 5th Meeting," "Delos Documents," Avery Classics Collection, Columbia University.
77. "Notes from the 5th Meeting."
78. "In speaking of reforming the environment of man, I include a surgeon's operations on the human body, for the latter is mobile environment of the brain.... I define industrialization as the extra-corporeal, organic metabolic regeneration of humanity. Industrialization consists of tools. All the tools are externalizations of originally integral functions of humans." Buckminster Fuller, "Letter to Doxiadis," *Main Currents in Modern Thought* 25 (March–April 1969): 87–97.
79. Buckminster Fuller, "Why I Am Interested in Ekistics," *Ekistics* 20 (October 1965): 180–1.
80. Buckminster Fuller in the discussions about energy, *Ekistics* 34 (October 1972): 241. The point was first made in Delos 4 in 1966.
81. "Communication via Humor," *Ekistics* 34 (October 1972): 282–83.
82. Letter from Marshall McLuhan to Tom and Dorothy Easterbrook, 1 August, 1972, in Molinaro et al., eds., *Letters of Marshall McLuhan*, 454; and letter from Marshall McLuhan to Gyorgy Kepes, 1 August 1972, in Molinaro et al., eds., *Letters of Marshall McLuhan*, 453. Tom Easterbrook was one of the group of four people, including Tyrwhitt, who carried out the key research with McLuhan from 1953 to 1955.
83. "Marshall McLuhan Executive Seminar," January 7–9, 1970, Grand Bahamas Hotel and Country Club.
84. As reported by Koichi Tonuma, *Ekistics* 52 (September–October 1985): 516.
85. Buckminster Fuller, "Accommodating Human Unsettlement," *Town Planning Review* 49 (January 1978): 51–60. Report of Fuller's report to the U.N. Habitat conference, September 20, 1976.

Afterword
The Demystifica-*hic*-tion of In-*hic*-formation

Thomas Keenan

Not much is said in this volume about the old medium of writing. But in a powerful text like the one I propose to read here, in conclusion, the question of writing opens up a profound critique of the contemporary politics of free media, old and new. Without naively imagining that writing is all that different from any other information or communications technology, without seeking to "return" from media or technology to the reassuring humanity of language, and without suggesting that the only thing writing—or any other medium—does is handle information, Nuruddin Farah's exemplary narrative *Sweet and Sour Milk* demonstrates that any media theory has much to learn from attending to the political aporias of language. Without relinquishing a claim for exposure and open media, for the value of public memory over political secrecy, and for the force of information, the novel asks about the limits of the ideology that sees writing—or media—as a simple instrument of demystification, and reminds us to be aware of the astonishing disruptions to which it can be subjected, in reading and in speaking.

In the prologue to Farah's 1979 novel, the first volume in what became a three-part series called "Variations on the Theme of an African Dictatorship," we meet—just as he is about to die—the technocrat-turned-dissident Soyaan. He has written a clandestine document called "Dionysius's Ear," a text of "eight typewritten pages" that the book generally refers to simply as "the Memorandum."[1]

This text, and all that the novel does with it, allegorizes and decomposes a powerful theory of media and writing, of information, a theory of free expression which in the quarter-century since Farah wrote the book has become practically hegemonic, for better and for worse, especially with reference to African and other dictatorships, and most especially in the age of a generalized writing that takes the form of television, global media, and the Internet. Farah's text, from deep within, undoes the paradigm that traces the passage from writing to emancipation as an ineluctable narrative, and it does so by systematically introducing the question of *reading*, and with it another temporality altogether. He does it in the name of a more profound relationship between media, new and old, and politics, not against their linkage but so as to open up some new possibilities, measure some risks and dangers, and insist—for political reasons—on a thought of the political that puts the reliability of meaning and grounds in question.[2]

Back to the Memo. Later we learn that the full title is a figure for the Somali state of General Mohammed Siad Barre, that peculiar Somalia which preceded the "starving Somalia" we have since come to know, a Somalia that mixed Islam and Stalinism, Italian restaurants and East German prisons and KGB-trained torturers, the knock at the door at dawn and the ear-splitting sound of MIGs overhead, the ninety-nine names of the General and the hard imperatives of the National Security Service, the skillful exploitation of clan politics and the pervasive influence of "the General's media" and the underground traffic in cassettes.[3] As Soyaan is dying, the radio is on, and Dulman, the country's most famous actress, is singing the psalms of the General's praise-names. Soyaan switches off the radio and, the novel tells us, "thus he strangled the singer in the midst of a syllable, made Dulman choke on the consonants of sycophancy" (10).

The radio and the microphone, what the novel calls "amplification" (119), are not simply what we now call "the media"—they are also the technical instruments of a society and a politics based on speech and its reproduction, on information and its transmission. Hence the title, and the project, of the Dionysius's Ear memo. The General's Somalia—and the novel denies the General any name more specific or particular than that, generality—is a society, thinks Soyaan, of informers, of "daily gatherers of spoken indiscretions," of "ear-servants" (9–10). Like the cave of Dionysius which echoes the whispers of its prisoners for the benefit of the tyrant,[4] the Somalia analyzed in the Memo is a dictatorship of the illiterate, administered by security forces "who neither read nor write, but report daily, report what they hear as they hear it, word by word." An "oral tradition" shies away from writing; it needs only amplifiers, a "society of planted ears." "Everything is done verbally." There are few files, reports, warrants, documentation, almost no writing and no paper trail. Instead, "the General ... has had an ear-service of tyranny constructed."

One of the authors of the memorandum explains it this way:

> "The Memo Soyaan and I worked on, the one my sister typed, if you want to know, is titled 'Dionysius's Ear.' It is not a long memo. Maximum eight typewritten pages. Dionysius's Ear.
>
> "Dionysius, the Syracusan tyrant?
>
> "That is it."
>
> [...]
>
> "But why Dionysius?"
>
> "The Syracusan tyrant had a cave built in the shape of a human ear which echoed to him in polysyllables whatever the prisoners whispered secretly to one another. Soyaan and I saw a similarity between this and the method the General uses so far. The Security Services in this country recruit their main corps from illiterates, men and women who belong to an oral tradition, and who neither read nor write but report daily, report what they hear as they hear it, word by word. They report verbatim what they think they heard when they walked into a shop. They need no warrant to arrest anybody. Everything is done verbally. Instructions are given on the phone: 'Before dawn arrest so-and-so.' Most of these are not even traceable to their origin, for there are no written warrants. [...] We've found that two-thirds of the prisoners have no files, that over two thirds of them are serving indeterminate prison sentences. We have indeed discovered that the only privileged prisoners are those with thin files of three-page reports concerning hour of arrest, reason for arrest (i.e., high treason), et cetera. We say in our Memo that the General has had an ear-service of tyranny constructed." (136–7)

The Memorandum, it seems (we never get to read it), offers documentation of the non-existent documentation, and is the first product of a small group of intellectuals and professionals opposed to the General. Farah's novel tells the story of this group, and motivates it by giving to Soyaan's twin brother Loyaan the task, the responsibility, of investigating his mysterious death—which is to say, of pursuing, discovering, and understanding the few traces he left behind. Loyaan wants to know "why": "what did he die of?" he asks himself, and reminds himself to "address yourself to the question, address yourself to the challenge." The death remains a mystery. As the story unfolds, Loyaan is recruited into Soyaan's dissident group, and this process of initiation, enlightenment, the entry into democracy and freedom and rights as initiation into writing and reading, is the apparent allegorical lesson and narrative vector of the novel. But along the way, the simplicity and perfection of the opposition between speech and writing becomes more complicated, just as the opposition between dictatorship and opposition begins to tremble, and the politics of information and writing turns and returns on itself, suffers from additions and subtractions, and most of all, from interruptions.

Faced with the ear, writing and publicity constitute the originary political acts of the opposition. They are the counter-media of demystification, information, revolution, memory, and enlightenment to be arrayed against the tyranny of talking, rumor, whisper, broadcast, disappearance, indeterminate detention, and the omnipresent oral informant. The memo is an exemplary, auto-exemplary, manifesto for a free media, for a politics of traces and memory, starting with writing.

The spoken word, then, leaves no traces, and the first step in opposing the society of listeners and the regime of disappearances is to write things down. Hence this memorandum on the ear, and the strategy of *memoranda* as such. Writing is memory and visibility, it remains and it shows and it can be shown, and in its persistence and reproducibility constitutes a force, a power, in itself. Against the fleeting consonants of sycophancy and the verbatim informers, the opposition offers the written word. That is, it seems, the politico-literary theory of these dissidents: a familiar, even stereotyped, theory, but no less forceful and persuasive for its typicality. It is, in fact, a politics of typicality—of letters, type, publication, code and reproduction and dissemination.

Though the words are not used, the practices of the dissidents constitute or imitate that of an embryonic human rights organization, a movement for democratic change, perhaps, but less a proto-political organization than a stereo-typical human rights reporting agency.[5] "We belonged to a clandestine group, Soyaan and I, [one of them says,] a clandestine movement of opposition which is composed of intellectuals and professionals who've taken an oath—*per modo di dire*—to serve not the interests of any superpower but this nation's" (139).

They aim to write. Their watchword, their code, is information and its free accessibility—they seek to document, research, recall, and publish what otherwise remains trapped invisibly in the oral network of rumor and silence.

> What did we ask of those ten whom we invited to our meeting? [. . .] "We want you to collect information for a common pool. We want you to research with us. We will disseminate the information received in that manner, we will eventually publish our findings, we will distribute them gratis in cyclostyled format, we will start with the General himself," we said. "We can foretell," we added, "that the written word, more powerful than the gun, will frighten them. In the chaos ensuing from all that, and just as they start their purge, we will announce our clandestinity and publish a leaflet of our intention, and you will see that more people will adhere to it. Then we will baptise it as a movement, give it a name." (140)

They do this at considerable cost—one of the scraps of paper Soyaan leaves behind cites a law authorizing the death penalty for distributing or publishing material critical of the state.

At the start, we know very little about the Memorandum and the force of the written word, only about the existence of "a strong political statement." In a flashback that interrupts the prologue narrating Soyaan's mysterious death, he returns to a clandestine afternoon at the beach. Perhaps it is that beach which looms large in the political mythology of Mogadishu, the beach where the corpses of so many of Siad Barre's political opponents were fed to the sharks at night that on some mornings the tide ran red, or perhaps the beach on which the humanitarian U.S. Marines of Operation Restore Hope were later met and outnumbered by the world's television cameras.[6] But in the novel, the beach is only the scene of a private meeting between lovers, at least a quasi-private rendezvous, an opportunity to discuss the secret text. Swimming in the surf, Soyaan's lover brings "it" up without warning, telling him that she's "enjoyed reading it," just "it," but that it is "dangerous stuff." Then, in a scene almost straight out of the Western canon, she writes. The writing of this unnamed woman proves decisive for the interpretation of writing as such, as political force, in the novel. She reads and writes.

> "I've enjoyed reading it," she said.
>
> "I was wondering how you found it." But he didn't wait to say any more.
>
> He made a dive, improvised and clumsy. She followed him in. She couldn't stay under for as long as he did. The sea spat her out. Up and out, and she panted. She waited for him to re-emerge.
>
> When he did, she asked: "Has anybody else seen it?"
>
> "Why?"
>
> "Dangerous stuff."
>
> "Do you think so?"
>
> "It certainly is a strong political statement."

They swim some more. And then she writes.

> She wrote his name on the sand. The sea washed away her writing. They silently watched the water recede. He wished he could read her message in the water receding. She wished she could make him see reason about the political statement he had made. Would he?
>
> "You haven't shown it to anybody else, have you?"
>
> He didn't reply. (7)

Perhaps you recognize this scene, not from Farah but from Edmund Spenser's version, dated 1594:

> One day I wrote her name upon the strand,
>   But came the waves and washed it away:
>   Again I wrote it with a second hand,
>   But came the tide, and made my pains his prey.
> Vain man, said she, that doest in vain assay,
>   A mortal thing so to immortalize,
>   For I myself shall like to this decay,
>   And eek my name be wiped out likewise.
> Not so, (quod I) let baser things devise
>   To die in dust, but you shall live by fame:
>   My verse your virtues rare shall eternize,
>   And in the heavens write your glorious name.
> Where whenas death shall all the world subdue,
>   Our love shall live, and later life renew.[7]

Having been set up by the poet's vain efforts in the sand, the female voice is made to speak for the fleeting materiality of all that is human, of necessity subject to being "wiped out" by time and certain "decay." Words, like the people they name, are all written in sand, are one day erased "like a face drawn in sand at the edge of the sea."[8] There is nothing not written in sand, says Spenser's she, in effect. Over against that bad corporeality, the poet-speaker can sort things out, and—on the basis of his inaugural mistake—explain how language in fact works, demonstrating that names are not reducible to their inscription in this or that sensible or mortal substance but are memorable, remarkable, precisely because they are immaterial, virtual, ideal, transcendent enough to outlive the rest of us. Inscription, yes, but in the heavens. Words, and what they bear, survive because they are not merely material. What matters is not the matter, but the outliving of mere life in language, the life beyond life of "verse" and "fame."

This is not exactly the case here. In Farah's rewriting, the roles are reversed: she writes, he can't read anything, and there are no promises about the future. Writing *his* name, destined for erasure, she aims to warn him about his writing, about the possibility of its being read and the risks of that possibility—which is of course precisely its intentional structure, we later learn. But the message of the erasure—not the erased message but the message to be read in the erasure itself—remains obscure to him. He wishes he could read, and so does she. Can he?

What does she write? His name, yes, but only insofar as it is washed away. Washed away, the name constitutes a message—that is when reading becomes necessary, in the advent of the tide which smoothes the disturbed surface of the sand, cleanses it and takes the name back with it. The predictability of the tide reminds us of the obvious: Erasure does not accidentally befall the name, but is anticipated in the act of inscription. The point is not the name but its effacement. She writes its silent, predictable, inevitable disappearance; she writes out its elimination, liquidation. What she writes is an *obliteration*; the name is, in the lexicon of another continent's dictatorships, disappeared. It names only its "recession": that is why the text specifies that the message is *in the water*, not the sand. Where does it go? "He wished he could read her message in the water receding." The name is not denominative, not descriptive, the utterance not a call or an appeal or even a performative. It is neither literal nor figural—it is, perhaps, *obliteral*, and it is that status which, in the text, triggers the demand for reading.

This too is a theory of writing, of the strong political statement, but it complicates the one proposed by the Memorandum and its writers, considerably, by recasting the questions of time and apparition, of message and information, of retention and recession. Without neutralizing the difference between speech and writing, and without discarding the practice of writing at all, the obliteral endangers the solidity of the principles of nomination and inscription, of information, in which the dissidents have put their trust. Something happens to the ground of reading, to the very surface of inscription, which does not defeat the project of writing but does complicate it a great deal.

"He wished he could read her message in the water receding." Without destroying the theory or the practice of the written word, Soyaan's lover—unnamed here, later called Margaritta—writes in another way, on another surface, and asks what happens when messages recede, when reading remains a question. She writes out the lesson about writing's recession: can he read what it has to say about political statements? Would he?

He makes no reply—to the question about the others, about the publicity of the message and about the "anybody else." But shortly afterwards, another message arrives—in the form of a mysterious silent child—that brings to the minds of the lovers "the name of a man," a name they both try immediately to forget, and that leads Soyaan to "wonder if the beach were really as private as he had thought." The answer is of course no, and the message incarnated by the child is the promise of Soyaan's disappearance, not merely that of his name. At least that's how he seems to understand it. In the wake of the child's silent arrival and departure, Soyaan simply proclaims himself "ready"—"when they finally come, having broken the pride of dawn, they will find me prepared." Sometime thereafter, he dies—perhaps poisoned. All that we finally learn about the fate of the Memo is that she has it hidden, and that after his death it is seized in a police raid. By novel's end, most of the group is arrested, tortured, dead, or turned—and Loyaan is dispatched to a diplomatic posting in, of all places, Belgrade.

The name remains, though. The moral of the story—the story of the name—is unusual. Soyaan the dissident-technocrat becomes, in death, the property of the State. After he dies, the General's State claims him as a hero of the Revolution—and writes his name everywhere. And the Minister, the same Minister who bears the name that Soyaan and his lover remember and try to forget in the aftermath of the silent child's apparition, is the official in charge of celebrating the name: "Soyaan is the hero the Revolution has knighted. Schools, squares, streets will be named after him" (184). Why? As Loyaan theorizes, again in a rather Spenserian vein:

> Hero-worship is a phenomenon as necessary as history itself. Every nation needs heroes in which to invest a past, heroes and legendary figures about whom one tells stories to children and future generations. (185)

Canonized, Soyaan's name goes public, and in the name begins the story, the narrative of nation. That is another move in the theory of language, literature, and the maintenance of state power. "Like it or not, Soyaan is the hero the Revolution has decided to knight," says the Minister. Soyaan's father, who collaborates with the State in the promotion of his son to posthumous hero status, says that "he will live longer than you or I. [...] I am responsible for the spiritual revival of his name" (93).

The name, then, the written name, has multiple futures and diverse fates. Its destiny is not given in advance, and it is open not just to interpretation but to appropriation. Perhaps this is what Margaritta sought to convey. Naming names, making lists, spreading information and fighting against secrecy and forgetting—these are of course critical activities for an opposition, but they are also equivocal ones, subject to unanticipated outcomes—whether it's erasure or identification, betrayal or repetition.

The novel, though, is not content to leave things at this. Names get erased, and they get remembered; denomination and information are obviously subject to both these dangers. The dissidents are aware of these risks, at least to some extent. But Farah asks us to consider the question in somewhat different terms: he asks about the search for information, for knowledge, for answers, and about the difficulty of reading what gets discovered.

The exemplary question is that of death—Loyaan pursues the question of how, why, and of what his brother died. Asked about this, early on, he offers an answer which in spite, or perhaps because, of its odd generality, its obliquity and opacity, persists across the novel. "What did he die of?" asks a friend. And Loyaan

> found that he was at a loss for an explanation. [...] What had Soyaan died of. Wasn't that why he had wanted a post-mortem examination? [...] But what had Soyaan died of? *Address yourself to the question. Come on.* A voice spoke to Loyaan from inside him. A voice? *Address yourself to the challenge.* And he improvised a response, safe, vague: "He died of complications." (29–30)

Not complications of something, but "of complications," pure and simple—except that complications are precisely not simple. Later, he is again asked, this time by the Minister: "Soyaan, what did he die of?"

> Just this very minute, since it was the second time anyone had seriously put this query to him, Loyaan admitted to himself that Soyaan's death actually stuck him as an activity peculiarly perverse. But he wouldn't openly admit a failure on his part to deal squarely with the question. He remembered the answer he had given [before]. "Complications." (42)

What are complications? Just before he dies, Soyaan lectures his twin brother—newly arrived on the scene—on his political theory, his activist theory of writing and publicity. He lays out the powerful conceptual apparatus which seems to inform the work of the dissidents, the strategy of the memoranda, and the challenge to the national security state with its rich hybrid of terror and censorship, cooptation, praise names and hero-worship. The signature of his impending death, and the decisively complicating factor in his discourse, is the violent punctuation of his "monologue" with... *hiccups*. Dying, he is reading Machiavelli: "there is nothing more difficult to take in hand, more perilous to conduct, or more uncertain in its success than to take the lead in the introduction of a new order of things." He is also hiccupping: "Soyaan hiccupped a series of involuntary spasms of breathlessness" (16).

Like the scene at the water's edge, the sentences which transcribe the first arrival of the hiccups introduce, and transform ruinously, a problematic of classical typicality into an altogether different understanding of politics and language, human rights and media, reading and responsibility.

The masses need to be informed, he explains, about "what was really happening, information being essential in a country where everything was censored... hiccup...."

He calls for:

> "The politics of confron-*hic*-tation."
>
> "I don't understand."
>
> "The demystifica-*hic*-tion of in-*hic*-formation. Tell the *hic* masses in the simplest *hic* of terms what is happening. Demystify *hic* politics. Empty those heads filled with tons of rhetoric. Uncover whether hiding *hic* behind pregnant letters such as KGB, CIA, or other *hic* wicked alphabet of mysteries *hic*. Do you *hic* understand now *hic*?"
>
> Soyaan's eyes were trained on Loyaan. "I am not sure if I do."

And Soyaan promises, with a smile and a hiccup: "you will in *hic* time" (15).

Easily recognizable in this progression of imperatives is the standard conceptual array of Enlightenment-inspired human rights discourse, in fact a kind of *summa* of the dissidents' theory of media and memory: the opposition between free expression and censorship, between information and rhetoric, the privilege of simplicity and directness of terms, the power of facts, the politics of demystification and unveiling and renewal, and the promise of a clarified, enlightened, informed future.[9]

Less easily understood are the hiccups. They complicate things.

> a. An involuntary spasm of the respiratory organs, consisting in a quick inspiratory movement of the diaphragm checked suddenly by closure of the glottis, and accompanied by a characteristic sound. Also, the affection consisting in a succession of such spasms.[10]

These are not Aristophanes' hiccups in Plato's *Symposium*, "which prevented him from speaking" and could only be cured—homeopathically, as it were—with an induced fit of sneezing.[11] These do not defer, delay, prevent or suspend speech—they mix with and riddle the discourse with their own singular alternation of inhalation and closure, breath and breathlessness, arriving in involuntary spasms to torture the speech and the speaker, to render understanding uncertain, and to make reading difficult.

They *irrupt*, in the dictionary sense: they do not just invade or burst in, but (as is said of populations) they also increase irregularly in number. From within the machinery of language they exceed and distort it. Hiccups are vocal: they come from the same place as speech but interrupt it. Their rhythms are otherwise, unsynchronized and unsyncopated, erratic. They are guided by no intention; they signify nothing but the straying of intention and the ruin of information, and here that loss of presence, of self-consciousness, of breath, even promises the imminent death of the intender. They are a sort of *parole involuntaire*. They separate the speaker from his words and the words from themselves.

Out of control, they are at the heart of this controlled discourse on linguistic control. Out of place, they take place in a speech on the need to sort things out, but without any possibility of being in control or in place, only ever displaced and displacing. Puncturing, perforating, yes, but also increasing, filling, a vast unconscious alphabetic swarm of letters and syllables threatening to overcrowd this discourse on the need for emptiness, to cover over this discourse on uncovering. Unconscious, spoken as if by an other speaker within the speaker, they are emitted with and in his words but irreducible to them. So what are hiccups? Or rather, what is their force here? Are they rhetorical, mystifying, obscuring, filling? Or are they informative, simple, clarifying, understandable? What do they do here?

"The demystifica-*hic*-tion of in-*hic*-formation." With this phrase, the promise of what information might do, what force it has, and hence an entire axiomatic that links human rights and democracy with free media and transparency, is threatened. Not destroyed, but challenged, and we are asked to read in its recession.

"'Do you *hic* understand now?' 'I am not sure if I do.'"

What do we "understand" here?—the speech about understanding, purged of its hiccups, or the speech as it's hiccupped? To answer these questions we would have to understand the difference between understanding and not understanding, between information and its hiccups, and that is precisely the security that Farah's text denies us. Not no, not yes, but "I am not sure I do" seems the best answer to the question of understanding and hiccupping. This lack of surety does not immobilize the demand for information or paralyze the brother, but—like Margaritta's experiment in reading and writing—it ups the ante, radicalizes the gesture. Just as reading becomes more urgent as the message recedes, here the question of understanding is posed with unprecedented radicality in the act of its disruption. Understanding cannot take its own possibility for granted: most compellingly when rights, responsibilities, and freedoms are at stake.

Can you understand a hiccup? This is not the question posed *in* the literary narrative but rather *by* it. It uses the inscription of the hiccup to challenge the discourse of information, not to abandon it or to replace it or to defer it indefinitely, but to twist it from some inassimilable "within," to open it up to an inarticulate force in speech—and in writing—that the direct opposition between information and rhetoric, understanding and mystification, closure and discovery, could never master. The hiccup does not inform, it deforms. "In-*hic*-formation." Let me repeat so as to be clear: the challenge to this conventional discourse on information and demystification, on the clarity of speech, is made within it—within the very flow of speech that enunciates the call for freedom of information—but from and with something that is not simply informative. And the hiccups are not just noise that threatens to drown out the signal, either; it would be too much to call them signals, or significant, in their own right, but they do something here. That is how we can start to read them.

Reading is not understanding, though. These hiccups force the question of politics and language to its breaking point. Indeed, they signal the death of the speaker—the death or the ruin of intention and self-presence, of his meaning-to-say-what-he-means, but also the death of Soyaan himself.

Hiccups are in fact said to be the cause of death, although perhaps only by metonymy—but that's the text's trope:

> Soyaan hiccupped a series of involuntary spasms of breathlessness. He looked all the more disturbed as he stretched out his hand to Loyaan, who took it and held it in his. He repeated and repeated and repeated Loyaan's name in between those spasms of breathlessness. First the warmth went out of Soyaan's hand. Then the brightness out of his eyes. Everything assumed an artificial quietness, for an unbroken fraction of a second. And Soyaan hiccupped his last. (16)

Does the text seek to undermine, to discredit, the dissidents, the oppositional politics, the struggle for rights of expression and political freedoms, with these hiccups? No, but it inscribes a fold, an irritant, a difficulty within that theory and practice of language and politics. From the receding water to the persistent hiccup, what gets removed is not the politics but the ground, the confidence and stability, the predictability ("we can foretell"), the certainty of the surface and the confidence in the future. Precisely what the activist cannot take for granted is the self-evidence of the text, the standpoint and the ground. The hardest lesson to learn, in writing as in politics, is this one, the one that another democracy movement and another writing will insist on and teach, as they do here: that they do not depend, finally or simply, on grounds, or information, or facts. Because grounds are difficult to ascertain, and are sometimes simply lacking, there is politics, reading and writing, media new and old.

Notes

1. The novel was first published in London by Allison and Busby Ltd. in 1979, and the following year in the African Writers Series at Heinemann. I have used the later American edition: *Sweet and Sour Milk*, Saint Paul: Graywolf Press, 1992.
2. For a brief but powerful reading of the novel, see Farah's interview with Maja Jaggi, "Bitter crumbs, Sweet and Sour Milk," *The Guardian* (London), 3 April 1993, 29. Jaggi writes: "in place of the General's rigid ideological certainties, Farah offers open-ended, ambiguous truths. There are no political panaceas, no platform. His characters are simply forced to make imperfect moral choices." Among the few interpretive considerations of the novel are: Barbara Turfan, "Opposing Dictatorship: a Comment on Nuruddin Farah's 'Variations on the Theme of an African Dictatorship,'" in Derek Wright, ed., *Emerging Perspectives on Nuruddin Farah*, Africa World Press, Trenton, NJ / Asmara, Eritrea, 2002, 265-281; and Josef Gugler, "African Literary Comment on Dictators: Wole Soyinka's Plays and Nuruddin Farah's Novels," *The Journal of Modern African Studies* 26:1, March 1988, 171-177.
3. See the report from Africa Watch, largely authored by Rakiya Omaar, *Somalia: A Government At War With Its Own People*, New York, January 1990.
4. On the cave of Dionysius and acoustic surveillance in general, see: Doerte Zbikowski, "The Listening Ear" in Thomas Levin, et al., ed., *CTRL [SPACE]: Rhetorics of Surveillance from Bentham to Big Brother*, Karlsruhe: ZKM, and Cambridge: MIT Press, 2002, 32–49.
5. For a beautiful account of the importance of filing in human rights work, see Lawrence Weschler, *A Miracle, A Universe*, New York: Pantheon, 1990.
6. Farah has told the story of the aftermath of the American intervention in *Links*, New York: Riverhead, 2004. At the time, he wrote a sharply-worded, and complicated, opinion piece: "Praise the Marines? I Suppose So," *The New York Times*, 28 December 1992, A15. On the beach, see Ahmed Omar Askar, *Sharks and Soldiers*, Järvenpää, Finland, 1992.
7. J.C. Smith and E. De Selincourt, eds., *The Poetical Works of Edmund Spenser*, London: Oxford University Press, 1912, 575 (Amoretti, Sonnet 75, spelling modernized).
8. Michel Foucault, *The Order of Things*, New York: Random House, 1970, 387.
9. I have tried to outline some elements of this conceptual array in a series of essays: "Live from ...,"in Elizabeth Diller and Ricardo Scofidio, *Back to the Front: Tourisms of War*, Caen: F.R.A.C. Basse-Normandie, 1994, 130-163; "Publicity and Indifference: media, surveillance, 'humanitarian intervention,'" in Levin et al, eds., *CTRL [SPACE]*, 544-561; and "Mobilizing shame," *South Atlantic Quarterly* 103, no. 2/3, Spring/Summer 2004, 435-449.
10. *Oxford English Dictionary*, Second Edition: "(*Hickop*, *hiccup*, appears, from its date, to be a variation of the earlier *hickock*, HICKET q.v. *Hiccough* was a later spelling, app. under the erroneous impression that the second syllable was *cough*, which has not affected the received pronunciation, and ought to be abandoned as a mere error.) [...] 1580 HOLLYBAND *Treas. Fr. Tong*, *Le hoquet*, the hickop, yexing. 1581 MULCASTER *Positions* x. (1887) 57 For the hikup. 1621 BURTON *Anat. Mel.* III. ii. VI. ii. (1651) 553 By some false accusation, as they do to such as have the hick-hop, to make them forget it."
11. Plato, *Symposium*, trans. W.R.M. Lamb, Loeb Classical Library, Cambridge: Harvard University Press and London: William Heinemann, 1961, 185c-d, 189a.

Contributors

Geoffrey Batchen is Professor of the History of Photography in the Doctoral Program in Art History at CUNY Graduate Center in New York. He has three books to his name: *Burning with Desire: The Conception of Photography* (MIT Press, 1997); *Each Wild Idea: Writing, Photography, History* (MIT Press, 2001) and *Forget Me Not: Photography and Remembrance* (Van Gogh Museum and Princeton Architectural Press, 2004).

Vannevar Bush (1890–1974) was Dean of the School of Engineering at Massachusetts Institute of Technology, president of the Carnegie Institution of Washington, DC, chairman of the President's National Defense Research Committee (1940), Director of the Office of Scientific Research and Development (1941–1947), Chairman of the National Advisory Committee for Aeronautics (1939–1941), founder of the National Science Foundation, and a central figure in the development of nuclear fission and the Manhattan Project. He wrote *Endless Horizons* (Public Affairs Press, 1946), *Modern Arms and Free Men* (Simon and Schuster, 1949), *Science Is Not Enough* (Morrow, 1967), *Pieces of the Action* (Morrow, 1970), and, perhaps most famously, "As We May Think" (*Atlantic Monthly*, July 1945).

Wendy Hui Kyong Chun is Associate Professor of Modern Culture and Media at Brown University. She is author of *Control and Freedom: Power and Paranoia in the Age of Fiber Optics* (MIT Press, 2005). She has been a fellow at the Radcliffe Institute for Advanced Study and a Henry Merritt Wriston Fellow at Brown. She is currently working on a book on the relationship between race and software entitled *Programmed Visions: Software, DNA, Race.*

Julian Dibbell is a contributing editor for *Wired* magazine and the author of *My Tiny Life: Crime and Passion in a Virtual World* (Fourth Estate, 1999). He has been writing about digital culture for fifteen years and currently is working on a book about the economies of online role-playing games. He lives in South Bend, Indiana.

Richard Dienst teaches critical theory at Rutgers University, where he is Associate Professor in the Department of English. He is the author of *Still Life in Real Time: Theory after Television* (Duke University Press, 1994) and the coeditor of *Reading the Shape of the World: Toward an International Cultural Studies* (Westview Press, 1996).

Mary Ann Doane is George Hazard Crooker Professor of Modern Culture and Media and of English at Brown University. She is the author of *The Emergence of Cinematic Time: Modernity,*

Contingency, the Archive (Harvard University Press, 2002), *Femmes Fatales: Feminism, Film Theory, Psychoanalysis* (Routledge, 1991), and *The Desire to Desire: The Woman's Film of the 1940s* (Indiana University Press, 1987).

Thomas Elsaesser is Professor in the Department of Media and Culture and Director of Research Film and Television at the University of Amsterdam. Among his most recent books as (co) editor are: *Cinema Futures: Cain, Abel or Cable?* (Amsterdam University Press, 1998), *The BFI Companion to German Cinema* (BFI, 1999), *The Last Great American Picture Show* (Amsterdam University Press, 2004) and *Harun Farocki — Working on the Sightlines* (Amsterdam University Press, 2004). His books as author include *Fassbinder's Germany: History, Identity, Subject* (Amsterdam University Press, 1996), *Weimar Cinema and After* (Routledge, 2000), *Metropolis* (BFI, 2000), *Studying Contemporary American Film* (Oxford University Press, 2002, with Warren Buckland), *Filmgeschichte und Frühes Kino* (text + kritik, 2002), *Terreur, Mythes et Representation* (Tausendaugen, 2005) and *European Cinema: Face to Face with Hollywood* (Amsterdam University Press, 2005).

Wolfgang Ernst is Professor of Media Theories at the Humboldt-University Berlin. He is author of *M.edium F.oucault. Weimarer Vorlesungen über Archive, Archeologie, Monumente und Medien* (VDG, Verlag und Datenbank für Geisteswissenschaften, 2000); *Das Rumoren der Archive. Ordnung aus Unordnung* (Merve, 2002); *Sammeln — Speichern — Erzählen. Infrastrukturelle Konfigurationen des deutschen Gedächtnisses* (Fink, 2003).

Alexander R. Galloway is Assistant Professor of Media Ecology at New York University. His first book, *Protocol: How Control Exists After Decentralization*, is published by the MIT Press.

Wolfgang Hagen is Privatdozent for media studies at the University of Basel and works as editor in chief of the cultural department of DeutschlandRadio Kultur in Berlin. He is author of *Gegenwartsvergessenheit* [Present Mindedness]: *Lazarsfeld - Adorno - Innis - Luhmann* (Merve, 2003).

Ken Hillis is Associate Professor of Media Studies, Department of Communication Studies, The University of North Carolina at Chapel Hill. He is author of *Digital Sensations: Space, Identity and Embodiment in Virtual Reality* (University of Minnesota Press, 1999). His coedited anthology on eBay and everyday life will be published by Routledge in 2006.

Lynne Joyrich is Associate Professor of Modern Culture and Media at Brown University. A member of the *Camera Obscura* editorial collective, she is the author of *Re-viewing Reception: Television, Gender, and Postmodern Culture* (Indiana University Press, 1996) and of journal articles and anthology chapters on film and television studies and feminist and queer cultural criticism.

Thomas Keenan teaches media theory, literature, and human rights at Bard College, where he is Associate Professor of Comparative Literature and directs the Human Rights Project. He is author of *Fables of Responsibility* (Stanford University Press, 1997), and editor of books on the museum and on the wartime journalism of Paul de Man. He is writing another book called *Live Feed: Crisis, Intervention, Media,* about the news media and contemporary conflicts.

Friedrich Kittler is Professor of Media History and Aesthetics in the Institute for Aesthetics at Humboldt University-Berlin. His books in English include: *Discourse Networks 1800 / 1900* (Stanford University Press, 1990), *Gramophone, Film, Typewriter* (Stanford University Press, 1999), and *Literature, Media, Information Systems: Essays* (G+B Arts International, 1997). His recent books include *Eine Kulturgeschichte der Kulturwissenschaft* [A Cultural History of Cultural Studies] (Fink

2000), *Vom Griechenland* [On Greece] (with Cornelia Vismann, Merve, 2001), and *Optische Medien* [Optical Media] (Merve, 2002).

Peter Krapp is Assistant Professor of Film and Media Studies at the University of California, Irvine. He is the author of *Déja Vu: Aberrations of Cultural Memory* (University of Minnesota Press, 2004) and edited *Medium Cool*, a collection on contemporary media theory (*South Atlantic Quarterly* 101:3, 2002).

Thomas Y. Levin is Associate Professor of German at Princeton University where he teaches media and cultural theory. He has translated and edited Seigfried Kracauer's *The Mass Ornament* (Harvard UP, 1995), and coedited *CTRL [SPACE]: Rhetorics of Surveillance from Bentham to Big Brother* (MIT Press, 2002), the catalogue of a major exhibition he curated at the Center for Art and Media (ZKM) in Karlsruhe, Germany. He is working on two small books, one on Rembrandt Media, and the other on the film-theoretical cinema of Guy Debord and the Situationist International.

Geert Lovink is a media theorist and director of the Institute for Network Cultures at Hogeschool/University of Amsterdam (HvA/UvA). He is the author of *Dark Fiber* (MIT Press, 2002), *Uncanny Networks* (MIT Press, 2002) and *My First Recession* (V2 Publishers/NAi Publishing, 2003).

Lev Manovich is Associate Professor of New Media at University of California, San Diego. His publications include *The Language of New Media* and *Soft Cinema: Navigating the Database* (both from MIT Press).

Tara McPherson is Associate Professor of Gender Studies and Critical Studies in USC'S School of Cinema-TV. She is the author of *Reconstructing Dixie: Race, Gender and Nostalgia in the Imagined South* (Duke University Press, 2003) and coeditor of the anthology *Hop on Pop: The Politics and Pleasures of Popular Culture* (Duke University Press, 2003). She is currently launching the multimedia journal, *Vectors*, and is working on a book on racial epistemologies in the electronic age.

Anders Michelsen is Assistant Professor and coordinator of visual culture studies in the Department of Art & Cultural Studies, University of Copenhagen. Most recent books: *Designmaskinen. Design af den moderne verden* [The Design Machine. Design of the Modern World] (Gyldendal, 1999); *Kunstteori. Positioner i nutidig kunstdebat* [Art Theory. Positions in Contemporary Art Discourse] (Borgen, 1999). He serves as advisory editor of *Atlantica Revista de Arte y Pensamiento* [Atlantic of Arts] (CAAM:Gran Canaria).

Nicholas Mirzoeff is Professor of Art and Art Professions at New York University. He is the author of *Watching Babylon: The War in Iraq and Global Visual Culture* (Routledge, 2004).

Lisa Nakamura is Assistant Professor of Communication Arts and Visual Culture Studies at the University of Wisconsin, Madison. She is the author of *Cybertypes: Race, Ethnicity, and Identity on the Internet* (Routledge, 2002) and coeditor of *Race in Cyberspace* (Routledge, 2000). She is working on a new book entitled *Visual Cultures of the Internet,* forthcoming from University of Minnesota Press, 2006.

Vicente L. Rafael is Professor of History at the University of Washington in Seattle. He is the author of *Contracting Colonialism* (Duke University Press, 1988, rev. 1993), *White Love and Other Events in Filipino History* (Duke University Press, 2000), and *The Promise of the Foreign: Nationalism and the Technics of Translation in the Spanish Philippines* (Duke University Press, 2005).

Arvind Rajagopal is author of *Politics After Television: Hindu Nationalism and the Reshaping of the Public in India* (Cambridge University Press, 2001), which won the Ananda Kentish Coomaraswamy Prize for the best book on South Asia from the South Asia Council of the Association of Asian Studies in 2003. He is editor of *Technologies of Perception and the Cultures of Globalization* (*Social Text* No. 68, 2001), and of *America and Its Others* (*Interventions: International Journal of Postcolonial Studies*, 2004). In 1998–1999, he was a member of the Institute for Advanced Study in Princeton, in their School of Social Science. He is Associate Professor of media studies at New York University.

Cornelia Vismann is Wissenschaftliche Mitarbeiterin at the Max-Planck-Institut for European Legal History in Frankfurt am Main, Germany. She has published *Akten: Medientechnik und Recht* [Filing Systems: Mediatechnique and Law]. (Fischer Taschenbuch Verlag, 2000; 2nd ed., 2001, an English translation is in preparation), and together with Friedrich Kittler, *Vom Griechenland* [On Greece] (Merve, 2002). She is currently working on a habilitation project concerning questions of information theory and the constitution.

McKenzie Wark teaches at Lang College, New School University. He is the author of *A Hacker Manifesto* (Harvard UP, 2004), *Dispositions* (Salt, 2002), *Celebrities, Culture and Cyberspace* (Pluto, 2000), *Virtual Republic* (Allen & Unwin, 1997), and *Virtual Geography* (Indiana University Press, 1994). He is coauthor of *Speed Factory* (Fremantle Arts Centre Press, 2001) and coeditor of *Readme!*, the nettime anthology (Autonomedia, 1998).

Mark Wigley is Dean of the Graduate School of Architecture, Preservation and Planning at Columbia University. He is the author of *The Architecture of Deconstruction: Derrida's Haunt* (MIT Press, 1993), *White Walls, Designer Dresses: The Fashioning of Modern Architecture* (MIT Press, 1995), and *Constant's New Babylon: The Hyper-Architecture of Desire* (010 Press, 1998). He coedited *The Activist Drawing: Retracing Situationist Architectures from Constant's New Babylon to Beyond* (MIT Press, 2001) and is currently preparing a pre-history of virtual space.

Index